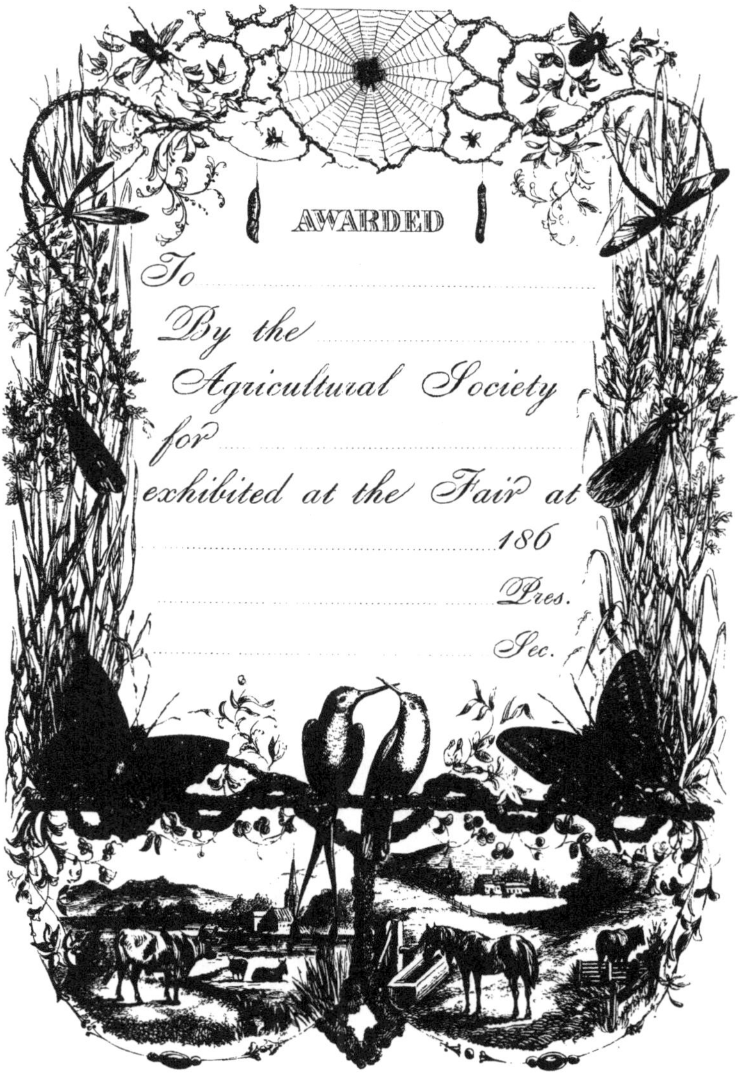

AWARDED

To ....

By the ....

Agricultural Society

for ....

exhibited at the Fair at

....186

....Pres.

....Sec.

A

TREATISE

ON SOME OF THE

INSECTS INJURIOUS TO VEGETATION.

BY

THADDEUS WILLIAM HARRIS, M. D.

THIRD EDITION.

BOSTON:
WILLIAM WHITE, PRINTER TO THE STATE.
1862.

Cambridge:
Welch, Bigelow, and Company,
Printers to the University.

EDITOR'S PREFACE.

BY a resolve of the Massachusetts Legislature, of 1859, chap. 93, I was directed to issue a new edition of Dr. Harris's admirable Report on Insects Injurious to Vegetation, with suitable additions and illustrations.

Arrangements were made to begin the work without delay, but it was found necessary to obtain extensive collections of insects in order to have fresh specimens for use in making the drawings. This required much time and labor, and, even with the utmost diligence, it was found impossible to secure all that were needed during the spring and summer of 1859.

It has been thought best to insert the additions contemplated in the resolve, in the form of foot-notes. No alterations have been made in the author's language, and the additional notes are enclosed in brackets to distinguish them from those in the former editions. Large additions to the text, however, have been made from the author's own manuscripts. These will be found exclusively in the chapter upon the butterflies. In giving a somewhat wider significance to the title, I have but carried out the plan adopted by the author in his last revision of the work.

Professor Louis Agassiz very kindly offered to supervise the drawings, comparing them with the original specimens before engraving. It is believed that very great scientific accuracy has thus been secured in the illustrations. Special acknowledgments are due to Professor Agassiz for this valuable service, and also for assistance rendered by way of suggestion and advice throughout.

Acknowledgments are also due to the following gentlemen, who have contributed notes on the subjects named: — Dr. John L.

Leconte, of Philadelphia, on the Coleoptera; Philip R. Uhler, Esq., of Baltimore, on the Orthoptera and Hemiptera; Dr. John G. Morris, of Baltimore, on the Lepidoptera; Edward Norton, Esq., of Farmington, Connecticut, on the Hymenoptera; and Baron R. Osten Sacken, Secretary of the Russian Legation at Washington, on the Diptera. These distinguished entomologists have made specialties of the orders on which they have had the kindness to furnish notes, and their contributions have done much to make the work complete. I am greatly indebted, also, to Mr. Alex. E. R. Agassiz for very valuable services, and to Mr. Francis G. Sanborn, whose enthusiasm in making collections, and otherwise promoting the progress of the work, has continued unabated from the first. Also to Messrs. James M. Barnard and Edward S. Rand, Jr., who have devoted much time and thought to the details of the work.

Many individuals have aided by presenting or lending specimens for illustration, or otherwise, among whom should be mentioned, in addition to the above, Messrs. S. H. Scudder, J. H. Treat, and J. O. Treat. To prevent any misconception, it should be stated that, in the specimens from which Figs. 109, 111, 112, 113, 115, 116, 117, 126, 127, 128, 129, and 130 were drawn, the *second* pair of feet were displayed instead of the *first*, and that in Fig. 114 the fore foot should have been omitted.

The drawings for the steel plates were made by Mr. Antoine Sonrel; those for the wood-cuts by the Messrs. Sonrel and J. Burckhardt. The engraving and coloring of the steel plates is the work of Mr. John H. Richard; the engraving on wood, that of Mr. Henry Marsh. The work of these artists needs no comment. The printing has been done by Messrs. Welch, Bigelow, & Co., of the University Press, Cambridge. This also speaks for itself.

No labor has been spared to secure the utmost accuracy and perfection in every respect, and it is hoped and believed that the objects of the Legislature in ordering a new edition of this valuable treatise have been fully accomplished.

CHARLES L. FLINT,
Secretary of the State Board of Agriculture.

BOSTON, December, 1861.

AUTHOR'S PREFACE.

THE first edition of this work was printed in the year 1841. It formed one of the scientific Reports, which were prepared and published by the Commissioners on the Zoölogical and Botanical Survey of Massachusetts, agreeably to an order of the General Court, and at the expense of the State. The Commission for this Survey bore the date of June 10th, 1837; and the following instructions from his Excellency, Governor Everett, accompanied it: —

"It is presumed to have been a leading object of the Legislature, in authorizing this Survey, to promote the agricultural benefit of the Commonwealth, and you will keep carefully in view the economical relations of every subject of your inquiry. By this, however, it is not intended that scientific order, method, or comprehension should be departed from. At the same time, that which is practically useful will receive a proportionally greater share of attention, than that which is merely curious; the promotion of comfort and happiness being the great human end of all science."

Upon a division of duties among the Commissioners, the department of Insects was assigned to me. Some idea of the extent of this department may be formed by an examination of my Catalogues of the Insects of Massachusetts, appended to the first and second editions of Professor Hitchcock's Report, in which above 2,300 species were enumerated; and these doubtless fall very far short of the actual number to be found within this Commonwealth.

In entering upon my duty, I was deterred from attempting to describe all these insects by the magnitude of the undertaking, and by the consideration that such a work, much as it might promote the cause of science, if well done, could not be expected to prove either interesting or particularly useful to the great body of the people. The subject and the plan of my Report were suggested by the instructions of the Governor, and by the want of a work, combining scientific and practical details on the natural history of our noxious insects. From among such of the latter as are injurious to plants, I selected for description chiefly those that were remarkable for their size, for the peculiarity of their structure and habits, or for the extent of their ravages; and these alone will be seen to constitute a formidable host. As they are found not only in Massachusetts, but throughout New England, and indeed in most parts of the United States, the propriety of giving to the work a more comprehensive title than it first bore, becomes apparent. This was accordingly done in the small impression that was printed at my own charge, while the original Report was passing through the press, and in which some other alterations were made to fit it for a wider circulation.

In the course of eight years, all the copies of the Report, and of the other impression, were entirely disposed of. Meanwhile, some materials for a new edition were collected, and these have been embodied in the present work, which I have been called upon to prepare and carry through the press.

Believing that the aid of science tends greatly to improve the condition of any people engaged in agriculture and horticulture, and that these pursuits form the basis of our prosperity, and are the safeguards of our liberty and independence, I have felt it to be my duty, in treating the subject assigned to me, to endeavor to make it useful and acceptable to those persons whose honorable employment is the cultivation of the soil.

T. W. H.

CAMBRIDGE, Mass., Oct. 15, 1852.

CONTENTS.

CHAPTER I.

INTRODUCTION.

INSECTS DEFINED. — BRAIN AND NERVES. — AIR-PIPES AND BREATHING-HOLES. — HEART AND BLOOD. — INSECTS ARE PRODUCED FROM EGGS. — METAMORPHOSES, OR TRANSFORMATIONS. — EXAMPLES OF COMPLETE TRANSFORMATION. — PARTIAL TRANSFORMATION. — LARVA, OR INFANT STATE. — PUPA, OR INTERMEDIATE STATE. — ADULT, OR WINGED STATE. — HEAD, EYES, ANTENNÆ, AND MOUTH. — THORAX OR CHEST, WINGS, AND LEGS. — ABDOMEN OR HIND-BODY, PIERCER, AND STING. — NUMBER OF INSECTS COMPARED WITH PLANTS. — CLASSIFICATION; ORDERS; COLEOPTERA; ORTHOPTERA; HEMIPTERA; NEUROPTERA; LEPIDOPTERA; HYMENOPTERA; DIPTERA; OTHER ORDERS AND GROUPS. — REMARKS ON SCIENTIFIC NAMES. . . . 1-22

CHAPTER II.

COLEOPTERA.

BEETLES. — SCARABÆIANS. — GROUND-BEETLES. — TREE-BEETLES. — COCKCHAFERS OR MAY-BEETLES. — FLOWER-BEETLES. — STAG-BEETLES. — BUPRESTIANS, OR SAW-HORNED BORERS. — SPRING-BEETLES. TIMBER-BEETLES. — WEEVILS. — CYLINDRICAL BARK-BEETLES. — CAPRICORN-BEETLES, OR LONG-HORNED BORERS. — LEAF-BEETLES. — CRIOCERIANS. — LEAF-MINING BEETLES. — TORTOISE-BEETLES. — CHRYSOMELIANS. — CANTHARIDES. 23-140

CHAPTER III.

ORTHOPTERA.

EARWIGS. — COCKROACHES. — MANTES, OR SOOTHSAYERS. — WALKING-LEAVES. — WALKING-STICKS, OR SPECTRES. — MOLE-CRICKET. — FIELD CRICKETS. — CLIMBING CRICKET. — WINGLESS CRICKET. — GRASSHOPPERS. — KATY-DID. — LOCUSTS. 141-191

CHAPTER IV.

HEMIPTERA.

BUGS. — SQUASH-BUG. — CHINCH-BUG. — PLANT-BUGS. — HARVEST-FLIES. — TREE-HOPPERS. — LEAF-HOPPERS. — VINE-HOPPER. — BEAN-HOPPER. — THRIPS. — PLANT-LICE. — AMERICAN BLIGHT. — ENEMIES OF PLANT-LICE. — BARK-LICE. 192-256

CHAPTER V.

LEPIDOPTERA.

CATERPILLARS. — BUTTERFLIES. — SKIPPERS. — HAWK-MOTHS. — ÆGERIANS OR BORING-CATERPILLARS. — GLAUCOPIDIANS. — MOTHS. — SPINNERS. — LITHOSIANS. — TIGER-MOTHS. — ERMINE-MOTHS. — TUSSOCK-MOTHS. — LACKEY-MOTHS. — LAPPET-MOTHS. — SATURNIANS. — CERATOCAMPIANS. — CARPENTER-MOTHS. — PSYCHIANS. — NOTODONTIANS. — OWL-MOTHS. — CUT-WORMS. — GEOMETERS, OR SPAN-WORMS, AND CANKER-WORMS. — DELTA-MOTHS. — LEAF-ROLLERS. — BUD-MOTHS. — FRUIT-MOTHS. — BEE-MOTHS. — CORN-MOTHS. — CLOTHES-MOTHS. — FEATHER-WINGED MOTHS. 257-511

CHAPTER VI.

HYMENOPTERA.

STINGERS AND PIERCERS. — HABITS OF SOME OF THE HYMENOPTERA. — SAW-FLIES AND SLUGS. — ELM SAW-FLY. — FIR SAW-FLY. — VINE SAW-FLY. — ROSE-BUSH SLUG. — PEAR-TREE SLUG. — HORN-TAILED WOOD-WASPS. — GALL-FLIES. — CHALCIDIANS. — BARLEY INSECT AND JOINT-WORM. 512-561

CHAPTER VII.

DIPTERA.

GNATS AND FLIES. — MAGGOTS, AND THEIR TRANSFORMATIONS. — GALL-GNATS. — HESSIAN FLY. — WHEAT-FLY. — REMARKS UPON AND DESCRIPTIONS OF SOME OTHER DIPTEROUS INSECTS. — RADISH-FLY. — TWO-WINGED GALL-FLIES, AND FRUIT-FLIES. — CONCLUSION. . 562-626

APPENDIX. — THE ARMY-WORM 627-630

INDEX 631-640

EXPLANATION OF PLATES.

PLATE I. (FRONTISPIECE.)

			Page
Fig.	1.	Nepa apiculata	12
"	2.	Agrion basalis	12
"	3.	Mutilla coccinea	15
"	4.	Asilus (Erax) æstuans, *Linn.*	17
"	5.	Cassida (Coptocycla) aurichalcea, *Fab.*	122
"	6.	Locusta (Œdipoda) sulphurea, *Fab.*	177
"	7.	Nymphalis Arthemis, *Drur.*	283

PLATE II. (PAGE 23.)

Fig.	1.	Eumolpus auratus, *Fab.*	134
"	2.	Chrysobothris (Trachypteris) Harrisii, *Hentz*	51
"	3.	Galeruca vittata, *Fab.*	124
"	4.	Coccinella novemnotata	246
"	5.	Haltica chalybea, *Illig.*	129
"	6.	Attelabus bipustulatus, *Fab.*	66
"	7.	Dicerca (Stenurus) divaricata, *Say*	48
"	8.	Sitophilus Oryzæ, *Linn.*	83
"	9.	Chrysomela trimaculata, *Fab.*	132
"	10.	Clytus flexuosus, *Fab.*	103
"	11.	Callidium antennatum, *Newm.*	100
"	12.	Hylotrupes bajulus, *Linn.*	100
"	13.	Saperda (Compsidea) tridentata, *Oliv.*	111
"	14.	Omaloplia (Serica) vespertina, *Gyll.*	33
"	15.	Clytus speciosus, *Say*	101
"	16.	Saperda candida, *Fab.*	107
"	17.	" " Larva	108
"	18.	Desmocerus cyaneus, *Fab.*	115
"	19.	Saperda vestita, *Say*	109
"	20.	Areoda (Cotalpa) lanigera, *Linn.*	24
"	21.	Saperda (Anaerea) calcarata, *Say*	106

PLATE III. (PAGE 141.)

Fig.	1.	Locusta (Chloealtis) curtipennis	184
"	2.	Locusta (Tragocephala) viridi-fasciata, *De Geer*	182

Fig. 3. Locusta (Œdipoda) Carolina, *Linn.* 176
" 4. Aphis mali 235
" 5. Tettigonia (Erythroneura) vitis 227
" 6. Clastoptera proteus 225
" 7. Cicada septendecim, *Linn.* 211
" 8. Chrysopa euryptera, *Burm.* 247
" 9. " " Larva and cocoon 247

PLATE IV. Page 257.)

Fig. 1. Vanessa (Grapta) comma, *Harr.* 300
" 2. " " " Vacant chrysalis 301
" 3. Thecla Humuli, *Harr.* 276
" 4. Papilio Asterias, *Fab.* ♂ 265
" 5. " " ♀ 265
" 6. " " Larva 263
" 7. " " Chrysalis 264

PLATE V. (Page 318.)

Fig. 1. Eudamus (Goniloba) Tityrus, *Smith* 310
" 2. Philampelus Satellitia, *Linn.* 325
" 3. Philampelus Achemon, *Drury* 326
" 4. Chœrocampa (Darapsa) pampinatrix, *Smith* 327
" 5. Ægeria (Trochilium) Pyri, *Harr.* 335
" 6. " " exitiosa, *Say* ♂ 331
" 7. " " " Vacant chrysalis 332
" 8. " " Cucurbitæ, *Harr.* 331

PLATE VI. (Page 340.)

Fig. 1. Lophocampa (Halesidota) Caryæ, *Harr.* Larva 361
" 2. " " " Cocoon 362
" 3. Deiopeia bella, *Drury* 342
" 4. Perophora Melsheimerii, *Harr.* Larva Case 415
" 5. " " ♀ 415, 417
" 6. Pygæra (Datana) ministra, *Drury* 430
" 7. Eudryas grata, *Fab.* Larva 427
" 8. " " Imago 427
" 9. Arctia (Spilosoma) acrea, *Drury* ♂ 354
" 10. " " " ♀ 354
" 11. Notodonta (Pygæra) concinna, *Smith* 426
" 12. Clostera Americana, *Harr.* 433

PLATE VII. (Page 376.)

Fig. 1. Orgyia leucostigma, *Smith.* Larva 366
" 2. " " ♀ after depositing eggs . . 367
" 3. " " " " " " 367
" 4. " " ♂ 367, 368
" 5. " " Cocoon and eggs 367

Fig. 6. Tinea granella. Larva 497
" 7. " " Wheat attacked by 497
" 8. Pyralis farinalis, *Harr.* 475
" 9. Gortyna Zeæ, *Harr.* 439
" 10. Hyphantria (Spilosoma) textor. Cocoon 358
" 11. " " " Pupa 358
" 12. " " " Young larva . . . 357, 358
" 13. Clisiocampa Americana, *Harr.* Larva 371
" 14. " " ♂ 372
" 15. " " Vacant cocoon 372
" 16. " " Cluster of eggs 370
" 17. " " ♀ 372
" 18. Clisiocampa silvatica, *Harr.* 376
" 19. " " Larva 375

PLATE VIII. (PAGE 512.)

Fig. 1. Tachina vivida, *Harr.* 612
" 2. Gasterophilus (Gastrus) Equi, *Linn.* 623
" 3. Lophyrus Abietis, *Harr.* ♂ 520
" 4. " " " antenna 520
" 5. " " ♀ 520
" 6. Cynips dichlocerus. Natural size 549
" 7. " " Magnified 549
" 8. " " Gall on Rose-bush 549
" 9. Cynips confluens. Galls on oak-leaf 546
" 10. " " 546, 547
" 11. Cimbex Ulmi. Cocoon 519
" 12. Cimbex Laportei. ♂ 518

NOTE. — The hair-line at the side of a cut shows its natural size.

INSECTS

INJURIOUS TO VEGETATION.

CHAPTER I.

INTRODUCTION.

INSECTS DEFINED. — BRAIN AND NERVES. — AIR-PIPES AND BREATHING-HOLES. — HEART AND BLOOD. — INSECTS ARE PRODUCED FROM EGGS. — METAMORPHOSES, OR TRANSFORMATIONS. — EXAMPLES OF COMPLETE TRANSFORMATION. — PARTIAL TRANSFORMATION. — LARVA, OR INFANT STATE. — PUPA, OR INTERMEDIATE STATE. — ADULT, OR WINGED STATE. — HEAD, EYES, ANTENNÆ, AND MOUTH. — THORAX OR CHEST, WINGS, AND LEGS. — ABDOMEN OR HIND-BODY, PIERCER, AND STING. — NUMBER OF INSECTS COMPARED WITH PLANTS. — CLASSIFICATION ; ORDERS ; COLEOPTERA ; ORTHOPTERA ; HEMIPTERA ; NEUROPTERA ; LEPIDOPTERA ; HYMENOPTERA ; DIPTERA ; OTHER ORDERS AND GROUPS. — REMARKS ON SCIENTIFIC NAMES.

THE benefits which we derive from insects, though neither few in number nor inconsiderable in amount, are, if we except those of the silk-worm, the bee, and the cochineal, not very obvious, and are almost entirely beyond our influence. On the contrary, the injuries that we suffer from them are becoming yearly more apparent, and are more or less within our control. A familiar acquaintance with our insect enemies and friends, in all their forms and disguises, will afford us much help in the discovery and proper application of the remedies for the depredations of the former, and will tend to remove the repugnance wherewith the latter are commonly regarded.

Destructive insects have their appointed tasks, and are limited in the performance of them ; they are exposed to many accidents through the influence of the elements, and they fall a prey to numerous animals, many of them also of the insect race, which, while they fulfil their own part in the economy of nature, contribute to prevent the undue increase of the noxious tribes. Too often, by an unwise interference with the plan of Providence, we defeat the very measures contrived for our protection. We not only suffer from our own carelessness, but through ignorance fall into many mistakes. Civilization and cultivation, in many cases, have destroyed the balance originally existing between plants and insects, and between the latter and other animals. Deprived of their natural food by the removal of the forest trees and shrubs, and the other indigenous plants that once covered the soil, insects have now no other resource than the cultivated plants that have taken the place of the original vegetation. The destruction of insect-eating animals, whether quadrupeds, birds, or reptiles, has doubtless tended greatly to the increase of insects. Colonization and commerce have, to some extent, introduced foreign insects into countries where they were before unknown. It is to such causes as these that we are to attribute the unwelcome appearance and the undue multiplication of many insects in our cultivated grounds, and even in our store-houses and dwellings. We have no reason to believe that any absolutely new insects are generated or created from time to time. The supposed new species, made known to us first by their unwonted depredations, may have come to us from other parts, or may have been driven by the hand of improvement from their native haunts, where heretofore the race had lived in obscurity, and thus had escaped the notice of man.

To understand the relations that insects bear to each other and to other objects, and to learn how best to check the ravages of the noxious tribes, we must make ourselves thoroughly acquainted with the natural history of these animals. This subject is particularly important to all persons who are interested in agricultural pursuits. For their use, chiefly, this account of the principal insects that are injurious to vegetation in New England, has been prepared. It has been thought best to prefix thereto some remarks on the structure and classification of insects, to serve as an introduction to the succeeding chapters, and, in some measure, to supply the want of a more general and complete work on this branch of natural history.

The word *Insect*, which, in the Latin language, from whence it was derived, means cut into or notched, was designed to express one of the chief characters of this group of animals, whose body is marked by several cross-lines or incisions. The parts between these cross-lines are called segments, or rings, and consist of a number of jointed pieces, more or less movable on each other.

Insects have a very small brain, and, instead of a spinal marrow, a kind of knotted cord, extending from the brain to the hinder extremity; and numerous small whitish threads, which are the nerves, spread from the brain and knots, in various directions. Two long air-pipes, within their bodies, together with an immense number of smaller pipes, supply the want of lungs, and carry the air to every part. Insects do not breathe through their mouths, but through little holes, called spiracles, generally nine in number, along each side of the body. Some, however, have the breathing-holes placed in the hinder extremity, and a few young water-insects breathe by means of gills. The heart is a long tube, lying under the skin of the back, having little holes on each side for the admission of the juices of the body, which are prevented from escaping again by valves or clappers, formed to close the holes within. Moreover, this tubular heart is divided into several chambers, by transverse partitions, in each of which there is a hole shut by a valve, which allows the blood to flow only from the hinder to the fore part of the heart, and prevents it from passing in the contrary direction. The blood, which is a colorless or yellow fluid, does not circulate in proper arteries and veins; but is driven from the fore part of the heart into the head, and thence escapes into the body, where it is mingled with the nutritive juices that filter through the sides of the intestines, and the mingled fluid penetrates the crevices among the flesh and other internal parts, flowing along the sides of the air-pipes, whereby it receives from the air that influence which renders it fitted to nourish the frame and maintain life.

Insects are never spontaneously generated from putrid animal or vegetable matter, but are produced from eggs. A few, such as some plant-lice, do not lay their eggs, but retain them within their bodies till the young are ready to escape. Others invariably lay their eggs where their young, as soon as they are hatched, will find a plentiful supply of food immediately within their reach.

Most insects, in the course of their lives, are subject to very great changes of form, attended by equally remarkable changes in their habits and propensities. These changes, transformations, or *metamorphoses*, as they are called, might cause the same insect, at different ages, to be mistaken for as many different animals. For example, a caterpillar, after feeding upon leaves till it is fully grown, retires into some place of concealment, casts off its caterpillar-skin, and presents itself in an entirely different form, one wherein it has neither the power of moving about, nor of taking food; in fact, in this its second or chrysalis state, the insect seems to be a lifeless oblong oval or conical body, without a distinct head, or movable limbs; after resting awhile, an inward struggle begins, the chrysalis-skin bursts open, and from the rent issues a butterfly or a moth, whose small and flabby wings soon extend and harden, and become fitted to bear away the insect in search of the honeyed juice of flowers and other liquids that suffice for its nourishment.

The little fish-like animals that swim about in vessels of stagnant water, and devour the living atoms that swarm in the same situations, soon come to maturity, cast their skins, and take another form, wherein they remain rolled up like a ball, and either float at the surface of the water, for the purpose of breathing through the two tunnel-shaped tubes on the top of their backs, or, if disturbed, suddenly uncurl their bodies, and whirl over and over from one side of the vessel to the other. In the course of a few days these little water-tumblers are ready for another transformation; the skin splits on the back between the breathing-tubes, the head, body, and limbs of a mosquito suddenly burst from the opening, the slender legs rest on the empty skin till the latter fills with water and sinks, when the insect abandons its native element, spreads its tiny wings, and flies away, piping its war-note, and thirsting for the blood which its natural weapons enable it to draw from its unlucky victims.

The full-fed maggot, that has rioted in filth till its tender skin seems ready to burst with repletion, when the appointed time arrives, leaves the offensive matters it was ordained to assist in removing, and gets into some convenient hole or crevice; then its body contracts or shortens, and becomes egg-shaped, while the skin hardens, and turns brown and dry, so that, under this form, the creature appears more like a seed than a living animal; after some time passed in this inactive and equivocal form, during which wonderful changes have taken place within the seed-like shell, one end of the shell is forced off, and from the inside comes forth a buzzing fly, that drops its former filthy habits with its cast-off dress, and now, with a more refined taste, seeks only to lap the solid viands of our tables, or sip the liquid contents of our cups.

Caterpillars, grubs, and maggots undergo a complete transformation in coming to maturity; but there are other insects, such as crickets, grasshoppers, bugs, and plant-lice, which, though differing a good deal in the young and adult states, are not subject to so great a change, their transformations being only partial. For instance, the young grasshopper comes from the egg a wingless insect, and consequently unable to move from place to place in any other way than by the use of its legs; as it grows larger it is soon obliged to cast off its skin, and, after one or two moultings, its body not only increases in size, but becomes proportionally longer than before, while little stump-like wings begin to make their appearance on the top of the back. After this, the grasshopper continues to eat voraciously, grows larger and larger, and hops about without any aid from its short and motionless wings, repeatedly casts off its outgrown skin, appearing each time with still longer wings, and more perfectly formed limbs, till at length it ceases to grow, and, shedding its skin for the last time, it comes forth a perfectly formed and mature grasshopper, with the power of spreading its ample wings, and of using them in flight.

Hence there are three periods in the life of an insect, more or less distinctly marked by corresponding changes in the form, powers, and habits. In the first, or period of infancy, an insect is technically called a *larva*, a word signifying a mask, because therein its future form is more or less masked or concealed. This name is not only applied to grubs, caterpillars, and maggots, and to other insects that undergo a complete transformation, but also to young and wingless grasshoppers, and bugs, and indeed to all young insects before the wings begin to appear. In this first period, which is generally much the longest, insects are always wingless, pass most of their time in eating, grow rapidly, and usually cast off their skins repeatedly.

The second period — wherein those insects that undergo a partial transformation retain their activity and their appetites for food, continue to grow, and acquire the rudiments of wings, while others, at this age, entirely lose their larva form, take no food, and remain at rest in a deathlike sleep — is called the *pupa* state, from a slight resemblance that some of the latter present to an infant trussed in bandages, as was the fashion among the Romans. The pupæ from caterpillars, however, are more commonly called chrysalids, because some of them, as the name implies, are gilt or adorned with golden spots; and grubs, after their first transformation, are often named nymphs, for what reason does not appear. At the end of the second period, insects again shed their skins, and come forth fully grown, and (with few exceptions) provided with wings. Thus they enter upon their last or adult state, wherein they no longer increase in size, and during which they provide for a continuation of their kind. This period usually lasts only a short time, for most insects die immediately after their eggs are laid. Bees, wasps, and ants, however, which live in society, and labor together for the common good of their communities, continue much longer in the adult state.

In winged or adult insects, two of the transverse incisions with which they are marked are deeper than the rest, so that the body seems to consist of three principal portions, the first whereof is the head, the second or middle portion the thorax, or chest, and the third or hindmost the abdomen, or hind-body. In some wingless insects these three portions are also to be seen; but in most young insects, or larvæ, the body consists of the head and a series of twelve rings or segments, the thorax not being distinctly separated from the hinder part of the body, as may be perceived in caterpillars, grubs, and maggots.

The eyes of adult insects, though apparently two in number, are compound, each consisting of a great number of single eyes closely united together, and incapable of being rolled in their sockets. Such also are the eyes of the larvæ, and of the active pupæ of those insects that undergo an imperfect transformation. Moreover, many winged insects have one, two, or three little single eyes, placed near each other on the crown of the head, and called *ocelli*, or eyelets. The eyes of grubs, caterpillars, and of other completely transforming larvæ, are not compound, but consist of five or six eyelets clustered together, without touching, on each side of the head; some, however, such as maggots, are totally blind. Near to the eyes are two jointed members, named *antennæ*, corresponding, for the most part, in situation, with the ears of other animals, and supposed to be connected with the sense of hearing, of touch, or of both united. The antennæ are very short in larvæ, and of various sizes and forms in other insects.

The mouth of some insects is made for biting or chewing, that of others for taking the food only by suction. The biting-insects have the parts of the mouth variously modified to suit the nature of the food; and these parts are, an upper and an under lip, two nippers or jaws on each side, moving sidewise, and not up and down, and four or six little jointed members, called *palpi* or feelers, whereof two belong to the lower lip, and one or two to each of the lower jaws. The mouth of sucking-insects consists essentially of these same parts, but so different in their shape and in the purposes for which they are designed, that the resemblance between them and those of biting-insects is not easily recognized. Thus the jaws of caterpillars are transformed to a spiral sucking-tube in butterflies and moths, and those of maggots to a hard proboscis, fitted for piercing, as in the mosquito and horse-fly, or to one of softer consistence, and ending with fleshy lips for lapping, as in common flies; while in bugs, plant-lice, and some other insects resembling them, the parts of the mouth undergo no essential change from infancy to the adult state, but are formed into a long, hard, and jointed beak, bent under the breast when not in use, and designed only for making punctures and drawing in liquid nourishment.

The parts belonging to the thorax are the wings and the legs. The former are two or four in number, and vary greatly in form and consistence, in the situation of the wing-bones or veins, as they are generally called, and in their position or the manner in which they are closed or folded when at rest. The under-side of the thorax is the breast, and to this are fixed the legs, which are six in number in adult insects, and in the larvæ and pupæ of those that are subject only to a partial transformation. The parts of the legs are the hip-joint, by which the leg is fastened to the body, the thigh, the shank (*tibia*), and the foot, the latter consisting sometimes of one joint only, more often of two, three, four, or five pieces (*tarsi*), connected end to end, like the joints of the finger, and armed at the extremity with one or two claws. Of the larvæ that undergo a complete transformation, maggots and some others are destitute of legs ; many grubs have six, namely, a pair beneath the under-side of the first three segments, and sometimes an additional fleshy prop-leg under the hindmost extremity ; caterpillars and false caterpillars have, besides the six true legs attached to the first three rings, several fleshy prop-like legs, amounting sometimes to ten or sixteen in number, placed in pairs beneath the other segments.

The abdomen, or hindmost, and, as to size, the principal part of the body, contains the organs of digestion, and other internal parts, and to it also belong the piercer and the sting with which many winged or adult insects are provided. The piercer is sometimes only a flexible or a jointed tube, capable of being thrust out of the end of the body, and is used for conducting the eggs into the crevices or holes where they are to be laid. In some other insects it consists of a kind of scabbard, containing a central borer, or instruments like saws, designed for making holes wherein the eggs are to be inserted. The sting, in like manner, consists of a sheath enclosing a sharp instrument for inflicting wounds, connected wherewith in the inside of the body is a bag of venom or poison. The parts belonging to the abdomen of larvæ are various, but are mostly designed to aid them in their motions, or to provide for their respiration.

An English entomologist has stated, that, on an average, there are six distinct insects to one plant. This proportion is probably too great for our country, where vast tracts are covered with forests, and the other original vegetable races still hold possession of the soil. There are above 1,200 flowering plants in Massachusetts, and it will be within bounds to estimate the species of insects at 4,800, or in the proportion of four to one plant. To facilitate the study of such an immense number, some kind of classification is necessary; it will be useful to adopt one, even in describing the few species now before us. The basis of this classification is founded upon the structure of the mouth, in the adult state, the number and nature of the wings, and the transformations. The first great divisions are called orders, of which the following seven are very generally adopted by naturalists.

1. Coleoptera (*Beetles*). Insects with jaws, two thick wing-covers meeting in a straight line on the top of the back, and two filmy wings, which are folded transversely. Transformation complete. Larvæ, called grubs, generally provided with six true legs, and sometimes also with a terminal prop-leg; more rarely without legs. Pupa with the wings and the legs distinct and unconfined.

Many of these insects, particularly in the larva state, are very injurious to vegetation. The tiger-beetles (*Cicindeladæ**), the predaceous ground-beetles (*Carabidæ*), the diving-beetles (*Dytiscidæ*), the lady-birds (*Coccinelladæ*), and some others, are eminently serviceable by preying upon caterpillars, plant-lice, and other noxious or destructive insects. The water-lovers (*Hydrophilidæ*), rove-beetles (*Staphylinidæ*), carrion-beetles (*Silphadæ*), skin-beetles (*Dermestadæ*, *Byrrhidæ*, and *Trogidæ*), bone-beetles (some of the *Nitiduladæ* and *Cleridæ*), and various kinds of dung-beetles (*Sphæridiadæ*, *Histeridæ*, *Geotrupidæ*,† *Coprididæ*,† and *Aphodiadæ*†), and clocks (*Pimeliadæ* and *Blaptidæ*), act the useful part of scavengers, by removing carrion, dung, and other filth, upon which alone they and their larvæ subsist. Many

Fig. 1.

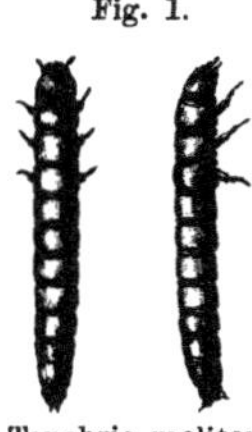

Tenebrio molitor.
(Meal-worm.)
Larva.

* See the Catalogue of Insects appended to Professor Hitchcock's Report on the Geology, Mineralogy, Botany, and Zoölogy of Massachusetts. 2d edit. 8vo. Amherst. 1835.

† All the Scarabæidæ of my Catalogue, from *Ateuchus* to *Geotrupes* inclusive, to which may be added many included in the genus *Scarabæus*.

Coleoptera (some *Staphylinidæ* and *Nitiduladæ*, *Diaperididæ*, some *Serropalpidæ*, *Mycetophagidæ*, *Erotylidæ*, and *Endomychidæ*) live altogether on agarics, mushrooms, and toadstools, plants of very little use to man, many of them poisonous, and in a state of decay often offensive; these fungus-eaters are therefore to be reckoned among our friends. There are others, such as the stag-beetles (*Lucanidæ*), some spring-beetles (*Elateridæ*), darkling-beetles (*Tenebrionidæ*), (Figs. 1 – 3,) and many bark-beetles (*Helopidæ*, *Cisteladæ*, *Serropalpidæ*, *Œdemeradæ*, *Cucujadæ*, and some *Trogositadæ*), which, living under the bark and in the trunks and roots of old trees, though they may occasionally prove injurious, must on the whole be considered as serviceable, by contributing to destroy and reduce to dust plants that have passed their prime and are fast going to decay. And, lastly, the blistering-beetles (*Cantharididæ*) have, for a long time, been employed with great benefit in the healing art.

Fig. 2.

Pupa.

Fig. 3.

Imago.

2. Orthoptera (*Cockroaches*, *Crickets*, *Grasshoppers*, *&c.*). Insects with jaws, two rather thick and opaque upper wings, overlapping a little on the back, and two larger, thin wings, which are folded in plaits, like a fan. Transformation partial. Larvæ and pupæ active, but wanting wings.

All of the insects of this order, except the camel-crickets (*Mantidæ*), which prey on other insects, are injurious to our household possessions, or destructive to vegetation.

3. Hemiptera (*Bugs*, *Locusts*, *Plant-lice*, *&c.*). Insects with a horny beak for suction, four wings, whereof the uppermost are generally thick at the base, with thinner extremities, which lie flat, and cross each other on the top of the back, or are of uniform thickness throughout, and slope at the sides like a roof. Transformation partial. Larvæ and pupæ nearly like the adult insect, but wanting wings.

The various kinds of field and house bugs give out a strong and disagreeable smell. Many of them (some *Pentatomadæ* and *Ly-*

gæidæ, Cimicidæ, Reduviadæ, Hydrometradæ, Nepadæ [Plate I. Fig. 1, Nepa apiculata], and *Notonectadæ*) live entirely on the juices of animals, and by this means destroy great numbers of noxious insects; some are of much service in the arts, affording us the costly cochineal, scarlet grain, lac, and manna; but the benefits derived from these are more than counterbalanced by the injuries committed by the domestic kinds, and by the numerous tribes of plant-bugs, locusts or cicadæ, tree-hoppers, plant-lice, bark-lice, mealy bugs, and the like, that suck the juices of plants, and require the greatest care and watchfulness on our part to keep them in check.

4. NEUROPTERA (*Dragon-flies, Lace-winged flies; May-flies, Ant-lion, Day-fly, White Ants, &c.*). Insects with jaws, four netted wings, of which the hinder ones are the largest, and no sting or piercer. Transformation complete, or partial. Larva and pupa various.

The white ants, wood-lice, and wood-ticks, (*Termitidæ* and *Psocidæ*,) the latter including also the little ominous death-watch, are almost the only noxious insects in the order, and even these do not injure living plants. The dragon-flies, or, as they are commonly called in this country, devil's-needles (*Libelluladæ*), (Figs. 4, 5,) (Plate I. Fig. 2, Agrion basalis,) prey upon gnats and mosquitoes; and their larvæ and pupæ, as well as those of the day-flies (*Ephemeradæ*), semblians (*Semblididæ*), and those of some of the May-flies, called caddis-worms (*Phryganeadæ*), (Fig. 6,) all of which live in the water, devour aquatic insects. The predaceous habits of the ant-lions (*Myrmeleontidæ*), (Fig. 7,) have been often described. The lace-winged flies (*Hemerobiadæ*), (Fig. 8,) in the larva state, live wholly on plant-lice, great numbers of which they destroy. The mantispians (*Mantispadæ*), and the scorpion-flies (*Panorpadæ*), are also predaceous insects.

5. LEPIDOPTERA (*Butterflies and Moths*). Mouth with a spiral sucking-tube; wings four, covered with branny scales. Transformation complete. The larvæ are caterpillars, and have six true legs, and from four to ten fleshy prop-legs.

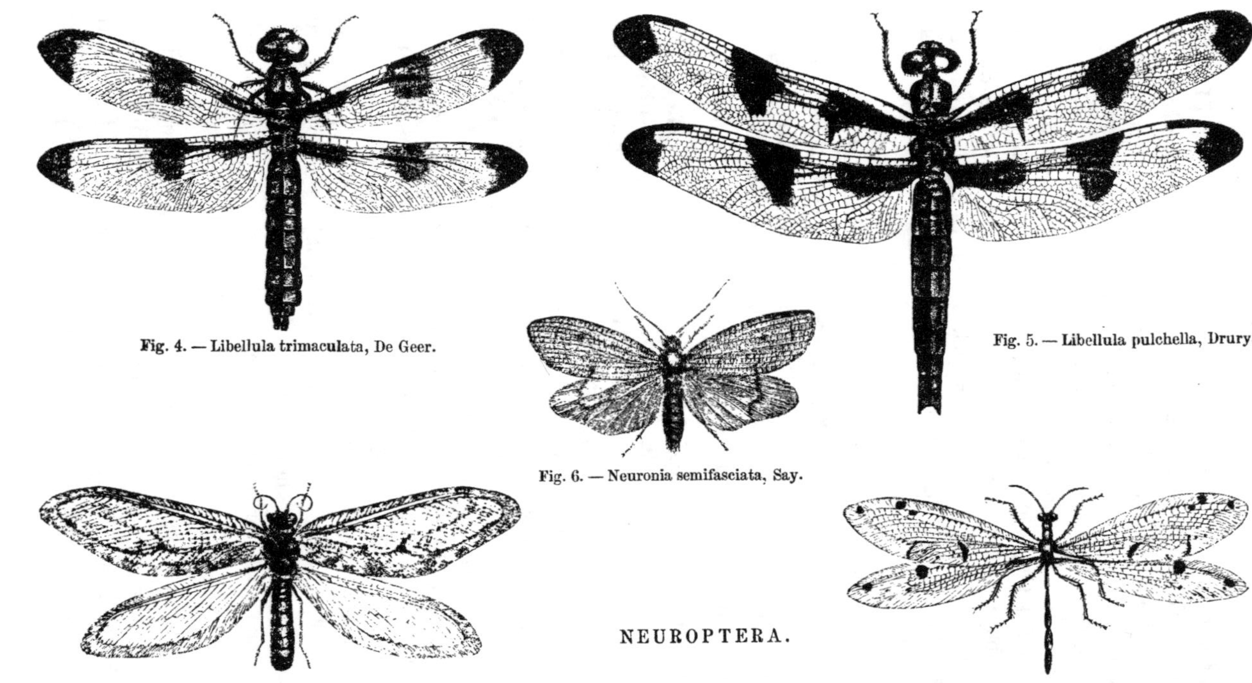

Fig. 4. — Libellula trimaculata, De Geer.

Fig. 5. — Libellula pulchella, Drury.

Fig. 6. — Neuronia semifasciata, Say.

Fig. 8. — Polystœchotes punctatus, Fab.

NEUROPTERA.

Fig. 7. — Myrmeleon obsoletus, Say.

Pupa with the cases of the wings and of the legs indistinct, and soldered to the breast.

Some kinds of caterpillars are domestic pests, and devour cloth, wool, furs, feathers, wax, lard, flour, and the like; but by far the greatest number live wholly on vegetable food, certain kinds being exclusively leaf-eaters, while others attack the buds, fruit, seeds, bark, pith, stems, and roots of plants.

6. HYMENOPTERA (*Saw-flies*, *Ants*, *Wasps*, *Bees*, *&c.*). Insects with jaws, four veined wings, in most species, the hinder pair being the smallest, and a piercer or sting at the extremity of the abdomen. Transformation complete. Larvæ mostly maggot-like, or slug-like; of some, caterpillar-like. Pupæ with the legs and wings unconfined.

In the adult state these insects live chiefly on the honey and pollen of flowers, and the juices of fruits. The larvæ of the saw-flies (*Tenthredinidæ*), under the form of false-caterpillars and slugs, are leaf-eaters, and are oftentimes productive of much injury to plants. The larvæ of the xiphydrians (*Xiphydriadæ*), and of the horn-tails (*Uroceridæ*), are borers and wood-eaters, and consequently injurious to the plants inhabited by them. Pines and firs suffer most from their attacks. Some of the warty excrescences on the leaves and stems of plants, such as oak-apples, gall-nuts, and the like, arise from the punctures of four-winged gall-flies (*Diplolepididæ*), and the irritation produced by their larvæ, which reside in these swellings. The injury caused by them is, comparatively, of very little importance, while, on the other hand, we are greatly indebted to these insects for the gall-nuts that are extensively used in coloring and in medicine, and form the chief ingredient in ink. We may, therefore, write down these insects among the benefactors of the human race. Immense numbers of caterpillars and other noxious insects are preyed upon by internal enemies, the larvæ of the ichneumon-flies (*Evaniadæ*, *Ichneumonidæ*, and *Chalcididæ*), which live upon the fat of their victims, and finally destroy them. Some of these ichneumon-flies (*Ichneumones ovulorum**) are extremely small, and confine their attacks

* Now placed among the *Proctotrupidæ*.

to the eggs of other insects, which they puncture, and the little creatures produced from the latter find a sufficient quantity of food to supply all their wants within the larger eggs they occupy. The ruby-tails (*Chrysididæ*) and the cuckoo-bees (*Hylæus, Sphecodes, Nomada, Melecta, Epeolus, Cœlioxys,* and *Stelis*) lay their eggs in the provisioned nests of other insects, whose young are robbed of their food by the earlier-hatched intruders, and are consequently starved to death. The wood-wasps (*Crabronidæ*), and numerous kinds of sand-wasps (*Larradæ, Bembicidæ, Sphegidæ, Pompilidæ,* and *Scoliadæ*), mud-wasps (*Pelopæus*), the stinging velvet-ants (*Mutilladæ*), (Plate I. Fig. 3, Mutilla coccinea,) and the solitary wasps (*Odynerus* and *Eumenes*), are predaceous in their habits, and provision their nests with other insects, which serve for food to their young.

The food of ants consists of animal and vegetable juices; and though these industrious little animals sometimes prove troublesome by their fondness for sweets, yet, as they seize and destroy many insects also, their occasional trespasses may well be forgiven. Even the proverbially irritable paper-making wasps and hornets (*Polistes* and *Vespa*) are not without their use in the economy of nature; for they feed their tender offspring not only with vegetable juices, but with the softer parts of other insects, great numbers of which they seize and destroy for this purpose. The solitary and social bees (*Andrenadæ* and *Apidæ*) live wholly on the honey and pollen of flowers, and feed their young with a mixture of the same, called bee-bread.

Various kinds of bees are domesticated for the sake of their stores of wax and honey, and are thus made to contribute directly to the comfort and convenience of man, in return for the care and attention afforded them. Honey and wax are also obtained from several species of wild bees (*Melipona, Trigona,* and *Tetragona*), essentially different from the domesticated kinds. While bees and other hymenopterous insects seek only the gratification of their own inclinations, in their frequent visits to flowers, they carry on their bodies the yellow dust or pollen from one blossom to another, and scatter it over the parts prepared to receive and be fertilized by it, whereby they render an important service to vegetation.

7. Diptera (*Mosquitoes, Gnats, Flies, &c.*). Insects with a horny or fleshy proboscis, two wings only, and two knobbed threads, called balancers or poisers, behind the wings. Transformation complete. The larvæ are maggots, without feet, and with the breathing-holes generally in the hinder extremity of the body. Pupæ mostly incased in the dried skin of the larvæ, sometimes, however, naked, in which case the wings and the legs are visible, and are found to be more or less free or unconfined.

The two-winged insects, though mostly of moderate or small size, are not only very numerous in kinds or species, but also extremely abundant in individuals of the same kind, often appearing in swarms of countless multitudes. Flies are destined to live wholly on liquid food, and are therefore provided with a proboscis, enclosing hard and sharp-pointed darts, instead of jaws, and fitted for piercing and sucking, or ending with soft and fleshy lips for lapping. In our own persons we suffer much from the sharp suckers and bloodthirsty propensities of gnats and mosquitoes (*Culicidæ*), and also from those of certain midges (*Ceratopogon* and *Simulium*), including the tormenting black-flies (*Simulium molestum*) of this country. The larvæ of these insects live in stagnant water, and subsist on minute aquatic animals. Horse-flies and the golden-eyed forest-flies (*Tabanidæ*), whose larvæ live in the ground, and the stinging stable-flies (*Stomoxys*), which closely resemble common house-flies, and in the larvæ state live in dung, attack both man and animals, goading the latter sometimes almost to madness by their severe and incessant punctures. The winged horse-ticks (*Hippoboscæ*), the bird-flies (*Ornithomyiæ*), the wingless sheep-ticks (*Melophagi*), and the spider-flies (*Nycteribiæ*), and bee-lice (*Braulæ*), which are also destitute of wings, are truly parasitical in their habits, and pass their whole lives upon the skin of animals. Bot-flies, or gad-flies (*Œstridæ*), as they are sometimes called, appear to take no food while in the winged state, and are destitute of a proboscis; the nourishment obtained by their larvæ, which, as is well known, live in the bodies of horses, cattle, sheep, and other animals, being sufficient to last these insects during the rest of their lives. Some flies, though apparently harmless in the winged state, deposit their eggs on plants, on the juices of which their young subsist, and are oftentimes productive of immense injury to vegetation ; among these the most notorious for their depredations are the gall-gnats (*Cecidomyiæ*), including the wheat-fly and Hessian fly, the root-eating maggots of some of the long-legged gnats (*Tipulæ*), those of the flower-flies (*Anthomyiæ*), and the two-winged gall-flies and fruit-flies (*Ortalides*). To this list of noxious flies are to be added the common house-flies (*Muscæ*), which pass through the maggot state in dung and other filth, the blue-bottle or blow-flies, and meat-flies (*Luciliæ* and *Calliphoræ*), together with the maggot-producing or viviparous flesh-flies (*Sarcophagæ* and *Cynomyiæ*), whose maggots live in flesh, the cheese-fly (*Piophila*), the parent of the well-known skippers, and a few others that in the larva state attack our household stores.

Some flies are harmless in all their states, and many are eminently useful in various ways. Even the common house-flies, and flesh-flies, together with others for which no names exist in our language, render important services by feeding while larvæ upon dung, carrion, and all kinds of filth, by which means, and by similar services rendered by various tribes of scavenger-beetles, these offensive matters speedily disappear, instead of remaining to decay slowly, thereby tainting the air and rendering it unwholesome. Those whose larvæ live in stagnant water, such as gnats (*Culicidæ*), feather-horned gnats (*Chironomus*, &c.), the soldier-flies (*Stratiomyadæ*), the rat-tailed flies (*Helophilus*), &c., &c., tend to prevent the water from becoming putrid, by devouring the decayed animal and vegetable matter it contains. The maggots of some flies (*Mycetophilæ* and various *Muscadæ*) live in mushrooms, toadstools, and similar excrescences growing on trees ; those of others (*Sargi*, *Xylophagidæ*, *Asilidæ*, *Thereveæ*, *Milesiæ*, *Xylotæ*, *Borbori*, &c., &c.), in rotten wood and bark, thereby joining with the grubs of certain beetles to hasten the removal of these dead and useless substances, and make room for new and more vigorous vegetation. Some of these wood-eating insects, with others, when transformed to flies, (*Asilidæ* [Plate I. Fig. 4, Asilus æstuans], *Rhagionidæ*, *Dolichopidæ*, and *Xylophagidæ*,) prey on other insects. Some (*Syrphidæ*), though not predaceous themselves in the winged state, deposit their eggs among plant-lice, upon the blood of which their young afterwards subsist. Many (*Conopidæ*, excluding *Stomoxys*, *Tachinæ*, *Ocypteræ*, *Phoræ*, &c.) lay their eggs on caterpillars, and on various other larvæ, within the bodies of which the maggots hatched from these eggs live till they destroy their victims. And finally others (*Anthracidæ* and *Volucellæ*) drop their eggs in the nests of insects, whose offspring are starved to death, by being robbed of their food by the offspring of these cuckoo-flies. Besides performing their various appointed tasks in the economy of nature, flies, and other insects, subserve another highly important purpose, for which an all-wise Providence has designed them, namely, that of furnishing food to numerous other animals. Not to mention the various kinds of insect-eating quadrupeds, such as bats, moles, and the like, many birds live partly or entirely on insects. The finest song-birds, nightingales and thrushes, feast with the highest relish on maggots of all kinds, as well as on flies and other insects, while the warblers, vireos, and especially the fly-catchers and swallows, devour these two-winged insects in great numbers.

The seven foregoing orders constitute very natural groups, relatively of nearly equal importance, and sufficiently distinct from each other, but connected at different points by various resemblances. It is impossible to show the mutual relations of these orders, when they are arranged in a continuous series, but these can be better expressed and understood by grouping the orders together in a cluster, so that each order shall come in contact with several others.

Besides these seven orders, there are several smaller groups, which some naturalists have thought proper to raise to the rank of independent orders. Upon the principal of these a few remarks will now be made.

The little order Strepsiptera of Kirby, or Rhipiptera of Latreille, consists of certain minute insects, which undergo their transformations within the bodies of bees and wasps. One of them, the *Xenos Peckii*, was discovered by Professor Peck in the common brown wasp (*Polistes fuscata*) of this country. The larva is maggot-like, and lives between the rings of the back of the wasp; the pupa resembles that of some flies, and is cased in the dried skin of the larva. The females never acquire wings, and never leave the bodies of the bees or wasps into which they penetrate while young. The males, in the adult state, have a pair of short, narrow, and twisted members, instead of fore-wings, and two very large hind-wings, folded lengthwise like a fan. The mouth is provided with a pair of slender, sharp-pointed jaws, better adapted for piercing than for biting. It is very difficult to determine the proper place of these insects in a natural arrangement. Latreille puts them between the Lepidoptera and Diptera, but thinks them most nearly allied to some of the Hymenoptera.[1]

The flea tribe (*Pulicidæ*) was placed among the bugs, or Hemiptera, by Fabricius. It constitutes the order APTERA of Leach, SIPHONAPTERA of Latreille, and APHANIPTERA of Kirby. Fleas are destitute of wings, in the place whereof there are four little scales, pressed closely to the sides of their bodies; their mouth is fitted for suction, and provided with several lancet-like pieces for making punctures; they undergo a complete transformation; their larvæ are worm-like and without feet; and their pupæ have the legs free. These insects, of which there are many different kinds, are intermediate in their characteristics between the Hemiptera and the Diptera, and seem to connect more closely these two orders.

The earwigs (*Forficuladæ*), of which also there are many kinds, were placed by Linnæus in the order Coleoptera, but most naturalists now include them among the Orthoptera; indeed, they seem to be related to both orders, but most

[[1] Systematic authors now consider the order of Strepsiptera as simply a family, though a very aberrant one, of Coleoptera. It is placed after the Rhipiphoridæ, under the name Stylopidæ, from its principal genus, Stylops, which is parasitic in certain genera of bees; a species of this genus has been discovered in Nova Scotia, and will probably be found hereafter in New England. — LEC.]

closely to the Orthoptera, with which they agree in their partial transformations, and active pupæ. They form the little order DERMAPTERA of Leach, or EUPLEXOPTERA of Westwood.

The spider-flies, bird-flies, sheep-tick, &c. (*Hippoboscadæ*), which, with Latreille and others, I have retained among the Diptera, form the order HOMALOPTERA of Leach, and the English entomologists.

The May-flies, or case-flies (*Phryganeadæ*), have been separated from the Neuroptera; and constitute the order TRICHOPTERA of Kirby. Latreille and most of the naturalists of the continent of Europe still retain them in Neuroptera, to which they seem properly to belong.

The *Thrips* tribe consists of minute insects more closely allied to Hemiptera than to any other order, but resembling in some respects the Orthoptera also. It forms the little order THYSANOPTERA of Haliday; but I propose to leave it, as Latreille has done, among the Hemiptera.

The English entomologists separate from Hemiptera the cicadas or harvest-flies, lantern-flies, frog-hoppers, plant-lice, bark-lice, &c., under the name of HOMOPTERA; but these insects seem too nearly to resemble the true Hemiptera to warrant the separation.

Burmeister, a Prussian naturalist, has subdivided the Neuroptera into the orders NEUROPTERA and DICTYOTOPTERA, the latter to include the species which undergo only a partial transformation. If Hemiptera is to be subdivided, as above mentioned, then this division of Neuroptera will be justifiable also.

Objections have often been raised against the study of natural history, and many persons have been discouraged from attempting it, on account of the formidable array of scientific names and terms which it presents to the beginner; and some men of mean and contracted minds have made themselves merry at the expense of naturalists, and have sought to bring the writings of the latter into contempt, because of the scientific language and names they were obliged to employ. Entomology, or the science that treats of insects, abounds in such names more than any other branch of natural history; for the different kinds of insects very far outnumber the species in every class of the animal, vegetable, and mineral kingdoms. It is owing to this excessive number of species, and to the small size and unobtrusive character of many insects, that comparatively very few have received any common names, either in our own, or in other modern tongues; and hence most of those that have been described in works of natural history are known only by their scientific names. The latter have the advantage over other names in being intelligible to all well-educated persons in all parts of the world; while the common names of animals and plants in our own and other modern languages are very limited in their application, and moreover are often misapplied.

For example, the name weevil is given, in this country, to at least six different kinds of insects, two of which are moths, two are flies, and two are beetles. Moreover, since nearly four thousand species of weevils have actually been scientifically named and described, when mention is made of "the weevil," it may well be a subject of doubt to which of these four thousand species the speaker or writer intends to refer; whereas, if the scientific name of the species in question were made known, this doubt would at once be removed. To give each of these weevils a short, appropriate, significant, and purely English name, would be very difficult, if not impossible, and there would be great danger of overburdening the memory with such a number of names; but, by means of the ingenious and simple method of nomenclature invented by Linnæus, these weevils are all arranged under three hundred and fifty-five generical, or surnames, requiring in addition only a small number of different words, like christian names, to indicate the various species or kinds. There is oftentimes a great convenience in the use of single collective terms for groups of animals and plants, whereby the necessity for enumerating all the individual contents or the characteristics of these groups is avoided. Thus the single word *Ruminantia* stands for camels, lamas, giraffes, deer, antelopes, goats, sheep, and kine, or for all the hoofed quadrupeds which ruminate or chew the cud, and have no front teeth in the upper jaw; *Lepidoptera* includes all the various kinds of butterflies, hawk-moths, and millers or moths, or insects having wings covered with branny scales, and a spiral tongue instead of jaws, and whose young appear in the form of caterpillars. It would be difficult to find or invent any single English words which would be at once so convenient and so expressive. This, therefore, is an additional reason why scientific names ought to be preferred to all others, at least in works of natural history, where it is highly important that the objects described should have names that are short, significant in themselves, and not liable to be mistaken or misapplied.

There is no art, profession, trade, or occupation, which can be taught or learned without the use of technical words or phrases belonging to each, and which, to the inexperienced and untaught, are as unintelligible as the terms of science. It is not at all more difficult to learn and remember the latter than the former, when the attention has been properly given to the subject. The seaman, the farmer, and the mechanic soon become familiar with the names and phrases peculiar to their several callings, uncouth, and without apparent signification, as many of them are. So, too, the terms of science lose their forbidding and mysterious appearance and sound by the frequency of their recurrence, and finally become as harmonious to the ear, as they are clear and definite in their application.

CHAPTER II.

COLEOPTERA.

BEETLES. — SCARABÆIANS. — GROUND-BEETLES. — TREE-BEETLES. — COCKCHAFERS OR MAY-BEETLES. — FLOWER-BEETLES. — STAG-BEETLES. — BUPRESTIANS, OR SAW-HORNED BORERS. — SPRING-BEETLES. —TIMBER-BEETLES. — WEEVILS. — CYLINDRICAL BARK-BEETLES. — CAPRICORN-BEETLES, OR LONG-HORNED BORERS. — LEAF-BEETLES. — CRIOCERIANS. — LEAF-MINING BEETLES. — TORTOISE-BEETLES. — CHRYSOMELIANS. — CANTHARIDES.

THE wings of beetles are covered and concealed by a pair of horny cases or shells, meeting in a straight line on the top of the back, and usually having a little triangular or semicircular piece, called the scutel, wedged between their bases. Hence the order to which these insects belong is called COLEOPTERA, a word signifying wings in a sheath. Beetles * are biting-insects, and are provided with two pairs of jaws moving sidewise. Their young are grubs, and undergo a complete transformation in coming to maturity.

At the head of this order Linnæus placed a group of insects, to which he gave the name of SCARABÆUS. It includes the largest and most robust animals of the beetle kind, many of them remarkable for the singularity of their shape, and the formidable horn-like prominences with which they are furnished, — together with others, which, though they do not present the same imposing appearance, require to be noticed, on account of the injury sustained by vegetation from their attacks. An immense number of Scarabæians (SCARABÆIDÆ), as they may be called, are now known, differing greatly from each other, not only in structure, but

* Beetle, in old English, *betl*, *bytl*, or *bitel*, means a biter, or insect that bites.

in their habits in the larva and adult states. They are all easily distinguished by their short movable horns, or antennæ, ending with a knob, composed of three or more leaf-like pieces, which open like the petals of a flower-bud. Another feature that they possess in common is the projecting ridge (*clypeus*) of the forehead, which extends more or less over the face, like the visor or brim of a cap, and beneath the sides of this visor the antennæ are implanted. Moreover, the legs of these beetles, particularly the first pair, are fitted for digging, being deeply notched or furnished with several strong teeth on the outer edges; and the feet are five-jointed. This very extensive family of insects is subdivided into several smaller groups, each composed of beetles distinguished by various peculiarities of structure and habits. Some live mostly upon or beneath the surface of the earth, and were, therefore, called ground-beetles by De Geer; some, in their winged state, are found on trees, the leaves of which they devour, — they are the tree-beetles of the same author; and others, during the same period of their lives, frequent flowers, and are called flower-beetles. The ground-beetles, including the earth-borers (*Geotrupidæ*), and dung-beetles (*Coprididæ* and *Aphodiadæ*), which, in all their states, are found in excrement, the skin-beetles (*Trogidæ*), which inhabit dried animal substances, and the gigantic Hercules-beetles (*Dynastidæ*), which live in rotten wood or beneath old dung-heaps, must be passed over without further comment. The other groups contain insects that are very injurious to vegetation, and therefore require to be more particularly noticed.

One of the most common, and the most beautiful of the tree-beetles of this country, is the *Areoda lanigera*,[2] or woolly Areoda, sometimes also called the goldsmith-beetle (Plate II. Fig. 20). It is about nine tenths of an inch in length, broad oval in shape, of a lemon-yellow color above, glittering

[[2] *Areoda lanigera*, now called *Cotalpa lanigera;* the genus Cotalpa, established by Burmeister, differs from the true Areoda by not having the last joint of the tarsi armed beneath with an angular projection. — LEC.]

like burnished gold on the top of the head and thorax; the under-side of the body is copper-colored, and thickly covered with whitish wool; and the legs are brownish yellow, or brassy, shaded with green. These fine beetles begin to appear in Massachusetts about the middle of May, and continue generally till the twentieth of June. In the morning and evening twilight they come forth from their retreats, and fly about with a humming and rustling sound among the branches of trees, the tender leaves of which they devour. Pear-trees are particularly subject to their attacks, but the elm, hickory, poplar, oak, and probably also other kinds of trees, are frequented and injured by them. During the middle of the day they remain at rest upon the trees, clinging to the under-sides of the leaves, and endeavor to conceal themselves by drawing two or three leaves together, and holding them in this position with their long unequal claws. In some seasons they occur in profusion, and then may be obtained in great quantities by shaking the young trees on which they are lodged in the daytime, as they do not attempt to fly when thus disturbed, but fall at once to the ground. The larvæ of these insects are not known; probably they live in the ground upon the roots of plants. The group to which the goldsmith-beetle belongs may be called Rutilians (RUTILIDÆ), from *Rutela*, or more correctly *Rutila*, signifying shining, the name of the principal genus included in it. The Rutilians connect the ground-beetles with the tree-beetles of the following group, having the short and robust legs of the former, with the leaf-eating habits of the latter.

Fig 9.

The spotted Pelidnota, *Pelidnota punctata* (Fig. 9), is also arranged among the Rutilians. This large beetle is found on the cultivated and wild grape-vine, sometimes in great abundance, during the months of July and August. It is of an oblong oval shape, and about an inch long. The wing-covers are tile-colored, or dull brownish-yellow, with three distinct black dots on each ; the thorax is darker and slightly bronzed, with a black dot on each side ; the body beneath, and the legs, are of a deep bronzed green color. These beetles fly by day ; but may also be seen at the same time on the leaves of the grape, which are their only food. They sometimes prove very injurious to the vine. The only method of destroying them is to pick them off by hand and crush them under foot. The larvæ live in rotten wood, such as the stumps and roots of dead trees ; and do not differ essentially from those of other Scarabæians.

Among the tree-beetles, those commonly called dors, chafers, May-bugs, and rose-bugs, are the most interesting to the farmer and gardener, on account of their extensive ravages, both in the winged and larva states. They were included by Fabricius in the genus *Melolontha*, a word used by the ancient Greeks to distinguish the same kind of insects, which were supposed by them to be produced from or with the flowers of apple-trees, as the name itself implies. These beetles, together with many others, for which no common names exist in our language, are now united in one family called MELOLONTHADÆ, or Melolonthians. The following are the general characters of these insects. The body is oblong oval, convex, and generally of a brownish color ; the antennæ are nine or more commonly ten jointed, the knob is much longer in the males than in the females, and consists generally of three leaf-like pieces, sometimes of a greater number, which open and shut like the leaves of a book ; the visor is short and wide ; the upper jaws are furnished at the base on the inner side with an oval space, crossed by ridges, like a millstone, for grinding ; the thorax is transversely square, or nearly so ; the wing-cases do not cover the whole of the body, the hinder extremity of which is exposed ; the legs are rather long, the first pair armed externally with two or three teeth ; and the claws are notched beneath, or are split at the end like the nib of a pen. The powerful and horny jaws are admirably fitted for cutting and grinding the leaves of plants, upon which these beetles subsist; their notched or double claws support them securely on the foliage; and their strong and jagged fore-legs, being formed for digging in the ground, point out the place of their transformations.

The habits and transformations of the common cockchafer of Europe have been carefully observed, and will serve to exemplify those of the other insects of this family, which, as far as they are known, seem to be nearly the same. This insect devours the leaves of trees and shrubs. Its duration in the perfect state is very short, each individual living only about a week, and the species entirely disappearing in the course of a month. After the sexes have paired, the males perish, and the females enter the earth to the depth of six inches or more, making their way by means of the strong teeth which arm the fore-legs; here they deposit their eggs, amounting, according to some writers, to nearly one hundred, or, as others assert, to two hundred from each female, which are abandoned by the parent, who generally ascends again to the surface, and perishes in a short time.

From the eggs are hatched, in the space of fourteen days, little whitish grubs, each provided with six legs near the head, and a mouth furnished with strong jaws. When in a state of rest, these grubs usually curl themselves in the shape of a crescent. They subsist on the tender roots of various plants, committing ravages among these vegetable substances, on some occasions of the most deplorable kind, so as totally to disappoint the best-founded hopes of the husbandman. During the summer they live under the thin coat of vegetable mould near the surface, but, as winter approaches, they descend below the reach of frost, and remain torpid until the succeeding spring, at which time they change their skins, and reascend to the surface for food. At the close of their third summer (or, as some say, of the fourth or fifth) they cease eating, and penetrate about two feet deep into the earth; there, by its motions from side to side, each grub forms an oval cavity, which is lined by some glutinous substance thrown from its mouth. In this cavity it is changed to a pupa by casting off its skin. In this state, the legs, antennæ, and wing-cases of the future beetle are visible through the transparent skin which envelops them, but appear of a yellowish-white color; and thus it remains until the month of February, when the thin film which encloses the body is rent, and three months afterwards the perfected beetle digs its way to the surface, from which it finally emerges during the night. According to Kirby and Spence, the grubs of the cockchafer sometimes destroy whole acres of grass by feeding on its roots. They undermine the richest meadows, and so loosen the turf that it will roll up as if cut by a turfing spade. They do not confine themselves to grass, but eat the roots of wheat, of other grains, and also those of young trees. About seventy years ago, a farmer near Norwich, in England, suffered much by them, and, with his man, gathered eighty bushels of the beetles. In the year 1785 many provinces in France were so ravaged by them, that a premium was offered by government for the best mode of destroying them. The Society of Arts in London, during many years, held forth a premium for the best account of this insect, and the means of checking its ravages, but without having produced one successful claimant.

In their winged state, these beetles, with several other species, act as conspicuous a part in injuring the trees as the grubs do in destroying the herbage. During the month of May they come forth from the ground, whence they have received the name of May-bugs, or May-beetles. They pass the greater part of the day upon trees, clinging to the undersides of the leaves, in a state of repose. As soon as evening approaches, they begin to buzz about among the branches, and continue on the wing till towards midnight. In their droning flight they move very irregularly, darting hither and thither with an uncertain aim, hitting against objects in their way with a force that often causes them to fall to the ground.

They frequently enter houses in the night, apparently attracted, as well as dazzled and bewildered, by the lights. Their vagaries, in which, without having the power to harm, they seem to threaten an attack, have caused them to be called dors, — that is, darers; while their seeming blindness and stupidity have become proverbial, in the expressions, "blind as a beetle," and "beetle-headed."

Besides the leaves of fruit-trees, they devour those of various forest-trees and shrubs, with an avidity not much less than that of the locust, so that, in certain seasons, and in particular districts, they become an oppressive scourge, and the source of much misery to the inhabitants. Mouffet relates that, in the year 1574, such a number of them fell into the river Severn as to stop the wheels of the water-mills; and, in the Philosophical Transactions, it is stated, that in the year 1688 they filled the hedges and trees of Galway, in such infinite numbers as to cling to each other like bees when swarming; and, when on the wing, darkened the air, annoyed travellers, and produced a sound like distant drums. In a short time the leaves of all the trees, for some miles round, were so totally consumed by them, that at midsummer the country wore the aspect of the depth of winter.

Another chafer, *Anomala vitis F.* is sometimes exceedingly injurious to the vine. It prevails in certain provinces of France, where it strips the vines of their leaves, and also devours those of the willow, poplar, and fruit-trees.

The animals and birds appointed to check the ravages of these insects are, according to Latreille, the badger, weasel, marten, bats, rats, the common dung-hill fowl, and the goat-sucker or night-hawk. To this list may be added the common crow, which devours not only the perfect insects, but their larvæ, for which purpose it is often observed to follow the plough. In "Anderson's Recreations" it is stated, that "a cautious observer, having found a nest of five young jays, remarked that each of these birds, while yet very young, consumed at least fifteen of these full-sized grubs in one day, and of course would require many more of a smaller size. Say that, on an average of sizes, they consumed twenty apiece, these for the five make one hundred. Each of the parents consume say fifty; so that the pair and family devour two hundred every day. This, in three months, amounts to twenty thousand in one season. But as the grub continues in that state four seasons, this single pair, with their family alone, without reckoning their descendants after the first year, would destroy eighty thousand grubs. Let us suppose that the half, namely, forty thousand, are females, and it is known that they usually lay about two hundred eggs each, it will appear, that no less than eight millions have been destroyed, or prevented from being hatched, by the labors of a single family of jays. It is by reasoning in this way, that we learn to know of what importance it is to attend to the economy of nature, and to be cautious how we derange it by our short-sighted and futile operations." Our own country abounds with insect-eating beasts and birds, and without doubt the more than abundant Melolonthæ form a portion of their nourishment.

We have several Melolonthians whose injuries in the perfect and grub state approach to those of the European cockchafer. *Phyllophaga* * *quercina* of Knoch, the May-beetle, as it is generally called here, is our common species. (Fig. 10.) It is of a chestnut-brown color, smooth, but finely punctured, that is, covered with little impressed dots, as if pricked with the point of a needle; each wing-case has two or

Fig. 10.

* A genus proposed by me in 1826. It signifies leaf-eater. Dejean subsequently called this genus *Ancylonycha*.[3]

[[3] The genus *Phyllophaga* was indeed proposed by Dr. Harris, but was not accompanied by any description; it must therefore yield to the name *Lachnosterna* of Hope, described in 1837. Burmeister has improperly adopted for the genus the name given by Dejean, but which was not sanctioned by a description until 1845. It is a very numerous genus, and many of the species resemble each other very closely. — LEC.]

three slightly elevated longitudinal lines; the breast is clothed with yellowish down. The knob of its antennæ contains only three leaf-like joints. Its average length is nine tenths of an inch. In its perfect state it feeds on the leaves of trees, particularly on those of the cherry-tree. It flies with a humming noise in the night, from the middle of May to the end of June, and frequently enters houses, attracted by the light. In the course of the spring, these beetles are often thrown from the earth by the spade and plough, in various states of maturity, some being soft and nearly white, their superabundant juices not having evaporated, while others exhibit the true color and texture of the perfect insect. The grubs devour the roots of grass and of other plants, and in many places the turf may be turned up like a carpet in consequence of the destruction of the roots. The grub* is a white worm with a brownish head, and, when fully grown, is nearly as thick as the little finger. It is eaten greedily by crows and fowls. The beetles are devoured by the skunk, whose beneficial foraging is detected in our gardens by its abundant excrement filled with the wing-cases of these insects.

A writer in the "New York Evening Post" says, that the beetles, which frequently commit serious ravages on fruit-trees, may be effectually exterminated by shaking them from the trees every evening. In this way two pailfuls of beetles were collected on the first experiment; the number caught regularly decreased until the fifth evening, when only two beetles were to be found. The best time, however, for shaking trees on which the May-beetles are lodged, is in the morning, when the insects do not attempt to fly. They are most easily collected in a cloth spread under the trees to receive them when they fall, after which they should be thrown into boiling water to kill them, and may then be given as food to swine.

* There is a grub, somewhat resembling this, which is frequently found under old manure-heaps, and is commonly called muck-worm. It differs, however, in some respects, from that of the May-beetle, or dor-bug, and is transformed to a dung-beetle called *Scarabæus relictus* by Mr. Say.

There is an undescribed kind of *Phyllophaga*, or leaf-eater, called, in my Catalogue of the Insects of Massachusetts,* *fraterna*, because it is nearly akin to the *quercina*, in general appearance. It differs from the latter, however, in being smaller, and more slender; the punctures on its thorax and wing-covers are not so distinct, and the three elevated lines on the latter are hardly visible. It measures thirteen twentieths of an inch in length. This beetle may be seen in the latter part of June and the beginning of July. Its habits are similar to those of the more abundant May-beetle or dor-bug.

Another common *Phyllophaga* has been described by Knoch and Say, under the name of *hirticula* (Fig. 11), meaning a little hairy. It is of a bay-brown color, the punctures on the thorax are larger and more distinct than in the *quercina*, and on each wing-cover are three longitudinal rows of short, yellowish hairs. It measures about seven tenths of an inch in length. Its time of appearance is in June and July.

Fig. 11.

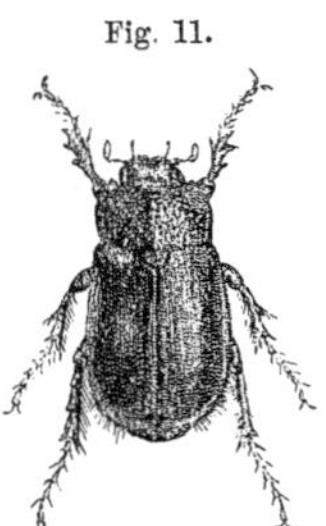

In some parts of Massachusetts the *Phyllophaga Georgicana* (Fig. 12) of Gyllenhall, or Georgian leaf-eater, takes the place of the *quercina*. It is extremely common, during May and June, in Cambridge, where the other species is rarely seen. It is of a bay-brown color, entirely covered on the upper side with very short, yellowish gray hairs, and measures seven tenths of an inch, or more, in length.

Fig. 12.

* In order to save unnecessary repetitions, it may be well to state, that the Catalogue above named, to which frequent reference will be made in the course of this treatise, was drawn up by me, and was published in Professor Hitchcock's Report on the Geology, Mineralogy, Botany, and Zoölogy of Massachusetts, and that two editions of it appeared with the Report, the first in 1833, and the second, with numerous additions, in 1835.

Phyllophaga pilosicollis (Fig. 13) of Knoch, or the hairy-necked leaf-eater, is a small chafer, of an ochre-yellow color, with a very hairy thorax. It is often thrown out of the ground by the spade, early in the spring; but it does not voluntarily come forth till the middle of May. It measures half an inch in length.

Fig. 13.

Hentz's *Melolontha variolosa*[4] (Fig. 14), or scarred Melolontha, differs essentially from the foregoing beetles in the structure of its antennæ, the knob of which consists of seven narrow, strap-shaped ochre-yellow leaves, which are excessively long in the males. This fine insect is of a light brown color, with irregular whitish blotches, like scars, on the thorax and wing-covers. It measures nine tenths of an inch, or more, in length. It occurs abundantly, in the month of July, at Martha's Vineyard, and in some other places near the coast; but is rare in other parts of Massachusetts.

Fig. 14.

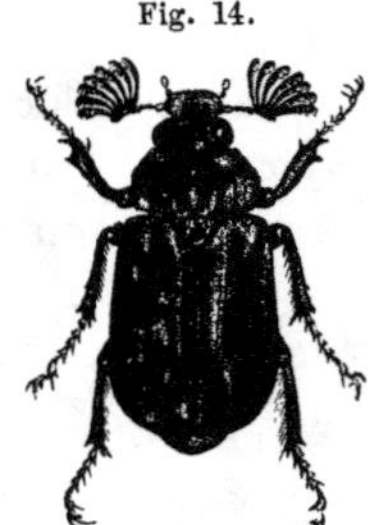

The foregoing Melolonthians are found in gardens, nurseries, and orchards, where they are more or less injurious to the fruit-trees, in proportion to their numbers in different seasons. They also devour the leaves of various forest-trees, such as the elm, maple, and oak.

Omaloplia[5] *vespertina* (Plate II. Fig. 14) of Gyllenhal, and *sericea* of Illiger, attack the leaves of the sweetbrier, or sweet-leaved rose, on which they may be found in profusion in the evening, about the last of June. They somewhat resemble the May-beetles in form, but are proportionally shorter and

[4 *Melolontha variolosa.* This insect belongs to the genus *Polyphylla*, proposed by Dr. Harris, and now adopted by all entomologists. — Lec.]

[5 *Omaloplia.* The species here mentioned, with all the other allied American species, belong rather to *Serica* of M'Leay, than to true *Omaloplia*, which is thus far confined to the other continent. — Lec.]

5

thicker, and much smaller in size. The first of them, the vespertine or evening Omaloplia, is bay-brown; the wing-covers are marked with many longitudinal shallow furrows, which, with the thorax, are thickly punctured. This beetle varies in length from three to four tenths of an inch. *Omaloplia sericea*, the silky Omaloplia, closely resembles the preceding in everything but its color, which is a very deep chestnut-brown, iridescent or changeable like satin, and reflecting the colors of the rainbow.

All these Melolonthians are nocturnal insects, never appearing, except by accident, in the day, during which they remain under shelter of the foliage of trees and shrubs, or concealed in the grass. Others are truly day-fliers, committing their ravages by the light of the sun, and are consequently exposed to observation.

One of our diurnal Melolonthians is supposed by many naturalists to be the *Anomala varians* (Fig. 15) of Fabricius; and it agrees very well with this writer's description of the *lucicola;* but Professor Germar thinks it to be an undescribed species, and proposes to name it *cœlebs*. It resembles the vine-chafer of Europe in its habits, and is found in the months of June and July on the cultivated and wild grape-vines, the leaves of which it devours. During the same period, these chafers may be seen in still greater numbers on various kinds of sumach, which they often completely despoil of their leaves. They are of a broad oval shape, and very variable in color. The head and thorax of the male are greenish black, margined with dull ochre or tile-red, and thickly punctured; the wing-covers are clay-yellow, irregularly furrowed, and punctured in the furrows; the legs are pale red, brown, or black. The thorax of the female is clay-yellow, or tile-red, sometimes with two oblique blackish spots on the top, and sometimes almost entirely black; the wing-covers resemble those of the male; the legs are clay-yellow,

Fig. 15.

or light red. The males are sometimes entirely black, and this variety seems to be the beetle called *atrata*, by Fabricius. The males measure nearly, and the females rather more than seven twentieths of an inch in length. In the year 1825, these insects appeared on the grape-vines in a garden in this vicinity; they have since established themselves on the spot, and have so much multiplied in subsequent years as to prove exceedingly hurtful to the vines. In many other gardens they have also appeared, having probably found the leaves of the cultivated grape-vine more to their taste than their natural food. Should these beetles increase in numbers, they will be found as difficult to check and extirpate as the destructive vine-chafers of Europe.

The rose-chafer, or rose-bug, as it is more commonly and incorrectly called, is also a diurnal insect. It is the *Melolontha subspinosa* (Fig. 16) of Fabricius, by whom it was first described, and belongs to the modern genus *Macrodactylus* of Latreille. Common as this insect is in the vicinity of Boston, it is, or was a few years ago, unknown in the northern and western parts of Massachusetts, in New Hampshire, and in Maine. It may, therefore, be well to give a brief description of it. This beetle measures seven twentieths of an inch in length. Its body is slender, tapers before and behind, and is entirely covered with very short and close ashen-yellow down; the thorax is long and narrow, angularly widened in the middle of each side, which suggested the name *subspinosa*, or somewhat spined; the legs are slender, and of a pale red color; the joints of the feet are tipped with black, and are very long, which caused Latreille to call the genus *Macrodactylus*, that is, long toe, or long foot.

Fig. 16.

The natural history of the rose-chafer, one of the greatest scourges with which our gardens and nurseries have been afflicted, was for a long time involved in mystery, but is at last fully cleared up.* The prevalence of this insect on the

* See my Essay in the Massachusetts Agricultural Repository and Journal

rose, and its annual appearance coinciding with the blossoming of that flower, have gained for it the popular name by which it is here known. For some time after they were first noticed, rose-bugs appeared to be confined to their favorite, the blossoms of the rose; but within forty years they have prodigiously increased in number, have attacked at random various kinds of plants in swarms, and have become notorious for their extensive and deplorable ravages. The grape-vine in particular, the cherry, plum, and apple trees, have annually suffered by their depredations; many other fruit-trees and shrubs, garden vegetables and corn, and even the trees of the forest and the grass of the fields, have been laid under contribution by these indiscriminate feeders, by whom leaves, flowers, and fruits are alike consumed. The unexpected arrival of these insects in swarms, at their first coming, and their sudden disappearance at the close of their career, are remarkable facts in their history. They come forth from the ground during the second week in June, or about the time of the blossoming of the damask rose, and remain from thirty to forty days. At the end of this period the males become exhausted, fall to the ground and perish, while the females enter the earth, lay their eggs, return to the surface, and, after lingering a few days, die also.

The eggs laid by each female are about thirty in number, and are deposited from one to four inches beneath the surface of the soil; they are nearly globular, whitish, and about one thirtieth of an inch in diameter, and are hatched twenty days after they are laid. The young larvæ begin to feed on such tender roots as are within their reach. Like other grubs of the Scarabæians, when not eating they lie upon the side, with the body curved, so that the head and tail

Vol. X. p. 8, reprinted in the New England Farmer, Vol. VI. p. 18, &c.; my Discourse before the Massachusetts Horticultural Society, p. 31, 8vo, Cambridge, 1832; Dr. Green's communication on this insect in the New England Farmer Vol. VI. pp. 41, 49, &c.; my Report on Insects Injurious to Vegetation, in Massachusetts House Document, No. 72, April, 1838, p. 70; and a communication in the New England Farmer, Vol. IX. p 1.

are nearly in contact; they move with difficulty on a level surface, and are continually falling over on one side or the other. They attain their full size in the autumn, being then nearly three quarters of an inch long, and about an eighth of an inch in diameter. They are of a yellowish-white color, with a tinge of blue towards the hinder extremity, which is thick, and obtuse or rounded; a few short hairs are scattered on the surface of the body; there are six short legs, namely, a pair to each of the first three rings behind the head, and the latter is covered with a horny shell of a pale rust color. In October they descend below the reach of frost, and pass the winter in a torpid state. In the spring they approach towards the surface, and each one forms for itself a little cell of an oval shape, by turning round a great many times, so as to compress the earth and render the inside of the cavity hard and smooth. Within this cell the grub is transformed to a pupa, during the month of May, by casting off its skin, which is pushed downwards in folds from the head to the tail. The pupa has somewhat the form of the perfected beetle; but it is of a yellowish-white color, and its short stump-like wings, its antennæ, and its legs are folded upon the breast; and its whole body is enclosed in a thin film, that wraps each part separately. During the month of June this filmy skin is rent, the included beetle withdraws from its body and its limbs, bursts open its earthen cell, and digs its way to the surface of the ground. Thus the various changes, from the egg to the full development of the perfected beetle, are completed within the space of one year.

Such being the metamorphoses and habits of these insects, it is evident that we cannot attack them in the egg, the grub, or the pupa state; the enemy in these stages is beyond our reach, and is subject to the control only of the natural but unknown means appointed by the Author of Nature to keep the insect tribes in check. When they have issued from their subterranean retreats, and have congregated upon our vines, trees, and other vegetable productions, in the complete enjoyment of their propensities, we must unite our efforts to seize and crush the invaders. They must indeed be crushed, scalded, or burned, to deprive them of life, for they are not affected by any of the applications usually found destructive to other insects. Experience has proved the utility of gathering them by hand, or of shaking them or brushing them from the plants into tin vessels containing a little water. They should be collected daily during the period of their visitation, and should be committed to the flames or killed by scalding water. The late John Lowell, Esq., states,* that in 1823 he discovered, on a solitary apple-tree, the rose-bugs "in vast numbers, such as could not be described, and would not be believed if they were described, or, at least, none but an ocular witness could conceive of their numbers. Destruction by hand was out of the question," in this case. He put sheets under the tree, and shook them down, and burned them.

Dr. Green, of Mansfield, whose investigations have thrown much light on the history of this insect, proposes protecting plants with millinet, and says that in this way only did he succeed in securing his grape-vines from depredation. His remarks also show the utility of gathering them. "Eighty-six of these spoilers," says he, "were known to infest a single rose-bud, and were crushed with one grasp of the hand." Suppose, as was probably the case, that one half of them were females; by this destruction, eight hundred eggs, at least, were prevented from becoming matured. During the time of their prevalence, rose-bugs are sometimes found in immense numbers on the flowers of the common white-weed, or ox-eye daisy (*Chrysanthemum leucanthemum*), a worthless plant, which has come to us from Europe, and has been suffered to overrun our pastures and encroach on our mowing-lands. In certain cases it may become expedient rapidly to mow down the infested white-weed in dry

* Massachusetts Agricultural Repository, Vol. IX. p. 145.

pastures, and consume it, with the sluggish rose-buds, on the spot.

Our insect-eating birds undoubtedly devour many of these insects, and deserve to be cherished and protected for their services. Rose-bugs are also eaten greedily by domesticated fowls; and when they become exhausted and fall to the ground, or when they are about to lay their eggs, they are destroyed by moles, insects, and other animals, which lie in wait to seize them. Dr. Green informs us, that a species of dragon-fly, or devil's-needle, devours them. He also says that an insect, which he calls the enemy of the cut-worm, probably the larva of a Carabus or predaceous ground-beetle, preys on the grubs of the common dor-bug. In France the golden ground-beetle (*Carabus auratus*) devours the female dor or chafer at the moment when she is about to deposit her eggs. I have taken one specimen of this fine ground-beetle in Massachusetts, and we have several other kinds, equally predaceous, which probably contribute to check the increase of our native Melolonthians.

Very few of the flower-beetles are decidedly injurious to vegetation. Some of them are said to eat leaves; but the greater number live on the pollen and the honey of flowers, or upon the sap that oozes from the wounds of plants. In the infant or grub state, most of them eat only the crumbled substance of decayed roots and stumps; a few live in the wounds of trees, and by their depredations prevent them from healing, and accelerate the decay of the trunk.

The flower-beetles belong chiefly to a group called CETONIADÆ, or Cetonians. They are easily distinguished from the other Scarabæians by their lower jaws, which are generally soft on the inside, and are often provided with a flat brush of hairs, that serves to collect the pollen and juices on which they subsist. Their upper jaws have no grinding plate on the inside. Their antennæ consist of ten joints, the last three of which form a three-leaved oval knob. The head is often square, with a large and wide visor, overhanging and entirely concealing the upper lip. The thorax is either rounded, somewhat square, or triangular. The wing-cases do not cover the end of the body. The fore legs are deeply notched on the outer edge; and the claws are equal and entire. These beetles are generally of an oblong oval form, somewhat flattened above, and often brilliantly colored and highly polished, sometimes also covered with hairs. Most of the bright-colored kinds are day-fliers; those of dark and plain tints are generally nocturnal beetles. Some of them are of immense size, and have been styled the princes of the beetle tribes; such are the Incas of South America, and the Goliah beetle (*Hegemon Goliatus*) of Guinea, the latter being more than four inches long, two inches broad, and thick and heavy in proportion.

Two American Cetonians must suffice as examples in this group. The first is the Indian Cetonia, *Cetonia Inda* * (Fig. 17), one of our earliest visitors in the spring, making its appearance towards the end of April or the beginning of May, when it may sometimes be seen in considerable numbers around the borders of woods, and in dry, open fields, flying just above the grass with a loud humming sound, like a humble-bee, for which perhaps it might at first sight be mistaken. Like other insects of the same genus, it has a broad body, very obtuse behind, with a triangular thorax, and a little wedge-shaped piece on each side between the hinder angles of the thorax and the shoulders of the wing-covers; the latter, taken together, form an oblong square, but are somewhat notched or widely scalloped on the middle of the outer edges. The head and thorax of this beetle are dark copper-brown, or almost black, and thickly covered with short greenish-yellow hairs; the wing-cases are light yellowish-

Fig. 17.

* *Scarabæus Indus* of Linnæus, *Cetonia barbata* of Say.[6]

[[6] *Cetonia Inda.* The old genus *Cetonia* has been divided recently into many genera, some of which have again been merged together by later investigators; our species belong to the one called *Euryomia*, as enlarged by Lacordaire. — Lec.]

brown, but changeable, with pearly and metallic tints, and spattered with numerous irregular black spots; the underside of the body, which is very hairy, is of a black color, with the edges of the rings and the legs dull red. It measures about six tenths of an inch in length. During the summer months the Indian Cetonia is not seen; but about the middle of September a new brood comes forth, the beetles appearing fresh and bright, as though they had just completed their last transformation. At this time they may be found on the flowers of the golden-rod, eating the pollen, and also in great numbers on corn-stalks, and on the trunks of the locust-tree, feeding upon the sweet sap of these plants. Fortunate would it be for us if they fed on these only; but their love of sweets leads them to attack our finest peaches, which, as soon as ripe, they begin to devour, and in a very few hours entirely spoil. I have taken a dozen of them from a single peach, into which they had burrowed so that nothing but the naked tips of their hind-body could be seen; and not a ripe peach remained unbitten by them on the tree. When touched, they leave a strong and disagreeable scent upon the fingers. On the approach of cold weather they disappear, but I have not been able to ascertain what becomes of them at this time, and only conjecture that they get into some warm and sheltered spot, where they pass the winter in a torpid state, and in the spring issue from their retreats, and finish their career by depositing their eggs for another brood. Those that are seen in the spring want the freshness of the autumnal beetles, a circumstance that favors my conjecture. Their hovering over and occasionally dropping upon the surface of the ground, is probably for the purpose of selecting a suitable place to enter the earth and lay their eggs. Hence I suppose that their larvæ or grubs may live on the roots of herbaceous plants.

The other Cetonian beetle to be described is the *Osmoderma scaber*,* or rough Osmoderma (Fig. 18). It is a large

* *Trichius scaber*, Palisot de Beauvois; *Gymnodus scaber*, Kirby.

insect, with a broad, oval, and flattened body; the thorax is nearly round, but wider than long; there are no wedge-shaped pieces between the corners of the thorax and the shoulders of the wing-cases, and the outer edges of the latter are entire. It is of a purplish-black color, with a coppery lustre; the head is punctured, concave or hollowed on the top, with the edge of the broad visor turned up in the males; nearly flat, and with the edge of the visor not raised in the females; the wing-cases are so thickly and deeply and irregularly punctured as to appear almost as rough as shagreen; the under-side of the body is smooth and without hairs; and the legs are short and stout. In addition to the differences between the sexes above described, it may be mentioned that the females are generally much larger than the males, and often want the coppery polish of the latter. They measure from eight tenths of an inch to one inch and one tenth in length. They are nocturnal insects, and conceal themselves during the day in the crevices and hollows of trees, where they feed upon the sap that flows from the bark. They have the odor of Russia leather, and give this out so powerfully that their presence can be detected, by the scent alone, at the distance of two or three yards from the place of their retreat. This strong smell suggested the name *Osmoderma*, that is, scented skin, given to these beetles by the French naturalists. They seem particularly fond of the juices of cherry and apple trees, in the hollows of which I have often discovered them. Their larvæ live in the hollows of these same trees, feeding upon the diseased wood, and causing it more rapidly to decay. They are whitish fleshy grubs, with a reddish hard-shelled head, and closely resemble the grubs of the common dor-beetle. In the autumn each one makes an oval cell or pod, of fragments of wood, strongly cemented with a kind

Fig. 18.

of glue; it goes through its transformation within this cell, and comes forth in the beetle form in the month of July.

We have another scented beetle, equal in size to the preceding, of a deep mahogany-brown color, perfectly smooth, and highly polished, and the male has a deep pit before the middle of the thorax. This species of *Osmoderma* is called *eremicola* * (Fig. 19), a name that cannot be rendered literally into English by any single word; it signifies wilderness-inhabitant, for which might be substituted hermit. I believe that this insect lives in forest-trees, but the larva is unknown to me.

Fig. 19.

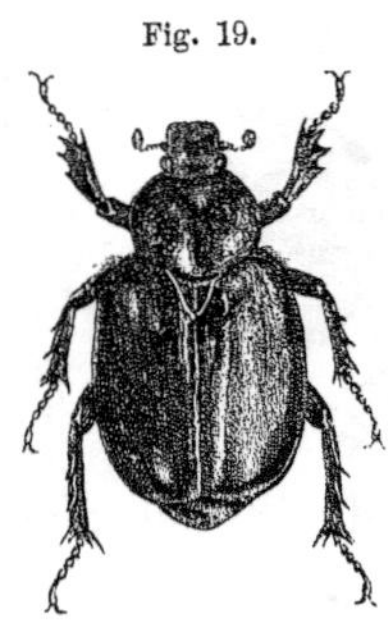

The family LUCANIDÆ, or Lucanians, so named from the Linnæan genus LUCANUS, must be placed next to the Scarabæians in a natural arrangement. This family includes the insects called stag-beetles, horn-bugs, and flying-bulls, names that they have obtained from the great size and peculiar form of their upper jaws, which are sometimes curved like the horns of cattle, and sometimes branched like the antlers of a stag. In these beetles the body is hard, oblong, rounded behind, and slightly convex; the head is large and broad, especially in the males; the thorax is short, and as wide as the abdomen; the antennæ are rather long, elbowed or bent in the middle, and composed of ten joints, the last three or four of which are broad, leaf-like, and project on the inside, giving to this part of the antennæ a resemblance to the end of a key; the upper jaws are usually much longer in the males than in the females, but even those of the latter extend considerably beyond the mouth; each of the under jaws is provided with a long hairy pencil or brush, which can be seen projecting beyond the mouth between the feelers; and the under lip has two shorter pencils of the same kind; the

* *Cetonia eremicola* of Knoch.

fore legs are oftentimes longer than the others, with the outer edge of the shanks notched into teeth; the feet are five-jointed, and the nails are entire and equal. These beetles fly abroad during the night, and frequently enter houses at that time, somewhat to the alarm of the occupants; but they are not venomous, and never attempt to bite without provocation. They pass the day on the trunks of trees, and live upon the sap, for procuring which the brushes of their jaws and lip seem to be designed. They are said also occasionally to bite and seize caterpillars and other soft-bodied insects, for the purpose of sucking out their juices. They lay their eggs in crevices of the bark of trees, especially near the roots, where they may sometimes be seen thus employed. The larvæ hatched from these eggs resemble the grubs of the Scarabæians in color and form, but they are smoother, or not so much wrinkled. The grubs of the large kinds are said to be six years in coming to their growth, living all this time in the trunks and roots of trees, boring into the solid wood, and reducing it to a substance resembling very coarse sawdust; and the injury thus caused by them is frequently very considerable. When they have arrived at their full size, they enclose themselves in egg-shaped pods, composed of gnawed particles of wood and bark stuck together and lined with a kind of glue; within these pods they are transformed to pupæ, of a yellowish-white color, having the body and all the limbs of the future beetle encased in a whitish film, which being thrown off in due time, the insects appear in the beetle form, burst the walls of their prison, crawl through the passages the larvæ had gnawed, and come forth on the outside of the trees.

The largest of these beetles in the New England States was first described by Linnæus, under the name of *Lucanus Capreolus* * (Fig. 20), signifying the young roebuck; but here it is called the horn-bug. Its color is a deep mahogany-

* *Lucanus Dama* of Fabricius.

brown; the surface is smooth and polished; the upper jaws of the male are long, curved like a sickle, and furnished internally beyond the middle with a little tooth; those of the female are much shorter, and also toothed; the head of the male is broad and smooth, that of the other sex narrower and rough with punctures. The body of this beetle measures from one inch to one inch and a quarter, exclusive of the jaws. The time of its appearance is in July and the beginning of August. The grubs live in the trunks and roots of various kinds of trees, but particularly in those of old apple-trees, willows, and oaks. All the foregoing beetles have, by some naturalists, been gathered into a single tribe, called lamellicorn or leaf-horned beetles, on account of the leaf-like joints wherewith the end of their antennæ is provided.

Fig. 20.

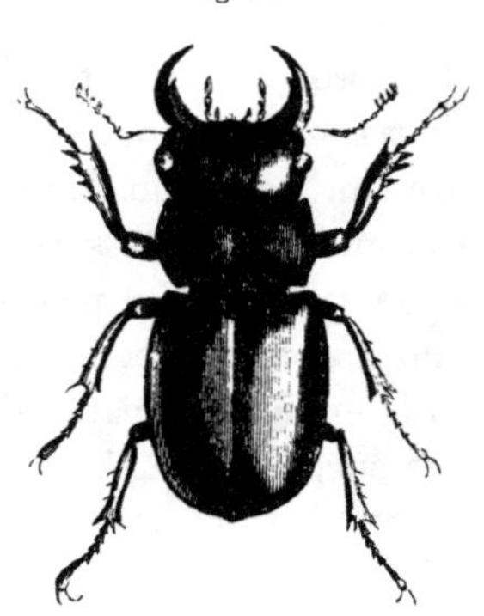

The beetles next to be described have been brought together into one great tribe, named serricorn or saw-horned beetles, because the tips of the joints of their antennæ usually project more or less on the inside, somewhat like the teeth of a saw. The beetles belonging to the family BUPRESTIDÆ, or the Buprestians, have antennæ of this kind. The *Buprestis* of the ancients, as its name signifies in Greek, was a poisonous insect, which, being swallowed with grass by grazing cattle, produced a violent inflammation, and such a degree of swelling as to cause the cattle to burst. Linnæus, however, unfortunately applied this name to the insects of the above-mentioned family, none of which are poisonous to animals, and are rarely, if ever, found upon the grass. It is in allusion to the original signification of the word *Buprestis*, that popular English writers on natural history sometimes give the name of burncow to the harmless Buprestians; while the French, with greater propriety, call them *richards*, on account of the rich and brilliant colors wherewith many of them are adorned. The Buprestians, then, according to the Linnæan application, or rather misapplication, of the name, are hard-shelled beetles, often brilliantly colored, of an elliptical or oblong oval form, obtuse before, tapering behind, and broader than thick, so that, when cut in two transversely, the section is oval. The head is sunk to the eyes in the fore part of the thorax; and the antennæ are rather short, and notched on one side like the teeth of a saw. The thorax is broadest behind, and usually fits very closely to the shoulders of the wing-covers. The legs are rather short, and the feet are formed for standing firmly, rather than for rapid motion; the soles being composed of four rather wide joints, covered with little spongy cushions beneath, and terminated by a fifth joint, which is armed with two claws. Most beetles, as already stated, have a little triangular piece, called the scutel, wedged between the bases of the wing-covers and the hinder part of the thorax, commonly of a triangular or semicircular form, and in the greater number of coleopterous insects quite conspicuous; in the Buprestians, however, the scutel is generally very small, and sometimes hardly perceptible. These beetles are frequently seen on the trunks and limbs of trees basking in the sun. They walk slowly, and, at the approach of danger, fold up their legs and antennæ and fall to the ground. Being furnished with ample wings, their flight is swift, and attended with a whizzing noise. They keep concealed in the night, and are in motion only during the day.

The larvæ are wood-eaters or borers. Our forests and orchards are more or less subject to their attacks, especially after the trees have passed their prime. The transformations of these insects take place in the trunks and limbs of trees. The larvæ that are known to me have a close resemblance to each other; a general idea of them can be formed from a description of that which attacks the pig-nut hickory (Fig. 21). It is of a yellowish-white color, very long, narrow, and depressed in form, but abruptly widened near the anterior extremity. The head is brownish, small, and sunk in the fore part of the first segment; the upper jaws are provided with three teeth, and are of a black color; and the antennæ are very short. The segment which receives the head is short and transverse; next to it is a large oval segment, broader than long, and depressed or flattened above and beneath. Behind this, the segments are very much narrowed, and become gradually longer; but are still flattened, to the last, which is terminated by a rounded tubercle or wart. There are no legs, nor any apparatus which can serve as such, except two small warts on the under-side of the second segment from the thorax. The motion of the grub appears to be effected by the alternate contractions and elongations of the segments, aided, perhaps, by the tubercular extremity of the body, and by its jaws, with which it takes hold of the sides of its burrow, and thus draws itself along. These grubs are found under the bark and in the solid wood of trees, and sometimes in great numbers. They frequently rest with the body bent sidewise, so that the head and tail approach each other. This posture those found under bark usually assume. They appear to pass several years in the larva state. The pupa bears a near resemblance to the perfect insect, but is entirely white, until near the time of its last transformation. Its situation is immediately under the bark, the head being directed outwards, so that, when the pupa-coat is cast off, the beetle has merely a thin covering of bark to perforate, before making its escape from the tree. The form of this perforation is oval, as is also a transverse section of the burrow, that shape being best adapted to the form, motions, and egress of the insect.

Fig. 21.

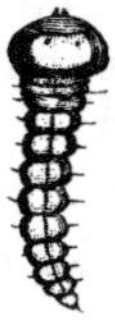

Larva of Buprestis.

Some of these beetles are known to eat leaves and flowers, and of this nature is probably the food of all of them. The injury they may thus commit is not very apparent, and cannot bear any comparison with the extensive ravages of their larvæ. The solid trunks and limbs of sound and vigorous trees are often bored through in various directions by these insects, which, during a long-continued life, derive their only nourishment from the woody fragments they devour. Pines and firs seem particularly subject to their attacks, but other forest-trees do not escape, and even fruit-trees are frequently injured by these borers. The means to be used for destroying them are similar to those employed against other borers, and will be explained in a subsequent part of this essay. It may not be amiss, however, here to remark, that woodpeckers are much more successful in discovering the retreats of these borers, and in dragging out the defenceless culprits from their burrows, than the most skilful gardener or nurseryman.

The largest of these beetles in this part of the United States is the *Buprestis* (*Chalcophora*) *Virginica* (Fig. 22) of Drury, or Virginian Buprestis. It is of an oblong oval form, brassy, or copper-colored; sometimes almost black, with hardly any metallic reflections. The upper side of the body is roughly punctured; the top of the head is deeply indented; on the thorax there are three polished black elevated lines; on each wing-cover are two small square impressed spots, a long elevated smooth black line near the outer, and another near the inner margin, with several short lines of the same kind between them; the underside of the body is sparingly covered with short whitish down. It measures from eight tenths of an inch to one inch or more in length. This beetle appears towards the end of May, and through the month of June, on pine-trees and on fences. In the larva state it bores into the trunks of the different kinds of pines, and is oftentimes very injurious to these trees.

Fig. 22.

The wild cherry-tree (*Prunus serotina*), and also the garden cherry and peach trees, suffer severely from the attacks of borers, which are transformed to the beetles called *Buprestis* (*Dicerca*) *divaricata* by Mr. Say, because the wing-covers divaricate or spread apart a little at the tips. (Plate II. Fig. 7.) These beetles are copper-colored, sometimes brassy above, and thickly covered with little punctures; the thorax is slightly furrowed in the middle; the wing-covers are marked with numerous fine irregular impressed lines and small oblong square elevated black spots; they taper very much behind, and the long and narrow tips are blunt-pointed; the middle of the breast is furrowed; and the males have a little tooth on the under-side of the shanks of the intermediate legs. They measure from seven to nine tenths of an inch. These beetles may be found sunning themselves upon the limbs of cherry and peach trees during the months of June, July, and August.

The borer of the hickory has already been described. It is transformed to a beetle which appears to be the *Buprestis (Dicerca) lurida* * (Fig. 23) of Fabricius. It is of a lurid or dull brassy color above, bright copper beneath, and thickly punctured all over; there are numerous irregular impressed lines, and several narrow elevated black spots on the wing-covers, the tip of each of which ends with two little points. It measures from about six to eight tenths of an inch in length. This kind of Buprestis appears during the greater part of the summer on the trunks and limbs of the hickory.

Fig. 23.

Buprestis (Chrysobothris) dentipes † (Fig. 24) of Germar, so named from the little tooth on the under-side of the thick fore legs, inhabits the trunks of oak-trees. It completes its transformations and comes out of the trees between the end of May and the first of July. It is oblong, oval, and flattened, of a bronzed brownish or purplish-black color above, copper-colored beneath, and rough like shagreen with

Fig. 24.

* *Buprestis obscura*, F., found in the Middle and Southern States, closely resembles the *lurida*.

† *Buprestis characteristica*, Harris. N. E. Farmer, Vol. VIII. p. 2.

7

numerous punctures ; the thorax is not so wide as the hinder part of the body, its hinder margin is hollowed on both sides to receive the rounded base of each wing-cover, and there are two smooth elevated lines on the middle ; on each wing-cover there are three irregular smooth elevated lines, which are divided and interrupted by large thickly punctured impressed spots, two of which are oblique ; the tips are rounded. Length from one half to six tenths of an inch.

Buprestis (*Chrysobothris*) *femorata* (Fig. 25) of Fabricius has the first pair of thighs toothed beneath, like the preceding, which it resembles also in its form and general appearance. It is of a greenish-black color above, with a brassy polish, which is very distinct in the two large transverse impressed spots on each wing-cover ; and the thorax has no smooth elevated lines on it. It measures from four tenths to above half of an inch in length. Its time of appearance is from the end of May to the middle of July, during which it may often be seen, in the middle of the day, resting upon or flying round the trunks of white-oak trees, and recently cut timber of the same kind of wood. I have repeatedly taken it upon and under the bark of peach-trees also. The grubs or larvæ bore into the trunks of these trees.

Fig. 25.

The *Buprestis* (*Chrysobothris*) *fulvoguttata** (Fig. 26), or tawny-spotted Buprestis, first described by me in the eighth volume of the "New England Farmer," is proportionally shorter and more convex than the two foregoing species. It is black and bronzed above, and brassy beneath ; the thorax is covered with very fine wavy transverse lines, and is some-

Fig. 26.

* Mr. Kirby has re-described and figured this insect under the name of *Buprestis* (*Trachypteris*) *Drummondi*, in the fourth volume of the "Fauna Boreali-Americana." [7]

[[7] *Buprestis* (*Chrysobothris*) *fulvoguttata* does not belong to *Chrysobothris*, but to *Melanophila*, Esch. The anterior thighs are not armed with a tooth, and the base of the thorax is truncate. — Lec.]

times copper-colored; the wing-covers are thickly punctured; and on each there are three small tawny yellow spots, with sometimes an additional one by the side of the first spot; the tips are rounded, and the fore legs are not toothed. It varies very much in size, measuring from about three to four tenths of an inch in length. I have taken this insect from the trunks of the white pine in the month of June, and have seen others that were found in the Oregon Territory.

Professor Hentz has described a small and broad beetle having the form of the above, under the name of *Buprestis* (*Chrysobothris*) *Harrisii.* (Plate II.Fig. 2.) It is entirely of a brilliant blue-green color, except the sides of the thorax, and the thighs, which in the male are copper-colored. It measures a little more than three tenths of an inch in length. The larvæ of this species inhabit the small limbs of the white pine, and young sapling trees of the same kind, upon which I have repeatedly captured the beetles about the middle of June.

These seven species form but a very small part of the Buprestians inhabiting Massachusetts and the other New England States. My knowledge of the habits of the others is not sufficiently perfect to render it worth while to insert descriptions of them here. The concealed situation of the grubs of these beetles, in the trunks and limbs of trees, renders it very difficult to discover and dislodge them. When trees are found to be very much infested by them, and are going to decay in consequence of the ravages of these borers, it will be better to cut them down, and burn them immediately, rather than to suffer them to stand until the borers have completed their transformations and made their escape.

Closely related to the Buprestians are the Elaters, or spring-beetles, (Elateridæ,) which are well known by the faculty they have of throwing themselves upwards with a jerk, when laid on their backs. On the under-side of the breast, between the bases of the first pair of legs, there is a short blunt spine, the point of which is usually concealed in a corresponding cavity behind it. When the insect, by any accident, falls upon its back, its legs are so short, and its back is so convex, that it is unable to turn itself over. It then folds its legs close to its body, bends back the head and thorax, and thus unsheathes its breast-spine; then, by suddenly straightening its body, the point of the spine is made to strike with force upon the edge of the sheath, which gives it the power of a spring, and reacts on the body of the insect, so as to throw it perpendicularly into the air. When it again falls, if it does not come down upon its feet, it repeats its exertions until its object is effected. In these beetles the body is of a hard consistence, and is usually rather narrow and tapering behind. The head is sunk to the eyes in the fore part of the thorax; the antennæ are of moderate length, and more or less notched on the inside like a saw. The thorax is as broad at the base as the wing-covers; it is usually rounded before, and the hinder angles are sharp and prominent. The scutel is of moderate size. The legs are rather short and slender, and the feet are five-jointed.

The larvæ or grubs of the Elaters live upon wood and roots, and are often very injurious to vegetation. Some are confined to old or decaying trees, others devour the roots of herbaceous plants. In England they are called wire-worms, from their slenderness and uncommon hardness. They are not to be confounded with the American wire-worm, a species of *Iulus*, which is not a true insect, but belongs to the class Myriapoda, a name derived from the great number of feet with which most of the animals included in it are furnished; whereas the English wire-worm has only six feet. The European wire-worm is said to live, in its feeding or larva state, not less than five years; during the greater part of which time it is supported by devouring the roots of wheat, rye, oats, and grass, annually causing a large diminution of the produce, and sometimes destroying whole crops. It is said to be particularly injurious in gardens recently converted from pasture lands. We have several grubs allied to this destructive insect, which are quite common in land newly broken up; but fortunately, as yet, their ravages are inconsiderable. We may expect these to increase in proportion as we disturb them and deprive them of their usual articles of food, while we continue also to persecute and destroy their natural enemies, the birds, and may then be obliged to resort to the ingenious method adopted by European farmers and gardeners for alluring and capturing these grubs. This method consists in strewing sliced potatoes or turnips in rows through the garden or field; women and boys are employed to examine the slices every morning, and collect the insects which readily come to feed upon the bait. Some of these destructive insects, which I have found in the ground among the roots of plants, were long, slender, worm-like grubs, closely resembling the common meal-worm; they were nearly cylindrical, with a hard and smooth skin, of a buff or brownish-yellow color, the head and tail only being a little darker; each of the first three rings was provided with a pair of short legs; the hindmost ring was longer than the preceding one, was pointed at the end, and had a little pit on each side of the extremity; beneath this part there was a short retractile wart, or prop-leg, serving to support the extremity of the body, and prevent it from trailing on the ground. Other grubs of Elaters differ from the foregoing in being proportionally broader, not cylindrical, but somewhat flattened, with a deep notch at the extremity of the last ring, the sides of which are beset with little teeth. Such grubs are mostly wood-eaters, devouring the woody parts of roots, or living under the bark and in the trunks of old trees.

After their last transformation, Elaters or spring-beetles make their appearance upon trees and fences, and some are found on flowers. They creep slowly, and generally fall to the ground on being touched. They fly both by day and night. Their food, in the beetle state, appears to be chiefly derived from flowers; but some devour the tender leaves of plants.

The largest of our spring-beetles is the *Elater* (*Alaus*) *oculatus* of Linnæus (Fig. 27). It is of a black color; the thorax is oblong-square, and nearly one third the length of the whole body, covered above with a whitish powder, and with a large oval velvet-black spot, like an eye, on each side of the middle, from which the insect derives its name, *oculatus*, or eyed; the wing-covers are marked with slender longitudinal impressed lines, and are sprinkled with numerous white dots; the under-side of the body, and the legs, are covered with a white mealy powder. This large beetle measures from one inch and a quarter to one inch and three quarters in length. It is found on trees, fences, and the sides of buildings, in June and July. It undergoes its transformations in the trunks of trees. I have found many of them in old apple-trees, together with their larvæ, which eat the wood, and from which I subsequently obtained the insects in the beetle state. These larvæ are reddish-yellow grubs, proportionally much broader than the other kinds, and very much flattened. One of them, which was found fully grown early in April, measured two inches and a half in length, and nearly four tenths of an inch across the middle of the body, and was not much narrowed at either extremity. The head was broad, brownish, and rough above; the upper jaws or nippers were very strong, curved, and pointed; the eyes were small and two in number, one being placed at the base of each of the short antennæ; the last segment of the body was blackish, rough with little sharp-pointed warts, with a deep semicircular notch at the end, and furnished around the sides with little teeth, the two hindmost of which were long, forked, and curved upwards like hooks; under this segment was a large retractile fleshy prop-foot, armed behind with little claws, and around the

Fig. 27.

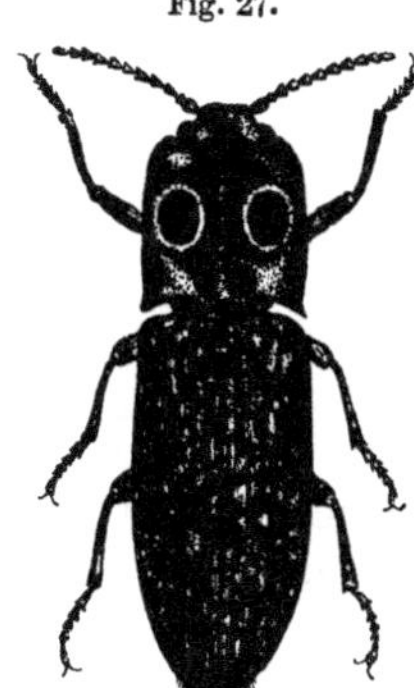

sides with short spines; the true legs were six, a pair to each of the first three rings; and were tipped with a single claw. Soon after this grub was found, it cast its skin and became a pupa, and in due time the latter was transformed to a beetle.

Elater (*Pyrophorus*) *noctilucus*, the night-shining Elater, is the celebrated *cucuio* or fire-beetle of the West Indies, from whence it is frequently brought alive to this country. It resembles the preceding insect somewhat in form, and is an inch or more in length. It gives out a strong light from two transparent eye-like spots on the thorax, and from the segments of its body beneath. It eats the pulpy substance of the sugar-cane, and its grub is said to be very injurious to this plant, by devouring its roots.

The next two common Elaters, together with several other species, are distinguished by their claws, which resemble little combs, being furnished with a row of fine teeth along the under-side. The thorax is short and rounded before, and the body tapers behind. They are found under the bark of trees, where they pass the winter, having completed their transformations in the previous autumn. Their grubs live in wood. The first of these beetles is the ash-colored Elater, *Elater* (*Melanotus*) *cinereus* of Weber (Fig. 28). It is about six tenths of an inch long, and is dark brown, but covered with short gray hairs, which give it an ashen hue; the thorax is convex, and the wing-covers are marked with lines of punctures, resembling stitches. It is found on fences, the trunks of trees, and in paths, in April and May.

Fig. 28.

Elater (*Melanotus*) *communis* of Schönherr, is, as its name implies, an exceedingly common and abundant species. It closely resembles the preceding, but is smaller, seldom exceeding half an inch in length; it is also rather lighter colored; the thorax is proportionally a little longer, not so convex, and has a slender longitudinal furrow in the middle.

This Elater appears in the same places as the *cinereus* in April, May, and June; and the recently transformed beetles can also be found in the autumn under the bark of trees, where they pass the winter.

Another kind of spring-beetle, which absolutely swarms in paths and among the grass during the warmest and brightest days in April and May, is the *Elater* (*Ludius*) *appressifrons* of Say. Its specific name probably refers to the front of the head or visor being pressed downwards over the lip. The body is slender and almost cylindrical, of a deep chestnut-brown color, rendered gray, however, by the numerous short yellowish hairs with which it is covered; the thorax is of moderate length, not much narrowed before, convex above, with very long and sharp-pointed hinder angles, and in certain lights has a brassy hue; the wing-covers are finely punctured, and have very slender impressed longitudinal lines upon them; the claws are not toothed beneath. This beetle usually measures from four to five tenths of an inch in length; but the females frequently greatly exceed these dimensions, and, being much more robust, with a more convex thorax, were supposed by Mr. Say to belong to a different species, named by him *brevicornis*, the short-horned. The larvæ are not yet known to me; but I have strong reasons for thinking that they live in the ground, upon the roots of the perennial grasses and other herbaceous plants.

Although above sixty different kinds of spring-beetles are now known to inhabit Massachusetts, I shall add to the foregoing a description of only one more species. This is the *Elater* (*Agriotes*) *obesus*[8] of Say (Fig. 29). It is a short and thick beetle, as the specific name implies; its real color is a dark brown, but it is covered with dirty yellowish-gray hairs, which on the wing-covers are arranged in longitudinal stripes; the head and

Fig. 29.

[[8] *Elater* (*Agriotes*) *obesus*. I am inclined to believe this species to be the *Elater mancus*, Say, and not his *E. obesus*, which is now entirely unknown. — LEC.]

thorax are thickly punctured, and the wing-covers are punctured in rows. Its length is about three tenths of an inch. This beetle closely resembles one of the kinds which, in the grub state, is called the wire-worm in Europe, and possibly it may be the same. This circumstance should put us on our guard against its depredations. It is found in April, May, and June, among the roots of grass, on the under-side of boards and rails on the ground, and sometimes also on fences.

The utility of a knowledge of the natural history of insects in the practical arts of life was never more strikingly and triumphantly proved than by Linnæus himself, who, while giving to natural science its language and its laws, neglected no opportunity to point out its economical advantages.* On one occasion this great naturalist was consulted by the King of Sweden upon the cause of the decay and destruction of the ship-timber in the royal dock-yards, and, having traced it to the depredations of insects, and ascertained the history of the depredators, by directing the timber to be sunk under water during the season when these insects made their appearance in the winged state, and were busied in laying their eggs, he effectually secured it from future attacks. The name of these insects is *Lymexylon navale*, the naval timber-destroyer. They have since increased to an alarming extent in some of the dock-yards of France, and in one of them, at least, have become very injurious, wholly in consequence of the neglect of seasonable advice given by a naval officer, who was also an entomologist, and pointed out the source of the injury, together with the remedy to be applied.

* See the Preface to Smith's "Introduction to Botany," and Pulteney's "View of the Writings of Linnæus," for several examples, one of which it may not be amiss to mention here. Linnæus was the first to point out the advantages to be derived from employing the *Arundo arenaria*, or beach-grass, in fixing the sands of the shore, and thereby preventing the encroachments of the sea. The Dutch have long availed themselves of his suggestion, and its utility has been tested to some extent in Massachusetts.

These destructive insects belong to a family called LYMEXYLIDÆ, which may be rendered timber-beetles. They cannot be far removed from the Buprestians and the spring-beetles in a natural arrangement. From the latter, however, the insects of this small group are distinguished by having the head broad before, narrowed behind, and not sunk into the thorax; they have not the breast-spine of the Elaters, and their legs are close together, and not separated from each other by a broad breast-bone as in the Buprestians; and the hip-joints are long, and not sunk into the breast. In the principal insects of this family the antennæ are short, and, from the third joint, flattened, widened, and saw-toothed on the inside; and the jaw-feelers of the males have a singular fringed piece attached to them. The body is long, narrow, nearly cylindrical, and not so firm and hard as in the Elaters. The feet are five-jointed, long, and slender.

The larvæ of *Lymexylon* and *Hylecœtus* are very odd-looking, long, and slender grubs. The head is small; the first ring is very much hunched; and on the top of the last ring there is a fleshy appendage, resembling a leaf in *Lymexylon*, and like a straight horn in *Hylecœtus*. They have six short legs near the head. These grubs inhabit oak-trees, and make long cylindrical burrows in the solid wood. They are also found in some other kinds of trees.

Fig. 30.

Only a few native insects of this family are known to me, and these fortunately seem to be rare in New England. I shall describe only two of them. The first was obtained by beating the limbs of some forest-tree. It may be called *Lymexylon sericeum* (Fig. 30), the silky timber-beetle. It is of a chestnut-brown color above, and covered with very short shining yellowish hairs, which give it a silky lustre. The head is bowed down beneath the fore part of the thorax; the eyes are very large, and almost meet above and below; the antennæ are brownish red, widened and compressed from the fourth to the last joint inclusive; the thorax is longer than wide, rounded before, convex above, and deeply indented on each side of the base; the wing-covers are convex, gradually taper behind, and do not cover the tip of the abdomen; the under-side of the body, and the legs, are brownish red. Its length is from four to six tenths of an inch. This insect was unknown to Mr. Say, and does not seem to have been described before.

The generical name *Hylecœtus*, given to some insects of this family, means a sleeper in the woods, or one who makes his bed in the forest. We have one hitherto undescribed species, which may be called *Hylecœtus Americanus*, the American timber-beetle. Its head, thorax, abdomen, and legs are light brownish red; the wing-covers, except at the base, where they are also red, and the breast, between the middle and hindmost legs, are black. The head is not bowed down under the fore part of the thorax; the eyes are small and black, and on the middle of the forehead there is one small reddish eyelet, a character unusual among beetles, very few of which have eyelets; the antennæ resemble those of *Lymexylon sericeum*, but are shorter; the thorax is nearly square, but wider than long; and on each wing-cover there are three slightly elevated longitudinal lines or ribs. This beetle is about four tenths of an inch long. It appears on the wing in July.

The foregoing beetles, though differing much in form and habits, possess one character in common; namely, their feet are five-jointed. Those that follow have four-jointed feet. In this great section of Coleopterous insects are arranged the Weevil tribe, the Capricorn beetles or long-horned borers, and various kinds of leaf-eating beetles, all of which are exceedingly injurious to vegetation.

So great is the extent of the Weevil tribe,* and so imperfectly known is the history of a large part of our native

* See page 21.

species, that I shall be obliged to confine myself to an account of a few only of the most remarkable weevils, and principally those that have become most known for their depredations. Mr. Köllar's excellent "Treatise on Insects injurious to Gardeners, Foresters, and Farmers," contains an account of several kinds of weevils that are unknown in this country; and indeed but few resembling them have hitherto been discovered here. Should future observations lead to the detection in our gardens and orchards of any like those which in Europe attack the vine, the plum, the apple, the pear, and the leaves and stems of fruit-trees, the work of Mr. Köllar may be consulted with great advantage.

Weevils, in the winged state, are hard-shelled beetles, and are distinguished from other insects by having the fore part of the head prolonged into a broad muzzle or a longer and more slender snout, in the end of which the opening of the mouth and the small horny jaws are placed. The flies and moths produced from certain young insects, called weevils by mistake, do not possess these characters, and their larvæ or young differ essentially from those of the true weevils. The latter belong to a group called RHYNCHOPHORIDÆ, literally, snout-bearers. These beetles are mostly of small size. Their antennæ are usually knobbed at the end, and are situated on the muzzle or snout, on each side of which there is generally a short groove to receive the base of the antennæ when the latter are turned backwards. Their feelers are very small, and, in most kinds, are concealed within the mouth. The abdomen is often of an oval form, and wider than the thorax. The legs are short, not fitted for running or digging, and the soles of the feet are short and flattened. These beetles are often very hurtful to plants, by boring into the leaves, bark, buds, fruit, and seeds, and feeding upon the soft substance therein contained. They are diurnal insects, and love to come out of their retreats and enjoy the sunshine. Some of them fly well; but others have no wings, or only very short ones, under the wing-cases, and are therefore unable to fly. They walk slowly, and being of a timid nature, and without the means of defence, when alarmed they turn back their antennæ under the snout, fold up their legs, and fall from the plants on which they live. They make use of their snouts not only in feeding, but in boring holes, into which they afterwards drop their eggs.

The young of these snout-beetles are mostly short fleshy grubs, of a whitish color, and without legs. The covering of their heads is a hard shell, and the rings of their bodies are very convex or hunched, by both of which characters they are easily distinguished from the maggots of flies. Their jaws are strong and horny, and with them they gnaw those parts of plants which serve for their food. It is in the grub state that weevils are most injurious to vegetation. Some of them bore into and spoil fruits, grain, and seeds; some attack the leaves and stems of plants, causing them to swell and become cankered; while others penetrate into the solid wood, interrupt the course of the sap, and occasion the branch above the seat of attack to wither and die. Most of these grubs are transformed within the vegetable substances upon which they have lived; some, however, when fully grown, go into the ground, where they are changed to pupæ, and afterwards to beetles.

In the spring of the year, we often find among seed-peas many that have holes in them; and, if the peas have not been exposed to the light and air, we see a little insect peeping out of each of these holes, and waiting apparently for an opportunity to come forth and make its escape. If we turn out the creature from its cell, we perceive it to be a small oval beetle, rather more than one tenth of an inch long, of a rusty black color, with a white spot on the hinder part of the thorax, four or five white dots behind the middle of each wing-cover, and a white spot shaped like the letter T on the exposed extremity of the body. This little insect is the *Bruchus Pisi* of Linnæus (Fig. 31), the pea-Bruchus, or pea-weevil, but is better known in America by the incorrect name of pea-bug. The original meaning of the word *Bruchus* is a devourer, and the insects to which it is applied well deserve this name, for, in the larva state, they devour the interior of seeds, often leaving but little more than the hull untouched. They belong to a family of the great weevil tribe called BRUCHIDÆ, and are distinguished from other weevils by the following characters. The body is oval, and slightly convex; the head is bent downwards, so that the broad muzzle, when the insects are not eating, rests upon the breast; the antennæ are short, straight, and saw-toothed within, and are inserted close to a deep notch in each of the eyes; the feelers, though very small, are visible; the wing-cases do not cover the end of the abdomen; and the hindmost thighs are very thick, and often notched or toothed on the under-side, as is the case in the pea-weevil. The habits of the Bruchians and their larvæ are similar to those of the pea-weevil, which remain to be described. It may be well, however, to state here, that these beetles frequent the leguminous or pod-bearing plants, such as the pea, Gleditschia, Robinia, Mimosa, Cassia, &c., during and immediately after the flowering season; they wound the skin of the tender pods of these plants, and lay their eggs singly in the wounds. Each of the little maggot-like grubs hatched therefrom perforates the pod and enters a seed, the pulp of which suffices for its food till fully grown.

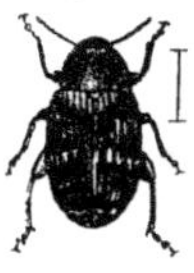

Fig. 31.

Few persons while indulging in the luxury of early green peas are aware how many insects they unconsciously swallow. When the pods are carefully examined, small discolored spots may be seen within them, each one corresponding to a similar spot on the opposite pea. If this spot in the pea be opened, a minute whitish grub, destitute of feet, will be found therein. It is the weevil in its larva form, which lives upon the marrow of the pea, and arrives at its full size by the time that the pea becomes dry. This larva or grub then bores a round hole from the hollow in the centre of the pea quite to the hull, but leaves the latter, and generally the germ of the future sprout, untouched. Hence these buggy peas, as they are called by seedsmen and gardeners, will frequently sprout and grow when planted. The grub is changed to a pupa within its hole in the pea in the autumn, and before the spring casts its skin again, becomes a beetle, and gnaws a hole through the thin hull in order to make its escape into the air, which frequently does not happen before the peas are planted for an early crop. After the pea-vines have flowered, and while the pods are young and tender, and the peas within them are just beginning to swell, the beetles gather upon them, and deposit their tiny eggs singly in the punctures or wounds which they make upon the surface of the pods. This is done mostly during the night, or in cloudy weather. The grubs, as soon as they are hatched, penetrate the pod and bury themselves in the opposite peas; and the holes through which they pass into the seeds are so fine as hardly to be perceived, and are soon closed. Sometimes every pea in a pod will be found to contain a weevil-grub; and so great has been the injury to the crop, in some parts of the country, that the inhabitants have been obliged to give up the cultivation of this vegetable.* These insects diminish the weight of the peas in which they lodge nearly one half, and their leavings are fit only for the food of swine. This occasions a great loss where peas are raised for feeding stock or for family use, as they are in many places. Those persons who eat whole peas in the winter after they are raised, run the risk of eating the weevils also; but if the peas are kept till they are a year old, the insects will entirely leave them.†

The pea-weevil is supposed to be a native of the United States. It seems to have been first noticed in Pennsylvania,

* See Kalm's Travels, (8vo, Warrington, 1770,) Vol. I. p. 173.

† See the "Boston Cultivator" for July 1, 1848, for an interesting account of the habits of these insects, by Mr. S. Deane.

many years ago, and has gradually spread from thence to New Jersey, New York, Connecticut, Rhode Island, and Massachusetts. It is yet rare in New Hampshire, and I believe has not appeared in the eastern parts of Maine. It is unknown in the North of Europe, as we learn from the interesting account given of it by Kalm, the Swedish traveller, who tells us of the fear with which he was filled on finding some of these weevils in a parcel of peas which he had carried home from America, having in view the whole damage which his beloved country would have suffered, if only two or three of these noxious insects had escaped him. They are now common in the South of Europe and in England, whither they may have been carried from this country. As the cultivated pea was not originally a native of America, it would be interesting to ascertain what plants the pea-weevil formerly inhabited. That it should have preferred the prolific exotic pea to any of our indigenous and less productive pulse, is not a matter of surprise, analogous facts being of common occurrence; but that, for so many years, a rational method for checking its ravages should not have been practised, is somewhat remarkable. An exceedingly simple one is recommended by Deane, but to be successful it should be universally adopted. It consists merely in keeping seed-peas in tight vessels over one year before planting them. Latreille and others recommend putting them, just before they are to be planted, into hot water for a minute or two, by which means the weevils will be killed, and the sprouting of the peas will be quickened. The insect is limited to a certain period for depositing its eggs; late-sown peas therefore escape its attacks. The late Colonel Pickering observed that those sown in Pennsylvania as late as the 20th of May were entirely free from weevils; and Colonel Worthington, of Rensselaer County, New York, who sowed his peas on the 10th of June, six years in succession, never found an insect in them during that period.

The crow black-bird is said to devour great numbers of

the beetles in the spring; and the Baltimore oriole or hang-bird splits open the green pods for the sake of the grubs contained in the peas, thereby contributing greatly to prevent the increase of these noxious insects. The instinct that enables this beautiful bird to detect the lurking grub, concealed, as the latter is, within the pod and the hull of the pea, is worthy our highest admiration; and the goodness of Providence, which has endowed it with this faculty, is still further shown in the economy of the insects also, which, through His prospective care, are not only limited in the season of their depredations, but are instinctively taught to spare the germs of the peas, thereby securing a succession of crops for our benefit and that of their own progeny.

The Attelabians (ATTELABIDÆ) are distinguished from the Bruchians by the form and greater length of the head, which is a little inclined, and ends with a snout, sometimes short and thick, and sometimes long, slender, and curved. The eyes also are round and entire, and the antennæ are usually implanted near the middle of the snout. The larvæ resemble those of most of the snout-beetles, being short, thick, whitish grubs, with horny heads, the rings of the body very much hunched, and deprived of legs, the place of which is supplied by fleshy warts along the under-side of the body. Some of the European insects of this family are known to be very injurious to the leaves, fruits, and seeds of plants.

The different kinds of *Attelabus* are said to roll up the edges of leaves, thereby forming little nests, of the shape and size of thimbles, to contain their eggs, and to shelter their young, which afterwards devour the leaves. The larvæ and habits of our native species are unknown to me. The most common one here is the *Attelabus analis* of Weber (Fig. 32), or the red-tailed Attelabus. It is one quarter of an inch long from the tip of the thick snout to the end of the body. The head, which is nearly cylindrical, the antennæ, legs, and

Fig. 32.

middle of the breast, are deep blue-black ; the thorax, wing-covers, and abdomen are dull red ; the wing-covers, taken together, are nearly square, and are punctured in rows. This beetle is found on the leaves of oak-trees in June and July.

The two-spotted Attelabus, *Attelabus bipustulatus* of Fabricius, (Plate II.Fig. 6,) is also found on oak-leaves during the same season as the preceding. It is of a deep blue-black color, with a square dull red spot on the shoulders of each wing-cover. It measures rather more than one eighth of an inch in length.

Two or three beetles of this family are very hurtful to the vine, in Europe, by nibbling the midrib of the leaves, so that the latter may be rolled up to form a retreat for their young. They also puncture the buds and the tender fruit of this and of other plants. In consequence of the damage caused by them and by their larvæ, whole vineyards are sometimes stripped of their leaves, and fruit-trees are despoiled of their foliage and fruits. These insects belong to the genus *Rynchites*, a name given to them in allusion to their snouts. I have not seen any of them on vines or fruit-trees in this country. The largest one found here is the *Rynchites bicolor* of Fabricius, or two-colored Rynchites. This insect is met with in June, July, and August, on cultivated and wild rose-bushes, sometimes in considerable numbers. That they injure these plants is highly probable, but the nature and extent of the injury is not certainly known. The whole of the upper side of this beetle is red, except the rather long and slender snout, which, together with the antennæ, legs, and under-side of the body, is black ; it is thickly covered with small punctures, and is slightly downy, and there are rows of larger punctures on the wing-covers. It measures one fifth of an inch from the eyes to the tip of the abdomen.

The grubs of many kinds of *Apion* destroy the seeds of plants. In Europe they do much mischief to clover in this way. They receive the above name from the shape of the beetles, which resembles that of a pear. Say's Apion, *Apion Sayi* * of Schönherr (Fig. 33), is a minute black species, not more than one tenth of an inch long, exclusive of the slender, sharp-pointed snout. Its grubs live in the pods of the common wild-indigo bush, *Baptisia tinctoria*, devouring the seeds. A smaller kind, somewhat like it, inhabits the pods and eats the seeds of the locust-tree, or *Robinia pseudacacia*.

Fig. 33.

Naturalists place here a little group of snout-beetles, called BRENTHIDÆ, or Brenthians, which differ entirely in their forms from the other weevils, both in the beetle and grub state. They have a long, narrow, and cylindrical body. The snout projects from the head in a straight line with the body, and varies in shape according to the sex of the insect, and even in individuals of the same sex. In the males it is broad and flat, sometimes as long as the thorax, sometimes much shorter, and it is widened at the tip, where are situated two strong nippers or upper jaws; in the females it is long, very slender, and not enlarged at the extremity, and the nippers are not visible to the naked eye. The feelers are too small to be seen. The antennæ are short, straight, slightly thickened towards the tip, and implanted before the prominent eyes, on the middle of the snout in the males, and at the base of it in the females. The legs are short, the first pair being the largest, and the hindmost unusually distant from the middle pair. These insects live under the bark and in the trunks of trees, but very little has been published respecting their habits; and the only description of their larvæ that has hitherto appeared is contained in my first Report on the Insects of Massachusetts, printed in the year 1838, in the seventy-second number of the "Documents of the House of Representatives."

The only beetle of this family known in the New England

* *Apion rostrum*, Say.

States is the *Brenthus* (*Arrhenodes*) *septemtrionis* * of Herbst (Fig. 34), the Northern Brenthus, so named because most of the other species are tropical insects. It is of a mahogany-brown color; the wing-cases are somewhat darker, ornamented with narrow tawny-yellow spots, and marked with deep furrows, the sides of which are punctured; the thorax is nearly egg-shaped, broadest behind the middle, and highly polished. The common length of this insect, including the snout, is six tenths of an inch; but much larger as well as smaller specimens frequently occur. The Northern Brenthus inhabits the white oak, on the trunks and under the bark of which it may be found in June and July, having then completed its transformations. The female, when about to lay her eggs, punctures the bark with her slender snout, and drops an egg in each hole thus made. The grub, as soon as it is hatched, bores into the solid wood, forming a cylindrical passage, which it keeps clear by pushing its castings out of the orifice of the hole, as fast as they accumulate. These castings or chips are like very fine sawdust; and the holes made by the insects are easily discovered by the dust around them. When fully grown, the grub measures rather more than an inch in length, and not quite one tenth of an inch in thickness. It is nearly cylindrical, being only a little flattened on the under-side, and is of a whitish color, except the last segment, which is dark chestnut-brown. Each of the first three segments is provided with a pair of legs, and there is a fleshy prop-leg under the hinder extremity of the body. The last segment is of a horny consistence, and is obliquely hollowed at the end, so as to form a kind of gouge or scoop, the edges of which are furnished with little notches or teeth. It is by means of this singular scoop that the grub shovels the minute grains of the wood out of its burrow. The pupa

Fig. 34.

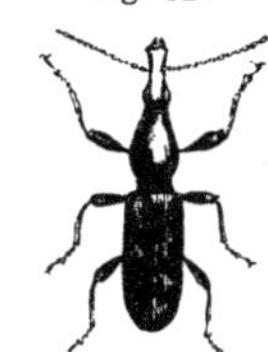

* A mistake undoubtedly for *septemtrionalis*. It is the *Brenthus maxillosus* of Olivier and Schönherr.

is met with in the burrow formed by the larva. It is of a yellowish-white color; the head is bent under the thorax, and the snout rests on the breast between the folded legs and wings; the back is furnished with transverse rows of little thorns or sharp teeth, and there are two larger thorns at the extremity of the body. These minute thorns probably enable the pupa to move towards the mouth of its burrow when it is about to be transformed, and may serve also to keep its body steady during its exertions in casting off its pupa skin. These insects are most abundant in trees that have been cut down for timber or fuel, which are generally attacked during the first summer after they are felled; it has also been ascertained that living trees do not always escape, but those that are in full vigor are rarely perforated by grubs of this kind. The credit of discovering the habits and transformations of the Northern Brenthus is due to the Rev. L. W. Leonard, of Dublin, New Hampshire, who has favored me with specimens in all their forms. This insect is now known to inhabit nearly all the States in the Union. I am inclined to think that the Brenthians ought to be placed at the end of the weevil tribe; but I have not ventured to alter the arrangement generally adopted.

The rest of the weevils are short and thick beetles, differing from all the preceding in their antennæ, which are bent or elbowed near the middle, the first joint being much longer than the rest. Their feelers are not perceptible. They belong to the family CURCULIONIDÆ, so called from the principal genus, *Curculio*, a name given by the Romans to the corn-weevil. The Curculionians vary in the form, length, and direction of their snouts. Those belonging to the old genus *Curculio* have short and thick snouts, at the extremity of which, and near to the sides of the mouth, the antennæ are implanted; those to which the name of *Rhynchœnus* was formerly applied have longer and more slender snouts, usually bearing the antennæ on or just behind the middle; and the third great genus, called *Calandra*, contains long-snouted beetles, whose antennæ are fixed just before the eyes at the base of the snout.

Fig. 35.

Curculio (*Pandeleteius*) *hilaris* of Herbst (Fig. 35), which we may call the gray-sided Curculio, is a little pale-brown beetle, variegated with gray upon the sides. Its snout is short, broad, and slightly furrowed in the middle; there are three blackish stripes on the thorax, between which are two of a light gray color; the wing-covers have a broad stripe of light gray on the outer side, edged within by a slender blackish line, and sending two short oblique branches almost across each wing-cover; and the fore-legs are much larger than the others. The length of this beetle varies from one eighth to one fifth of an inch. The larva lives in the trunks of the white oak, on which the beetles may be found about the last of May and the beginning of June.

Fig. 36.

The Pales weevil, *Curculio* (*Hylobius*) *Pales* of Herbst (Fig. 36), is a beetle of a deep chestnut-brown color, having a line and a few dots of a yellowish-white color on the thorax, and many small yellowish-white spots sprinkled over the wing-covers. All the thighs are toothed beneath, and the snout is slender, cylindrical, inclined, and nearly as long as the thorax. On account of the length of the snout this insect has been placed in the genus *Rhynchænus* by some naturalists; but the antennæ are implanted before the middle of the snout, and not far from the sides of the mouth. This beetle measures from two to three eighths of an inch in length, exclusive of the snout. It may be found in great abundance, in May and June, on board-fences, the sides of new wooden buildings, and on the trunks of pine-trees. I have discovered them, in considerable numbers, under the bark of the pitch-pine. The larvæ, which do not materially differ from those of other weevils, inhabit these and probably other kinds of pines, doing sometimes immense injury to them. Wilson, the ornithologist, describes the depredations of these insects, in his account* of the ivory-billed woodpecker, in the following words: "Would it be believed that the larvæ of an insect, or fly, no larger than a grain of rice, should silently, and in one season, destroy some thousand acres of pine-trees, many of them from two to three feet in diameter and a hundred and fifty feet high! Yet whoever passes along the high road from Georgetown to Charleston, in South Carolina, about twenty miles from the former place, can have striking and melancholy proofs of the fact. In some places the whole woods, as far as you can see around you, are dead, stripped of the bark, their wintry-looking arms and bare trunks bleaching in the sun, and tumbling in ruins before every blast, presenting a frightful picture of desolation. Until some effectual preventive or more complete remedy can be devised against these insects, and their larvæ, I would humbly suggest the propriety of protecting, and receiving with proper feelings of gratitude, the services of this and the whole tribe of woodpeckers, letting the odium of guilt fall to its proper owners." Some years ago Mr. Nuttall kindly procured for me, near the place above mentioned, specimens of the destructive insects referred to by Wilson. They were of three kinds. Those in greatest abundance were the Pales weevil. One of the others was a larger, darker-colored weevil, without white spots on it, and named *Hylobius picivorus* by Germar and Schönherr, or the pitch-eating weevil; it is seldom found in Massachusetts. The third was the white-pine weevil, to be next described. It is said that these beetles puncture the buds and the tender bark of the small branches, and feed upon the juice, and that the young shoots are often so much injured by them as to die and break off at the wounded part. But it is in the larva state that they are found to be most hurtful to the pines. The larvæ live under

* American Ornithology, Vol. IV. p. 21.

the bark, devouring its soft inner surface, and the tender, newly formed wood. When they abound, as they do in some of our pine forests, they separate large pieces of bark from the wood beneath, in consequence of which the part perishes, and the tree itself soon languishes and dies.

The white-pine weevil, *Rhynchænus* (*Pissodes*) *Strobi** of Professor Peck (Fig. 37), unites with the two preceding insects in destroying the pines of this country, as above described. But it employs also another mode of attack on the white pine, of which an interesting account is given by the late Professor Peck, the first describer of the insect, in the fourth volume of the "Massachusetts Agricultural Repository and Journal," accompanied by figures of the insect. The lofty stature of the white pine, and the straightness of its trunk, depend, as Professor Peck has remarked, upon the constant health of its leading shoot, for a long succession of years; and if this shoot be destroyed, the tree becomes stunted and deformed in its subsequent growth. This accident is not uncommon, and is caused by the ravages of the white-pine weevil.

Fig 37.

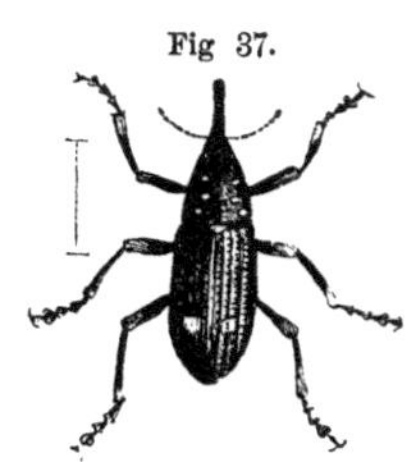

This beetle is oblong oval, rather slender, of a brownish color, thickly punctured, and variegated with small brown, rust-colored, and whitish scales. There are two white dots on the thorax; the scutel is white; and on the wing-covers, which are punctured in rows, there is a whitish transverse band behind the middle. The snout is longer than the thorax, slender, and a very little inclined. The length of this insect, exclusive of its snout, varies from one fifth to three tenths of an inch. Its eggs are deposited on the leading shoot of the pine, probably immediately under the outer bark. The larvæ, hatched therefrom, bore into the shoot in various directions, and probably remain in the wood more than one year. When the feeding state is passed, but before

* *Pissodes nemorensis* of Germar.

the insect is changed to a pupa, it gnaws a passage from the inside quite to the bark, which, however, remaining untouched, serves to shelter the little borers from the weather. After they have changed to beetles, they have only to cut away the outer bark to make their escape. They begin to come out early in September, and continue to leave the wood through that month and a part of October. The shoot at this time will be found pierced with small round holes on all sides; sometimes thirty or forty may be counted on one shoot. Professor Peck has observed that an unlimited increase is not permitted to this destructive insect; and that if it were, our forests would not produce a single mast. One of the means appointed to restrain the increase of the white-pine weevil is a species of ichneumon-fly, endued with sagacity to discover the retreat of the larva, the body of which it stings, and therein deposits an egg. From the latter a grub is hatched, which devours the larva of the weevil, and is subsequently transformed to a four-winged fly, in the habitation prepared for it. The most effectual remedy against the increase of these weevils is to cut off the shoot in August, or as soon as it is perceived to be dead, and commit it, with its inhabitants, to the fire.

Such is the substance of Professor Peck's history of this insect; to which may be added, that the beetles are found in great numbers, in April and May, on fences, buildings, and pine-trees; that they probably secrete themselves during the winter in the crevices of the bark, or about the roots of the trees, and deposit their eggs in the spring; or they may not usually leave the trees before spring.

Perhaps the method used for decoying the pine-eating beetles in Europe may be practised here with advantage. It consists in sticking some newly-cut branches of pine-trees in the ground, in an open place, during the season when the insects are about to lay their eggs. In a few hours these branches will be covered with the beetles, which may be shaken into a cloth and burned.

There are some of the long-snouted weevils which inhabit nuts of various kinds. Hence they are called nut-weevils, and belong chiefly to the modern genus *Balaninus*, a name that signifies living or being in a nut. The common nut-weevil of Europe lays her eggs in the hazelnut and filbert, having previously bored a hole for that purpose with her long and slender snout, while the fruit is young and tender, and dropping only one egg in each nut thus pricked. A little grub is soon hatched from the egg, and begins immediately to devour the soft kernel. Notwithstanding this, the nut continues to increase in size, and, by the time that it is ripe and ready to fall, its little inhabitant also comes to its growth, gnaws a round hole in the shell, through which it afterwards makes its escape, and burrows in the ground. Here it remains unchanged through the winter, and in the following summer, having completed its transformations, it comes out of the ground a beetle.

In this country weevil-grubs are very common in hazel-nuts, chestnuts, and acorns; but I have not hitherto been able to rear any of them to the beetle state. The most common of the nut-weevils known to me appears to be the *Rhynchœnus* (*Balaninus*) *nasicus* of Say (Fig. 38), the long-snouted nut-weevil. Its form is oval, and its ground color dark brown; but it is clothed with very short rust-yellow flattened hairs, which more or less conceal its original color, and are disposed in spots on its wing-covers. The snout is brown and polished, longer than the whole body, as slender as a bristle, of equal thickness from one to the other, and slightly curved; it bears the long elbowed antennæ, which are as fine as a hair, just behind the middle. This beetle measures nearly three tenths of an inch in length, exclusive of the snout. Specimens have been found paired upon the hazel-nut-tree in July, at which time probably the eggs are laid.

Fig. 38.

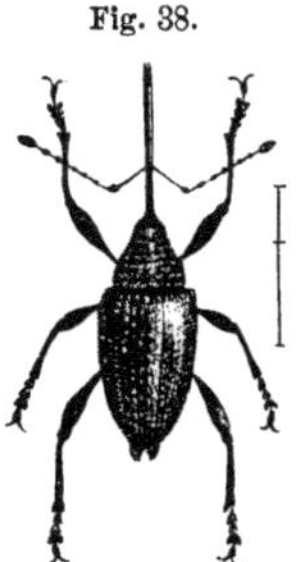

Others appear in September and October, and must pass the winter concealed in some secure place. From its size and resemblance to the nut-weevil of Europe, this is supposed to be the species which attacks the hazelnut here.

It is now well known that the falling of unripe plums is caused by little whitish grubs, which bore into the fruit. The loss occasioned by insects of this kind is frequently very great; and in some of our gardens and orchards the crop of plums is often entirely ruined by the depredations of the grubs, which have been ascertained to be the larvæ or young of a small beetle of the weevil tribe, called *Rhynchænus* (*Conotrachelus*) *Nenuphar*,* (Figs. 39 and 40,) the Nenuphar or plum-weevil. This weevil, or *curculio*, as it is often called, is a little rough, dark-brown, or blackish beetle, looking like a dried bud when it is shaken from the trees, which resemblance is increased by its habit of drawing up its legs and bending its snout close to the lower side of its body, and remaining for a time without motion, and seemingly lifeless. It is from three twentieths to one fifth of an inch long, exclusive of the curved snout, which is rather longer than the thorax, and is bent under the breast, between the fore legs, when at rest. Its color is a dark brown, variegated with spots of white, ochre-yellow, and black. The thorax is uneven; the wing-covers have several short ridges upon them, those on the middle of the back forming two considerable humps, of a black color, behind which there is a wide band of ochre-yellow and white. Each of the thighs has two little teeth on the under-side. I have found these beetles as early as the 30th of March, and as late as the 10th of June, and at various intermediate times, according with the for-

Fig. 39. Fig. 40.

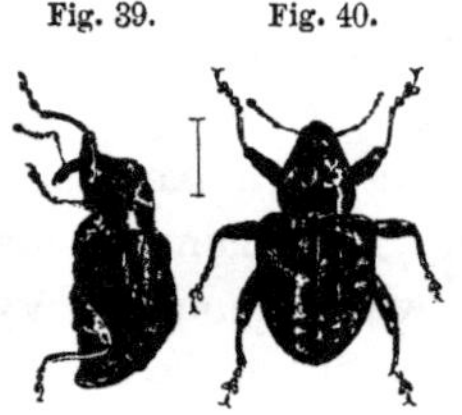

* First described by Herbst, in 1797, under the name of *Curculio Nenuphar;* Fabricius redescribed it under that of *Rhynchænus Argula;* and Dejean has named it *Conotrachelus variegatus.*

wardness or backwardness of vegetation in the spring, and have frequently caught them flying in the middle of the day. They begin to sting the plums as soon as the fruit is set, and continue their operations to the middle of July, or, as some say, till the first of August. In doing this, the beetle first makes a small crescent-shaped incision, with its snout, in the skin of the plum, and then, turning round, inserts an egg in the wound. From one plum it goes to another, until its store of eggs is exhausted; so that, where these beetles abound, not a plum will escape being stung. Very rarely is there more than one incision made in the same fruit; and the weevil lays only a single egg therein. The insect hatched from this egg is a little whitish grub, destitute of feet, and very much like a maggot in appearance, except that it has a distinct, rounded, light-brown head. It immediately burrows obliquely into the fruit, and finally penetrates to the stone. The irritation, arising from the wounds and from the gnawings of the grubs, causes the young fruit to become gummy, diseased, and finally to drop before it is ripe. Meanwhile, the grub comes to its growth, and, immediately after the falling of the fruit, quits the latter and burrows in the ground. This may occur at various times between the middle of June and of August; and, in about three weeks afterwards, the insect completes its transformations, and comes out of the ground in the beetle form.

The earliest account of the habits of the plum weevil, that I have seen, was written by Dr. James Tilton, of Wilmington, Delaware. It will be found, under the article *Fruit*, in Dr. James Mease's edition of Willich's "Domestic Encyclopædia," published at Philadelphia in 1803. The same account has been reprinted in the "Georgic Papers for 1809" of the Massachusetts Agricultural Society, and in other works. According to Dr. Tilton, this insect attacks not only nectarines, plums, apricots, and cherries, but also peaches, apples, pears, and quinces, the truth of which has been abundantly confirmed by later writers. I have myself ascertained that the *cherry-worm*, so called, which is very common in this fruit when gathered from the tree, produces, at maturity, the same curculio as that of the plum; but, unlike the latter, it rarely causes the stung cherry to drop prematurely to the ground. The late Dr. Joel Burnett, of Southborough, the author of two interesting articles on the plum-weevil,* sent to me, in the summer of 1839, some specimens of the insect, in the chrysalis state, which were raised from the small grubs in apples; and, since that time, I have seen the same grubs in apples, pears, and quinces, in this vicinity. They are not to be mistaken for the more common *apple-worms*, from which they are easily distinguished by their inferior size, and by their want of feet. In 1831, Mr. Thomas Say, in a note on the plum-weevil, stated that it "depredates on the plum and peach and other stone-fruits;" and that his "kinsman, the late excellent William Bartram, informed him it also destroys the English walnut in this country." †

Observers do not agree concerning some points in the economy of this insect, such as the time required for it to complete its transformations, the condition and place wherein it passes the winter, and the agency of the curculio in producing the warts or excrescences on plum and cherry trees. The average time passed by the insect in the ground, during the summer, has appeared to me to be about three weeks; but the transformation may be accelerated or retarded by temperature and situation. It has also been my impression that the late broods remained in the ground all winter, and that from them are produced the beetles which sting the fruit in the following spring. Dr. Burnett's observations coincide with this opinion. According to him, the insect "undergoes transformation in about fifteen or twenty days, in the month of June or fore part of July; but all the larvæ, (as

* New England Farmer, Vol. XVIII. p. 304, March 11, 1840; and Hovey's Magazine of Horticulture, Vol. IX. p. 281, August, 1843, reprinted in the New England Farmer, Vol. XXII. p. 49, August 16, 1843, and in the Transactions of the Massachusetts Horticultural Society, for 1843 – 1846, p. 18.

† Descriptions of Curculionites, p. 19 (8vo, New Harmony, 1831).

far as he had observed,) that go into the earth as late as the 20th of July, do not ascend that season, but remain there in the pupa stage until next spring." Dr. Tilton, in his account of the curculio, stated that "it remains in the earth, in the form of a grub, during the winter, ready to be metamorphosed into a beetle as the spring advances." According to M. H. Simpson, Esq., of Saxonville, the larvæ, or grubs, "go through their chrysalis state in three weeks after going into the ground, and remain in a torpid state through the season, unless the earth is disturbed." * Dr. E. Sanborn, of Andover, has come to entirely different conclusions, from a series of experiments made upon these insects. It is his opinion that they do not remain in the ground, during the winter, either in the grub or in the beetle state; but that, under all conditions of place and temperature, "in about six weeks" after they have entered the earth "they return to the surface perfectly finished, winged, and equipped for the work of destruction"; and that, "as neither the curculio nor its grub burrows in the ground during the winter, the common practice of guarding against its ravages, by various operations in the soil, rests upon a false theory, and is productive of no valuable results."† If these conclusions be correct, these insects must pass the winter above ground, in the beetle state, and the place of their concealment, during this season, remains to be discovered.

In July, 1818, Professor W. D. Peck obtained, from the warty excrescences of the cherry-tree, the same insects that he "had long known to occasion the fall of peaches, apricots, and plums, before they had acquired half their growth"; and, not aware that this species had already received a scientific name, he called it *Rhynchænus Cerasi*, the cherry-weevil. His account of it, with a figure, may be seen in the fifth volume of the "Massachusetts Agricultural Repository and

* Hovey's Magazine, Vol. XVI. p. 257, June, 1850.

† See Dr. Sanborn's interesting communications on the Plum Curculio, in the Boston Cultivator, for May 19, 1849, and July 13, 1850, and in the Puritan Recorder for May 2, and the Cambridge Chronicle for May 30, 1850.

Journal." The grubs, found by Professor Peck in the tumors of the cherry-tree, went into the ground on the 6th of July, and on the 30th of the same month, or twenty-four days from their leaving the bark, the perfect insects began to rise, and were soon ready to deposit their eggs.

The plum, still more than the cherry tree, is subject to a disease of the small limbs, that shows itself in the form of large irregular warts, of a black color. Professor Peck referred this disease, as well as that of the cherry-tree, to the agency of insects, but was uncertain whether to attribute it to his cherry-weevil "or to another species of the same genus." It was his opinion, that "the seat of the disease is in the bark. The sap is diverted from its regular course, and is absorbed entirely by the bark, which is very much increased in thickness; the cuticle bursts, the swelling becomes irregular, and is formed into black lumps, with a cracked, uneven, granulated surface. The wood, besides being deprived of its nutriment, is very much compressed, and the branch above the tumor perishes." Dr. Burnett rejected the idea of the insect origin of this disease, which he considered as a kind of fungus, arising in the alburnum, from an obstruction of the vessels, and bursting through the bark, which became involved in the disease. These tumors appear to me to begin between the bark and wood. They are at first soft, cellular, and full of sap, but finally become hard and woody. But whether caused by vitiated sap, as Dr. Burnett supposed, or by the irritating punctures of insects, which is the prevailing opinion, or whatever be their origin and seat, they form an appropriate bed for the growth of numerous little parasitical plants or *fungi*, to which botanists give the name of *Sphæria morbosa.* These plants are the minute black granules that cover the surface of the wart, and give to it its black color. When fully matured, they are filled with a gelatinous fluid, and have a little pit or depression on their summit. They come to their growth, discharge their volatile seed, and die in the course of a single summer; and with them perishes the tumor whence they sprung. It is worthy of remark, that they are sure to appear on these warts in due time, and that they are never found on any other part of the tree.

Insects are often found in the warts of the plum-tree, as well as in those of the cherry-tree. The larvæ of a minute *Cynips*, or gall-fly, are said to inhabit them,* but have never fallen under my observation. The naked caterpillars of a minute moth are very common in the warts of the plum-tree, in which also are sometimes found other insects, among them little grubs from which genuine plum-weevils have been raised. This is a very interesting fact in the economy of the plum-weevil. It may be questioned, however, whether it be a mere mistake of instinct that leads the curculio to lay its eggs in the warts of the plum-tree, or a special provision of a wise Providence to secure thereby a succession of the species in unfruitful seasons.

The following, among other remedies that have been suggested, may be found useful in checking the ravages of the plum-weevil. Let the trees be briskly shaken or suddenly jarred every morning and evening during the time that the insects appear in the beetle form, and are engaged in laying their eggs. When thus disturbed, they contract their legs and fall; and, as they do not immediately attempt to fly or crawl away, they may be caught in a sheet spread under the tree, from which they should be gathered into a large wide-mouthed bottle, or other tight vessel, and be thrown into the fire. Keeping the fruit covered with a coat of whitewash, which is to be applied with a syringe as often as necessary, has been much recommended of late to repel the attacks of the curculio. A little glue, added to the whitewash, causes it to stick better and last longer. We may succeed by this remedy in securing a crop of plums; but as we cannot apply it to cherries and apples, they will be sure to suffer more than ever, and hence no check will

* Schweinitz, Synopsis Fungorum; in Transactions of the American Philosophical Society, Philadelphia, New Series, Vol. IV. p. 204.

be given to the increase of the weevil. All the fallen fruit should be immediately gathered and thrown into a tight vessel, and after it is boiled or steamed to kill the enclosed grubs, it may be given as food to swine. Many of the grubs will be found in the bottom of the vessel in which the fallen fruit has been deposited. Not one of these should be allowed to escape to the ground, but they should all be killed before they have time to complete their transformations. The diseased excrescences on the trees should be cut out, and, as they often contain insects, they should be burnt. If the wounds are washed with strong brine, the formation of new warts will be checked. The moose plum-tree (*Prunus Americana*) seems to be free from warts, even when growing in the immediate vicinity of diseased foreign trees. It would, therefore, be the best of stocks for budding or ingrafting upon. It can be easily raised from the stone, and grows rapidly, but does not attain a great size.

Among the many insects that have been charged with being the cause of the wide-spread pestilence, commonly called the potato-rot, there is a kind of weevil that lives in the stalk of the potato. The history of this little insect was first made known by Miss Margaretta H. Morris, of Germantown, Pennsylvania. In August, 1849, her attention was called to this subject by Mr. Wilkinson, the principal of the Mount Airy Agricultural Institute, "who discovered small grubs in the potato-vines on his farm, and naturally feared injurious consequences." On the 28th of the same month and year, Miss Morris sent to me some specimens of the insects in a piece of the potato-stalk, wherein they underwent their transformations. They proved to be the beetles described by Mr. Say under the name of *Baridius trinotatus* (Fig. 41), so called from their having three black dots on their backs. This kind of beetle is about three twentieths of an inch long. Its body is covered with short whitish hairs, which give to it a gray appearance.

Fig. 41.

One of the black dots is on the scutel, and the others are on the hinder angles of the thorax; and by these it can be readily distinguished from other species. According to Miss Morris, it lays its eggs singly on the plant at the base of a leaf. The grubs burrow into and consume the inner substance of the stalk, proceeding downwards towards the root. In many fields in the neighborhood of Germantown every stem was found to be infested by these insects, causing the premature decay of the vines, and giving to them the appearance of having been scalded. The insects undergo all their transformations in the stalks. Their pupa state lasts from fourteen to twenty days, and they take the beetle form during the last of August and beginning of September. These insects, though common enough in the Middle States, I have never found in New England, in the course of thirty years of observation, and have failed to discover them here since my attention was called to their depredations by Miss Morris. That they may become very injurious to the potato crop where they abound, will be readily admitted; but, as they do not occur in all places, either here or in Europe, where the potato-rot has prevailed, they cannot be justly said to produce this disease.*

The most pernicious of the Rhynchophorians, or snout-beetles, are the insects properly called grain-weevils, belonging to the old genus *Calandra.* These insects must not be confounded with the still more destructive larvæ of the corn-moth (*Tinea granella*), which also attacks stored grain, nor with the orange-colored maggots of the wheat-fly (*Cecidomyia Tritici*), which are found in the ears of growing wheat. Although the grain-weevils are not actually injurious to vegetation, yet as the name properly belonging to them has often been misapplied in this country, thereby creating no little confusion, some remarks upon them may tend to prevent future mistakes.

* See my communication on this insect, &c., in the New England Farmer, for June 22, 1850, Vol. II. p. 204.

The true grain-weevil or wheat-weevil of Europe, *Calandra (Sitophilus) granaria*, or *Curculio granarius* of Linnæus, in its perfected state is a slender beetle of a pitchy-red color, about one eighth of an inch long, with a slender snout slightly bent downwards, a coarsely punctured and very long thorax, constituting almost one half the length of the whole body, and wing-covers that are furrowed and do not entirely cover the tip of the abdomen. This little insect, both in the beetle and grub state, devours stored wheat and other grains, and often commits much havoc in granaries and brewhouses. Its powers of multiplication are very great, for it is stated that a single pair of these destroyers may produce above six thousand descendants in one year. The female deposits her eggs upon the wheat after it is housed, and the young grubs hatched therefrom immediately burrow into the wheat, each individual occupying alone a single grain, the substance of which it devours, so as often to leave nothing but the hull; and this destruction goes on within while no external appearance leads to its discovery, and the loss of weight is the only evidence of the mischief that has been done to the grain. In due time the grubs undergo their transformations, and come out of the hulls, in the beetle state, to lay their eggs for another brood. These insects are effectually destroyed by kiln-drying the wheat; and grain that is kept cool, well ventilated, and is frequently moved, is said to be exempt from attack.

Rice is attacked by an insect closely resembling the wheat-weevil, from which, however, it is distinguished by having two large red spots on each wing-cover; it is also somewhat smaller, measuring only about one tenth of an inch in length, exclusive of the snout. This beetle, the *Calandra (Sitophilus) Oryzæ*,* or rice-weevil (Plate II. Fig. 8), is not entirely confined to rice, but depredates upon wheat, and also on Indian corn. In the Southern States it is called *the black weevil*, to distinguish it from other insects that in-

* *Curculio Oryzæ* of Linnæus.

fest grain. I am not aware that these weevils attack wheat in New England; but I have seen stored Southern corn swarming with them; and, should they multiply and extend in this section of the country, they will become a source of serious injury to one of the most valuable of our staple productions. It is said that this weevil lays its eggs on the rice in the fields, as soon as the grain begins to swell. If this indeed be true, we have very little to fear from it here, our Indian corn being so well protected by the husks that it would probably escape from any injury, if attacked. On the contrary, if the insects multiply in stored grain, then our utmost care will be necessary to prevent them from infesting our own garners. The parent beetle bores a hole into the grain, and drops therein a single egg, going from one grain to another till all her eggs are laid. She then dies, leaving, however, the rice well seeded for a future harvest of weevil-grubs. In due time the eggs are hatched, the grubs live securely and unseen in the centre of the rice, devouring a considerable portion of its substance, and when fully grown they gnaw a little hole through the end of the grain, artfully stopping it up again with particles of rice-flour, and then are changed to pupæ. This usually occurs during the winter; and in the following spring the insects are transformed to beetles, and come out of the grain. By winnowing and sifting the rice in the spring, the beetles can be separated, and then should be gathered immediately and destroyed.

The sudden change of the temperature that generally occurs in the early part of May, brings out great numbers of insects from their winter quarters, to enjoy the sunshine and the ardent heat which are congenial to their natures. While a continued hum is heard, among the branches of the trees, from thousands of bees and flies, drawn thither by the fragrance of the bursting buds and the tender foliage, and the very ground beneath our feet seems teeming with insect life, swarms of little beetles of various kinds come forth to try their wings, and, with an uncertain and heavy flight, launch into the air. Among these beetles there are many of a dull red or fox color, nearly cylindrical in form, tapering a very little before, obtusely rounded at both extremities, and about one quarter of an inch in length. They are seen slowly creeping upon the sides of wooden buildings, resting on the tops of fences, or wheeling about in the air, and every now and then suddenly alighting on some tree or wall, or dropping to the ground. If we go to an old pine-tree we may discover from whence they have come, and what they have been about during the past period of their lives. Here they will be found creeping out of thousands of small round holes which they have made through the bark for their escape. Upon raising a piece of the bark, already loosened by the undermining of these insects, we find it pierced with holes in every direction, and even the surface of the wood will be seen to have been gnawed by these little miners. After enjoying themselves abroad for a few days, they pair, and begin to lay their eggs. The pitch-pine is most generally chosen by them for this purpose, but they also attack other kinds of pines. They gnaw little holes here and there through the rough bark of the trunk and limbs, drop their eggs therein, and, after this labor is finished, they become exhausted and die. In the autumn the grubs hatched from these eggs will be found fully grown. They have a short, thick, nearly cylindrical body, wrinkled on the back, are somewhat curved, and of a yellowish-white color, with a horny darker-colored head, and are destitute of feet. They devour the soft inner substance of the bark, boring through it in various directions for this purpose, and, when they have come to their full size, they gnaw a passage to the surface for their escape after they have completed their transformations. These take place deep in their burrows late in the autumn, at which time the insects may be found, in various states of maturity, within the bark. Their depredations interrupt the descent of the sap, and prevent the formation of new wood; the bark becomes loosened from the wood, to a greater or less extent, and the tree languishes and prematurely decays. The name of this insect is *Hylurgus terebrans*,* the boring Hylurgus (Fig. 42); the generical name signifying a carpenter, or worker in wood. It belongs to the family SCOLYTIDÆ, including various kinds of destructive insects, which may be called cylindrical bark-beetles. The insects of this family may be recognized by the following characters. The body is nearly cylindrical, obtuse before and behind, and generally of some shade of brown. The head is rounded, sunk pretty deeply in the fore part of the thorax, and does not end with a snout; the antennæ are short, more or less crooked or curved in the middle, and end with an oval knob; the feelers are very short. The thorax is rather long, and as broad as the following part of the body. The wing-covers are frequently cut off obliquely, or hollowed at the hinder extremity. The legs are short and strong, with little teeth on the outer edge or extremity of the shanks, and the feet are not wide and spongy beneath.

Fig. 42.

Though these cylindrical bark-beetles are of small size, they multiply very fast, and where they abound are productive of much mischief, particularly in forests, which are often greatly injured by their larvæ, and the wood is rendered unfit for the purposes of art. In the year 1780, an insect of this family made its appearance in the pine-trees of one of the mining districts of Germany, where it increased so rapidly that in three years afterwards whole forests had disappeared beneath its ravages, and an end was nearly put to the working of the extensive mines in this range of country, for the want of fuel to carry on the operations. Pines and firs are the most subject to their attacks, but there are some kinds which infest other trees. The premature decay of the elm in some parts of Europe is occasioned by the ravages of the *Scolytus destructor*, of which an interesting

* *Scolytus terebrans* of Olivier.

account was written in 1824, by Mr. Macleay. An abstract of his paper may be found in the fifth volume of the "New England Farmer." * The larvæ or grubs of these bark-beetles resemble those of the *Hylurgus terebrans*, or pine bark-beetle already described. Like the grubs of the weevils, they are short and thick, and destitute of legs.

The red cedar is inhabited by a very small bark-beetle, named by Mr. Say *Hylurgus dentatus*, the toothed Hylurgus. It is nearly one tenth of an inch in length, and of a dark-brown color; the wing-cases are rough with little grains, which become more elevated towards the hinder part, and are arranged in longitudinal rows, with little furrows between them. The tooth-like appearance of these little elevations suggested the name given to this species. The female bores a cylindrical passage beneath the bark of the cedar, dropping her eggs at short intervals as she goes along, and dies at the end of her burrow when her eggs are all laid. The grubs hatched from these proceed in feeding nearly at right angles, forming on each side numerous parallel furrows, smaller than the central tube of the female. They complete their transformations in October, and eat their way through the bark, which will then be seen to be perforated with thousands of little round holes, through which the beetles have escaped.

Under the bark of the pitch-pine I have found, in company with the pine bark-beetle, a more slender bark-beetle, of a dark chestnut-brown color, clothed with a few short yellowish hairs, with a long, almost egg-shaped thorax, which is very rough before, and short wing-covers, deeply punctured in rows, hollowed out at the tip like a gouge, and beset around the outer edge of the hollow with six little teeth on each side. This beetle measures one fifth of an inch, or rather more, in length. It arrives at maturity in the autumn, but does not come out of the bark till the following spring, at which time it lays its eggs. It is the *Tomicus exesus*, or excavated Tomicus; the specific name, signifying eaten out

* Page 169.

or excavated, was given to it by Mr. Say on account of the hollowed and bitten appearance of the end of its wing-covers. Its grubs eat zigzag and wavy passages, parallel to each other, between the bark and the wood. They are much less common in the New England than in the Middle and Southern States, where they abound in the yellow pines.

Another bark-beetle is found here, closely resembling the preceding, from which it differs chiefly in the inferiority of its size, being but three twentieths of an inch in length, and in having only three or four teeth at the outer extremity of each wing-cover. It is the *Tomicus Pini* of Mr. Say (Fig. 43). The grubs of this insect are very injurious to pine-trees. I have found them under the bark of the white and pitch pine, and they have also been discovered in the larch. The beetles appear during the month of August.

Fig. 43.

There is another small bark-beetle, the *Tomicus liminaris*[9] of my Catalogue, which has been found, in great numbers, by Miss Morris, under the bark of peach-trees, affected with the disease called *the yellows*, and hence supposed by her to be connected with this malady.* I have found it under the bark of a diseased elm; but have nothing more to offer, from my own observations, concerning its history, except that it completes its transformation in August and September. It is of a dark-brown color; the thorax is punctured, and the wing-covers are marked with deeply punctured furrows, and are beset with short hairs. It does not average one tenth of an inch in length.

The pear-tree in New England has been found to be subject to a peculiar malady, which shows itself during midsummer by the sudden withering of the leaves and fruit, and the discoloration of the bark of one or more of the limbs,

[[9] This species differs from the others known in this country by having the last three joints of the antennæ dilated laterally, forming a lamellate club like that of the Scarabæidæ; it therefore belongs to the genus Phloiotribus. — Lec.]

* See Miss Morris on the Yellows, in Downing's Horticulturist, Vol. IV. p. 502.

followed by the immediate death of the part affected. This kind of blight, as it has been called, being oftenest confined to a single branch, or to the extremity of a branch, seems to be a local affection only. It ends with the death of the branch, down to a certain point, but does not extend below the seat of attack, and does not affect the health of other parts of the tree. In June, 1816, the Hon. John Lowell, of Roxbury, discovered a minute insect in one of the affected limbs of a pear-tree; afterwards, he repeatedly detected the same insects in blasted limbs, and his discoveries have been confirmed by Mr. Henry Wheeler and the late Dr. Oliver Fiske, of Worcester, and by many other persons. Mr. Lowell submitted the limb and the insect contained therein to the examination of Professor Peck, who gave an account and figure of the latter, in the fourth volume of the "Massachusetts Agricultural Repository and Journal."

From this account, and from the subsequent communication by Mr. Lowell, in the fifth volume of the "New England Farmer," it appears that the grub or larva of the insect eats its way inward through the alburnum or sap-wood into the hardest part of the wood, beginning at the root of a bud, behind which probably the egg was deposited, following the course of the eye of the bud towards the pith, around which it passes, and part of which it also consumes; thus forming, after penetrating through the alburnum, a circular burrow or passage in the heart-wood, contiguous to the pith which it surrounds. By this means the central vessels, or those which convey the ascending sap, are divided, and the circulation is cut off. This takes place when the increasing heat of the atmosphere, producing a greater transpiration from the leaves, renders a large and continued flow of sap necessary to supply the evaporation. For the want of this, or from some other unexplained cause, the whole of the limb above the seat of the insect's operations suddenly withers, and perishes during the intense heat of midsummer. The larva is changed to a pupa, and subsequently to a little beetle, in the bottom of its burrow, makes its escape from the tree in the latter part of June, or beginning of July, and probably deposits its eggs before August has passed.

This insect, which may be called the *blight-beetle*, from the injury it occasions, attacks also apple, apricot, and plum trees, though less frequently than pear-trees. In the latter part of May, 1843, a piece of the blighted limb of an apple-tree was sent to me for examination. It was twenty-eight inches in length, and three quarters of an inch in diameter at the lower end. Its surface bore the marks of twenty buds, thirteen of which were perforated by the insects; and from the burrows within I took twelve of the blight-beetles in a living and perfect condition, the thirteenth insect having previously been cut out. On the 9th of July, 1844, the Hon. M. P. Wilder sent to me a piece of a branch from a plum-tree, which contained, within the space of one foot, four nests or branching burrows, in each of which several insects in the grub and chrysalis state were found, and also one that had completed its transformations. Soon afterwards I caught one of the blight-beetles on a plum-tree, probably about to lay her eggs. In the following month of August, I received a blighted branch of an apricot-tree, one inch in diameter at the largest end, and containing, within the short distance of six inches, seven or eight perfect blight-beetles, each in a separate burrow, and vestiges of other burrows that had been destroyed in cutting the branch.*

This little beetle, which is only one tenth of an inch in length; was named *Scolytus Pyri*, the pear-tree Scolytus, by Professor Peck. It is of a deep brown color, with the antennæ and legs of the color of iron-rust. The thorax is short, very convex, rounded and rough before; the wing-covers are minutely punctured in rows, and slope off very suddenly and obliquely behind; the shanks are widened and flattened towards the end, beset with a few little teeth

* See my communications on these insects in the Massachusetts Ploughman for June 17, 1843. Also Downing's Horticulturist for February, 1848, Vol. II. p. 365.

externally, and end with a short hook; and the joints of the feet are slender and entire. This insect cannot be retained in the genus *Scolytus*, as defined by modern naturalists, but is to be placed in the genus *Tomicus*. The minuteness of the insect, the difficulty attending the discovery of the precise seat of its operations before it has left the tree, and the small size of the aperture through which it makes its escape from the limb, are probably the reasons why it has eluded the researches of those persons who disbelieve in its existence as the cause of the blasting of the limbs of the pear-tree. It is to be sought for at or near the lowest part of the diseased limbs, and in the immediate vicinity of the buds situated about that part. The remedy, suggested by Mr. Lowell and Professor Peck, to prevent other limbs and trees from being subsequently attacked in the same way, consists in cutting off the blasted limb *below* the seat of injury, and burning it *before* the *perfect* insect has made its escape. It will therefore be necessary carefully to examine our pear-trees daily, during the month of June, and watch for the first indication of disease, or the remedy may be applied too late to prevent the dispersion of the insects among other trees.

There are some other beetles, much like the preceding in form, whose grubs bore into the solid wood of trees. They were formerly included among the cylindrical bark-beetles, but have been separated from them recently, and now form the family BOSTRICHIDÆ, or Bostrichians. Some of these beetles are of large size, measuring more than an inch in length, and, in the tropical regions where they are found, must prove very injurious to the trees they inhabit. The body in these beetles is hard and cylindrical, and generally of a black color. The thorax is bulging before, and the head is sunk and almost concealed under the projecting fore part of it. The antennæ are of moderate length, and end with three large joints, which are saw-toothed internally. The larvæ are mostly wood-eaters, and are whitish fleshy grubs, wrinkled on the back, furnished with six legs, and resemble in form the grubs of some of the small Scarabæians.

The shagbark or walnut tree is sometimes infested by the grubs of the red-shouldered Apate, or *Apate basillaris* of Say, an insect of this family. The grubs bore diametrically through the trunks of the walnut to the very heart, and undergo their transformations in the bottom of their burrows. Several trees have fallen under my observation which have been entirely killed by these insects. The beetles are of a deep black color, and are punctured all over. The thorax is very convex and rough before; the wing-covers are not excavated at the tip, but they slope downwards very suddenly behind, as if obliquely cut off, the outer edge of the cut portion is armed with three little teeth on each wing-cover, and on the base or shoulders there is a large red spot. This insect measures one fifth of an inch or more in length.

The most powerful and destructive of the wood-eating insects are the grubs of the long-horned or Capricorn-beetles (CERAMBYCIDÆ), called *borers* by way of distinction. There are many kinds of borers which do not belong to this tribe. Some of them have already been described, and others will be mentioned under the orders to which they belong. Those now under consideration differ much from each other in their habits. Some live altogether in the trunks of trees, others in the limbs; some devour the wood, others the pith; some are found only in shrubs, some in the stems of herbaceous plants, and others are confined to roots. Certain kinds are limited to plants of one species, others live indiscriminately upon several plants of one natural family; but the same kind of borer is not known to inhabit plants differing essentially from each other in their natural characters. As might be expected from these circumstances, the beetles produced from these borers are of many different kinds. Nearly one hundred species have been found in Massachusetts, and probably many more remain to be discovered.

The Capricorn-beetles agree in the following respects.

The antennæ are long and tapering, and generally curved like the horns of a goat, which is the origin of the name above given to these beetles. The body is oblong, approaching to a cylindrical form, a little flattened above, and tapering somewhat behind. The head is short, and armed with powerful jaws. The thorax is either square, barrel-shaped, or narrowed before; and is not so wide behind as the wing-covers. The legs are long; the thighs thickened in the middle; the feet four-jointed, not formed for rapid motion, but for standing securely, being broad and cushioned beneath, with the third joint deeply notched. Most of these beetles remain upon trees and shrubs during the daytime, but fly abroad at night. Some of them, however, fly by day, and may be found on flowers, feeding on the pollen and the blossoms. When annoyed or taken into the hands, they make a squeaking sound by rubbing the joints of the thorax and abdomen together. The females are generally larger and more robust than the males, and have rather shorter antennæ. Moreover, they are provided with a jointed tube at the end of the body, capable of being extended or drawn in like the joints of a telescope, by means of which they convey their eggs into the holes and chinks of the bark of plants.

The larvæ hatched from these eggs are long, whitish, fleshy grubs, with the transverse incisions of the body very deeply marked, so that the rings are very convex or hunched both above and below. The body tapers a little behind, and is blunt-pointed. The head is much smaller than the first ring, slightly bent downwards, of a horny consistence, and is provided with short but very powerful jaws, by means whereof the insect can bore, as with a centre-bit, a cylindrical passage through the most solid wood. Some of these borers have six very small legs, namely, one pair under each of the first three rings; but most of them want even these short and imperfect limbs, and move through their burrows by alternate extension and contraction of their bodies, on each or on most of the rings of which, both above and below, there is an oval space covered with little elevations, somewhat like the teeth of a fine rasp; and these little oval rasps, which are designed to aid the grubs in their motions, fully make up to them the want of proper feet.

Some of these borers always keep one end of their burrows open, out of which, from time to time, they cast their chips, resembling coarse sawdust; others, as fast as they proceed, fill up the passages behind them with their castings, well known here by the name of powder-post. These borers live from one year to three or perhaps more years before they come to their growth. They undergo their transformations at the furthest extremity of their burrows, many of them previously gnawing a passage through the wood to the inside of the bark, for their future escape. The pupa is at first soft and whitish, and it exhibits all the parts of the future beetle under a filmy veil which inwraps every limb. The wings and legs are folded upon the breast, the long antennæ are turned back against the sides of the body, and then bent forwards between the legs. When the beetle has thrown off its pupa-skin, it gnaws away the thin coat of bark that covers the mouth of its burrow, and comes out of its dark and confined retreat, to breathe the fresh air, and to enjoy for the first time the pleasure of sight, and the use of the legs and wings with which it is provided.

The Capricorn-beetles have been divided into three families, corresponding with the genera *Prionus*, *Cerambyx*, and *Leptura* of Linnæus. Those belonging to the first family are generally of a brown color, have flattened and saw-toothed or beaded antennæ of a moderate length, projecting jaws, and kidney-shaped eyes. Those in the second have eyes of the same shape, more slender or much longer antennæ, and smaller jaws; and are often variegated in their colors. The beetles belonging to the third family are readily distinguished by their eyes, which are round and prominent. These three families are divided into many smaller groups and genera, the peculiarities of which cannot be particularly pointed out in a work of this kind.

The Prionians, or PRIONIDÆ, derive their name from a Greek word signifying a saw, which has been applied to them either because the antennæ, in most of these beetles, consists of flattened joints, projecting internally somewhat like the teeth of a saw, or on account of their upper jaws, which sometimes are very long and toothed within. It is said that some of the beetles thus armed can saw off large limbs by seizing them between their jaws, and flying or whirling sidewise round the enclosed limb, till it is completely divided. The largest insects of the Capricorn tribe belong to this family, some of the tropical species measuring five or six inches in length, and one inch and a half or two inches in breadth. Their larvæ are broader and more flattened than the grubs of the other Capricorn-beetles, and are provided with six very short legs. When about to be transformed, they collect a quantity of their chips around them, and make therewith an oval pod or cocoon, to enclose themselves.

Our largest species is the broad-necked Prionus (Fig. 44), *Prionus laticollis* * of Drury, its first describer. It is of a long oval shape and of a pitchy-black color. The jaws, though short, are very thick and strong; the antennæ are stout and saw-toothed in the male, and more slender in the other sex; the thorax is short and wide, and armed on the lateral edges with three teeth; the wing-covers have three slightly elevated lines on each of them, and are rough with a multitude of large punctures, which run together irregularly. It measures from one inch and one

Fig. 44.

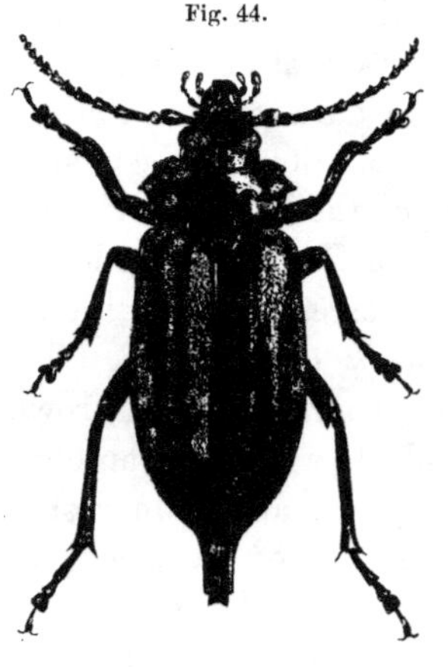

* *Prionus brevicornis* of Fabricius.

eighth to one inch and three quarters in length; the females being always much larger than the males. The grubs of this beetle, when fully grown, are as thick as a man's thumb. They live in the trunks and roots of the balm of gilead, Lombardy poplar, and probably in those of other kinds of poplar also. The beetles may frequently be seen upon, or flying round, the trunks of these trees in the month of July, even in the daytime, though the other kinds of Prionus generally fly only by night.

The one-colored Prionus, *Prionus unicolor* * [10] of Drury (Fig. 45), inhabits pine-trees. Its body is long, narrow, and flattened, of a light bay-brown color, with the head and antennæ darker. The thorax is very short, and armed on each side with three sharp teeth; the wing-covers are nearly of equal breadth throughout, and have three slightly elevated ribs on each of them. This beetle measures from one inch and one quarter to one inch and a half in length, and about three or four tenths of an inch in breadth. It flies by night, and frequently enters houses in the evening, from the middle of July to September.

Fig. 45.

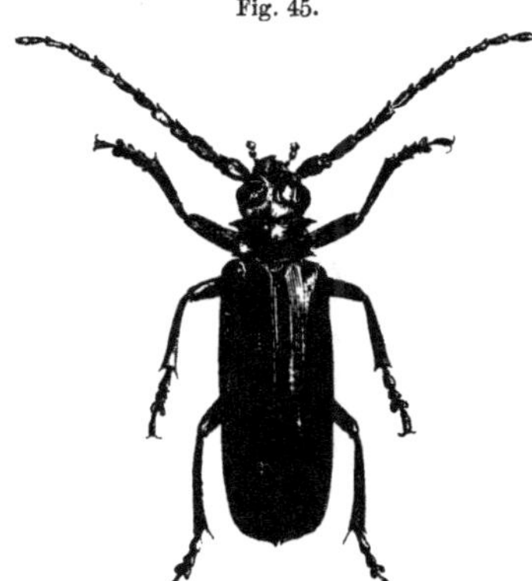

The second family of the Capricorn-beetles may be allowed to retain the scientific name, CERAMBYCIDÆ, of the tribe to which it belongs. The Cerambycians have not the very prominent jaws of the Prionians; their eyes are always kidney-shaped or notched for the reception of the first joint of the antennæ, which are not saw-toothed, but generally

* *P. cylindricus* of Fabricius.

[[10] This species was very properly separated by Serville as a distinct genus *Orthosoma*. — LEC.]

slender and tapering, sometimes of moderate length, sometimes excessively long, especially in the males; the thorax is longer and more convex than in the preceding family, not thin-edged, but often rounded at the sides.

Some of these beetles, distinguished by their narrow wing-covers, which are notched or armed with two little thorns at the tip, and by the great length of their antennæ, belong to the genus *Stenocorus*, a name signifying narrow or straitened. One of them, which is rare here, inhabits the hickory, in its larva state forming long galleries in the trunk of this tree in the direction of the fibres of the wood. This beetle is the *Stenocorus* (*Cerasphorus*) *cinctus*,* or banded Stenocorus (Fig 46). It is of a hazel color, with a tint of gray, arising from the short hairs with which it is covered; there is an oblique ochre-yellow band across each wing-cover; and a short spine or thorn on the middle of each side of the thorax. The antennæ of the males are more than twice the length of the body, which measures from three quarters of an inch to one inch and one quarter in length.

Fig 46.

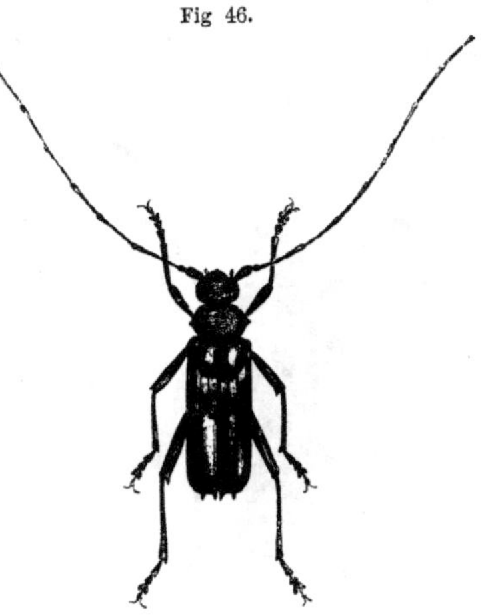

The ground beneath black and white oaks is often observed to be strewn with small branches, neatly severed from these trees as if cut off with a saw. Upon splitting open the cut end of a branch, in the autumn or winter after it has fallen, it will be found to be perforated to the extent of six or eight inches in the course of the pith, and a slender grub, the author of the mischief, will be discovered therein. In

* *Cerambyx cinctus*, Drury; *Stenocorus garganicus*, Fabricius.

the spring this grub is transformed to a pupa, and in June or July it is changed to a beetle, and comes out of the branch.

Fig. 47.

The history of this insect was first made public by Professor Peck,* who called it the oak-pruner, or *Stenocorus* (*Elaphidion*) *putator* (Fig 47).[11] In its adult state it is a slender long-horned beetle, of a dull brown color, sprinkled with gray spots, composed of very short close hairs; the antennæ are longer than the body in the males, and equal to it in length in the other sex, and the third and fourth joints are tipped with a small spine or thorn; the thorax is barrel-shaped, and not spined at the sides; and the scutel is yellowish-white. It varies in length from four and a half to six tenths of an inch. It lays its eggs in July. Each egg is placed close to the axilla or joint of a leaf-stalk or of a small twig, near the extremity of a branch. The grub (Fig. 48) hatched from it penetrates at that spot to the pith, and then continues its course towards the body of the tree, devouring the pith, and thereby forming a cylindrical burrow, several inches in length, in the centre of the branch. Having reached its full size, which it does towards the end of the summer, it divides the branch at the lower end of its burrow (Fig 49, pupa), by gnawing away the wood transversely from within, leaving only the ring of bark untouched. It then retires backwards, stops up the end of its hole, near the transverse section, with fibres of the wood, and awaits the fall of the branch, which is usually broken off and precipitated to the ground by the autumnal winds.

Fig. 48.

Larva.

Fig. 49.

Pupa.

* Massachusetts Agricultural Repository and Journal, Vol. V., with a plate.

[11 This species was previously described by Fabricius as *Stenocorus villosus*, which specific name must therefore be preserved. — LEC.]

The leaves of the oak are rarely shed before the branch falls, and thus serve to break the shock. Branches of five or six feet in length and an inch in diameter are thus severed by these insects, a kind of pruning that must be injurious to the trees, and should be guarded against if possible. By collecting the fallen branches in the autumn, and burning them before the spring, we prevent development of the beetles, while we derive some benefit from the branches as fuel.

It is somewhat remarkable that, while the pine and fir tribes rarely suffer to any extent from the depredations of caterpillars and other leaf-eating insects, the resinous odor of these trees, offensive as it is to such insects, does not prevent many kinds of borers from burrowing into and destroying their trunks. Several of the Capricorn-beetles, while in the grub state, live only in pine and fir trees, or in timber of these kinds of wood. They belong chiefly to the genus *Callidium*, a name of unknown or obscure origin. Their antennæ are of moderate length; they have a somewhat flattened body; the head nods forward, as in *Stenocorus;* the thorax is broad, nearly circular, and somewhat flattened or indented above; and the thighs are very slender next to the body, but remarkably thick beyond the middle. The larvæ are of moderate length, more flattened than the grubs of the other Capricorn-beetles, have a very broad and horny head, small but powerful jaws, and are provided with six extremely small legs. They undermine the bark, and perforate the wood in various directions, often doing immense injury to the trees, and to new buildings, in the lumber composing which they may happen to be concealed. Their burrows are wide and not cylindrical, are very winding, and are filled up with a kind of compact sawdust as fast as the insects advance. The larva state is said to continue two years, during which period the insects cast their skins several times. The sides of the body in the pupa are thin-edged, and finely notched, and the tail is forked.

One of the most common kinds of *Callidium* found here is a flattish, rusty-black beetle, with some downy whitish spots across the middle of the wing-covers; the thorax is nearly circular, is covered with fine whitish down, and has two elevated polished black points upon it; and the wing-covers are very coarsely punctured. It measures from four tenths to three quarters of an inch in length. This insect is the *Callidium bajulus* (Plate II. Fig. 12); the second name, meaning a porter, was given to it by Linnæus, on account of the whitish patch which it bears on its back. It inhabits fir, spruce, and hemlock wood and lumber, and may often be seen on wooden buildings and fences in July and August. We are informed by Kirby and Spence, that the grubs sometimes greatly injure the wood-work of houses in London, piercing the rafters of the roofs in every direction, and, when arrived at maturity, even penetrating through sheets of lead which covered the place of their exit. One piece of lead, only eight inches long and four broad, contained twelve oval holes made by these insects, and fragments of the lead were found in their stomachs. As this insect is now common in the maritime parts of the United States, it was probably first brought to this country by vessels from Europe.

The violet Callidium, *Callidium violaceum*,* [12] (Plate II. Fig. 11,) is of a Prussian blue or violet color; the thorax is transversely oval, and downy, and sometimes has a greenish tinge; and the wing-covers are rough with thick irregular punctures. Its length varies from four to six tenths of an inch. It may be found in great abundance on piles of pine wood, from the middle of May to the first of June; and the larvæ and pupæ are often met with in splitting the wood. They live mostly just under the bark, where their broad and winding tracks may be traced by the hardened sawdust with

* *Cerambyx violaceus* of Linnæus.

[12 Our species is considered different from the European *Callidium violaceum*, under the name *C. antennatum*, Newman. — Lec.]

which they are crowded. Just before they are about to be transformed, they bore into the solid wood to the depth of several inches. They are said to be very injurious to the sapling pines in Maine. Professor Peck supposed this species of Callidium to have been introduced into Europe in timber exported from this country, as it is found in most parts of that continent that have been much connected with North America by navigation. Thus Europe and America seem to have interchanged the porter and violet Callidium, which, by means of shipping, have now become common to the two continents.

From the regularity of its form, and the noble size it attains, the sugar-maple is accounted one of the most beautiful of our forest-trees, and is esteemed as one of the most valuable, on account of its many useful properties. This fine tree suffers much from the attacks of borers, which in some cases produce its entire destruction. We are indebted to the Rev. L. W. Leonard, of Dublin, N. H., for the first account of the habits and transformations of these borers. In the summer of 1828, his attention was called to some young maples, in Keene, which were in a languishing condition. He discovered the insect in its beetle state under the loosened bark of one of the trees, and traced the recent track of the larva three inches into the solid wood. In the course of a few years, these trees, upon the cultivation of which much care had been bestowed, were nearly destroyed by the borers. The failure, from the same cause, of several other attempts to raise the sugar-maple, has since come to my knowledge. The insects are changed to beetles, and come out of the trunks of the trees in July. In the vicinity of Boston, specimens have been repeatedly taken, which were undoubtedly brought here in maple logs from Maine. The beetle was first described in 1824, in the Appendix to Keating's "Narrative of Long's Expedition," by Mr. Say, who called it *Clytus speciosus;* that is, the beautiful Clytus. (Plate II. Fig. 15.) It was afterwards inserted, and accurately represented by the pencil of Lesueur, in Say's "American Entomology," and, more recently, a description and figure of it have appeared in Griffith's translation of Cuvier's "Animal Kingdom," under the name of *Clytus Hayii.*

The beautiful Clytus, like the other beetles of the genus to which it belongs, is distinguished from a Callidium by its more convex form, its more nearly globular thorax, which is neither flattened nor indented, and by its more slender thighs. The head is yellow, with the antennæ and the eyes reddish black; the thorax is black, with two transverse yellow spots on each side; the wing-covers, for about two thirds of their length, are black, the remaining third is yellow, and they are ornamented with bands and spots arranged in the following manner: a yellow spot on each shoulder, a broad yellow curved band or arch, of which the yellow scutel forms the key-stone, on the base of the wing-covers, behind this a zigzag yellow band forming the letter W, across the middle another yellow band arching backwards, and on the yellow tip a curved band and a spot of a black color; the legs are yellow; and the under side of the body is reddish yellow, variegated with brown. It is the largest known species of Clytus, being from nine to eleven tenths of an inch in length, and three or four tenths in breadth. It lays its eggs on the trunk of the maple in July and August. The grubs burrow into the bark as soon as they are hatched, and are thus protected during the winter. In the spring they penetrate deeper, and form, in the course of the summer, long and winding galleries in the wood, up and down the trunk. In order to check their devastations, they should be sought for in the spring, when they will readily be detected by the sawdust that they cast out of their burrows; and, by a judicious use of a knife and stiff wire, they may be cut out or destroyed before they have gone deeply into the wood.

Many kinds of Clytus frequent flowers, for the sake of the

pollen, which they devour. During the month of September, the painted Clytus, *Clytus pictus*,* (Plate II. Fig. 10,) is often seen in abundance, feeding by day upon the blossoms of the golden-rod. If the trunks of our common locust-tree, *Robinia pseudacacia*, are examined at this time, a still greater number of these beetles will be found upon them, and most often paired. The habits of this insect seem to have been known, as long ago as the year 1771, to Dr. John Reinhold Foster, who then described it under the name of *Leptura Robiniæ*, the latter being derived from the tree which it inhabits. Drury, however, had previously described and figured it, under the specific name here adopted, which, having the priority, in point of time, over all the others that have been subsequently imposed, must be retained. This Capricorn-beetle has the form of the beautiful maple Clytus. It is velvet-black, and ornamented with transverse yellow bands, of which there are three on the head, four on the thorax, and six on the wing-covers, the tips of which are also edged with yellow. The first and second bands on each wing-cover are nearly straight; the third band forms a V, or, united with the opposite one, a W, as in the *speciosus;* the fourth is also angled, and runs upwards on the inner margin of the wing-cover towards the scutel; the fifth is broken or interrupted by a longitudinal elevated line; and the sixth is arched, and consists of three little spots. The antennæ are dark brown; and the legs are rust-red. These insects vary from six tenths to three quarters of an inch in length.

In the month of September these beetles gather on the locust-trees, where they may be seen glittering in the sunbeams with their gorgeous livery of black velvet and gold, coursing up and down the trunks in pursuit of their mates, or to drive away their rivals, and stopping every now and then to salute those they meet with a rapid bowing of the shoulders, accompanied by a creaking sound, indicative of

* *Leptura picta*, Drury; *Clytus flexuosus*, Fabricius.

recognition or defiance. Having paired, the female, attended by her partner, creeps over the bark, searching the crevices with her antennæ, and dropping therein her snow-white eggs, in clusters of seven or eight together, and at intervals of five or six minutes, till her whole stock is safely stored. The eggs are soon hatched, and the grubs immediately burrow into the bark, devouring the soft inner substance that suffices for their nourishment till the approach of winter, during which they remain at rest in a torpid state. In the spring they bore through the sap-wood, more or less deeply into the trunk, the general course of their winding and irregular passages being in an upward direction from the place of their entrance. For a time they cast their chips out of their holes as fast as they are made, but after a while the passage becomes clogged and the burrow more or less filled with the coarse and fibrous fragments of wood, to get rid of which the grubs are often obliged to open new holes through the bark. The seat of their operations is known by the oozing of the sap and the dropping of the sawdust from the holes. The bark around the part attacked begins to swell, and in a few years the trunks and limbs will become disfigured and weakened by large porous tumors, caused by the efforts of the trees to repair the injuries they have suffered. According to the observations of General H. A. S. Dearborn, who has given an excellent account* of this insect, the grubs attain their full size by the 20th of July, soon become pupæ, and are changed to beetles and leave the trees early in September. Thus the existence of this species is limited to one year.

Whitewashing, and covering the trunks of the trees with grafting composition, may prevent the female from depositing her eggs upon them; but this practice cannot be carried to any great extent in plantations or large nurseries of the trees. Perhaps it will be useful to head down young trees to the ground, with the view of destroying the grubs con-

* Massachusetts Agricultural Repository and Journal, Vol. VI. p. 272.

tained in them, as well as to promote a more vigorous growth. Much evil might be prevented by employing children to collect the beetles while in the act of providing for the continuation of their kind. A common black bottle, containing a little water, would be a suitable vessel to receive the beetles as fast as they were gathered, and should be emptied into the fire in order to destroy the insects. The gathering should be begun as soon as the beetles first appear, and should be continued as long as any are found on the trees, and furthermore should be made a general business for several years in succession. I have no doubt, should this be done, that, by devoting one hour every day to this object, we may, in the course of a few years, rid ourselves of this destructive insect.

The largest Capricorn-beetle, of the Cerambycian family, found in New England, is the *Lamia* (*Monohammus titillator*) of Fabricius (Fig. 50), or the tickler, so named probably on

Fig. 50.

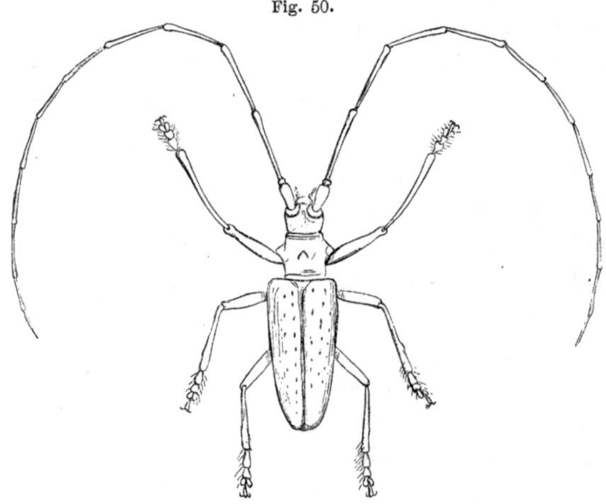

account of the habit which it has, in common with most of the Capricorn-beetles, of gently touching now and then the surface on which it walks with the tips of its long antennæ. Three or four of these beetles may sometimes be seen together in June and July, on logs or on the trunks of trees in the woods, the males paying their court to the females, or contending with their rivals, waving their antennæ, and showing the eagerness of the contest or pursuit by their rapid creaking sounds.

The head of the Lamias is vertical or perpendicular; the antennæ of the males are much longer than the body, and taper to the end; the thorax is cylindrical before and behind, and is armed on the middle of each side with a very large pointed wart or tubercle; the tips of the wing-covers are rounded; and the fore legs are longer than the rest, with broad hairy soles in the males.

The *titillator* is of a brownish color, variegated or mottled with spots of gray, and the wing-covers, which are coarsely punctured, have also several small tufted black spots upon them; the middle legs are armed with a small tooth on the upper edge; the antennæ of the male are twice as long as the body, and those of the other sex equal the body in length, which measures from one inch and one eighth to one inch and one quarter. What kind of tree the grub of this insect inhabits is unknown to me.

Trees of the poplar tribe, both in Europe and America, are subject to the attacks of certain kinds of borers, differing essentially from all the foregoing when arrived at maturity. They belong to the genus *Saperda*. In the beetle state the head is vertical, the antennæ are about the length of the body in both sexes, the thorax is cylindrical, smooth, and unarmed at the sides, and the fore legs are shorter than the others. Our largest kind is the *Saperda calcarata* of Say (Plate II. Fig. 21), or the spurred Saperda, so named because the tips of the wing-covers end with a little sharp point or spur. It is covered all over with a short and close nap, which gives it a fine blue-gray color, it is finely punctured with brown, there are four ochre-yellow lines on the head, and three on the top of the thorax, the scutel is also ochre-yellow, and there are several irregular lines and spots of the same color on the wing-covers. It is from one inch to an inch and a quarter in length. This beetle closely resembles the European *Saperda carcharias*, which inhabits the poplar; and the grubs of our native species, with those of the broad-necked Prionus, have almost entirely destroyed the Lombardy poplar in this vicinity. They live also in the trunks of our American poplars. They are of a yellowish-white color, except the upper part of the first segment, which is dark buff. When fully grown they measure nearly two inches in length. The body is very thick, rather larger before than behind, and consists of twelve segments separated from each other by deep transverse furrows. The first segment is broad, and slopes obliquely downwards to the head; the second is very narrow; on the upper and under sides of each of the following segments, from the third to the tenth inclusive, there is a transverse oval space, rendered rough like a rasp by minute projections. These rasps serve instead of legs, which are entirely wanting. The beetles may be found on the trunks and branches of the various kinds of poplars, in August and September; they fly by night, and sometimes enter the open windows of houses in the evening.

The borers of the apple-tree have become notorious, throughout the New England and Middle States, for their extensive ravages. They are the larvæ of a beetle called *Saperda bivittata* * by Mr. Say, the two-striped, or the brown and white striped Saperda (Plate II. Fig. 16); the upper side of its body being marked with two longitudinal white stripes between three of a light-brown color, while the face, the antennæ, the under side of the body, and the legs are white. This beetle varies in length from a little more than one half to three quarters of an inch. It comes forth from the trunks of the trees, in its perfected state, early in June, making its escape in the night, during which time only it uses its ample wings in going from tree to tree in search

* *Saperda candida?* Fabricius.

of companions and food. In the daytime it keeps at rest among the leaves of the plants which it devours.

The trees and shrubs principally attacked by this borer are the apple-tree, the quince, mountain ash, hawthorn and other thorn bushes, the June-berry or shad-bush, and other kinds of *Amelanchier* and *Aronia*. Our native thorns and Aronias are its natural food; for I have discovered the larvæ in the stems of these shrubs, and have repeatedly found the beetles upon them, eating the leaves, in June and July. It is in these months that the eggs are deposited, being laid upon the bark near the root, during the night. The larvæ hatched therefrom are fleshy whitish grubs, nearly cylindrical, and tapering a little from the first ring to the end of the body. (Plate II. Fig. 17.) The head is small, horny, and brown; the first ring is much larger than the others, the next two are very short, and, with the first, are covered with punctures and very minute hairs; the following rings, to the tenth inclusive, are each furnished, on the upper and under side, with two fleshy warts situated close together, and destitute of the little rasp-like teeth, that are usually found on the grubs of the other Capricorn-beetles; the eleventh and twelfth rings are very short; no appearance of legs can be seen, even with a magnifying glass of high power.

The grub, with its strong jaws, cuts a cylindrical passage through the bark, and pushes its castings backwards out of the hole from time to time, while it bores upwards into the wood. The larva state continues two or three years, during which the borer will be found to have penetrated eight or ten inches upwards in the trunk of the tree, its burrow at the end approaching to, and being covered only by, the bark. Here its transformation takes place. The pupa does not differ much from other pupæ of beetles; but it has a transverse row of minute prickles on each of the rings of the back, and several at the tip of the abdomen. These probably assist the insect in its movements, when casting off its pupa-skin. The final change occurs about the first of June, soon after which, the beetle gnaws through the bark that covers the end of its burrow, and comes out of its place of confinement in the night.

Notwithstanding the pains that have been taken by some persons to destroy and exterminate these pernicious borers, they continue to reappear in our orchards and nurseries every season. The reasons of this are to be found in the habits of the insects, and in individual carelessness. Many orchards suffer deplorably from the want of proper attention; the trees are permitted to remain, year after year, without any pains being taken to destroy the numerous and various insects that infest them; old orchards, especially, are neglected, and not only the rugged trunks of the trees, but even a forest of unpruned suckers around them, are left to the undisturbed possession and perpetual inheritance of the Saperda.

On the means that have been used to destroy this borer, a few remarks only need to be made; for it is evident that they can be fully successful only when generally adopted. Killing it by a wire thrust into the holes it has made, is one of the oldest, safest, and most successful methods. Cutting out the grub, with a knife or gouge, is the most common practice; but it is feared that these tools have sometimes been used without sufficient caution. A third method, which has more than once been suggested, consists in plugging the holes with soft wood. If a little camphor be previously inserted, this practice promises to be more effectual; but experiments are wanting to confirm its expediency.

The coated Saperda, or *Saperda vestita* (Plate II. Fig. 19), described by Mr. Say in the Appendix to Keating's Narrative of Major Long's Expedition, resembles the foregoing species in form. It measures from six to eight tenths of an inch in length; it is entirely covered with a close greenish-yellow down or nap, and has two or three small black dots near the middle of each wing-cover. Mr. Say discovered it near the southern extremity of Lake Michigan, and states that it is also sometimes found in Pennsylvania; but he does not appear to have known anything of its history. It is also found in Massachusetts, but has been rarely seen until within a few years. One of my specimens was taken in Milton about twenty years ago, and several others were taken in Cambridge, during the summers of 1843 and 1844, upon the European lindens, from the trunks and branches of which they had just come forth. A knowledge of the habits of this insect might have led to its more frequent discovery. One of the lindens above named was a noble and venerable tree, with a trunk measuring eight feet and five inches in circumference, three feet from the ground. A strip of the bark, two feet wide at the bottom, and extending to the top of the trunk, had been destroyed, and the exposed surface of the wood was pierced and grooved with countless numbers of holes, wherein the borers had been bred, and whence swarms of the beetles must have issued in past times. Some of the large limbs and a portion of the top of the tree had fallen, apparently in consequence of the ravages of these insects; and it is a matter of surprise that this fine linden should have withstood and outlived the attacks of such a host of miners and sappers.

The lindens of Philadelphia have suffered much more severely from these borers. Dr. Paul Swift, in a letter written in May, 1844, gave to me the following interesting account of them. "The trees in Washington and Independence Squares were first observed to have been attacked about seven years ago. Within two years, it has been found necessary to cut down forty-seven European lindens in the former square alone, where there now remain only a few American lindens, and these a good deal eaten." "Many of the beetles were found upon the small branches and leaves on the 28th day of May, and it is said that they come out as early as the first of the month, and continue to make their way through the bark of the trunk and large branches during the whole of the warm season. They immediately fly into the top of the tree, and there feed upon the epidermis of the tender twigs, and the petioles of the leaves, often wholly denuding the latter, and causing the leaves to fall. They deposit their eggs, two or three in a place, upon the trunk and branches, especially about the forks, making slight incisions or punctures, for their reception, with their strong jaws. As many as ninety eggs have been taken from a single beetle. The grubs, hatched from these eggs, undermine the bark to the extent of six or eight inches, in sinuous channels, or penetrate the solid wood an equal distance. It is supposed that three years are required to mature the insect. Various expedients have been tried to arrest their course, but without effect. A stream, thrown into the tops of the trees from the hydrant, is often used with good success to dislodge other insects; but the borer-beetles, when thus disturbed, take wing and hover over the trees till all is quiet, and then alight and go to work again. The trunks and branches of some of the trees have been washed over with various preparations without benefit. Boring the trunk near the ground, and putting in sulphur and other drugs, and plugging, have been tried with as little effect."

This beetle I have taken in Massachusetts only in June, mostly between the 1st and 17th, and none after the 20th day of the month. The grub closely resembles that of the apple-tree borer. Figures of the insect, in all its stages, may be seen in the tenth volume of Hovey's Magazine, page 330.

There is another destructive *Saperda*, whose history remains to be written. It is the *Saperda tridentata* (Plate II. Fig. 13), so named by Olivier on account of the tridentate or three-toothed red border of its wing-covers. This beetle is of a dark brown color, with a tint of gray, owing to a thin coating of very short down. It is ornamented with a curved line behind the eyes, two stripes on the thorax, and a three-toothed or three-branched stripe on the outer edge of each wing-cover, of a rusty red color. There are also six black dots on the thorax, two above, and two on the sides; and each of the angles between the branches and the lateral stripes of the wing-covers is marked with a blackish spot. The two hinder branches are oblique, and extend nearly or quite to the suture; the anterior branch is short and hooked. Its average length is about half an inch; but it varies from four to six tenths of an inch. The males are smaller than the females, but have longer antennæ.

This pretty beetle has been long known to me, but its habits were not ascertained till the year 1847. On the 19th of June, in that year, Theophilus Parsons, Esq. sent me some fragments of bark and insects which were taken by Mr. J. Richardson from the decaying elms on Boston Common; and, among the insects, I recognized a pair of these beetles in a living state. My curiosity was immediately excited to learn something more concerning these beetles and their connection with the trees, but was not satisfied by a partial examination made in the course of the summer. It was not till the following winter, that an opportunity was afforded for a thorough search, with the permission of the Mayor, the Hon. Josiah Quincy, Jun., and with the help of the Superintendent of the Common.

The trees were found to have suffered terribly from the ravages of these insects. Several of them had already been cut down, as past recovery; others were in a dying state, and nearly all of them were more or less affected with disease or premature decay. Their bark was perforated, to the height of thirty feet from the ground, with numerous holes, through which insects had escaped; and large pieces had become so loose, by the undermining of the grubs, as to yield to slight efforts, and come off in flakes. The inner bark was filled with the burrows of the grubs, great numbers of which, in various stages of growth, together with some in the pupa state, were found therein; and even the surface of the wood, in many cases, was furrowed with their irregular tracks. Very rarely did they seem to have penetrated far into the wood itself; but their operations were mostly confined to the inner layers of the bark, which thereby became loosened from the wood beneath. The grubs rarely exceed three quarters of an inch in length. They have no feet, and they resemble the larvæ of other species of *Saperda*, except in being rather more flattened. They appear to complete their transformations in the third year of their existence.

The beetles probably leave their holes in the bark during the month of June and in the beginning of July; for, in the course of thirty years, I have repeatedly taken them at various dates, from the 5th of June to the 10th of July. It is evident, from the nature and extent of their depredations, that these insects have alarmingly hastened the decay of the elm-trees on Boston Mall and Common, and that they now threaten their entire destruction. Other causes, however, have probably contributed to the same end. It will be remembered that these trees have greatly suffered, in past times, from the ravages of canker-worms. Moreover, the impenetrable state of the surface-soil, the exhausted condition of the subsoil, and the deprivation of all benefit from the decomposition of accumulated leaves, which, in a state of nature, the trees would have enjoyed, but which a regard for neatness has industriously removed, have doubtless had no small influence in diminishing the vigor of the trees, and thus made them fall unresistingly a prey to insect-devourers. The plan of this work precludes a more full consideration of these and other topics connected with the growth and decay of these trees; and I can only add, that it may be prudent to cut down and burn all that are much infested by the borers.

The tall blackberry, *Rubus villosus*, is sometimes cultivated among us for the sake of its fruit, which richly repays the care thus bestowed upon it. It does not seem to be known that this plant and its near relation, the raspberry, suffer from borers that live in the pith of the stems. These borers differ somewhat from the preceding, being cylindrical in the middle, and thickened a little at each end. The head is proportionally larger than in the other borers; the first three rings of the body are short, the second being the widest, and each of them is provided beneath with a pair of minute sharp-pointed warts or imperfect legs; the remaining rings are smooth, and without tubercles or rasps; the last three are rather thicker than those which immediately precede them, and the twelfth ring is very obtusely rounded at the end. The beetles from these borers are very slender, and of a cylindrical form, and their antennæ are of moderate length and do not taper much towards the end.

The species which attacks the blackberry appears to be the *Saperda (Oberea) tripunctata* of Fabricius (Fig. 51). It is of a deep black color, except the fore part of the breast and the top of the thorax, which are rusty yellow, and there are two black elevated dots on the middle of the thorax, and a third dot on the hinder edge close to the scutel; the wing-covers are coarsely punctured, in rows on the top, and irregularly on the sides and tips, each of which is slightly notched and ends with two little points. The two black dots on the middle of the thorax are sometimes wanting. This beetle varies from three tenths to half an inch in length. It finishes its transformations towards the end of July, and lays its eggs early in August, one by one, on the stems of the blackberry and raspberry, near a leaf or small twig. The grubs burrow directly into the pith, which they consume as they proceed, so that the stem, for the distance of several inches, is completely deprived of its pith, and consequently withers and dies before the end of the summer. In Europe one of these slender Saperdas attacks the hazel-bush, and another the twigs of the pear-tree, in the same way.

Fig. 51.

The Lepturians, or LEPTURADÆ, constitute the third family of the Capricorn-beetles. In most of them the body is narrowed behind, which is the origin of the name applied to them, signifying really narrow tail. They differ from the other Capricorn-beetles in the form of their eyes, which are not deeply notched, but are either oval or rounded and prominent, and the antennæ are more distant from them, and are implanted near the middle of the forehead. Moreover, the head is not deeply sunk in the fore part of the thorax, but is connected with it by a narrowed neck. The thorax varies somewhat in shape, but is generally narrowed before and widened behind. The Lepturians are often gayly colored, and fly about by day, visiting flowers for the sake of the pollen and tender leaves, which they eat. Their grubs live in the trunks and stumps of trees, are rather broad and somewhat flattened, and are mostly furnished with six extremely short legs.

The largest and finest of these beetles in New England is the *Desmocerus palliatus*,* (Plate II. Fig. 18,) which appears on the flowers and leaves of the common elder towards the end of June and until the middle of July. It is of a deep violet or Prussian-blue color, sometimes glossed with green, and nearly one half of the fore part of the wing-covers is orange-yellow, suggesting the idea of a short cloak of this color thrown over the shoulders, which the name *palliatus*, that is, cloaked, was designed to express. The head is narrow. The thorax has nearly the form of a cone cut off at the top, being narrow before and wide behind; it is somewhat uneven, and has a little sharp projecting point on each side of the base. The antennæ have the third and the three following joints abruptly thickened at the extremity, giving them the knotty appearance indicated by the generical name *Desmocerus*, which signifies knotty horn. The larvæ live in the lower part of the stems of the elder, and devour the pith; they have hitherto escaped my researches, but I have found the beetles in the burrows made by them.

The bark of the pitch-pine is often extensively loosened by the grubs of Lepturians at work beneath it, in consequence

* *Cerambyx palliatus* of Forster; *Stenocorus cyaneus*, Fabricius.

of which it falls off in large flakes, and the tree perishes. These grubs live between the bark and the wood, often in great numbers together, and, when they are about to become pupæ, each one surrounds itself with an oval ring of woody fibres, within which it undergoes its transformations. The beetle is matured before winter, but does not leave the tree until spring. It is the ribbed Rhagium, or *Rhagium lineatum*,* (Fig. 52,) so named because it has three elevated longitudinal lines or ribs on each wing-cover; and it measures from four and a half to seven tenths of an inch in length. The head and thorax are gray, striped with black, and thickly punctured; the antennæ are about as long as the two forenamed parts of the body together; the thorax is narrow, cylindrical before and behind, and swelled out in the middle by a large pointed wart or tubercle on each side; the wing-covers are wide at the shoulders, gradually taper behind, and are slightly convex above; they are coarsely punctured between the smooth elevated lines, and are variegated with reddish ash-color and black, the latter forming two irregular transverse bands; the under side of the body, and the legs, are variegated with dull red, gray, and black. The gray portions on this beetle are occasioned by very short hairs, forming a close kind of nap, which is easily rubbed off.

Fig. 52.

The Buprestians and the Capricorn-beetles seem evidently allied in their habits, both being borers during the greater part of their lives, and living in the trunks and limbs of trees, to which they are more or less injurious in proportion to their numbers. Some of the beetles in these two groups resemble each other closely in their forms and habits. The resemblance between the slender cylindrical Saperdas and some of the cylindrical Buprestians belonging to the genus *Agrilus*, is indeed very remarkable, and cannot fail to strike a common observer. Their larvæ also are not only very similar in

* *Stenocorus lineatus* of Olivier.

their forms, but they have the same habits; living in the centre of stems, and devouring the pith.

The insects that have passed under consideration in the foregoing part of this treatise spend by far the greater portion of their lives, namely, that wherein they are larvæ only, in obscurity, buried in the ground, or concealed within the roots, the stems, or the seeds of plants, where they perform their appointed tasks unnoticed and unknown. Thus the work of destruction goes secretly and silently on, till it becomes manifest by its melancholy consequences; and too late we discover the hidden foes that have disappointed the hopes of the husbandman, and ruined those spontaneous productions of the soil that constitute so important a source of our comfort and prosperity.

There still remain several groups of beetles to be described, consisting almost entirely of insects that spend the whole, or the principal part, of their lives upon the leaves of plants, and which, as they derive their nourishment, both in the larva and adult states, from leaves alone, may be called leaf-beetles, or, as they have recently been named, phyllophagous, that is, leaf-eating insects. When, as in certain seasons, they appear in considerable numbers, they do not a little injury to vegetation, and, being generally exposed to view on the leaves that they devour, they soon attract attention. But the power possessed by most plants of renewing their foliage, enables them soon to recover from the attacks of these devourers; and the injury sustained, unless often repeated, is rarely attended by the ruinous consequences that follow the hidden and unsuspected ravages of those insects that sap vegetation in its most vital parts. Moreover, the leaf-eaters are more within our reach, and it is not so difficult to destroy them, and protect plants from their depredations. The leaf-beetles are generally distinguished by the want of a snout, by their short legs and broad cushioned feet, and their antennæ of moderate length, often thickened a little towards the end, or not distinctly tapering. Some of them have an oblong body and a narrow or cylindrical thorax, and resemble very much some of the Lepturians, with which Linnæus included them. Others, and indeed the greater number, have the body oval, broad, and often very convex.

The oblong leaf-beetles, called Criocerians (CRIOCERIDIDÆ), have some resemblance to the Capricorn-beetles. They are distinguished by the following characters. The eyes are prominent and nearly round; the antennæ are of moderate length, composed of short, nearly cylindrical or beaded joints, and are implanted before the eyes; the thorax is narrow and almost cylindrical or square; the wing-covers, taken together, form an oblong square, rounded behind, and much wider than the thorax; and the thighs of the hind legs are often thickened in the middle.

The three-lined leaf-beetle, *Crioceris trilineata* of Olivier,[13] (Fig. 53,) will serve to exemplify the habits of the greater part of the insects of this family. This beetle is about one quarter of an inch long, of a rusty buff or nankin-yellow color, with two black dots on the thorax, and three black stripes on the back, namely, one on the outer side of each wing-cover, and one in the middle on the inner edges of the same; the antennæ (except the first joint), the outside of the shins, and the feet are dusky. The thorax is abruptly narrowed or pinched in on the middle of each side. When held between the fingers, these insects make a creaking sound like the Capricorn-beetles. They appear early in June on the leaves of the potato-vines, having at that time recently come out of the ground, where they pass the winter in the pupa state. Within a few years, these insects have excited some attention, on account of their prevalence in some parts of the country, and from a mistaken notion that they were the cause of the *potato-rot*. They eat the leaves

Fig. 53.

[[13] The genus *Crioceris* as now restricted contains only species indigenous to the other continent, although one of them, *C. asparagi*, has been recently introduced from Europe, and is found abundantly near Brooklyn, New York. The species above mentioned belongs to *Lema*. — LEC.]

of the potato, gnawing large and irregular holes through them; and, in the course of a few days, begin to lay their oblong oval golden-yellow eggs, which are glued to the leaves, in parcels of six or eight together. The grubs, which are hatched in about a fortnight afterwards, are of a dirty yellowish or ashen-white color, with a darker-colored head, and two dark spots on the top of the first wing. They are rather short, approaching to a cylindrical form, but thickest in the middle, and have six legs, arranged in pairs beneath the first three rings. After making a hearty meal upon the leaves of the potato, they cover themselves with their own filth. The vent is situated on the upper side of the last ring, so that their dung falls upon their backs, and, by motions of the body, is pushed forwards, as fast as it accumulates, towards the head, until the whole of the back is entirely coated with it. This covering shelters their soft and tender bodies from the heat of the sun, and probably serves to secure them from the attacks of their enemies. When it becomes too heavy or too dry, it is thrown off, but replaced again by a fresh coat in the course of a few hours. In eating, the grubs move backwards, never devouring the portion of the leaf immediately before the head, but that which lies under it. Their numbers are sometimes very great, and the leaves are then covered and nearly consumed by these filthy insects. When about fifteen days old, they throw off their loads, creep down the plant, and bury themselves in the ground. Here each one forms for itself a little cell of earth, cemented and varnished within by a gummy fluid discharged from its mouth, and when this is done, it changes to a pupa. In about a fortnight more the insect throws off its pupa skin, breaks open its earthen cell, and crawls out of the ground. The beetles come out towards the end of July or early in August, and lay their eggs for a second brood of grubs. The latter come to their growth and go into the ground in the autumn, and remain there in the pupa form during the winter.

The only method that occurs to me, by means of which we may get rid of them, when they are so numerous as to be seriously injurious to plants, is to brush them from the leaves into shallow vessels containing a little salt and water or vinegar.

The habits of the Hispas, little leaf-beetles, forming the family HISPADÆ, were first made known by me in the year 1835, in the "Boston Journal of Natural History," * where a detailed account of them, with descriptions of three native species, and figures of the larvæ and pupæ, may be found. The upper side of the beetles is generally rough, as the generical name implies. The larvæ burrow under the skin of the leaves of plants, and eat the pulpy substance within, so that the skin, over and under the place of their operations, turns brown and dries, and has somewhat of a blistered appearance, and within these blistered spots the larvæ or grubs, the pupæ, or the beetles may often be found. The eggs of these insects are little rough blackish grains, and are glued to the surface of the leaves, sometimes singly, and sometimes in clusters of four or five together. The grubs of our common species are about one fifth of an inch in length, when fully grown. The body is oblong, flattened, rather broader before than behind, soft, and of a whitish color, except the head and the top of the first ring, which are brown, or blackish, and of a horny consistence. It has a pair of legs to each of the first three rings; the other rings are provided with small fleshy warts at the sides, and transverse rows of little rasp-like points above and beneath. The pupa state lasts only about one week, soon after which the beetles come out of their burrows.

The leaves of the apple-tree are inhabited by some of these little mining insects, which in the beetle state are probably the *Hispa rosea* † of Weber, or the rosy Hispa (Fig. 54). They are of a deep or tawny reddish-yellow color above, marked with little deep red lines and spots. The head is

* Vol. I. p. 141.

† *Hispa quadrata*, Fabricius; *H. marginata*, Say.

small; the antennæ are short, thickened towards the end, and of a black color; the thorax is narrow before and wide behind, rough above, striped with deep red on each side; the wing-covers taken together form an oblong square; there are three smooth longitudinal lines or ribs on each of them, spotted with blood-red, and the spaces between these lines are deeply punctured in double rows; the under side of the body is black, and the legs are short and reddish. They measure about one fifth of an inch in length. These beetles may be found on the leaves of the apple-tree, and very abundantly on those of the shad-bush (*Amelanchier ovalis*), and choke-berry (*Pyrus arbutifolia*), during the latter part of May and the beginning of June.

Fig. 54.

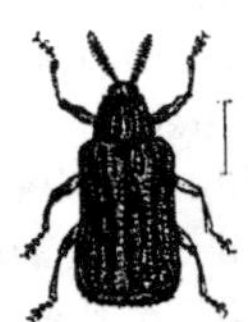

In the middle of June, another kind of *Hispa* may be found pairing and laying eggs on the leaves of the locust-tree. The grubs appear during the month of July, and are transformed to beetles in August. They measure nearly one quarter of an inch in length, are of a tawny yellow color, with a black longitudinal line on the middle of the back, partly on one and partly on the other wing-cover, the inner edges of which meet together and form what is called the *suture;* whence this species was named *Hispa suturalis* by Fabricius; the head, antennæ, body beneath, and legs are black; and the wing-covers are not so square behind as in the rosy Hispa.

The tortoise-beetles, as they are familiarly called from their shape, are leaf-eating insects, belonging to the family CASSIDADÆ. This name, derived from a word signifying a helmet, is applied to them because the fore part of the semicircular thorax generally projects over the head like the front of a helmet. In these beetles the body is broad oval or rounded, flat beneath, and slightly convex above. The antennæ are short, slightly thickened at the end, and inserted close together on the crown of the head. The latter is small, and concealed under, or deeply sunk into, the thorax. The legs are very short, and hardly seen from above. These insects are often gayly colored or spotted, which increases their resemblance to a tortoise; they creep slowly, and fly by day. Their larvæ and pupæ resemble those of the following species in most respects.

Cassida aurichalcea (Plate I. Fig. 5), so named by Fabricius on account of the brilliant brassy or golden lustre it assumes, is found during most of the summer months on the leaves of the bitter-sweet (*Solanum dulcamara*), and in great abundance on various kinds of Convolvulus, such as our large-flowered *Convolvulus sepium*, the morning-glory, and the sweet-potato vine. The leaves of these plants are eaten both by the beetles and their young. The former begin to appear during the months of May and June, having probably survived the winter in some place of shelter and concealment, and their larvæ in a week or two afterwards. The larvæ are broad oval, flattened, dark-colored grubs (Fig. 55), with a kind of fringe, composed of stiff prickles, around the thin edges of the body, and a long forked tail. This fork serves to hold the excrement when voided; and a mass of it half as large as the body of the insect is often thus accumulated. The tail, with the loaded fork, is turned over the back, and thus protects the insect from the sun, and probably also from its enemies. The first broods of larvæ arrive at their growth and change to pupæ early in July, fixing themselves firmly by the hinder part of their bodies to the leaves, when this change is about to take place. The pupa remains fastened to the cast-skin of the larva. It is broad oval, fringed at the sides, and around the fore part of the broad thorax, with large prickles. Soon afterwards the beetles come forth, and lay their eggs for a second brood of grubs, which, in turn, are changed to beetles in the course of the autumn. In June, 1824, the late Mr. John Lowell sent me

Fig. 55.

specimens of this little beetle, which he found to be injurious to the sweet-potato vine, by eating large holes through the leaves. This beetle is very broad oval in shape, and about one fifth of an inch in length. When living, it has the power of changing its hues, at one time appearing only of a dull yellow color, and at other times shining with the splendor of polished brass or gold, tinged sometimes also with the variable tints of pearl. The body of the insect is blackish beneath, and the legs are dull yellow. It loses its brilliancy after death. The wing-covers, the parts which exhibit the change of color, are lined beneath with an orange-colored paint, which seems to be filled with little vessels; and these are probably the source of the changeable brilliancy of the insect.

The Chrysomelians (CHRYSOMELADÆ) compose an extensive tribe of leaf-eating beetles, formerly included in the old genus *Chrysomela*. The meaning of this word is golden beetle, and many of the insects to which it was applied by Linnæus are of brilliant and metallic colors. They differ, however, so much in their essential characters, their forms, and their habits, that they are now very properly distributed into four separate groups or families. The first of these, called GALERUCADÆ, or Galerucians, consists mostly of dull-colored beetles; having an oblong oval, slightly convex body; a short, and rather narrow, and uneven thorax; slender antennæ, more than half the length of the body, and implanted close together on the forehead; slender legs, which are nearly equal in size; and claws split at the end. They fly mostly by day, and are by nature either very timid or very cunning, for, when we attempt to take hold of them, they draw up their legs, and fall to the ground. They sometimes do great injury to plants, eating large holes in the leaves, or consuming entirely those that are young and tender. The larvæ are rather short cylindrical grubs, generally of a blackish color, and are provided with six legs. They live and feed together in swarms, and sometimes appear in very great numbers on the leaves of plants, committing ravages, at these times, as extensive as those of the most destructive caterpillars. This was the case in 1837 at Sevres, in France, and in 1838 and 1839 in Baltimore and its vicinity, where the elm-trees were entirely stripped of their leaves during midsummer by swarms of the larvæ of *Galeruca Calmariensis;* and, in the latter place, after the trees had begun to revive, and were clothed with fresh leaves, they were again attacked by new broods of these noxious grubs. These insects, which were undoubtedly introduced into America with the European elm, are as yet unknown in the New England States. The eggs of the Galerucians are generally laid in little clusters or rows along the veins of the leaves, and those of the elm Galeruca are of a yellow color. The pupa state of some species occurs on the leaves, of others in the ground; and some of the larvæ live also in the ground on the roots of plants.

One of the most common kinds is the *Galeruca vittata,** or striped Galeruca, (Plate II. Fig. 3,) generally known here by the names of striped bug, and cucumber-beetle. This destructive insect is of a light-yellow color above, with a black head, and a broad black stripe on each wing-cover, the inner edge or suture of which is also black, forming a third narrower stripe down the middle of the back; the abdomen, the greater part of the fore legs, and the knees and feet of the other legs, are black. It is rather less than one fifth of an inch long. Early in the spring it devours the tender leaves of various plants. I have found it often on those of our Aronias, *Amelanchier botryapium* and *ovalis*, and *Pyrus arbutifolia*, towards the end of April. It makes its first appearance, on cucumber, squash, and melon vines, about the last of May and first of June, or as soon as the leaves begin to expand; and, as several broods are produced in the course of the summer, it may be found at various times on these plants, till the latter are destroyed by frost. Great

* *Crioceris vittata* of Fabricius.

numbers of these little beetles may be obtained in the autumn from the flowers of squash and pumpkin vines, the pollen and germs of which they are very fond of. They get into the blossoms as soon as the latter are opened, and are often caught there by the twisting and closing of the top of the flower; and, when they want to make their escape, they are obliged to gnaw a hole through the side of their temporary prison. The females lay their eggs in the ground, and the larvæ probably feed on the roots of plants, but they have hitherto escaped my researches.

Various means have been suggested and tried to prevent the ravages of these striped cucumber-beetles, which have become notorious throughout the country for their attacks upon the leaves of the cucumber and squash. Dr. B. S. Barton, of Philadelphia, recommended sprinkling the vines with a mixture of tobacco and red pepper, which he stated to be attended with great benefit. Watering the vines with a solution of one ounce of Glauber's salts in a quart of water, or with tobacco-water, an infusion of elder, of walnut-leaves, or of hops, has been highly recommended. Mr. Gourgas, of Weston, has found no application so useful as ground plaster of Paris; and a writer in the "American Farmer" extols the use of charcoal dust. Deane recommended sifting powdered soot upon the plants when they are wet with the morning dew, and others have advised sulphur and Scotch snuff to be applied in the same way. As these insects fly by night, as well as by day, and are attracted by lights, burning splinters of pine knots or of staves of tar-barrels, stuck into the ground during the night, around the plants, have been found useful in destroying these beetles. The most effectual preservative, both against these insects and the equally destructive black flea-beetles which infest the vines in the spring, consists in covering the young vines with millinet stretched over small wooden frames. Mr. Levi Bartlett, of Warner, N. H., has described a method for making these frames expeditiously and economically, and his directions may be found in the second volume of the "New England Farmer,"* and in Fessenden's "New American Gardener,"† under the article *Cucumber*.

The cucumber flea-beetle above mentioned, a little, black, jumping insect, well known for the injury done by it, in the spring, to young cucumber plants, belongs to another family of the Chrysomelian tribe, called HALTICADÆ. The following are the chief peculiarities of the beetles of this family. The body is oval and very convex above; the thorax is short, nearly or quite as wide as the wing-covers behind, and narrowed before; the head is pretty broad; the antennæ are slender, about half the length of the body, and are implanted nearly on the middle of the forehead; the hindmost thighs are very thick, being formed for leaping; hence these insects have been called flea-beetles, and the scientific name *Haltica*, derived from a word signifying to leap, has been applied to them. The surface of the body is smooth, generally polished, and often prettily or brilliantly colored. The claws are very thick at one end, are deeply notched towards the other, and terminate with a long curved and sharp point, which enables the insect to lay hold firmly upon the leaves of the plants on which they live. These beetles eat the leaves of vegetables, preferring especially plants of the cabbage, turnip, mustard, cress, radish, and horse-radish kind, or those which, in botanical language, are called cruciferous plants, to which they are often exceedingly injurious. The turnip-fly, or more properly turnip flea-beetle, is one of these *Halticas*, which lays waste the turnip-fields in Europe, devouring the seed-leaves of the plants as soon as they appear above the ground, and continuing their ravages upon new crops throughout the summer. Another small flea-beetle is often very injurious to the grape-vines in Europe, and a larger species attacks the same plant in this country. The flea-beetles conceal themselves during the winter, in dry places, under stones, in tufts of withered grass and moss,

* Page 305. † Sixth edition, p. 91.

and in chinks of walls. They lay their eggs in the spring, upon the leaves of the plants upon which they feed. The larvæ, or young, of the smaller kinds burrow into the leaves, and eat the soft pulpy substance under the skin, forming therein little winding passages, in which they finally complete their transformations. Hence the plants suffer as much from the depredations of the larvæ, as from those of the beetles, a fact that has too often been overlooked. The larvæ of the larger kinds are said to live exposed upon the surface of the leaves which they devour, till they have come to their growth, and to go into the ground, where they are changed to pupæ, and soon afterwards to beetles. The mining larvæ, the only kinds which are known to me from personal examination, are little slender grubs, tapering towards each end, and provided with six legs. They arrive at maturity, turn to pupæ, and then to beetles in a few weeks. Hence there is a constant succession of these insects, in their various states, throughout the summer. The history of the greater part of our Halticas or flea-beetles is still unknown; I shall, therefore, only add, to the foregoing general remarks, descriptions of two or three common species, and suggest such remedies as seem to be useful in protecting plants from their ravages.

The most destructive species in this vicinity is that which attacks the cucumber plant as soon as the latter appears above the ground, eating the seed-leaves, and thereby destroying the plant immediately. Supposing this to be an undescribed insect, I formerly named it *Haltica Cucumeris*, the cucumber flea-beetle (Fig. 56); but Mr. Say subsequently informed me that it was the *pubescens* of Illiger, so named because it is very slightly pubescent or downy. Count Dejean, who gave to it the specific name of *fuscula*, considered it as distinct from the *pubescens*; and it differs from the descriptions of the latter in the color of its thighs, and in never having the tips and shoulders of the wing-covers yel-

Fig. 56.

lowish; so that it may still bear the name given to it in my Catalogue. It is only one sixteenth of an inch long, of a black color, with clay-yellow antennæ and legs, except the hindmost thighs, which are brown. The upper side of the body is covered with punctures, which are arranged in rows on the wing-cases; and there is a deep transverse furrow across the hinder part of the thorax. During the summer, these pernicious flea-beetles may be found, not only on cucumber-vines, but on various other plants having fleshy and succulent leaves, such as beans, beets, the tomato, and the potato. They injure all these plants, more or less, according to their numbers, by nibbling little holes in the leaves with their teeth; the functions of the leaves being thereby impaired in proportion to the extent of surface and amount of substance destroyed. The edges of the bitten parts become brown and dry by exposure to the air, and assume a rusty appearance. Since the prevalence of the disease commonly called *the potato-rot*, attention has been particularly directed to various insects that live upon the potato-plant; and, as these flea-beetles have been found upon it in great numbers, in some parts of the country, they have been charged with being the cause of the disease. The same charge has also been made against several other kinds of insects, some of which will be described in the course of this work. In my own opinion, the origin, extension, and continued reappearance of this wide-spread pestilence are not due to the depredations of insects of any kind. Mr. Phanuel Flanders, of Lowell, where the flea-beetles have appeared in unusual numbers, showed to me, in August, 1851, some potato-leaves that were completely riddled with holes by them, so that but little more than the ribs and veins remained untouched. He thinks that their ravages may be prevented by watering the leaves with a solution of lime, a remedy long ago employed in England, with signal benefit, in preserving the turnip crop from the attacks of the turnip flea-beetle.

The wavy-striped flea-beetle, *Haltica striolata** (Fig. 57), may be seen in great abundance on the horse-radish, various kinds of cresses, and on the mustard and turnip, early in May, and indeed at other times throughout the summer. It is very injurious to young plants, destroying their seed-leaves as soon as the latter expand. Should it multiply to any extent, it may in time become as great a pest as the European turnip flea-beetle, which it closely resembles in its appearance, and in all its habits. Though rather larger than the cucumber flea-beetle, and of a longer oval shape, it is considerably less than one tenth of an inch in length. It is of a polished black color, with a broad wavy buff-colored stripe on each wing-cover, and the knees and feet are reddish yellow. Specimens are sometimes found having two buff-yellow spots on each wing-cover instead of the wavy stripe. These were not known by Fabricius to be merely varieties of the *striolata*, and accordingly he described them as distinct, under the name of *bipustulata*,† the two-spotted.

Fig. 57.

The steel-blue flea-beetle, *Haltica chalybea* of Illiger, (Fig. 58, and Plate II. Fig. 5,) or the grape-vine flea-beetle, as it might be called on account of its habits, is found in almost all parts of the United States, on wild and cultivated grape-vines, the buds and leaves of which it destroys. Though it has received the specific name of *chalybea*, meaning steel-blue, it is exceedingly variable in its color, specimens being often seen on the same vine of a dark purple, violet, Prussian blue, greenish blue, and deep green color. The most common tint of the upper side is a glossy, deep, greenish blue; the under side is dark green; and the antennæ and feet are dull black. The body is oblong-oval, and the hinder part of the thorax is marked with a transverse furrow. It measures rather more than three twentieths of an inch in length. In this part of the

Fig. 58.

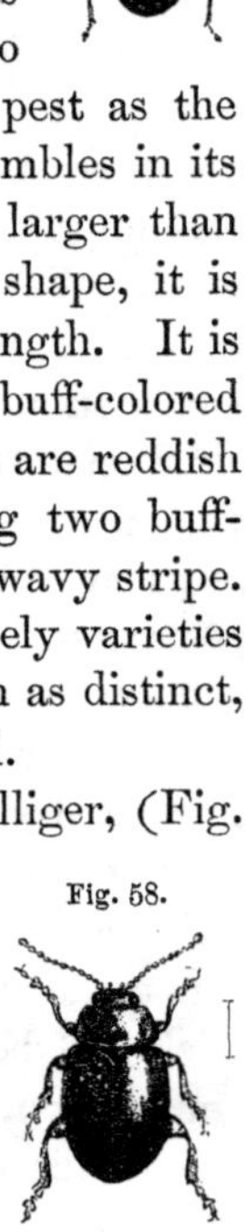

* *Crioceris striolata*, Fabricius. † *Crioceris bipustulata*, Fabricius.

country these beetles begin to come out of their winter quarters towards the end of April, and continue to appear till the latter part of May. Soon after their first appearance they pair, and probably lay their eggs on the leaves of the vine, and perhaps on other plants also. A second brood of the beetles is found on the grape-vines towards the end of July. I have not had an opportunity to trace the history of these insects any further, and consequently their larvæ are unknown to me. Mr. David Thomas has given an interesting account of their habits and ravages in the twenty-sixth volume of Silliman's "American Journal of Science and Arts." These brilliant insects were observed by him, in the spring of 1831, in Cayuga County, N. Y., creeping on the vines, and destroying the buds, by eating out the central succulent parts. Some had burrowed even half their length into the buds. When disturbed, they jump rather than fly, and remain where they fall for a time without motion. During the same season these beetles appeared in unusually great numbers in New Haven, Conn., and its vicinity, and the injury done by them was "wholly unexampled." "Some vines were entirely despoiled of their fruit buds, so as to be rendered, for that season, barren." Mr. Thomas found the vine-leaves were infested, in the years 1830 and 1831, by "small chestnut-colored smooth worms," and suspecting these to be the larvæ of the beetle (which he called *Chrysomela vitivora*), he fed them in a tumbler, containing some moist earth, until they were fully grown, when they buried themselves in the earth. "After a fortnight or so," some of the beetles were found in the tumbler. Hence there is no doubt that the former were the larvæ of the beetles, and that they undergo their transformations in the ground. A good description of the larvæ, and a more full account of their habits, seasons, and changes, are still wanted.

In England, where the ravages of the turnip flea-beetle have attracted great attention, and have caused many and various experiments to be tried with a view of checking them, it is thought that "the careful and systematic use of lime will obviate, in a great degree, the danger which has been experienced" from this insect. From this and other statements in favor of the use of lime, there is good reason to hope that it will effectually protect plants from the various kinds of flea-beetles, if dusted over them, when wet with dew, in proper season. Watering plants with alkaline solutions, it is said, will kill the insects without injuring the plants. The solution may be made by dissolving one pound of hard soap in twelve gallons of the soap-suds left after washing. This mixture should be applied twice a day with a water-pot. Köllar very highly recommends watering or wetting the leaves of plants with an infusion or tea of wormwood, which prevents the flea-beetles from touching them. Perhaps a decoction of walnut-leaves might be equally serviceable. Great numbers of the beetles may be caught by the skilful use of a deep bag-net of muslin, which should be swept over the plants infested by the beetles, after which the latter may be easily destroyed. This net cannot be used with safety to catch the insects on very young plants, on account of the risk of bruising or breaking their tender leaves.

The Chrysomelians, CHRYSOMELADÆ, properly so called, form the third family of the tribe to which I have given the same name, because these insects hold the chief place in it, in respect to size, beauty, variety, and numbers. These leaf-beetles are mostly broad oval, sometimes nearly hemispherical, in their form, or very convex above and flat beneath. The head is rather wide, and not concealed under the thorax. The latter is short, and broad behind. The antennæ are about half the length of the body, and slightly thickened towards the end, and arise from the sides of the head, between the eyes and the corners of the mouth; being much further apart than those of the Galerucians and flea-beetles. The legs are rather short, nearly equal in length, and the hindmost thighs are not thicker than the others, and are not fitted for leaping. The colors of these beetles are often rich and brilliant, among which blue and green, highly polished, and with a golden or metallic lustre, are the most common tints. The larvæ are soft-bodied, short, thick, and slug-shaped grubs, with six legs before, and a prop-leg behind. They live exposed on the leaves of plants, which they eat, and to which most of them fasten themselves by the tail, when about to be transformed. Some, however, go into the ground when about to change to pupæ. Many of these insects, both in the larva and beetle state, have been found to be very injurious to vegetation in other countries; but I am not aware that any of them have proved seriously injurious to cultivated or other valuable plants in this country. There are some, it is true, which may hereafter increase so as to give us much trouble, unless effectual means are taken to protect and cherish their natural enemies, the birds.

The largest species in New England inhabits the common milk-weed, or silk-weed (*Asclepias Syriaca*), upon which it may be found, in some or all of its states, from the middle of June till September. Its head, thorax, body beneath, antennæ, and legs are deep blue, and its wing-covers orange, with three large black spots upon them, namely, one on the shoulder, and another on the tip of each, and the third across the base of both wing-covers. Hence it was named *Chrysomela trimaculata* by Fabricius, or the three-spotted Chrysomela (Plate II. Fig. 9). It is nearly three eighths of an inch long, and almost hemispherical. Its larvæ and pupæ are orange-colored, spotted with black, and pass through their transformations on the leaves of the Asclepias.

The most elegant of our Chrysomelians is the *Chrysomela scalaris* of Leconte, literally the ladder Chrysomela (Fig. 59). It is about three tenths of an inch long, and of a narrower and more regularly oval shape than the preceding. The head, thorax, and under side of its body are dark green, the wing-covers silvery white, ornamented with small green

Fig. 59.

spots on the sides, and a broad jagged stripe along the suture or inner edges; the antennæ and legs are rust-red, and the wings are rose-colored. It is a most beautiful object when flying, with its silvery wing-covers, embossed with green, raised up, and its rose-red wings spread out beneath them. These beetles inhabit the lime or linden (*Tilia Americana*), and the elm, upon which they may be found in April, May, and June, and a second brood of them in September and October. They pass the winter in holes, and under leaves and moss. The trees on which they live are sometimes a good deal injured by them and by their larvæ (Fig. 60). The latter are hatched from eggs laid by the beetles on the leaves in the spring, and come to their growth towards the end of June. They are then about six tenths of an inch long, of a white color, with a black line along the top of the back, and a row of small square black spots on each side of the body; the head is horny and of an ochre-yellow color. Like the grubs of the preceding species, these are short, and very thick, the back arching upwards very much in the middle. I believe that they go into the ground to turn to pupæ. Should they become so numerous as seriously to injure the lime and elm trees, it may be found useful to throw decoctions of tobacco or of walnut-leaves on the trees by means of a garden or fire engine, a method which has been employed with good effect for the destruction of the larvæ of *Galeruca Calmariensis*.

Fig. 60.

The most common leaf-beetle of the family under consideration is the blue-winged Chrysomela, or *Chrysomela cœruleipennis* of Say (Fig. 61), an insect hardly distinct from the European *Chrysomela Polygoni*, and like the latter it lives in great numbers on the common knot-grass (*Polygonum aviculare*), which it completely strips of its leaves two or three times in the course of the summer. This little

Fig. 61.

beetle is about three twentieths of an inch long. Its head, wing-covers, and body beneath are dark blue; its thorax and legs are dull orange-red; the upper side of its abdomen is also orange-colored; and the antennæ and feet are blackish. The females have a very odd appearance before they have laid their eggs, their abdomen being enormously swelled out like a large orange-colored ball, which makes it very difficult for them to move about. I have found these insects on the knot-grass in every month from April to September inclusive. The larvæ eat the leaves of the same plant.

Having described the largest, the most elegant, and the most common of our Chrysomelians, I must omit all the rest, except the most splendid, which was called *Eumolpus auratus* by Fabricius, that is, the gilded Eumolpus (Plate II. Fig. 1). It is of a brilliant golden green color above, and of a deep purplish green below; the legs are also purple-green; but the feet and the antennæ are blackish. The thorax is narrower behind than the wing-covers, and the rest of the body is more oblong oval than in the foregoing Chrysomelians. It is about three eighths of an inch long. This splendid beetle may be found in considerable numbers on the leaves of the dog's-bane (*Apocynum Androsæmifolium*), which it devours, during the months of July and August. The larvæ are unknown to me.

The fourth family of the leaf-eating Chrysomelians consists of the Cryptocephalians (CRYPTOCEPHALIDÆ), so named from the principal genus *Cryptocephalus*, a word signifying concealed head. These insects somewhat resemble the beetles of the preceding family; but they are of a more cylindrical form, and the head is bent down, and nearly concealed in the fore part of the thorax. Their larvæ are short, cylindrical, whitish grubs, which eat the leaves of plants. Each one makes for itself a little cylindrical or egg-shaped case, of a substance sometimes resembling clay, and sometimes like horn, with an opening at one end, within which the grub lives, putting out its head and fore legs when it wishes to eat or to move.

When it is fully grown, it stops up the open end of its case, and changes to a pupa, and afterwards to a beetle within it, and then gnaws a hole through the case, in order to escape. As none of these insects have been observed to do much injury to plants in this country, I shall state nothing more respecting them, than that *Clythra dominicana*[14] inhabits the sumach, *C. quadriguttata*[15] oak-trees, *Chlamys gibbosa* low whortleberry bushes, *Cryptocephalus luridus* the wild indigo-bush, and most of the other species may be found on different kinds of oaks.

Although the blistering beetles, or Cantharides (CANTHARIDIDÆ), have been enumerated among the insects directly beneficial to man, on account of the important use made of them in medical practice, yet it must be admitted that they are often very injurious to vegetation. The green Cantharides, or Spanish flies, as they are commonly called, are found in the South of Europe, and particularly in Spain and Italy, where they are collected in great quantities for exportation. In these countries they sometimes appear in immense swarms, on the privet, lilac, and ash; so that the limbs of these plants bend under their weight, and are entirely stripped of their foliage by these leaf-eating beetles. In like manner our native Cantharides devour the leaves of plants, and sometimes prove very destructive to them.

The Cantharides are distinguished from all the preceding insects by their feet, the hindmost pair of which have only four joints, while the first and middle pairs are five-jointed. In this respect they agree with many other beetles, such as clocks or darkling beetles, meal-beetles, some of the mushroom-beetles, flat bark-beetles, and the like, with which they form a large and distinct section of Coleopterous insects.

[14 *Clythra (Coscinoptera) dominicana.* — LEC.]
[15 *Clythra (Babia) quadriguttata.* — LEC.]

The following are the most striking peculiarities of the family to which the blistering beetles belong. The head is broad and nearly heart-shaped, and it is joined to the thorax by a narrow neck. The antennæ are rather long and tapering, sometimes knotted in the middle, particularly in the males. The thorax varies in form, but is generally much narrower than the wing-covers. The latter are soft and flexible, more or less bent down at the sides of the body, usually long and narrow, sometimes short and overlapping on their inner edges. The legs are long and slender ; the soles of the feet are not broad, and are not cushioned beneath ; and the claws are split to the bottom, or double, so that there appear to be four claws to each foot. The body is quite soft, and when handled, a yellowish fluid, of a disagreeable smell, comes out of the joints. These beetles are timid insects, and when alarmed they draw up their legs and feign themselves dead. Nearly all of them have the power of raising blisters when applied to the skin, and they retain it even when dead and perfectly dry. It is chiefly this property that renders them valuable to physicians. Four of our native Cantharides have been thus successfully employed, and are found to be as powerful in their effects as the imported species. For further particulars relative to their use, the reader is referred to my account of them published in 1824, in the first volume of "The Boston Journal of Philosophy and the Arts," and in the thirteenth volume of "The New England Medical and Surgical Journal."

Occasionally potato-vines are very much infested by two or three kinds of Cantharides, swarms of which attack and destroy the leaves during midsummer. One of these kinds has thereby obtained the name of the potato-fly. It is the *Cantharis vittata*,* or striped Cantharis. It is of a dull tawny yellow or light yellowish-red color above, with two

* *Lytta vittata*, Fabricius.[16]

[[16] The name *Lytta* is now adopted by most entomologists in preference to that of *Cantharis* for these insects. — Lec.]

black spots on the head, and two black stripes on the thorax and on each of the wing-covers. The under side of the body, the legs, and the antennæ are black, and covered with a grayish down. Its length is from five to six tenths of an inch. In this and the three following species the thorax is very much narrowed before, and the wing-covers are long and narrow, and cover the whole of the back. The striped Cantharis is comparatively rare in New England; but in the Middle and Western States it often appears in great numbers, and does much mischief in potato-fields and gardens, eating up, not only the leaves of the potato, but those of many other vegetables. It is one of the insects to which the production of the *potato-rot* has been ascribed. The habits of this kind of Cantharis are similar to those of the following species.

There is a large blistering beetle which is very common on the virgin's bower (*Clematis Virginiana*), a trailing plant, which grows wild in the fields, and is cultivated for covering arbors. I have sometimes seen this plant completely stripped of its leaves by these insects, during the month of August. They are very shy, and when disturbed fall immediately from the leaves, and attempt to conceal themselves among the grass. They most commonly resort to the low branches of the Clematis, or those that trail upon the ground, and more rarely attack the upper parts of the vine. They also eat the leaves of various kinds of Ranunculus or buttercups, and, in the Middle and Southern States, those of *Clematis viorna* and *crispa*. This beetle is the *Cantharis marginata* of Olivier, or margined Cantharis (Fig. 62). It measures six or seven tenths of an inch in length. Its head and thorax are thickly covered with short gray down, and have a black spot on the upper side of each; the wing-covers are black, with a very narrow gray edging; and the under side of the body and the legs are also gray.

Fig. 62.

The most destructive kind of Cantharis found in Massachusetts is of a more slender form than the preceding, and measures only from five and a half to six tenths of an inch in length. Its antennæ and feet are black, and all the rest of its body is ashen gray, being thickly covered with a very short down of that color. Hence it is called *Cantharis cinerea*,*[17] or the ash-colored Cantharis (Fig. 63). When the insect is rubbed, the ash-colored substance comes off, leaving the surface black. It begins to appear in gardens about the 20th of June, and is very fond of the leaves of the English bean, which it sometimes entirely destroys. It is also occasionally found in considerable numbers on potato-vines ; and in Cambridge, Massachusetts, it has repeatedly appeared in great profusion upon hedges of the honey-locust, which have been entirely stripped of foliage by these voracious insects. They are also found on the wild indigo-weed. In the night, and in rainy weather, they descend from the plants, and burrow in the ground, or under leaves and tufts of grass. Thither also they retire for shelter during the heat of the day, being most actively engaged in eating in the morning and evening. About the 1st of August they go into the ground and lay their eggs, and these are hatched in the course of one month. The larvæ are slender, somewhat flattened grubs, of a yellowish color, banded with black, with a small reddish head, and six legs. These grubs are very active in their motions, and appear to live upon fine roots in the ground ; but I have not been able to keep them till they arrived at maturity, and therefore know nothing further of their history.

Fig. 63.

About the middle of August, and during the rest of this and the following month, a jet-black Cantharis may be seen on potato-vines, and also on the blossoms and leaves of vari-

* *Lytta cinerea*, Fabricius.

[17 As this specific name was previously applied by Forster to the species mentioned on the previous page as *Cantharis* or *Lytta marginata*, and has priority over that name, I have changed the name of the present species to *Lytta Fabricii*. — Lec.]

ous kinds of golden-rod, particularly the tall golden-rod (*Solidago altissima*), which seems to be its favorite food. In some places it is as plentiful in potato-fields as the striped and the margined Cantharis, and by its serious ravages has often excited attention. These three kinds, in fact, are often confounded under the common name of potato-flies; and it is still more remarkable, that they are collected for medical use, and are sold in our shops by the name of *Cantharis vittata*, without a suspicion of their being distinct from each other. The black Cantharis, or *Cantharis atrata** (Fig. 64), is totally black, without bands or spots, and measures from four tenths to half of an inch in length. I have repeatedly taken these insects, in considerable quantities, by brushing or shaking them from the potato-vines into a broad tin pan, from which they were emptied into a covered pail containing a little water, which, by wetting their wings, prevented their flying out when the pail was uncovered. The same method may be employed for taking the other kinds of Cantharides, when they become troublesome and destructive from their numbers; or they may be caught by gently sweeping the plants they frequent with a deep muslin bag-net. They should be killed by throwing them into scalding water, for one or two minutes, after which they may be spread out on sheets of paper to dry, and may be made profitable by selling them to the apothecaries for medical use.

Fig. 64.

There are some blistering beetles, belonging to another genus, which seem deserving of a passing notice, not on account of any great injury committed by them, but because they can be used in medicine like the foregoing, and are considered by some naturalists as forming one of the links connecting the orders Coleoptera and Orthoptera. These insects belong to the genus *Meloe*, so named, it is supposed, because they are of a black, or deep blue-black

* *Lytta atrata*, Fabricius.

color. They are called oil-beetles in England, on account of the yellowish liquid which oozes from their joints in large drops when they are handled. Their head is large, heart-shaped, and bent down, as in the other blistering beetles. Their thorax is narrowed behind, and very small in proportion to the rest of the body. The latter is egg-shaped, pointed behind, and so enormously large that it drags on the ground when the beetle attempts to walk. The wings are wanting, and of course these insects are unable to fly, although they have a pair of very short oval wing-covers, which overlap on their inner edges, and do not cover more than one third of the abdomen. These beetles eat the leaves of various kinds of buttercups.

Our common species is the *Meloe angusticollis* of Say, or narrow-necked oil-beetle. (Fig. 65 represents the female, and the antenna of the male at her left.) It is of a dark indigo-blue color; the thorax is very narrow, and the antennæ of the male are curiously twisted and knotted in the middle. It measures from eight tenths of an inch to one inch in length. It is very common on buttercups in the autumn, and I have also found it eating the leaves of potato-vines.

Fig. 65.

The foregoing insects are but a small number of those, belonging to the order Coleoptera, which are injurious to vegetation. Those only have been selected that are the most remarkable for their ravages, or would best serve to illustrate the families and genera to which they belong. The orders Orthoptera, Hemiptera, Lepidoptera, Hymenoptera, and Diptera remain to be treated in the same way, in carrying out the plan upon which this treatise has been begun, and to which it is limited.

CHAPTER III.

ORTHOPTERA.

Earwigs. — Cockroaches. — Mantes, or Soothsayers. — Walking-Leaves. — Walking-Sticks, or Spectres. — Mole-Cricket. — Field Crickets. — Climbing Cricket. — Wingless Cricket. — Grasshoppers. — Katy-did. — Locusts.

THE destructive insects popularly known in this country by the name of grasshoppers, but which in our version of the Bible, and in other works in the English language, are called locusts, have, from a period of very high antiquity, attracted the attention of mankind by their extensive and lamentable ravages. It should here be remarked, that in America the name of locust is very improperly given to the *Cicada* of the ancients, or the harvest-fly of English writers, some kinds of which will be the subject of future remark in this treatise. The name of locust will here be restricted to certain kinds of grasshoppers; while the popularly named locust, which, according to common belief, appears only once in seventeen years, must drop this name, and take the more correct one of Cicada or harvest-fly. The very frequent misapplication of names, by persons unacquainted with natural history, is one of the greatest obstacles to the progress of science, and shows how necessary it is that things should be called by their right names, if the observations communicated respecting them are to be of any service. Every intelligent farmer is capable of becoming a good observer, and of making valuable discoveries in natural history; but if he be ignorant of the proper names of the objects examined, or if he give to them names which previously have been applied by other persons to entirely different objects, he will fail to make the result of his observations intelligible and useful to the community.

The insects which I here call locusts, together with other grasshoppers, earwigs, crickets, spectres or walking-sticks, and walking-leaves, soothsayers, cockroaches, &c., belong to an order called ORTHOPTERA, literally straight wings; for their wings, when not in use, are folded lengthwise in narrow plaits like a fan, and are laid straight along the top or sides of the back. They are also covered by a pair of thicker wing-like members, which, in the locusts and grasshoppers, are long and narrow, and lie lengthwise on the sides of the body, sloping outwards on each side like the roof of a house; in the cockroaches, these upper wings or wing-covers are broader, almost oval, and lie horizontally on the top of the back, overlapping on their inner edges; and in the crickets, the wing-covers, when closed, are placed like those of cockroaches, but have a narrow outer border, which is folded perpendicularly downwards so as to cover the sides of the body also.

All the Orthopterous insects are provided with transversely movable jaws, more or less like those of beetles, but they do not undergo a complete transformation in coming to maturity. The young, in fact, often present a close resemblance to the adult insects in form, and differ from them chiefly in wanting wings. They move about and feed precisely like their parents, but change their skins repeatedly before they come to their full size. The second stage in the progress of the Orthopterous insects to maturity is not, like that of beetles, a state of inactivity and rest, in which the insect loses the grub-like or *larva* form which it had when hatched from the egg, and becomes a *pupa* or *chrysalis*, more nearly resembling the form of a beetle, but soft, whitish, and with its undeveloped wings and limbs incased in a thin transparent skin which impedes all motion. On the contrary, the Orthoptera in the pupa state do not differ from the young and from the old insects, except in having the rudiments of wings and wing-covers projecting, like little scales, from the back near the thorax. These pupæ are active and voracious, and increase greatly in size, which is not the case with the insects that are subject to a complete transformation, for such never eat or grow in the pupa state. When fully grown, they cast off their skins for the sixth or last time, and then appear in the adult or perfect state, fully provided with all their members, with the exception of a few kinds which remain wingless throughout their whole lives. The slight changes to which the Orthoptera are subject consist of nothing more than a successive series of moultings, during which their wings are gradually developed. These changes may receive the name of imperfect or incomplete transformation, in contradistinction to the far greater changes exhibited by those insects which pass through a complete transformation in their progress to maturity.

Cockroaches are general feeders, and nothing comes amiss to them, whether of vegetable or animal nature; the Mantes or soothsayers are predaceous and carnivorous, devouring weaker insects, and even those of their own kind occasionally; but by far the greater part of the Orthopterous insects subsist on vegetable food, grass, flowers, fruits, the leaves, and even the bark of trees; whence it follows, in connection with their considerable size, their great voracity, and the immense troops or swarms in which they too often appear, that they are capable of doing great injury to vegetation.

The Orthoptera may be divided into four large groups: —

1. Runners (*Orthoptera cursoria**), including earwigs and cockroaches, with all the legs fitted for rapid motion;

2. Graspers (*Orthoptera raptoria*), such as the Mantes, or soothsayers, with the shanks of the fore legs capable of being doubled upon the under side of the thigh, which, moreover, is armed with teeth, and thus forms an instrument for seizing and holding their prey;

* These are the four divisions proposed by Mr. Westwood in his "Introduction," who, however, applies to them their Latin names only.

3. Walkers (*Orthoptera ambulatoria*), like the spectres or walking-sticks, having weak and slender legs, which do not admit of rapid motion; and

4. Jumpers (*Orthoptera saltatoria*), such as crickets, grasshoppers, and locusts, in which the thighs of the hind legs are much larger than the others, and are filled and moved with powerful muscles, which enable these insects to leap with facility.

I. RUNNERS. (*Orthoptera Cursoria.*)

In English works on gardening, earwigs are reckoned among obnoxious insects, various remedies are suggested to banish them from the garden, and even traps and other devices are described for capturing and destroying them. They have a rather long and somewhat flattened body, which is armed at the hinder end with a pair of slender sharp-pointed blades, opening and shutting horizontally like scissors, or like a pair of nippers, which suggested the name of *Forficula*, literally little nippers, applied to them by scientific writers. Although no well authenticated instances are on record of their entering the human ear, yet, during the daytime, they creep into all kinds of crevices for the sake of concealment, and come out to feed chiefly by night. It is common with English gardeners to hang up, among the flowers and fruit-trees subject to their attacks, pieces of hollow reeds, lobster claws, and the like, which offer enticing places of retreat for these insects on the approach of daylight, and by means thereof great numbers of them are obtained in the morning. The little creeping animal, with numerous legs, commonly but erroneously called earwig in America, is not an insect; but of the true earwig we have several species, though they are by no means common, and certainly never appear in such numbers as to prove seriously injurious to vegetation. Nevertheless, it seemed well to give to this kind of insect a passing notice in its proper place among the Orthoptera, were it only for its notoriety in other countries.

Of cockroaches (*Blatta*) we have also several kinds; those which are indigenous I believe are found exclusively in woods, under stones and leaves, while the others, and particularly the Oriental cockroach (*Blatta orientalis*), (Fig. 66,) which is supposed to have originated in Asia, whence it has spread to Europe, and thence to America, and has multiplied and become established in most of our maritime commercial towns, are domestic species, and are found in houses, under kitchen hearths, about ovens, and in dark and warm closets, whence they issue at night, and prowl about in search of food. But, as these disgusting and ill-smelling insects confine themselves to our dwellings, and do not visit our gardens and fields, they will require no further remarks than the mention of a method which has sometimes been found useful in destroying them. Mix together a table-spoonful of red-lead and of Indian meal with molasses enough to make a thick batter, and place the mixture at night on a plate or piece of board in the closets or on the hearths frequented by the cockroaches. They will eat it and become poisoned thereby. The dose is to be repeated for several nights in succession. Dr. F. H. Horner* recommends the following preparation to destroy cockroaches. Mix one teaspoonful of powdered arsenic with a table-spoonful of mashed potato, and crumble one third of it, every night, at bedtime, about the kitchen hearth, or where the insects will find and devour it. As both of these preparations are very poisonous, great care should be taken in the use of them, and of any portions that may be left by the insects.

Fig. 66.

* Downing's Horticulturist, Vol. II. p. 343 (Jan. 1848).

II. GRASPERS. (*Orthoptera raptoria.*)

These, which consist of the *Mantes*, called praying mantes and soothsayers, from their singular attitudes and motions, and camel-crickets, from the great length of the neck, are chiefly tropical insects, though some of them are occasionally found in this country. Moreover, they are exclusively predaceous insects, seizing, with their singular fore legs, caterpillars, and other weaker insects, which they devour. They are, therefore, to be enumerated among the insects that are beneficial to mankind, by keeping in check those that subsist on vegetable food.

III. WALKERS. (*Orthoptera ambulatoria.*)

To this division belong various insects, mostly found in warm climates, and displaying the most extraordinary forms. Some of them are furnished with wings, which, by their shape, and the branching veins with which they are covered, exactly represent leaves, either green, or dry and withered; such are the walking-leaves, as they are called (*Phyllium pulchrifolium*, *siccifolium*, &c.). Others are wingless, of a long and cylindrical shape, resembling a stick with the bark on it, while the slender legs, standing out on each side, give to these insects almost precisely the appearance of a little branching twig, whence is derived the name of walking-sticks, generally applied to them. The South American *Bacteria arumatia*, *rubispinosa*, and *phyllina*, and two species of *Diapheromera?*[1] described and figured in Say's "American Entomology," under the names of *Spectrum femoratum* (Fig. 67, male) and *bivittatum*, are of the latter description. These insects are very sluggish and inactive, are found among trees

[[1] Two species of *Phasma* are noticed. The first is *Bacunculus femoratus*, Say, which has also received the name of *Bacunculus Sayi*, Burm., and under which name it is best known to European authors. The latter was long ago figured by Stoll, in his great work upon the Orthoptera, and his name preoccupied that of Say and should be retained for it; it is *Amisomorpha Buprestoides*. The former has been found in most of the States east of the Mississippi, while the latter is peculiar to Florida and some of the Southern States. — UHLER.]

and bushes, on which they often remain motionless for a long time, or walk slowly over the leaves and young shoots, which

Fig. 67.

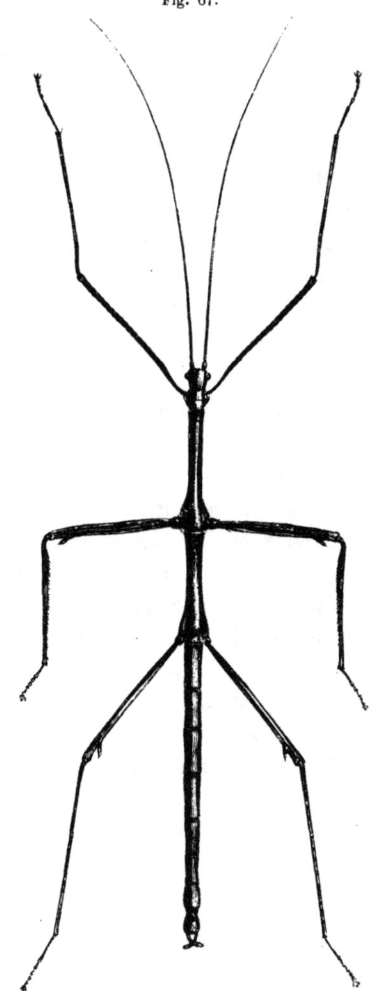

are their appropriate food. The American species are not so numerous, and have not proved so injurious as particularly to attract attention.

IV. JUMPERS. (*Orthoptera saltatoria.*)

These are by far the most abundant and prolific, and the most destructive of the Orthopterous insects. They were all included by Linnæus in his great genus *Gryllus*, in separate divisions, however, three of which correspond to the families *Achetadæ*,* *Grylliadæ*,† and *Locustiadæ*,‡ in my "Catalogue of the Insects of Massachusetts," and may retain the synonymous English names of Crickets, Grasshoppers, and Locusts. These three families may thus be distinguished from each other.

1. Crickets (ACHETADÆ); with the wing-covers horizontal, and furnished with a narrow, deflexed outer border; antennæ long and tapering; feet three-jointed (except Œcanthus, which has four joints to the hind feet); two tapering, downy bristles at the end of the body, between which, in most of the females, there is a long spear-pointed piercer.

2. Grasshoppers (GRYLLIDÆ); with the wing-covers sloping downwards at the sides of the body, or roofed, and not bordered; antennæ long and tapering; feet with four joints; end of the body, in the females, with a projecting sword or sabre-shaped piercer.

3. Locusts (LOCUSTADÆ); with the wing-covers roofed, and not bordered; antennæ rather short, and in general not tapering at the end; feet with only three joints; female without a projecting piercer.

1. CRICKETS. (*Achetadæ.*)

There may sometimes be seen in moist and soft ground, particularly around ponds, little ridges or hills of loose fresh earth, smaller than those which are formed by moles. They cover little burrows, that usually terminate beneath a stone or clod of turf. These burrows are made and inhabited by mole-crickets, which are among the most extraordinary of the cricket kind. The common mole-cricket of this country

* *Gryllus Acheta*, Linnæus. † *Gryllus Tettigonia*, L. ‡ *Gryllus Locusta*, L.

(Fig. 68) is, when fully grown, about one inch and a quarter in length, of a light bay or fawn color, and covered with a very short and velvet-like down. The wing-covers are not half the length of the abdomen, and the wings are also short, their tips, when folded, extending only about one eighth of an inch beyond the wing-covers. The fore legs are admirably adapted for digging, being very short, broad, and strong; and the shanks, which are excessively broad, flat, and three-sided, have the lower side divided by deep notches into four finger-like projections, that give to this part very much the appearance and the power of the hand of a mole. From this similarity in structure, and from its burrowing habits, this insect receives its scientific name of *Gryllotalpa*, derived from *Gryllus*, the ancient name of the cricket, and *Talpa*, a mole; and our common species has the additional name of *brevipennis*,* or short-winged, to distinguish it from the European species, which has much longer wings. Mole-crickets avoid the light of day, and are active chiefly during the night. They live on the tender roots of plants, and in Europe, where they infest moist gardens and meadows, they often do great injury by burrowing under the turf, and cutting off the roots of the grass, and by undermining and destroying, in this way, sometimes whole beds of cabbages, beans, and flowers. In the West Indies, extensive ravages have been committed in the plantations of the sugar-cane by another species, *Gryllotalpa didactyla*, which has only two

Fig. 68.

* Serville, "Orthoptères," p. 308.[2]

[[2] It was previously described by Burmeister, under the name *G. borealis*, and this name must be applied to it and retained. It was known to Catesby, who figures it in his "Natural History of Carolina." — UHLER.]

finger-like projections on the shin. The mole-cricket of Europe lays from two to three hundred eggs, and the young do not come to maturity till the third year; circumstances both contributing greatly to increase the ravages of these insects. It is observed, that, in proportion as cultivation is extended, destructive insects multiply, and their depredations become more serious. We may, therefore, in process of time, find mole-crickets in this country quite as much a pest as they are in Europe, although their depredations have hitherto been limited to so small an extent as not to have attracted much notice. Should it hereafter become necessary to employ means for checking them, poisoning might be tried, such as placing, in the vicinity of their burrows, grated carrots or potatoes mixed with arsenic. It is well known that swine will eat almost all kinds of insects, and that they are very sagacious in rooting them out of the ground. They might, therefore, be employed with advantage to destroy these and other noxious insects, if other means should fail.

We have no house-crickets in America;[3] our species inhabit gardens and fields, and enter our houses only by accident. Crickets are, in great measure, nocturnal and solitary insects, concealing themselves by day, and coming from their retreats to seek their food and their mates by night. There are some species, however, which differ greatly from the others in their social habits. These are not unfrequently seen during the daytime in great numbers in paths, and by the roadside; but the other kinds rarely expose themselves to the light of day, and their music is heard only at night. With crickets, as with grasshoppers, locusts, and harvest-flies, the males only are musical; for the females are not provided with the instruments from which the sounds emitted

[[3] This language may apply to the particular district in which Dr. Harris made his observations, but it would be gratuitous to say that we have no house-crickets in America, for nothing is better known to the country-people of Maryland than the "cricket on the hearth," and in some sections of the West they are also well known to inhabit the chimney-places and first-floor apartments of the dwellings. — UHLER.]

by these different insects are produced. In the male cricket these make a part of the wing-covers, the horizontal and overlapping portion of which, near the thorax, is convex, and marked with large, strong, and irregularly curved veins. When the cricket shrills, (we cannot say sings, for he has no vocal organs,) he raises the wing-covers a little, and shuffles them together lengthwise, so that the projecting veins of one are made to grate against those of the other. The English name cricket, and the French *cri-cri*, are evidently derived from the creaking sounds of these insects. Mr. White of Selborne says that "the shrilling of the field-cricket, though sharp and stridulous, yet marvellously delights some hearers, filling their minds with a train of summer ideas of everything that is rural, verdurous, and joyous"; sentiments in which few persons, if any, in America will participate; for with us the creaking of crickets does not begin till summer is gone, and the continued and monotonous sounds, which they keep up during the whole night, so long as autumn lasts, are both wearisome and sad. Where crickets abound, they do great injury to vegetation, eating the most tender parts of plants, and even devouring roots and fruits, whenever they can get them. Melons, squashes, and even potatoes, are often eaten by them, and the quantity of grass that they destroy must be great, from the immense numbers of these insects which are sometimes seen in our meadows and fields. They may be poisoned in the same way as mole-crickets. Crickets are not entirely confined to a vegetable diet; they devour other insects whenever they can meet with and can overpower them. They deposit their eggs, which are numerous, in the ground, making holes for their reception, with their long, spear-pointed piercers. The eggs are laid in the autumn, and do not appear to be hatched till the ensuing summer. The old insects for the most part, die on the approach of cold weather; but a few survive the winter, by sheltering themselves under stones, or in holes secure from the access of water.

The scientific name of the genus that includes the cricket is *Acheta*, and our common species is the *Acheta abbreviata* (Fig. 69), so named from the shortness of its wings, which do not extend beyond the wing-covers. It is about three quarters of an inch in length, of a black color, with a brownish tinge at the base of the wing-covers, and a pale line on each side above the deflexed border. The pale line is most distinct in the female, and is oftentimes entirely wanting in the male.

Fig. 69.

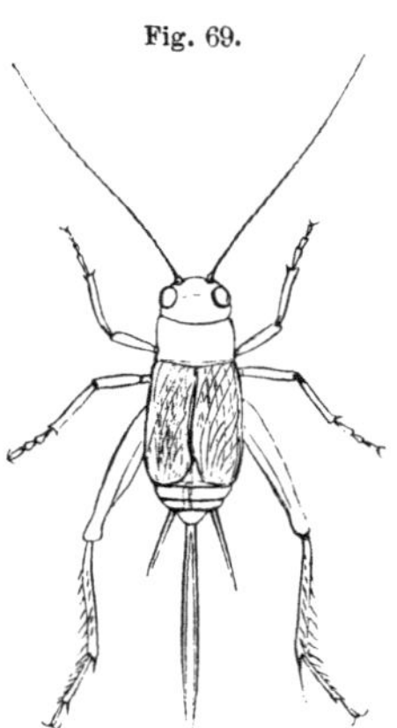

We have another species with very short or abortive wings; it is entirely of a black color, and measures six tenths of an inch in length from the head to the end of the body. It may be called *Acheta nigra*,[4] the black cricket.

A third species, differing from these two in being entirely destitute of wings, and in having the wing-covers proportionally much shorter, and the last joint of the feelers (*palpi*) almost twice the length of the preceding joint, is furthermore distinguished from them by its greatly inferior size, and its different coloring. It measures from three to above four tenths of an inch in length, and varies in color from dusky brown to rusty black, the wing-covers and hindmost thighs being always somewhat lighter. In the brownish-colored varieties three longitudinal black lines are distinctly visible on the top of the head, and a black line on each side of the thorax, which is continued along the sides of the wing-covers to their tips. This black line on the wing-covers is never wanting, even in the darkest varieties. The hindmost thighs have, on the outside, three rows of short oblique black lines, presenting somewhat of a twilled appearance.

[[4] It is *A. Pennsylvanica*, Burm. Priority of nomenclature requires this name to be retained. — UHLER.]

This is one of the social species, which, associated together in great swarms, and feeding in common, frequent our meadows and road-sides, and, so far from avoiding the light of day, seem to be quite as fond of it as others are of darkness. It may be called *Acheta vittata*,* (Fig. 70,) the striped cricket.

Fig. 70.

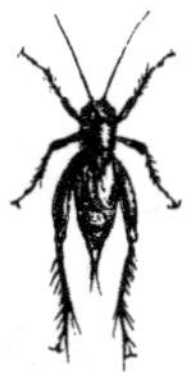

These kinds of crickets live upon the ground, and among the grass and low herbage; but there is another kind which inhabits the stems and branches of shrubs and trees, concealing itself during the daytime among the leaves, or in the flowers of these plants. Some Isabella grape-vines, which were trained against one side of my house, were much resorted to by these delicate and noisy little crickets. The males begin to be heard about the middle of August, and do not leave us until after the middle of September. Their shrilling is excessively loud, and is produced, like that of other crickets, by the rubbing of one wing-cover against the other; but they generally raise their wing-covers much higher than other crickets do while they are playing. These wing-covers, in the males, are also very large, and as long as the wings; they are exceedingly thin, and perfectly transparent, and have the horizontal portion divided into four unequal parts by three oblique raised lines, two of which are parallel and form an angle with the anterior line. The antennæ and legs are both very long and slender, the hinder thighs being much smaller in proportion than those of other crickets, and the hindmost feet have four instead of three joints. The two bristle-formed appendages at the end of the body are as long as the piercer, and the latter is only about half the length of the body, while, in the ground-crickets, the piercer is usually as long as the body, or longer. These insects have, therefore, been separated from the other crickets, under the generical name of *Œcanthus*, a word which means inhabiting flowers. They

* It belongs to M. Serville's new genus *Nemobius*.

20

may be called climbing crickets, from their habit of mounting upon plants and dwelling among the leaves and flowers. According to M. Salvi,* the female makes several perforations in the tender stems of plants, and in each perforation thrusts two eggs quite to the pith. The eggs are hatched about midsummer, and the young immediately issue from their nests and conceal themselves among the thickest foliage of the plant. When arrived at maturity the males begin their nocturnal serenade at the approach of twilight, and continue it with little or no intermission till the dawn of day. Should one of these little musicians get admission to the chamber, his incessant and loud shrilling will effectually banish sleep. Of three species which inhabit the United States, one only is found in Massachusetts. It is the *Œcanthus niveus* (Fig. 71), or white climbing cricket. The male is ivory-white, with the upper side of the first joint of the antennæ, and the head between the eyes, of an ochre-yellow color; there is a minute black dot on the under sides of the first and second joints of the antennæ; and in some individuals the extremities of the feet and the under sides of the hindmost thighs are ochre-yellow. The body is about half an inch long, exclusive of the wing-covers. The female (Fig. 72) is usually rather longer, but the wing-covers are much narrower than those of the male, and there is a great diversity of coloring in this sex; the body being sometimes almost white, or pale greenish-yellow, or dusky, and blackish beneath. There are three dusky stripes on the head and thorax, and the legs, antennæ, and piercer are more or less dusky or blackish. The wing-covers and wings are yellowish-white,

Fig. 71.

Fig. 72.

* Memorie intorno le Locuste grillaiole. 8vo, Verona, 1750.

sometimes with a tinge of green, and the wings are rather longer than the covers. Some of these insects have been sent to me by a gentleman who found them piercing and laying eggs in the branches of a peach-tree. Another correspondent, who is interested in the tobacco culture in Connecticut, informed me that they injured the plant by eating holes in the leaves.

2. Grasshoppers. (*Gryllidæ.*)

Grasshoppers, properly so called, as before stated, are those jumping orthopterous insects which have four joints to all their feet, long bristle-formed antennæ, and in which the females are provided with a piercer, flattened at the sides, and somewhat resembling a sword or cimeter in shape. The wing-covers slope downwards at the sides of the body, and overlap only a little on the top of the back near the thorax. This overlapping portion, which forms a long triangle, is traversed, in the males, by strong projecting veins, between which, in many of them, are membranous spaces as transparent as glass. The sounds emitted by the males, and varying according to the species, are produced by the friction of these overlapping portions together.

In Massachusetts there is one kind of grasshopper which forms a remarkable exception to the other native insects of this family; and, as it does not seem to have been named or described by any author, although by no means an uncommon insect, it may receive a passing notice here. It is found only under stones and rubbish in woods, has a short thick body, and remarkably stout hind thighs, like a cricket, but is entirely destitute of wing-covers and wings, even when arrived at maturity. It belongs to M. Serville's genus *Phalangopsis*, and I propose to call it *Phalangopsis maculata*,*

* *Gryllus maculatus*, Harris. Catalogue of the Insects of Massachusetts.[5]

[[5] According to the authority of Erichson, it was previously described with the name *Phalangopsis lapidicola*, Burm. — Uhler.]

(Fig. 73,) the spotted wingless cricket. Its body is of a pale yellowish-brown color, darker on the back, which is covered with little light-colored spots, and the outside of the hindmost thighs is marked with numerous short oblique lines, disposed in parallel rows, like those on the thighs of *Acheta vittata*. It varies in length from one half to more than three quarters of an inch, exclusive of the piercer and legs. The body is smooth and shining, and the back is arched.

Fig. 73.

Most grasshoppers are of a green color, and are furnished with wings and wing-covers, the latter frequently resembling the leaves of trees and shrubs, upon which, indeed, many of these insects pass the greater part of their lives. Their leaf-like form and green color evidently seem to have been designed for their better concealment. They are nocturnal insects, or at least more active by night than by day. When taken between the fingers, they emit from their mouths a considerable quantity of dark-colored fluid, as do also the locusts or diurnal grasshoppers. They devour the leaves of plants, and lead a solitary life, or at least do not associate and migrate from place to place in great swarms, like some of the crickets and the locusts. There is a remarkable difference in their habits, which does not appear to have been described hitherto. Some of these grasshoppers live upon grass and other herbaceous or low plants in fields and meadows. The piercer of the females is often straight, or only slightly curved. They commit their eggs to the earth, thrusting them into holes made therein with the piercer. They lay a large number of eggs at a time, and cover them with a kind of varnish, which, when dry, forms a thin film that completely encloses them. These eggs are elongated, and nearly of an ellipsoidal form. Other green *Grylli* live upon trees and shrubs. Their wing-covers and wings are broader, and their piercer is shorter and often more curved, than in the foregoing kinds. They do not lay their eggs in the ground, but deposit them upon branches and twigs, in regular rows. My attention was first directed to the eggs of the tree-grylli by Mr. F. C. Hill, late of Philadelphia.

Some of these grasshoppers have the front of the head obtuse, and others have it conical, or prolonged to a point between the antennæ. Among the former is the insect which, from its peculiar note, is called the katy-did. Its body is of a pale green color, the wing-covers and wings being somewhat darker. Its thorax is rough like shagreen, and has somewhat the form of a saddle, being curved downwards on each side, and rounded and slightly elevated behind, and is marked by two slightly transverse furrows. The wings are rather shorter than the wing-covers, and the latter are very large, oval, and concave, and enclose the body within their concavity, meeting at the edges above and below, somewhat like the two sides or valves of a pea-pod. The veins are large, very distinct, and netted like those of some leaves, and there is one vein of larger size running along the middle of each wing-cover, and simulating the midrib of a leaf. The musical organs of the male consist of a pair of taborets. They are formed by a thin and transparent membrane stretched in a strong half-oval frame in the triangular overlapping portion of each wing-cover. During the daytime these insects are silent, and conceal themselves among the leaves of trees; but at night they quit their lurking-places, and the joyous males begin the tell-tale call with which they enliven their silent mates. This proceeds from the friction of the taboret frames against each other when the wing-covers are opened and shut, and consists of two or three distinct notes almost exactly resembling articulated sounds, and corresponding with the number of times that the wing-covers are opened and shut; and the notes are repeated at intervals of a few minutes, for hours together. The mechanism of the taborets, and the concavity of the wing-covers, reverberate and increase the sound to such a degree, that it may be heard in the stillness of the night, at the distance of a quarter of a mile. At the approach of twilight the katy-did mounts to the upper branches of the tree in which he lives, and, as soon as the shades of evening prevail, begins his noisy babble, while rival notes issue from the neighboring trees, and the groves resound with the call of "katy-did, she-did" the live-long night. Of this insect I have met with no scientific description except my own, which was published in 1831 in the eighth volume of the "Encyclopædia Americana," page 42. It is the *Platyphyllum** *concavum*,† (Fig. 74,) and measures, from the head to the end of the wing-covers, rather more than one inch and a half, the body alone being one inch in length. The piercer is broad, laterally compressed, and curved like a cimeter; and there are, in both sexes, two little thorn-like projections from the middle of the breast between the fore legs. The katy-did is found in the perfect state during the months of September and October, at which time the female lays her eggs. These are slate-colored, and are rather more than

Fig. 74.

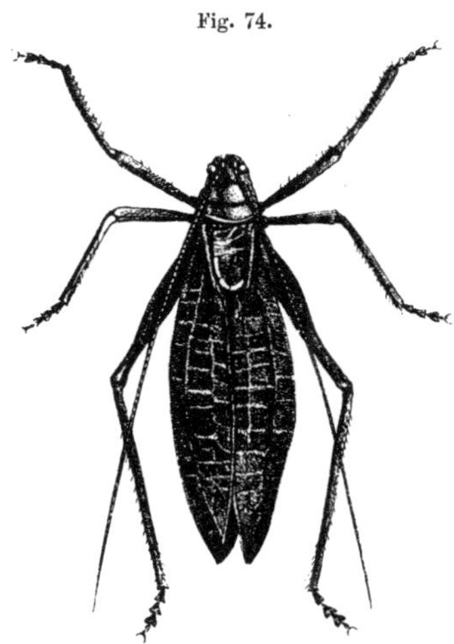

* *Platyphyllum* means broad-wing.

† Can this be the *Locusta perspicillata* of Fabricius? [6]

[[6] This is *Cyrtophyllus perspicillatus*, Burm. = *Locusta perspicillata*, Fab. Dr. Harris's generic name has priority over that of Burmeister, and hence this insect must be called *Platyphyllum perspicillatum*, Fab. The insect called katy-did in the Southern States is entirely different from this one, although its habit of sitting upon the trees and issuing this shrill note has induced some persons to mistake it for the true one from New England. The Southern katy-did belongs to the genus *Phylloptera*, and from the ovipositor being shaped somewhat like that of *Locusta curvicauda*, De Geer, Dr. Harris supposed it to be that species. — UHLER.]

one eighth of an inch in length. They resemble tiny oval bivalve shells in shape. The insect lays them in two contiguous rows along the surface of a twig, the bark of which is previously shaved off or made rough with her piercer. Each row consists of eight or nine eggs, placed somewhat obliquely, and overlapping each other a little, and they are fastened to the twig with a gummy substance. In hatching, the egg splits open at one end, and the young insect creeps through the cleft. I am indebted to Miss Morris for specimens of these eggs.

We have another broad-winged green grasshopper, differing from the katy-did, in having the wing-covers narrower, flat and not concave, and shorter than the wings, the thorax smooth, flat above, and abruptly bent downwards at a right angle on each side, and the breast without any projecting spines in the middle. The piercer has the same form as that of the katy-did. The musical organ of the left wing-cover, which is the uppermost, is not transparent, but is green and opaque, and is traversed by a strong curved vein; that of the right wing-cover is semi-transparent in the middle. This insect is the *Phylloptera oblongifolia*,* (Fig. 75,) or ob-

Fig. 75.

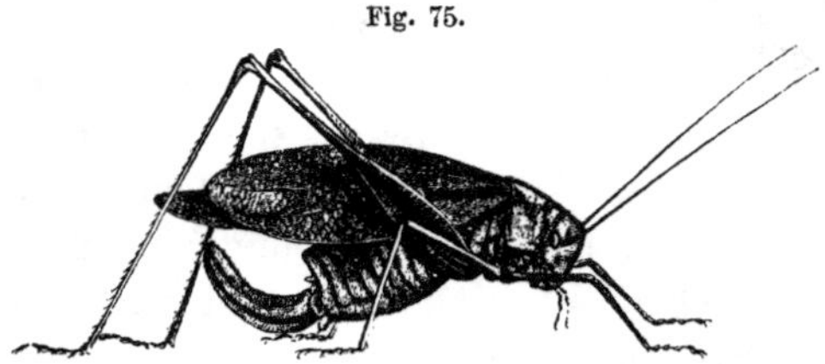

long leaf-winged grasshopper. Its body measures about an inch in length, and from the head to the tips of the wings, from an inch and three quarters to three inches. It is found in its perfect state during the months of September and October, upon trees, and, when it flies, makes a whizzing noise somewhat like that of a weaver's shuttle. The notes

* *Locusta oblongifolia* of De Geer, a different species from the *laurifolia* of Linnæus, with which it has been confounded by many naturalists.

of the male, though grating, are comparatively feeble. The females lay their eggs in the autumn on the twigs of trees and shrubs, in double rows, of seven or eight eggs in each row. These eggs, in form, size, and color, and in their arrangement on the twig, strikingly resemble those of the katy-did. The Rev. Thomas Hill, of Waltham, had the kindness to procure some of them for me from Philadelphia.

A third species, also of a green color, with still narrower wing-covers, which are of almost equal width from one end to the other, but are rounded at the tips, and are shorter than the wings, has the head, thorax, musical organs, and breast like those of the preceding species, but the piercer is

Fig. 76.

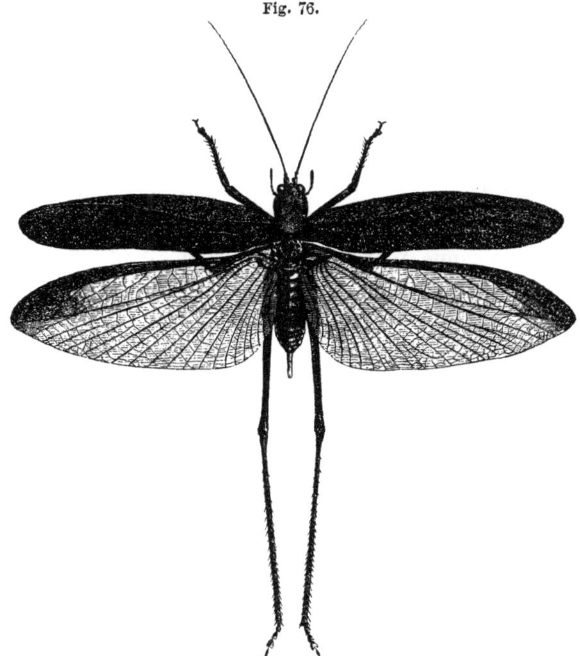

much shorter, and very much more crooked, being bent vertically upwards from near its base. The male has a long tapering projection from the under side of the extremity of the body, curved upwards like the piercer of the female. This grasshopper belongs to the genus *Phaneroptera*, so named, probably, because the wings are visible beyond the tips of the wing-covers; and, as it does not appear to have been described before, I propose to call it *angustifolia*,* (Fig. 76,) the narrow-leaved. It measures from the forehead to the end of the abdomen about three quarters of an inch, and to the tips of the wings from an inch and a half to an inch and three quarters. Its habits appear to be the same as those of the *oblongifolia*. It comes to maturity some time in the latter part of August or the beginning of September.

From the middle till the end of summer, the grass in our meadows and moist fields is filled with myriads of little grasshoppers, of different ages, and of a light green color, with a brown stripe on the top of the head, extending to the tip of the little smooth and blunt projection between the antennæ, and a broader brown stripe bounded on each side by deeper brown on the top of the thorax. The antennæ, knees, and shanks are green, faintly tinted with brown, and the feet are dusky. When come to maturity, they measure three quarters of an inch or more, from the forehead to the end of the body, or one inch to the ends of the wing-covers. The latter are abruptly narrowed in the middle, and taper thence to the tip, which, however, is rounded, and extends as far back as the wings. The color of the wing-covers is green, but they are faintly tinged with brown on the overlapping portion, and have the delicacy and semi-transparency of the

* I formerly mistook this insect for the *Locusta curvicauda* of De Geer, which is found in the Middle and Southern States, but not in Massachusetts, is a larger species, with wing-covers broadest in the middle, and different organs in the male, and belongs to the genus *Phylloptera*.[7]

[[7] This is the true *curvicauda*; it was figured by Drury as *P. myrtifolia*, but he unfortunately confounded it with a species somewhat resembling it from South America, which has caused some authors to refer his figure to the one described by Linnæus; but that is a different insect, belonging to the genus *Phylloptera*. The synonymy of this species is, *Phaneroptera curvicauda*, De Geer = *P. myrtifolia*, Drury = *P. septentrionalis*, Serv. = *P. angustifolia*, Harris. — UHLER.]

skin of an onion. The shrilling organs in the males consist of a transparent glassy spot, bounded and traversed by strong veins, in the middle of the overlapping portion of each wing-cover, which part is proportionally much larger and longer than in the other grasshoppers; but the transparent spot is rather smaller on the left than on the right wing-cover. The male is furthermore distinguished by having two small black spots or short dashes, one behind the other, on each wing-cover, on the outside of the transparent spot. The wings are green on their front margins, transparent, and reflecting a faint pink color behind. The piercer of the female is cimeter-shaped, being curved, and pointed at the end, and is about three tenths of an inch long. The hindmost thighs, in both sexes, are smooth and not spinous beneath; there are two little spines in the middle of the breast; and the antennæ are very long and slender, and extend, when turned back, considerably beyond the end of the hind legs. During the evening, and even at other times in shady places, the males make a sharp clicking noise, somewhat like that produced by snapping the point of a pen against the thumb-nail, but much louder. This kind of grasshopper very much resembles the *Locusta agilis* of De Geer, which is found in Pennsylvania and the Southern States, but does not inhabit Massachusetts, and is distinguished from our species by having the wings nearly one tenth of an inch longer than the wing-covers, the antennæ excessively long (two inches or more), and the piercer not quite so much curved as in our species, besides other differences which it is unnecessary to record here. As our species does not appear to have been named, or described by any previous writer, I propose to call it *Orchelimum vulgare* (Fig. 77), the common meadow-grasshopper, the generical name signifying literally, I dance in the meadow.

Fig. 77.

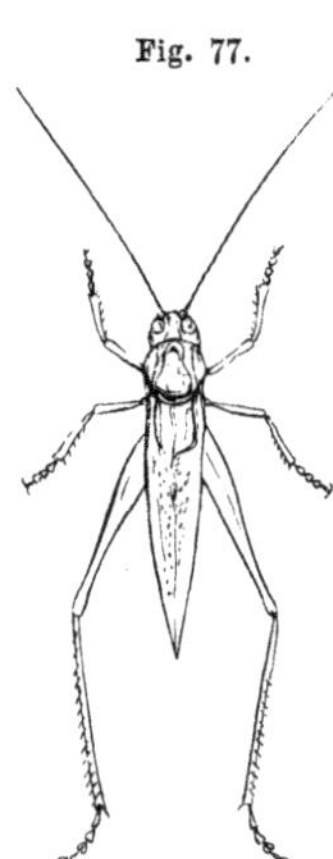

With this species another one is also found, bearing a considerable resemblance to it in color and form, but measuring only four or five tenths of an inch from the head to the end of the body, or from seven to eight tenths to the tips of the wings, which are a little longer than the wing-covers. The latter are narrow and taper to the end, which is rounded, but the overlapping portion is not so large as in the common species, and the male has not the two black spots on each wing-cover. The upper part of the abdomen is brown, with the edges of the segments greenish-yellow, and the piercer, which is nearly three tenths of an inch long, is brown and nearly straight. This little insect comes very near to *Locusta fasciata* of De Geer, who, however, makes no mention of the broad brown stripe on the head and thorax. I therefore presume that our species is not the same, and propose to call it *Orchelimum gracile* (Fig. 78), the slender meadow-grasshopper. M. Serville, by whom this genus was instituted, has described three species, two of which are stated to be North American, and the remaining one is probably also from this country; but his descriptions do not answer for either of our species. Both of these kinds of meadow-grasshoppers are eaten greedily by fowls of all kinds.

Fig. 78.

One more grasshopper remains to be described. It is distinguished from all the preceding species by having the head conical, and extending to a blunt point between the eyes. It belongs to the genus *Conocephalus*, a word expressive of the conical form of the head, and, in my Catalogue of the Insects of Massachusetts, bears the specific name of *ensiger* (Fig. 79, male), the sword-bearer, from the long, straight, sword-shaped piercer of the female. It measures an inch or more from the point of the head to the end of the body, and from one inch and three quarters to two inches to the end of the wing-covers. It is pale green, with the head whitish, or only faintly tinted with green, and the legs and abdomen are pale brownish-green. A little tooth

Fig. 79.

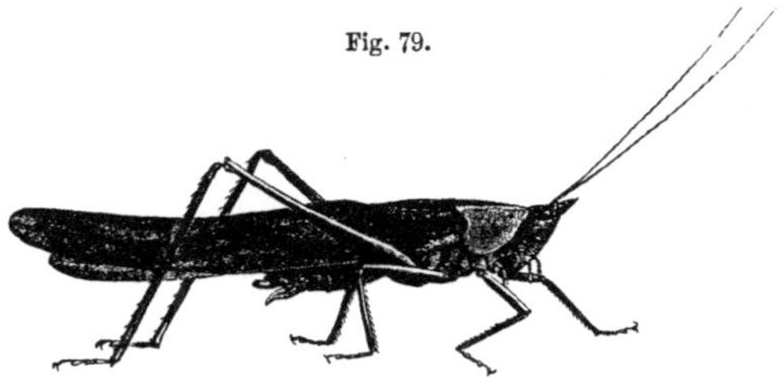

projects downwards from the under side of the conical part of the head, which extends between the antennæ, and immediately before this little tooth is a black line bent backwards on each side like the letter U. The hindmost thighs have five or six exceedingly minute spines on the inner ridge of the under side. The shrilling organ of the male on the left wing-cover is green and opaque, but that on the right has a space in the middle that is transparent like glass. The piercer of the female is above an inch long, very slightly bent near the body, and is perfectly straight from thence to the tip, which ends in a point. The color of this grasshopper is very apt to change after death to a dirty brown. It comes very near to the *dissimilis* described by M. Serville, but appears to be a different species.*

* In the collection belonging to the Boston Society of Natural History, there is an insect which I suppose to be the *Conocephalus dissimilis* of Serville. It was taken in North Carolina by Professor Hentz. The conical projection of the head is shorter and more obtuse than in the *ensiger*, the sides of the thorax are brownish, the hindmost thighs have a double row of black dots on the under side, and the spines on this part are more numerous and rather larger. Professor Hentz has sent to me from Alabama another species distinct from both of these, about the same in length, but considerably broader. The conical part of the head between the eyes is broader, flattened above, and, as well as the thorax, rough like shagreen. There is a projecting tubercle beneath, but the curved black line is wanting, and the tip of the cone has a minute point abruptly bent downwards, and forming a hook. The sides of the thorax are bent down suddenly so as to make an angular ridge on each side of the middle. The wing-covers are dotted with black around their edges, and have also an irregular row of larger and more distinct spots along

3. LOCUSTS. (*Locustadæ.*)

The various insects included under the name of locusts nearly all agree in having their wing-covers rather long and narrow, and placed obliquely along the sides of the body, meeting, and even overlapping for a short distance, at their upper edges, which together form a ridge on the back like a sloping roof. Their antennæ are much shorter than those of most grasshoppers, and do not taper towards the end, but are nearly of equal thickness at both extremities. Their feet have really only three joints; but as the under side of the first joint is marked by one or two cross lines, the feet, when seen only from below, seem to be four or five jointed. The females have not a long projecting piercer, like the crickets and grasshoppers, but the extremity of their body is provided with four short, wedge-like pieces, placed in pairs above and below, and opening and shutting opposite to each other, thus forming an instrument like a pair of nippers, only with four short blades instead of two. When one of these insects is about to lay her eggs, she drives these little wedges into the earth; these, being then opened and withdrawn, enlarge the orifice; upon which the insect inserts them again, and drives them down deeper than before, and repeats the operation above described until she has formed a perforation large and deep enough to admit nearly the whole of her abdomen.

The males, though capable of producing sounds, have not the cymbals and tabors of the crickets and grasshoppers; their instruments may rather be likened to violins, their hind legs being the bows, and the projecting veins of their wing-covers the strings. But besides these, they have on each side of the body, in the first segment of the abdomen, just above and a little behind the thighs, a deep cavity, closed by a thin piece of skin stretched tightly across it. These proba-

the middle. The hindmost thighs have a double row of strong spines beneath, and the piercer is straight and only about six tenths of an inch long. This insect may be called *Conocephalus uncinatus*, from the hook on the tip of the head.

bly act in some measure to increase the reverberation of the sound, like the cavity of a violin. When a locust begins to play, he bends the shank of one hind leg beneath the thigh, where it is lodged in a furrow designed to receive it, and then draws the leg briskly up and down several times against the projecting lateral edge and veins of the wing-cover. He does not play both fiddles together, but alternately, for a little time, first upon one, and then on the other, standing meanwhile upon the four anterior legs and the hind leg which is not otherwise employed. It is stated that, in Spain, people of fashion keep these insects, which they call *grillo*, in cages, for the sake of their music.

Locusts leap much better than grasshoppers, for the thighs of their hind legs, though shorter, are much thicker, and consequently more muscular within. The back part of the shanks of these legs, from a little below the knee to the end, is armed with strong sharp spines, arranged in two rows. These may serve as means of defence, but the lower ones also help to fix the legs firmly against the ground when the insect is going to leap. The power of flight in locusts is, in general, much greater than that of grasshoppers; for the wing-covers, being narrow, do not, like the much wider ones of grasshoppers, so much impede their passage through the air; while their wings, which are ample, except in a few species, and when expanded together form half of a circle, have very strong joints, and are moved by very powerful muscles within the chest. From the shoulders of the wings several stout ribs or veins pass towards the hinder margin, spreading apart, when the wings are opened, like the sticks of a fan, and are connected and strengthened by little crossing veins, which form a kind of network. The same structure exists in the wings of grasshoppers, but in them the longitudinal ribs are not so strong, and the network is much more delicate. Hence the flight of grasshoppers is short and unsteady, while that of locusts is longer and better sustained. Many locusts, when they fly, make a loud whizzing noise, the source of which does not seem to be understood. Those of our native locusts, whose flight is the most noisy, are the coral-winged, the yellow-winged, and the broad-winged species. But as these are comparatively small insects, and never assemble in such great swarms as the much larger migrating locusts of Asia and Africa, the noise of their flight bears no comparison to that of the latter. When a large number of these take flight together, it is said that the noise is like the rushing of a whirlwind; and hence we read, of the symbolical locusts of the Apocalypse, that the sound of their wings was as the sound of chariots of many horses running to battle;* and of others, that their coming is like the noise of chariots on the tops of mountains, or the crackling of stubble when overrun and consumed by a flame of fire.†

The East seems to have suffered severely at various times from the irruptions of immense swarms of locusts, darkening the sky during their passage, stripping the surface of the earth, where they alight, of all vestiges of vegetation, and thus reducing, in an inconceivably short time, the most fertile regions to barren wastes. The ground over which they have passed presents the appearance of having been scorched by fire; and hence the name of locust, which is derived from the Latin,‡ and means a burnt place, is highly expressive of the desolation occasioned by their ravages. Famine and pestilence have sometimes followed their appearance, as we find recorded by various writers. In the Scriptures § frequent mention is made of the destructive powers of locusts, and these accounts are fully confirmed by the testimony of numerous travellers in Asia and Africa, some of whom have been eyewitnesses of the devastations of these insects. Among

* Revelation ix. 9. † Joel ii. 5.

‡ *Locus* and *ustus.*

§ For an explanation of the various passages in which allusion is made to locusts, and for much interesting matter relating to the history of these insects, as contained in the Bible and elucidated by the accounts of historians and travellers, the reader is referred to the article *Locust* in the learned and instructive work of my father, entitled, "The Natural History of the Bible, by Thaddeus Mason Harris," 8vo, Boston, 1820.

the later accounts, that contained in Olivier's "Travels" does not seem to have been quoted by English writers. The following is a free translation of the passage. Olivier, at the time of writing it, was in Syria. "After a burning south wind had prevailed for some time, there came, from the interior of Arabia and from the southern parts of Persia, clouds of locusts, whose ravages in these countries are as grievous and as sudden as the destruction occasioned in Europe by the most severe hail-storm. Of these my companion, M. Brugières, and myself were twice witnesses. It is difficult to describe the effect produced on us by the sight of the whole atmosphere filled, on all sides, to a vast height, with a countless multitude of these insects, which flew along with a slow and even motion, and with a noise like the dashing of a shower of rain. The heavens were darkened by them, and the light of the sun was sensibly diminished. In a moment the roofs of the houses, the streets, and all the fields were completely covered with these insects, and in two days they almost entirely devoured the foliage of every plant. Fortunately, however, they continued but a short time, and seemed to have emigrated only for the purpose of providing for a continuation of their kind. In fact, nearly all of them which we saw on the next day were paired, and in a day or two afterwards the ground was covered with their dead bodies." * These were not the still more celebrated and destructive migratory locusts (*Locusta migratoria*), but consisted of the species called *Acrydium peregrinum*.

Although the ravages of locusts in America are not followed by such serious consequences as in the Eastern continent, yet they are sufficiently formidable to have attracted attention, and not unfrequently have these insects laid waste considerable tracts, and occasioned no little loss to the cultivator of the soil. Our salt-marshes, which are accounted among the most productive and valuable of our natural meadows, are frequented by great numbers of the small red-

* Olivier, Voyage dans l'Empire Ottoman, l'Égypte et la Perse, Tom. II. p. 424.

legged species (*Acrydium femur-rubrum*), (Fig. 80, p. 174,) intermingled occasionally with some larger kinds. These, in certain seasons, almost entirely consume the grass of these marshes, from whence they then take their course to the uplands, devouring, in their way, grass, corn, and vegetables, till checked by the early frosts, or by the close of the natural term of their existence. When a scanty crop of hay has been gathered from the grounds which these puny pests have ravaged, it becomes so tainted with the putrescent bodies of the dead locusts contained in it, that it is rejected by horses and cattle. In this country locusts are not distinguished from grasshoppers, and are generally, though incorrectly, comprehended under the same name, or under that of flying grasshoppers. They are, however, if we make allowance for their inferior size, quite as voracious and injurious to vegetation during the young or larva and pupa states, when they are not provided with wings, as they are when fully grown. In our newspapers I have sometimes seen accounts of the devastations of grasshoppers, which could only be applicable to some of our locusts.

At various times they have appeared in great abundance in different parts of New England. It is stated that, in Maine, "during dry seasons, they often appear in great multitudes, and are the greedy destroyers of the half-parched herbage." "In 1749 and 1754 they were very numerous and voracious; no vegetables escaped these greedy troops; they even devoured the potato tops; and in 1743 and 1756 they covered the whole country and threatened to devour everything green. Indeed, so great was the alarm they occasioned among the people, that days of fasting and prayer were appointed," * on account of the threatened calamity. The southern and western parts of New Hampshire, the northern and eastern parts of Massachusetts, and the southern part of Vermont, have been overrun by swarms of these

* See Williamson's History of Maine, Vol. I. pp. 102, 103, and compare with p. 172 of the same work.

miscalled grasshoppers, and have suffered more or less from their depredations.

Among the various accounts which I have seen, the following, extracted from the Travels of the late President Dwight,* seems to be the most full and circumstantial. "Bennington (Vermont), and its neighborhood, have for some time past been infested by grasshoppers (locusts) of a kind with which I had before been wholly unacquainted. At least, their history, as given by respectable persons, is in a great measure novel. They appear at different periods, in different years; but the time of their continuance seems to be the same. This year (1798) they came four weeks earlier than in 1797, and disappeared four weeks sooner. As I had no opportunity of examining them, I cannot describe their form or their size. Their favorite food is clover and maize. Of the latter they devour the part which is called the silk, the immediate means of fecundating the ear, and thus prevent the kernel from coming to perfection. But their voracity extends to almost every vegetable; even to the tobacco plant and the burdock. Nor are they confined to vegetables alone. The garments of laborers, hung up in the field while they are at work, these insects destroy in a few hours; and with the same voracity they devour the loose particles which the saw leaves upon the surface of pine boards, and which, when separated, are termed sawdust. The appearance of a board fence, from which the particles had been eaten in this manner, and which I saw, was novel and singular; and seemed the result, not of the operations of the plane, but of attrition. At times, particularly a little before their disappearance, they collect in clouds, rise high in the atmosphere, and take extensive flights, of which neither the cause nor the direction has hitherto been discovered. I was authentically informed that some persons, employed in raising the steeple of the church in Williamstown, were, while standing near the vane, covered by them, and saw, at the same time, vast swarms of

* Travels in New England and New York, by Timothy Dwight, Vol. II. p. 403.

them flying far above their heads. It is to be observed, however, that they customarily return, and perish on the very grounds which they have ravaged." Through the kindness of the Rev. L. W. Leonard, of Dublin, New Hampshire, I have been favored with specimens of the destructive locusts which occasionally appear in that part of New England, and which, most probably, are of the same species as the insects mentioned by President Dwight. They prove to be the little red-legged locusts, whose ravages on our salt-marshes I have already recorded.

In the summer of 1838, the vicinity of Baltimore, Maryland, was infested by insects of this kind ; and I was informed by a young gentleman from that place, then a student in Harvard College, that they were so thick and destructive in the garden and grounds of his father, that the negroes were employed to drive them from the garden with rods; and in this way they were repeatedly whipped out of the grounds, leaping and flying before the extended line of castigators like a flock of fowls. Some of these insects were brought to me by the same gentleman, on his return to the University, at the end of the summer vacation, and they turned out to be specimens of the red-legged locusts already mentioned.

It is not to be supposed that these are the only depredatory locusts in this country. Massachusetts alone produces a large number of species, some of which have never been described; and the habits of many of them have not been fully investigated. The difficulty which I have met with in ascertaining, from mere verbal reports, or from the accounts that occasionally appear in our public prints, the scientific names of the noxious insects which are the subjects of such remarks, and the impossibility, without this knowledge of their names, of fixing upon the true culprits, has induced me to draw up, in this treatise, brief descriptions of all our locusts, as a guide to other persons in their investigations.

All the locusts of Massachusetts that are known to me

may be included in three large groups or genera; viz. *Acrydium* (of Geoffroy and Latreille), *Locusta* (*Gryllus Locusta* of Linnæus), and *Tetrix* (of Latreille). These three genera may be distinguished from each other by the following characters.*

1. *Acrydium.* The thorax (*prothorax* of Kirby) and the wing-covers of ordinary dimensions; a projecting spine in the middle of the breast; and a little projecting cushion between the nails of all the feet.

2. *Locusta.* The thorax, and usually the wing-covers also, of ordinary dimensions; no projecting spine in the middle of the breast; cushions between the nails of the feet.

3. *Tetrix.* The thorax (*prothorax*) greatly prolonged, tapering to a point behind, and covering the whole of the back to the extremity of the abdomen; wing-covers exceedingly minute, consisting only of a little scale on each side of the body; fore part of the breast forming a projection, like a cravat or stock, to receive the lower part of the head; no spine in the middle of the breast; no cushions between the nails.

* I have not considered it necessary to give, in addition to these, the characters that distinguish them from the other genera of American locusts, which are not found in Massachusetts, but add the characteristics of these genera in this note.

Opsomala. Body slender and cylindrical; head long and conical, extending with an obtuse point between the antennæ; eyes oblong oval and oblique; antennæ short, flattened, and more or less enlarged toward the base, and tapering toward the point; a pointed tubercle between the fore legs on the breast; wing-covers narrow and pointed; face sloping down toward the breast, and forming an acute angle with the top of the head.

Truxalis. Body rather thicker; head shorter, but ending in a blunt cone between the antennæ; eyes oval and oblique; antennæ short, flattened, enlarged near the base, and tapering to a point; no tubercle between the fore legs; wing-covers wider and not so pointed; face sloping toward the breast, and forming an angle of forty-five degrees with the top of the head; thorax flat above, and marked with three longitudinal elevated lines.

Xiphicera. Robust; head not conical, but with a projection between the antennæ; face vertical; antennæ rather short, flattened more or less, and tapering at the end; a spine between the fore legs on the breast; wing-covers about as long as the abdomen, obtuse or notched at the end; thorax with three elevated crested lines, which are frequently notched.

Romalea. Very thick and short; head obtuse; face vertical; antennæ short, of equal thickness to the end, seventeen or eighteen jointed; thorax with a some-

I. ACRYDIUM. *Spine-breasted Locusts.*

This word, which is nearly the same as one of the Greek names of a locust, has been variously applied by different entomologists. I have followed Latreille and Serville in confining it to those locusts which have a projecting spine or tubercle in the middle of the fore part of the breast between the fore legs. To this genus belong the following native species.

1. *Acrydium alutaceum.* Leather-colored Locust.

Dirty brownish yellow; a paler yellow stripe on the top of the head and thorax; a slightly elevated longitudinal line on the top of the thorax; wing-covers semi-transparent, with irregular brownish spots; wings transparent, uncolored, netted with dirty yellow; abdomen with transverse rows of minute blackish dots; hindmost thighs whitish within and without, the white portion bounded by a row of minute distant black dots, and crossed, herring-bone fashion, by numerous brown lines; hindmost shanks reddish, with yellowish-white spines, which are tipped with black. Length, to the end of the abdomen, $1\frac{3}{4}$ inch; the wing-covers expand over 3 inches.

This insect was brought to me, from Martha's Vineyard, by Mr. Robert Treat Paine. It bears a close resemblance in form to *Acrydium Americanum* of De Geer,[8] a much larger and more showy Southern species.

2. *Acrydium flavo-vittatum.*[9] Yellow-striped Locust.

Dull green or olive-colored, with a yellowish line on each side from the forehead to the tips of the wing-covers; hind-

what elevated crest; a spine between the fore legs on the breast; wing-covers and wings much shorter than the abdomen.

The first two of these genera seem to connect the cone-headed grasshoppers with the locust family, while the last two approach nearer to the genus *Acrydium*; many foreign genera, however, are interposed between them.

[8 This reference to De Geer is incorrect, no such species being found in his works; it may refer to Drury. Illustrations I. pl. 49, f. 2. — UHLER.]

[9 This insect was previously described by Say, who calls it *A. bivittatus*. The difference between the species, as found in New England and that of the most shanks and feet blood-red, the spines tipped with black; wings transparent, faintly tinged with pale green, and netted with greenish-brown lines. The abdomen of the male is very obtuse and curves upwards at the end, and is furnished, on each side of the tip, with a rather large oblong square appendage, which has a little projecting angle in the middle of the lower side. Length, to tip of the abdomen, from 1 inch to 1¼; expands from 1¼ inch to 2 inches.

This and the following species probably belong to the subgenus *Oxya* of Serville. The yellow-striped locust is one of our most common insects. It is readily known by its color, and by the two yellowish lines on the thorax, extending, when the insect acquires wings, along the inner margin of the wing-covers. It is very troublesome in gardens, climbing upon the stems of beans, peas, and flowers, devouring the leaves and petals, and defiling them with its excrement. The young begin to appear in June, and they come to their growth and acquire their wings by the first of August. When about to moult, like other locusts, they cling to the stem of some plant, till the skin bursts and the insect withdraws its body and legs from it, and leaves the cast-skin still fastened to the plant.

3. *Acrydium femur-rubrum.*[10] Red-legged Locust. (Fig. 80.)

Grizzled with dirty olive and brown; a black spot extending from the eyes along the sides of the thorax; an oblique yellow line on each side of the body beneath the wings; a row of dusky brown spots along the middle of the wing-covers; and the hindmost shanks and feet blood-red, with black spines. The wings

Fig. 80.

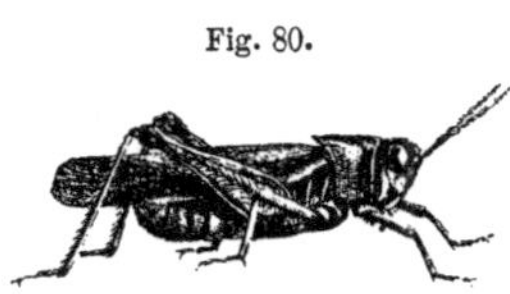

western sections of the Union, consists only in the color of the legs and greater depth of tint upon the thorax, &c. In the latter, the synonymy stands as follows: *A.* (*Caloptenus*) *bivittatus*, Say = *A.* (*Caloptenus*) *femoratus*, Burm. = *A. Milberti*, Serv. = *A. flavo-vittatum*, Harris. — UHLER.]

[10 This is also a *Caloptenus*. — UHLER.]

are transparent, with a very pale greenish-yellow tint next to the body, and are netted with brown lines. The hindmost thighs have two large spots, on the upper side, and the extremity, black; but are red below, and yellow on the inside. The appendages at the tip of the body in the male are of a long triangular form. Length from $\frac{3}{4}$ inch to 1 inch; exp. $1\frac{1}{4}$ to $1\frac{3}{4}$ inch.

The red-legged locust was first described by De Geer from specimens sent to him from Pennsylvania, and I have retained the scientific name which he gave to it. It is the *Gryllus* (*Locusta*) *erythropus* of Gmelin, and the *Acrydium femorale* of Olivier. It appears to be very generally diffused throughout the United States, and sometimes so greatly abounds in certain places as to be productive of great injury to vegetation. I have already described its prevalence on our salt-marshes; and it seems to constitute those large migrating swarms whose flight has been observed and recorded in various parts of this country. It comes to maturity with us by the latter part of July; some broods, however, a little earlier, and others later. It is most plentiful and destructive during the months of August and September, and does not disappear till some time in October.

II. Locusta. *Locusts Proper.*

With the English entomologists, I apply the name *Locusta* to that genus which includes the celebrated migrating locust, or *Gryllus Locusta migratoria* of Linnæus. By the older French entomologists the insects contained in it were united to the genus *Acrydium;* but Latreille afterwards separated them from *Acrydium* under the generical name of *Œdipoda* (which means swelled leg), and he is followed in this by Serville, the latest writer on the Orthoptera. In the insects of this genus the breast is not armed with a blunt spine or tubercle, a character which distinguishes the genus *Acrydium* from it. In other respects these two genera are much alike.

1. *Locusta Carolina.** [11] Carolina Locust. (Plate III. Fig. 3.)

Pale yellowish brown, with small dusky spots; wings black, with a broad yellow hind margin, which is covered with dusky spots at the tip. Length from 1 to 1½ inch; exp. 2¾ to above 3½ inches.

A more detailed description of this large, common, and well-known species is unnecessary. The Carolina locust is found in abundance by the road-side, from the middle to the end of summer. It generally makes use of its large and handsome wings in moving from place to place. It is frequently found in company with the red-legged locust in the vicinity of salt-marshes, but it generally prefers warm and dry situations. Pairing takes place with this species in the months of September and October, immediately after which the female prepares to lay her eggs. These are deposited at the bottom of a cylindrical hole in the ground, made in the manner already described, and are not hatched till the following spring. The abdomen of the female admits of being greatly extended in length; hence she frequently deposits her eggs at the depth of nearly two inches beneath the surface of the soil.

2. *Locusta corallina.* Coral-winged Locust.

Light brown; spotted with dark brown on the wing-covers; wings light vermilion or coral-red, with an external dusky border, which is wide and paler at the tip, narrowed and darker behind; hind shanks yellow with black-tipped spines. Length 1 to 1¼ inch; exp. 2¼ to 2½ inches.

This species closely resembles the *Acridium tuberculatum* of Palisot de Beauvois, which seems to be the *Œdipoda discoidea* of Serville, found in the Southern States, of a much larger size than the coral-winged locust, and having the wings of a much deeper and duller red color, and the black-

* *Gryllus Locusta Carolinus*, Linnæus.

[[11] *L. Carolina* must be referred to *Œdipoda*. — UHLER.]

ish border not so much narrowed behind. It cannot be mistaken for the *fenestralis*, which M. Serville describes as having the antennæ nearly as long as the body, whereas in this species they are not half that length. The coral-winged locust is the first that makes its appearance with wings in the spring, being found flying about in warm and dry pastures as early as the middle of April or the first of May, and is rendered very conspicuous by its bright-colored wings, and the loud noise which it makes in flying. It probably passes the winter in the pupa state, and undergoes its last transformation in the spring; but its history is not yet fully known to me, and this opinion is the result only of conjecture.

3. *Locusta sulphurea.*[12] Yellow-winged Locust. (Plate I. Fig. 6.)

Dusky brown; thorax slightly keeled in the middle; wing-covers ash-colored at their extremities, more or less distinctly spotted with brown; wings deep yellow next to the body, dusky at tip, the yellow portion bounded beyond the middle by a broad dusky brown band, which curves and is prolonged on the hind margin, but does not reach the angle next to the extremity of the body; hindmost thighs blackish at the end, and with two black and two whitish bands on the inside; hindmost shanks and their spines black, with a broad whitish ring just below the knees. Length $\frac{8}{10}$ to $1\frac{1}{5}$ inch; exp. $1\frac{3}{4}$ to $2\frac{1}{4}$ inches.

This insect agrees tolerably well with the brief description given by Fabricius of his *Gryllus sulphureus*, except that the wings are not sulphur-yellow, but of a deeper tint. It is also described and figured by Palisot de Beauvois under the name of *Acridium sulphureum*. It is a rare species in this vicinity. I have taken it, though sparingly, in its perfect state, in May and in September. The elevated ridge on the top of the thorax is higher than in any other species found in Massachusetts.

[12 *L. sulphurea* must be referred to *Œdipoda.* — UHLER.]

4. *Locusta Maritima.*[13] Maritime Locust.

Ash-gray; face variegated with white; wing-covers sprinkled with minute brownish spots, and semi-transparent at tip; wings transparent, faintly tinted with yellow next the body, uncolored at tip, with a series of irregular blackish spots forming a curved band across the middle; hindmost shanks and feet pale yellow, with the extreme points of the spines black. Length $\frac{3}{4}$ to $1\frac{1}{4}$ inch; exp. $1\frac{1}{10}$ inch to $2\frac{3}{4}$ inches.

This species comes very near to Mr. Kirby's description of the *Locusta leucostoma*; but is evidently distinct from it, and does not appear to have been described before. I have received it from Sandwich, and have found it in great abundance among the coarse grass which grows near the edges of our sandy beaches, but have never seen it except in the immediate vicinity of the sea. It comes to maturity and lays its eggs about the middle of August or a little later.

5. *Locusta æqualis.*[14] Barren-ground Locust.

Ash-gray, mottled with dusky brown and white; wing-covers semi-transparent at tip, with numerous dusky spots which run together so as to form three transverse bands; wings light yellow on their basal half, transparent with dusky veins and a few spots at the tip, with an intermediate broad black band, which, curving and becoming narrower on the hind margin, is continued to the inner angle of the wing; hindmost shanks coral-red, with a broad white ring below the knees, and the spines tipped with black. Length $1\frac{1}{4}$ inch; exp. $2\frac{1}{4}$ inches.

Mr. Say, to whom I sent a specimen of this handsome locust, informed me that it was his *Gryllus equalis*, probably intended for *æqualis*. It is found, during the months of July

[13 *L. maritima* must be referred to *Œdipoda*. — UHLER.]

[14 *L. æqualis* and *latipennis* are merely to be separated as races of one species, and cannot remain as separate species. They must be referred to the genus *Œdipoda*. — UHLER.]

and August, on dry barren hills and on sandy plains, upon the scanty herbage intermingled with the reindeer moss.

6. *Locusta latipennis.*[14] Broad-winged Locust.

Ash-colored, mottled with black and gray; wing-covers semi-transparent beyond the middle, with numerous blackish spots which run together at the base, and form a band across the middle; wings broad, light yellow on the basal half, the remainder dusky but partially transparent, with black net-work, and deep black at tip, and an intermediate irregular band, formed by a contiguous series of black spots, reaching only to the hind margin, but not continued towards the inner angle; hindmost shanks pale yellow, with a black ring below the knees, a broader one at the extremity, and a blackish spot behind the upper part of the shank. Length $\frac{9}{10}$ inch; exp. $1\frac{7}{10}$ inch.

It is possible that this may be a variety of the preceding species, from which it differs especially in the form and width of the wings and in the colors of the hindmost shanks. It is found in the same places, and at the same time, as the barren-ground locust.

7. *Locusta marmorata.*[15] Marbled Locust. (Fig. 81.)

Ash-colored, variegated with pale yellow and black; thorax suddenly narrowed before the middle, and the slightly elevated longitudinal line on the top is cut through in the middle by a transverse fissure; wing-covers marbled with large whitish and black spots, and semi-transparent at the end; wings light yellow on the half next to the body, transparent near the end, with two black spots on the tip, and a broad intermediate black band, which, narrowed and curving inwards on the hind margin, nearly reaches the inner angle; hindmost thighs pale yellow, black at the extremity, and nearly

Fig. 81.

[[15] *L. marmorata* must be referred to *Œdipoda*. — UHLER.]

surrounded by two broad black bands; hind shanks coral-red, with a black ring immediately below the knee, and followed by a white ring, black at the lower extremity also, with the tips of the spines black. In some individuals there is an additional black ring below the white one on the shanks. Length from $\frac{7}{10}$ to above $\frac{9}{10}$ inch; exp. $1\frac{4}{10}$ to $1\frac{8}{10}$ inch.

The marbled locust, which is one of our prettiest species, is found in the open places contiguous to or within pitch-pine woods, flying over the scanty grass and reindeer moss which not unfrequently grow in these situations. It is marked on the wings somewhat like the barren-ground locust, but is invariably smaller, with the thorax much more contracted before the middle. It appears, in the perfect state, from the middle of July to the middle of October.

8. *Locusta eucerata.*[16] Long-horned Locust.

Ash-colored, variegated with gray and dark brown; antennæ nearly as long as the body, and with flattened joints; thorax very much pinched or compressed laterally before the middle, with a slightly elevated longitudinal line, which is interrupted by two notches; wing-covers and wings long and narrow; the former variegated with dusky spots, and semi-transparent at tip; wings next to the body yellow, sometimes pale, sometimes deep and almost orange-colored, at other times uncolored and semi-transparent; with a broad black band across the middle, which is narrowed and prolonged on the hinder margin, and extends quite to the inner angle; beyond the band the wings are transparent, with the tips black or covered with blackish spots; hindmost shanks whitish, with a black ring at each end, a broad one of the same color just above the middle, and the spines tipped with black. Length $\frac{1}{2}$ inch to $\frac{7}{10}$ inch; exp. $1\frac{3}{10}$ inch to more than $1\frac{1}{2}$ inch.

The wings of this species are very variable in color at the base. The *fenestralis* described by M. Serville has the base

[16 *L. eucerata* must be referred to *Œdipoda.* — UHLER.]

of the wings vermilion-red, but in other respects it approaches to this species. The long-horned locust is found oftentimes in company with the marbled species, and also near sea-beaches with the maritime locust, from the last of July to the middle of October.

9. *Locusta nebulosa.*[17] Clouded Locust.

Dusky brown; thorax with a slender keel-like elevation, which is cut across in the middle by a transverse fissure; wing-covers pale, clouded, and spotted with brown; wings transparent, dusky at tip, with a dark brown line on the front margin; hindmost shanks brown, with darker spines, and a broad whitish ring below the knees. Length from $\frac{8}{10}$ inch to more than $1\frac{2}{10}$ inch; exp. from $1\frac{1}{2}$ inch to more than 2 inches.

A very common species, and easily known by its clouded wing-covers and colorless wings. It abounds in pastures, and even in corn-fields and gardens, during the months of September and October, at which time it is furnished with wings and may often be seen paired or busied in laying eggs. It does not appear to have been described before.

The three following locusts differ from the preceding in having the antennæ shorter than the thorax, and slightly thickened towards the end, and the face somewhat oblique, the mouth being nearer the breast than in our other species of Locusta; and they seem to constitute a distinct group or sub-genus, which may receive the name of *Tragocephala*,[18] or goat-headed locusts.

10. *Locusta* (*Tragocephala*) *infuscata.* Dusky Locust.

Dusky brown; thorax with a slender keel-like elevation; wing-covers faintly spotted with brown; wings transparent, pale greenish yellow next to the body, with a large dusky

[17 *L. nebulosa* must be referred to *Œdipoda*. — UHLER.]

[18 *Tragocephala* is synonymous with *Gomphocerus*, and *L infuscata*, *L. viridifasciata*, and *L. radiata* must be referred to it. — UHLER.]

cloud near the middle of the hind margin, and a black line on the front margin; hind thighs pale, with two large black spots on the inside; hind shanks brown, with darker spines, and a broad whitish ring below the knees. Length $\frac{3}{4}$ inch; exp. above $1\frac{1}{2}$ inch.

This somewhat resembles the clouded locust, from which, however, it is easily distinguished by its much shorter antennæ and the dusky cloud on the hinder margin of the wings. I have captured it in pastures, in the perfect state, from the middle of May to near the end of July. I believe that it has never been described before.

11. *Locusta* (*Tragocephala*) *viridi-fasciata*. Green-striped Locust. (Plate III. Fig. 2.)

Green; thorax keeled above; wing-covers with a broad green stripe on the outer margin extending from the base beyond the middle and including two small dusky spots on the edge, the remainder dusky but semi-transparent at the end; wings transparent, very pale greenish yellow next to the body, with a large dusky cloud near the middle of the hind margin, and a black line on the front margin; antennæ, fore and middle legs reddish; hind thighs green, with two black spots in the furrow beneath; hind shanks blue-gray, with a broad whitish ring below the knees, and the spines whitish, tipped with black. Length about 1 inch; exp. from more than $1\frac{3}{4}$ to nearly 2 inches.

This insect is the *Acrydium viridi-fasciatum* of De Geer, who was the first describer of it, the *Gryllus Virginianus* of Fabricius, the *Gryllus Locusta chrysomelas* of Gmelin, the *Acrydium marginatum* of Olivier, and the *Acridium hemipterum* of Palisot de Beauvois. It is remarkable that a species so strongly marked as this is should have been so profusely named. Palisot de Beauvois seems to have selected the most appropriate name for it; for the green portion of the wing-covers is thick and opaque, and the dusky portion thin and semi-transparent, as in the wing-covers of Hemipterous insects. It is very common in pastures and mowing lands from the first of June to the middle of August, being found in various states of maturity throughout this period. The young also appear still earlier, and are readily known by their green color, and large compressed thorax, which is arched and crested or keeled above, and by their very short and flattened antennæ. These locusts are sometimes very troublesome in gardens, living upon the leaves of vegetables and flowers, and attacking the buds and half-expanded petals. The larvæ or young survive the winter, sheltered among the roots of grass and under leaves.

12. *Locusta* (*Tragocephala*) *radiata.* Radiated Locust.

Rust-brown; thorax keeled above; wing-covers entirely brown, but semi-transparent at the end; wings transparent, with brown network, and the principal longitudinal veins black; they are very faintly tinted with green next to the body, have a large dusky cloud near the middle of the hind margin, and a brown streak on the front margin; hind shanks reddish brown, a little paler below the knees, and the spines tipped with black. Length about 1 inch; exp. from $1\frac{3}{4}$ to 2 inches.

This species is now for the first time described. It seems to be rare. I captured one specimen in Cambridge on the 1st of July, and have received another from Dr. D. S. C. H. Smith of Sutton, Massachusetts. It is found in North Carolina as early as the month of May in the perfect state.

The following species have the face still more oblique than the foregoing, but the antennæ are much longer, particularly in the males, in which they nearly equal the body in length, and are not enlarged towards the end. The eyes are oval and oblique, and there is a deep hollow before each of them for the reception of the first joint of the antennæ. The thorax is not crested or keeled, but is flattened above, with three slender threadlike elevated lines, and the hind margin is very nearly transverse, or not much (if at all) angulated behind. The wing-covers and wings are extremely short. The hind legs are long and slender. I propose therefore to separate these species from the other locusts under a subgenus by the name of *Chloëaltis*, derived from the Greek, and signifying a grasshopper.

13. *Locusta* (*Chloëaltis*) *conspersa*. Sprinkled Locust.

Light bay, sprinkled with black spots; a black line on the head behind each eye, extending on each side of the thorax on the lateral elevated line; wing-covers oblong-oval, pale yellowish brown, with numerous small darker brown spots; wings about three twentieths of an inch long, transparent, with dusky lines at the tip; hind shanks pale red, with the spines black at the end. Length nearly $\frac{9}{10}$ inch.

This may be merely a variety of the following species, though very differently colored.

14. *Locusta* (*Chloëaltis*) *abortiva*. Abortive Locust.

Brown; wing-covers with dark brown veins and confluent spots, covering two thirds of the abdomen; wings three twentieths of an inch long, transparent, with dusky lines at the tip; hind margin of the thorax straight; hind shanks coral-red, whitish just below the knees, the spines tipped with black. Length nearly $\frac{9}{10}$ inch.

This and the preceding locust have much the appearance of pupæ or young insects; nevertheless I believe that their wings and wing-covers never become larger, and Mr. Leonard informs me that they are found paired. I have captured the abortive locust in pastures near the end of July.

15. *Locusta* (*Chloëaltis*) *curtipennis*. Short-winged Locust.
(Plate III. Fig. 1.)

Olive-gray above, variegated with dark gray and black; legs and body beneath yellow; a broad black line extends from behind each eye on the sides of the thorax; wing-covers, in the male, as long as the abdomen, in the female, covering two thirds of the abdomen; wings rather shorter than the wing-covers, transparent, and faintly tinged with yellow; hinder knees black; spines on the hind shanks tipped with black. Length from $\frac{1}{2}$ to more than $\frac{8}{10}$ inch; exp. from $\frac{7}{10}$ to nearly 1 inch.

The flight of the short-winged locust is noiseless and short, but it leaps well. Great numbers of these insects are found in our low meadows, in the perfect state, from the first of August till the middle of October. They are easily distinguished from other locusts by their short and narrow wings, by the yellow color of the body beneath, and by the yellow legs and black knees.

III. Tetrix. *Grouse-locust.*

The Greeks applied the name of *Tetrix* to some kind of grouse, probably the heath-cock of Europe, and Latreille adopted it for a genus of locusts in which, perhaps, he fancied some resemblance to the bird in question. Linnæus placed these locusts in a division of his genus Gryllus, which he called *Bulla*, a name that ought to have been retained for them. The principal distinguishing characters of the genus have already been given, and I will only add that the body is broadest between the middle legs, narrows gradually to a point behind, and very abruptly to the head, which is much smaller than in the other locusts. The wings are large, forming nearly the quadrant of a circle, thin and delicate, and scalloped on the edge; when not in use they are folded beneath the projecting thorax. The four boring appendages of the females are notched on their edges with fine teeth, like a saw. Latreille and Serville have stated that the antennæ consist of only thirteen or fourteen joints; but some of our native species have twenty-two joints in the antennæ. Upon this variation I would arrange those now to be described in two groups.

I. *Antennæ* 14-*jointed; eyes very prominent, with a projecting ridge between them, formed by a horizontal extension of the flat top of the head; thorax prolonged beyond the extremity of the body.*

1. *Tetrix ornata.* Ornamented Grouse-locust.

Dark ash-colored; a large white patch between four black spots on the top of the thorax; a white spot on the top of the hind thighs; thorax nearly or quite as long as the wings. Length $\frac{11}{20}$ to $\frac{6}{10}$ inch to the apex of the thorax.

This species varies in wanting the white spot on the top of the thorax sometimes. It was first described by Mr. Say, under the name of *Acrydium ornatum.**

2. *Tetrix dorsalis.* Red-spotted Grouse-locust.

Rusty black, with ochre-yellow spots on the sides and legs, and a large rusty-red spot on the top of the thorax; wings extending beyond the apex of the thorax. Length $\frac{1}{2}$ inch.

3. *Tetrix quadrimaculata.* Four-spotted Grouse-locust.

Ash-colored or dark gray above, variegated with black; four velvet-black spots on the top of the thorax; wings projecting beyond the extremity of the thorax. Length from $\frac{4}{10}$ to $\frac{5}{10}$ of an inch.

This is a shorter and thicker species than the *ornamented grouse-locust.* It is not uncommon in pastures from the first of May to the first of June.

4. *Tetrix bilineata.* Two-lined Grouse-locust.

Ash-colored; thorax paler, with a narrow angular whitish line, on each side, extending from the head beyond the middle; the angular portion including a long blackish patch on each side; wings, in the male, rather shorter than the thorax, in the female longer. Length from $\frac{7}{20}$ to more than $\frac{9}{20}$ inch.

* American Entomology, Vol. I. Plate 5.

5. *Tetrix sordida.* Sordid Grouse-locust.

Yellowish ash-colored; thorax with minute elevated black points; wings, in both sexes, rather longer than the thorax. Length from $\frac{9}{20}$ inch to nearly $\frac{1}{2}$ inch.

I have taken this species both in May and September, and have received a specimen from Dr. D. S. C. H. Smith, of Sutton, Massachusetts.

II. *Antennæ 22-jointed; eyes hardly prominent, top of the head not horizontal between them, but curving towards the front, with a very slightly projecting ridge; wings smaller than in those of the preceding group.*

6. *Tetrix lateralis.* Black-sided Grouse-locust.

Pale brown; sides of the body blackish; thorax yellowish clay-colored, shorter than the wings, but longer than the body; wing-covers with a small white spot at the tips; male with the face and the edges of the lateral margins of the thorax yellow. Length from $\frac{9}{20}$ to $\frac{6}{10}$ of an inch.

This species was first described by Mr. Say under the name of *Acrydium laterale.** I have taken it from the middle of April to the middle of May. It varies in being darker above sometimes.

7. *Tetrix parvipennis.* (Fig. 82.) Small-winged Grouse-locust.

Dark brown; sides blackish; thorax clay-colored or pale brown, about as long as the body; wing-covers with a small white spot at the tips; wings much shorter than the thorax; male with the face and the edges of the lateral margins of the thorax yellow. Length from $\frac{7}{20}$ to more than $\frac{9}{20}$ inch.[19]

Fig. 82.

This species is much shorter and thicker than the *Tetrix lateralis.* I have taken it in April and May, in the perfect state, and have found the pupæ near the end of July.

* American Entomology, Vol. I. Plate 5.

[19 Color and style of marking is of very little value in separating the species of *Tetrix*, and the species described by Dr. Harris are probably all referable to the two species of Say. — UHLER.]

The habits of the grouse-locusts are said to be absolutely the same as those of other locusts. They seem, however, to be more fond of heat, being generally found in grassy places, on banks, by the sides of the road, and even on the naked sands, exposed to the full influence of the sun throughout the day. They are extremely agile, and consequently very difficult to capture, for they leap to an astonishing distance, considering their small size, being moreover aided in this motion by their ample wings. The young, which are deprived of wings, are generally found about midsummer, and are readily distinguished by the thorax, which is somewhat like a reversed boat, being furnished with a longitudinal ridge or keel from one end to the other. These little locusts are analogous to the insects belonging to the genus *Membracis* in the order HEMIPTERA, which also are distinguished by a very large thorax covering the whole of the upper side of the body, small wing-covers, and have the faculty of making great leaps. Indeed, these two kinds of insects very naturally connect the orders Orthoptera and Hemiptera together.

After so much space has been devoted to an account of the ravages of grasshoppers and locusts, and to the descriptions of the insects themselves, perhaps it may be expected that the means of checking and destroying them should be fully explained. The naturalist, however, seldom has it in his power to put in practice the various remedies which his knowledge or experience may suggest. His proper province consists in examining the living objects about him with regard to their structure, their scientific arrangement, and their economy or history. In doing this, he opens to others the way to a successful course of experiments, the trial of which he is generally obliged to leave to those who are more favorably situated for their performance.

In the South of France the people make a business, at certain seasons of the year, of collecting locusts and their eggs, the latter being turned out of the ground in little masses cemented and covered with a sort of gum in which they are enveloped by the insects. Rewards are offered and paid for their collection, half a franc being given for a kilogramme (about 2 lb. 3¼ oz. avoirdupois) of the insects, and a quarter of a franc for the same weight of their eggs. At this rate twenty thousand francs were paid in Marseilles, and twenty-five thousand in Arles, in the year 1613; in 1824, five thousand five hundred and forty-two, and in 1825, six thousand two hundred francs were paid in Marseilles. It is stated that an active boy can collect from six to seven kilogrammes (or from 13 lb. 3 oz. 13.22 dr. to 15 lb. 7 oz. 2.09 dr.) of eggs in one day. The locusts are taken by means of a piece of stout cloth, carried by four persons, two of whom draw it rapidly along, so that the edge may sweep over the surface of the soil, and the two others hold up the cloth behind at an angle of forty-five degrees.* This contrivance seems to operate somewhat like a horse-rake, in gathering the insects into winrows or heaps, from which they are speedily transferred to large sacks.

A somewhat similar plan has been successfully tried in this country, as appears by an account extracted from the "Portsmouth Journal," and published in the "New England Farmer."† It is there stated that, in July, 1826, Mr. Arnold Thompson, of Epsom, New Hampshire, caught, in one evening, between the hours of eight and twelve, in his own and his neighbor's grain-fields, five bushels and three pecks of grasshoppers, or more properly locusts. "His mode of catching them was by attaching two sheets together, and fastening them to a pole, which was used as the front part of the drag. The pole extended beyond the width of the sheets, so as to admit persons at both sides to draw it forward. At the sides of the drag, braces extended from the pole to raise the back part considerably from the ground, so that the grasshoppers could not escape. After running the drag about a dozen rods with rapidity, the braces were taken out, and

* See Annales de la Société Entomologique de France, Vol. II. pp. 486 – 489.

† Vol. V. p. 5.

the sheets doubled over; the grasshoppers were then swept from each end towards the centre of the sheet, where was left an opening to the mouth of a bag which held about half a bushel; when deposited and tied up, the drag was again opened and ready to proceed. When this bag was filled so as to become burdensome (their weight is about the same as that of the same measure of corn), the bag was opened into a larger one, and the grasshoppers received into a new deposit. The drag can be used only in the evening, when the grasshoppers are perched on the top of the grain. His manner of destroying them was by dipping the large bags into a kettle of boiling water. When boiled, they had a reddish appearance, and made a fine feast for the farmer's hogs."

When these insects are very prevalent on our salt-marshes, it will be advisable to mow the grass early, so as to secure a crop before it has suffered much loss. The time for doing this will be determined by data furnished in the foregoing pages, where it will be seen that the most destructive species come to maturity during the latter part of July. If, then, the marshes are mowed about the first of July, the locusts, being at that time small and not provided with wings, will be unable to migrate, and will consequently perish on the ground for the want of food, while a tolerable crop of hay will be secured, and the marshes will suffer less from the insects during the following summer. This, like all other preventive measures, must be generally adopted, in order to prove effectual; for it will avail a farmer but little to take preventive measures on his own land, if his neighbors, who are equally exposed and interested, neglect to do the same.

Among the natural means which seem to be appointed to keep these insects in check, violent winds and storms may be mentioned, which sometimes sweep them off in great swarms, and cast them into the sea. Vast numbers are drowned by the high tides that frequently inundate our marshes. They are subject to be attacked by certain thread-like brown or blackish worms (*Filaria*), resembling in appearance those called horse-hair eels (*Gordius*). I have taken three or four of these animals out of the body of a single locust. They are also much infested by little red mites, belonging apparently to the genus *Ocypete;* these so much weaken the insects, by sucking the juices from their bodies, as to hasten their death. Ten or a dozen of these mites will frequently be found pertinaciously adhering to the body of a locust, beneath its wing-covers and wings. A kind of sand-wasp preys upon grasshoppers, and provisions her nest with them. Many birds devour them, particularly our domestic fowls, which eat great numbers of grasshoppers, locusts, and even crickets. Young turkeys, if allowed to go at large during the summer, derive nearly the whole of their subsistence from these insects.

CHAPTER IV.

HEMIPTERA.

Bugs. — Squash-Bug. — Chinch-Bug. — Plant-Bugs. — Harvest-Flies. — Tree-Hoppers. — Leaf-Hoppers. — Vine-Hopper. — Bean-Hopper. — Thrips. — Plant-Lice. — American Blight. — Enemies of Plant-Lice. — Bark-Lice.

THE word bug seems originally to have been used for any frightful object, whether real or imaginary, whose appearance was to be feared at night. It was applied in the same sense as bugbear, and also as a term of contempt for something disagreeable or hateful. In later times it became, with the common people, a general name for insects, which, being little known, were viewed with dislike or terror. At present, however, we can say, with L'Estrange, though "we have a horror for uncouth monsters, upon experience all these bugs grow familiar and easy to us." We would except from this remark those domestic nocturnal species to which the name is now applied by way of pre-eminence; the real, by an easy transition in the use of language, having assumed the name of the imaginary objects of terror and disgust by night.

Entomologists now use the word bug for various kinds of insects, all, like the bed-bug, having the mouth provided with a slender beak, which, when not in use, is bent under the body, and lies upon the breast between the legs. This instrument consists of a horny sheath, containing, in a groove along its upper surface, three stiff bristles as sharp as needles. Bugs have no jaws, but live by sucking the juices of animals and plants, which they obtain by piercing them with their beaks. Although the domestic kinds above mentioned are without wing-covers and wings, yet most bugs have both, and, with the former, belong to an order called HEMIPTERA, literally half-wings, on account of the peculiar construction of their wing-covers, the hinder half of which is thin and filmy like the wings, while the fore part is thick and opaque. There are, however, other insects provided with the same kind of beak, but having the wing-covers sometimes entirely transparent, and sometimes more or less opaque, and these, by most entomologists, are also classed among Hemipterous insects, because they come much nearer to them than to any other insects, in structure and habits. Bugs, like other insects, undergo three changes, but they retain nearly the same form in all their stages; for the only transformation to which they are subject, from the young to the adult state, is occasioned by the gradual development of their wing-covers and wings, and the growth of their bodies, which make it necessary for them repeatedly to throw off their skins, to allow of their increase in size. Young, half-grown, and mature, all live in the same way, and all are equally active. The young come forth from the egg without wing-covers and wings, which begin to appear in the form of little scales on the top of their backs as they grow older, and increase in size with each successive moulting of the skin, till they are fully developed in the full-grown insect.

The Hemiptera are divided into two groups, distinguished by the following characters.

1. BUGS, or TRUE HEMIPTERA, (*Hemiptera heteroptera*,) in which the wing-covers are thick and opaque at the base, but thin and more or less transparent and wing-like at the tips, are laid horizontally on the top of the back, and cross each other obliquely at the end, so that the thin part of one wing-cover overlaps the same part of the other; the wings are also horizontal, and are not plaited; the head is more or less horizontal, and the beak issues from the fore part of it, and is abruptly bent backwards beneath the under side of the head

and the breast. Some of the insects belonging to this division live on animal, and others on vegetable juices.

2. HARVEST-FLIES, PLANT-LICE, and BARK-LICE, (*Hemiptera homoptera*,) in which the wing-covers are, as the scientific name implies, of one texture throughout, and are either entirely thin and transparent, like wings, or somewhat thicker and opaque; they are not horizontal, and do not cross each other at their extremities, but, together with the wings, are more or less inclined at the sides of the body, like the wing-covers of locusts; the face is either vertical, or slopes obliquely under the body, so that the beak issues from the under side of the head close to the breast. All the insects included in this division live on vegetable juices.

I. BUGS. (*Hemiptera heteroptera.*)

The hemipterous insects belonging to this division are various kinds of bugs, properly so called, such as squash-bugs, bed-bugs, fruit-bugs, water-bugs, water-boatmen, and many others, for which there are no common names in our language. In my Catalogue of the Insects of Massachusetts, the scientific names of ninety-five native species are given; but, as the mere description of these insects, unaccompanied by any details respecting their economy and habits, would not interest the majority of readers, and as I am not sufficiently prepared to furnish these details at present, I shall confine my remarks to two or three species only.

Fig. 83.

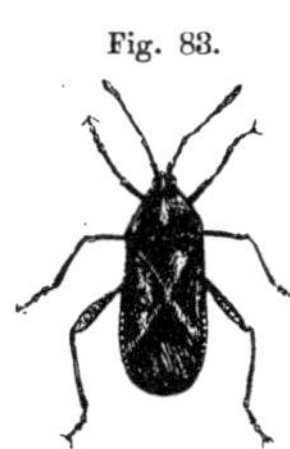

The common squash-bug, *Coreus tristis* (Fig. 83), so well known for the injurious effects of its punctures on the leaves of squashes, is one of the most remarkable of these insects. It was first described by De Geer, who gave it the specific name of *tristis*, from its sober color, which Gmelin unwarrantably changed to *mœstus*, having, however, the same meaning. Fabricius called it *Coreus rugator*, the latter word signifying one who wrinkles, which was probably applied to this insect because its punctures cause the leaves of the squash to become wrinkled. Mr. Say, not being aware that this insect had already been three times named and described, re-described it under the name of *Coreus ordinatus*. Of these four names, however, that of *tristis*, being the first, is the only one which it can retain. *Coreus*, its generical name, was altered by Fabricius from *Coris*, a word used by the Greeks for some kind of bug.

About the last of October squash-bugs desert the plants upon which they have lived during the summer, and conceal themselves in crevices of walls and fences, and other places of security, where they pass the winter in a torpid state. On the return of warm weather, they issue from their winter quarters, and when the vines of the squash have put forth a few rough leaves, the bugs meet beneath their shelter, pair, and immediately afterwards begin to lay their eggs. This usually happens about the last of June or beginning of July, at which time, by carefully examining the vines, we shall find the insects on the ground or on the stems of the vines, close to the ground, from which they are hardly to be distinguished on account of their dusky color. This is the place where they generally remain during the daytime, apparently to escape observation; but at night they leave the ground, get beneath the leaves, and lay their eggs in little patches, fastening them with a gummy substance to the under sides of the leaves. The eggs are round, and flattened on two sides, and are soon hatched. The young bugs are proportionally shorter and more rounded than the perfect insects, are of a pale ash-color, and have quite large antennæ, the joints of which are somewhat flattened. As they grow older and increase in size, after moulting their skins a few times, they become more oval in form, and the under side of their bodies gradually acquires a dull ochre-yellow color. They live together at first in little swarms or families beneath the leaves upon which they were hatched, and which, in consequence of the numerous punctures of the insects, and the quantity of sap imbibed by them, soon wither, and eventually become brown, dry, and wrinkled; when the insects leave them for fresh leaves, which they exhaust in the same way. As the eggs are not all laid at one time, so the bugs are hatched in successive broods, and consequently will be found in various stages of growth through the summer. They, however, attain their full size, pass through their last transformation, and appear in their perfect state, or furnished with wing-covers and wings, during the months of September and October. In this last state the squash-bug measures six tenths of an inch in length. It is of a rusty black color above, and of a dirty ochre-yellow color beneath, and the sharp lateral edges of the abdomen, which project beyond the closed wing-covers, are spotted with ochre-yellow. The thin overlapping portion of the wing-covers is black; the wings are transparent, but are dusky at their tips; and the upper side of the abdomen, upon which the wings rest when not in use, is of a deep black color, and velvety appearance.

The ground-color of this insect is really ochre-yellow, and the rusty black hue of the head, thorax, thick part of the wing-covers, and legs, is occasioned by numerous black punctures, that, on the head, are arranged in two broad black longitudinal lines, between which, as well as on the margin of the thorax, the yellow is distinctly to be seen. On the back part of the head of this bug, and rather behind the eyes, are two little glassy elevated spots, which are called eyelets, and which are supposed to enable the insect to see distant objects above it, while the larger eyes at the sides of the head are for nearer objects around it. Eyelets are also to be found in grasshoppers, locusts, and many other insects. In some of our species of *Coreus* there is a little thorn at the base of the antennæ, the legs are also thorny on the under side, and the hindmost thighs are much thicker than the others; but none of these characters are found in squash-bugs.* When handled, and still more when crushed, the latter give out an odor

* They appear to belong to the genus *Gonocerus* of Burmeister.

precisely similar to that of an over-ripe pear, but far too powerful to be agreeable.

In order to prevent the ravages of these insects, they should be sought and killed when they are about to lay their eggs; and if any escape our observation at this time, their eggs may be easily found and crushed. With this view the squash-vines must be visited daily, during the early part of their growth, and must be carefully examined for the bugs and their eggs. A very short time spent in this way every day, in the proper season, will save a great deal of vexation and disappointment afterwards. If this precaution be neglected or deferred till the vines have begun to spread, it will be exceedingly difficult to exterminate the insects, on account of their numbers; and if at this time dry weather should prevail, the vines will suffer so much from the bugs and drought together, as to produce but little if any fruit. Whatever contributes to bring forward the plants rapidly, and to promote the vigor and luxuriance of their foliage, renders them less liable to suffer by the exhausting punctures of the young bugs. Water drained from a cow-yard, and similar preparations, have, with this intent, been applied with benefit.

The wheat-fields and corn-fields of the South and West often suffer severely from the depredations of certain minute bugs, long known there by the name of chinch-bugs, which fortunately have not yet been observed in New England.* It is not improbable, however, that they may spread in this direction, and attack our growing grain and other crops. In anticipation of such a sad event, and to gratify a curiosity that has been expressed concerning these offensive insects, I venture to offer a few remarks upon them. Attention seems early to have been directed to them. They are mentioned in the eleventh volume of Young's "Annals of Agriculture," published, I believe, about 1788. From this work Messrs. Kirby and Spence probably obtained the following account,

* While this sheet is passing through the press, I have to record the discovery of one of these bugs in my own garden, on the 17th of June, 1852.

contained in the first volume of their interesting "Introduction to Entomology." "America suffers in its wheat and maize from the attack of an insect, which, for what reason I know not, is called the chinch-bug fly. It appears to be apterous, and is said in scent and color to resemble the bed-bug. They travel in immense columns from field to field, like locusts, destroying everything as they proceed; but their injuries are confined to the States south of the 40th degree of north latitude. From this account," add Kirby and Spence, "the depredator here noticed should belong to the tribe *Geocorisæ*, Latr.; but it seems very difficult to conceive how an insect that lives by suction, and has no mandibles, could destroy these plants so totally."

Fig. 84

I have ascertained, from an examination of living specimens, that the chinch-bug is the *Lygæus Leucopterus* (Fig. 84), or white-winged Lygæus, described by Mr. Say, in December, 1831, in a rare little pamphlet on the "Heteropterous Hemiptera of North America." It appears, moreover, to belong to the modern genus *Rhyparochromus*. In its perfect state it is not apterous, but is provided with wings, and then measures about three twentieths of an inch in length. It is readily distinguished by its white wing-covers, upon each of which there is a short central line and a large marginal oval spot of a black color. The rest of the body is black and downy, except the beak, the legs, the antennæ at base, and the hinder edge of the thorax, which are reddish yellow, and the fore part of the thorax, which has a grayish lustre. The young and wingless individuals are at first bright red, changing with age to brown and black, and are always marked with a white band across the back. It is a mistake that these insects are confined to the States south of the 40th degree; for I have been favored with them by Professor Lathrop, of Beloit College, Wisconsin, and by Dr. Le Baron, of Geneva, Illinois. The latter gentleman had no difficulty in obtaining a sufficient number without going out of his own garden. The eggs of the chinch-bug are laid in the ground, in which the young have been found, in great abundance, at the depth of an inch or more. They make their appearance on wheat about the middle of June, and may be seen in their various stages of growth on all kinds of grain, on corn, and on herds-grass, during the whole summer. Some of them continue alive through the winter in their places of concealment. A very good account of these destructive bugs, with an enlarged figure, will be found in the "Prairie Farmer," for December, 1845. In the same publication, for September, 1850, there is an excellent description of the chinch-bug, by Dr. Le Baron, who, not being aware that it had been previously named by Mr. Say, called it *Rhyparochromus devastator*.

During the summer of 1838, and particularly in the early part of the season, which, it will be recollected, was very dry, our gardens and fields swarmed with immense numbers of little bugs, that attacked almost all kinds of herbaceous plants. My attention was first drawn to them in consequence of the injury sustained by a few dahlias, marigolds, asters, and balsams, with which I had stocked a little border around my house. In the garden of my friends the Messrs. Hovey, at Cambridge Port, I observed, about the same time, that these insects were committing sad havoc, and was informed that various means had been tried to destroy or expel them without effect. On visiting my potato-patch shortly afterwards, I found the insects there also in great numbers on the vines; and, from information worthy of credit, am inclined to believe that these insects contributed, quite as much as the dry weather of that season, to diminish the produce of the potato-fields in this vicinity. They principally attacked the buds, terminal shoots, and most succulent growing parts of these and other herbaceous plants, puncturing them with their beaks, drawing off the sap, and, from the effects subsequently visible, apparently poisoning the parts attacked.

These shortly afterwards withered, turned black, and in a few days dried up ; or curled, and remained permanently stunted in their growth. Early in the morning the bugs would be found buried among the little expanding leaves of the growing extremities of the plants, at which time it was not very difficult to catch them ; but, after being warmed by the sun, they became exceedingly active, and, on the approach of the fingers, would loose their hold, and either drop suddenly or fly away. Sometimes, too, when on the stem of a plant, they would dodge round to the other side, and thus elude our grasp. In July, 1851, some of these insects were sent to me by a gentleman, who brought them from St. Johnsbury, Vt., where they were confidently believed to be the cause of the *potato-rot.*

This kind of bug is the *Phytocoris lineolaris* [1] (Fig. 85), a variety of which was first described and figured by Palisot de Beauvois, under the specific name above given, and was doubtingly referred by him to the genus *Coreus;* and it was subsequently described by Mr. Say, who called it *Capsus oblineatus.* All the insects belonging to the genus *Phytocoris* * (which means plant-bug) are found on plants, and subsist on their juices, which they obtain by suction through their sharp beaks. They are easily distinguished from other bugs by the following characters. Eyelets wanting; antennæ four-jointed, with the first and second joints much thicker than the last two, which are very slender and threadlike ; the head short and triangular; the body oval, flattened, and soft ; the thorax in the form of a broad triangle, with the tip of the anterior angle cut off, and the broadest side applied to the base of the wing-covers ; the latter, when folded, cover the whole of the abdomen, and their thin portions have only one

[[1] Dr. Harris misquotes Beauvois for this *Phytocoris;* the name applied by that author is *P. linearis,* not *lineolaris.* — UHLER.]

* This new genus, or sub-genus, was instituted by Fallén, and is not noticed by Latreille and Laporte. It differs ftom *Capsus* chiefly in having a smaller head, and the thorax wider behind, and narrower before, than in the latter genus.

or two little veins; the legs are slender, and the shanks are bristled with little points. There are, in Massachusetts, a good many species belonging to this genus; but, in my Catalogue of the insects of this Commonwealth, they are included among the species of *Capsus*, which, indeed, they closely resemble.

Fig. 85.

The *Phytocoris lineolaris* (Fig. 85), or little-lined plant-bug, measures one fifth of an inch, or rather more, in length. It is an exceedingly variable species. The males are generally much darker than the females, being very deep livid brown or almost black above. The head is yellowish, with three narrow longitudinal reddish stripes; the first joint of the antennæ, the terminal half of the second, and the last two joints are blackish; the beak is more than one third the whole length of the body, when folded beneath the breast, extends to the middle pair of legs, and is of a yellowish color, ringed with black; the thorax, or that part of the body that comes immediately behind the head, is thickly covered with punctures, has a yellow margin, and five longitudinal yellow lines upon it, which often disappear on the back part; the scutel, or escutcheon, a small triangular piece behind the thorax, and interposed between the bases of the wing-covers, is also margined with yellow, and has a yellow spot upon it in the form of the letter V, which is often imperfect, so that only three small yellow spots are visible in the place of the three extremities of the letter; the thick part of the wing-covers is brown, with the outer edge and the longitudinal veins sometimes pale or yellowish, and behind this thick part there is a large yellowish spot, on the posterior tip of which is a small black point; the thin or membranous part of the wing-covers is shaded with dusky clouds; the under side of the body is marked with a yellowish line or a longitudinal series of yellow spots on each side of the middle; the legs are dirty brownish yellow, the thighs blackish at base, and with two black rings near the tip, and the extremities of the feet are blackish. The females are most often of a pale olive-green, or of a dirty greenish-yellow color; the thorax spotted and more or less distinctly striped with black, and the thick part of the wing-covers also variegated with dusky or brownish lines and clouds. In both sexes, however, the yellow V, or the three spots on the thorax, and the large yellow spot tipped with black on the wing-covers, are conspicuous characters, which readily afford the means of identifying the species. I have taken this insect in the spring, as early as the 20th of April, and in the autumn, as late as the middle of October; from which I infer that it passes the winter in the perfect state in some place of security. It is most abundant during the months of June and July. Specimens have been sent to me from Maine, New York, North Carolina, and Alabama, and Mr. Say records its occurrence in Pennsylvania, Indiana, the Northwest Territory, and Missouri. It seems, therefore, to be very generally diffused throughout the Union.

The history of this species is yet imperfect. We know not where and when the eggs are laid; the young have not been observed; and the insects, during the early periods of their existence, have escaped notice, and are only known to us after they have completed their final transformations. It is possible that further information upon the history of these insects may afford some aid in devising proper remedies against their ravages. Upon a limited scale, as on plants growing in our gardens, may be tried the effect of sprinkling them with alkaline solutions, such as strong soap-suds, or potash-water, or with decoctions of tobacco and of walnut-leaves, or of dusting the plants with air-slacked lime or sulphur. But in field husbandry such applications would be impracticable. I am inclined to believe that nothing will prove so effectual as thorough irrigation, or copious and frequent showers of rain, which will bring forward the plants with such rapidity, that they will soon become so strong and vigorous as to withstand the attacks of these little bugs. The great increase of these and other noxious insects may fairly be attributed to the exterminating war which has wantonly been waged upon our insect-eating birds, and we may expect the evil to increase unless these little friends of the farmer are protected, or left undisturbed to multiply, and follow their natural habits. Meanwhile, some advantage may be derived from encouraging the breed of our domestic fowls. A flock of young chickens or turkeys, if suffered to go at large in a garden, while the mother is confined within their sight and hearing, under a suitable crate or cage, will devour great numbers of destructive insects; and our farmers should be urged to pay more attention than heretofore to the rearing of chickens, young turkeys, and ducks, with a view to the benefits to be derived from their destruction of insects.

II. HARVEST-FLIES, &c. (*Hemiptera Homoptera.*)

By many entomologists this division is raised to the rank of a separate order, under the name of HOMOPTERA; but the insects arranged in it are, as already stated, much more like the true HEMIPTERA, or bugs, than they are to the insects in any other order, which shows the propriety of keeping these two divisions together, and that separately they hold only a subordinate importance compared with other orders.

The insects belonging to this division are divided by naturalists into three large groups, or tribes.

1. Harvest-flies, or Cicadians (CICADADÆ); having short antennæ, which are awl-shaped or tipped with a little bristle; wings and wing-covers, in both sexes, inclined at the sides of the body; three joints to their feet; firm and hard skins; and in which the females have a piercer, lodged in a furrow beneath the extremity of the body.

2. Plant-lice (APHIDIDÆ); having antennæ longer than the head, and threadlike or tapering from the root to the end; wing-covers and wings frequently wanting in the females; feet two-jointed; the body very soft, generally furnished with two little tubercles at the end; no piercer in the females.

3. Bark-lice (COCCIDÆ); having threadlike or tapering antennæ, longer than the head; the males alone provided with wings, which lie horizontally on the top of the back; no beak in this sex; females wingless, but furnished with beaks; the feet with only one joint, terminated by a single claw; skins tolerably firm and hard; two slender threads at the extremity of the body; no piercer in the females.

1. HARVEST-FLIES. (*Cicadadæ.*)

The most remarkable insects in this group are those to which naturalists now apply the name of *Cicada.* They are readily distinguished by their broad heads, the large and very convex eyes on each side, and the three eyelets on the crown; by the transparent and veined wing-covers and wings; and by the elevation on the back part of the thorax in the form of the letter X. The males have a peculiar organization, which enables them to emit an excessively loud buzzing kind of sound, which, in some species, may be heard at the distance of a mile; and the females are furnished with a curiously contrived piercer, for perforating the limbs of trees, in which they place their eggs. Without attempting a detailed description of the complicated mechanism of these parts, which could only be made intelligible by means of figures, I shall merely give a brief and general account of them, which may suffice for the present occasion. The musical instruments of the male consist of a pair of kettle-drums, one on each side of the body, and these, in the *seventeen-year Cicada* (or locust as it is generally but improperly called in America), are plainly to be seen just behind the wings. These drums are formed of convex pieces of parchment, gathered into numerous fine plaits, and, in the species above named, are lodged in cavities on the sides of the body behind the thorax. They are not played upon with sticks, but by muscles or cords fastened to the inside of the drums. When these muscles contract and relax, which they do with great rapidity, the drum-heads are alternately tightened and loosened, recovering their natural convexity by their own elasticity. The effect of this rapid alternate tension and relaxation is the production of a rattling sound, like that caused by a succession of quick pressures upon a slightly convex and elastic piece of tin plate. Certain cavities within the body of the insect, which may be seen on raising two large valves beneath the belly, and which are separated from each other by thin partitions having the transparency and brilliancy of mica, or of thin and highly polished glass, tend to increase the vibrations of the sounds, and add greatly to their intensity. In most of our species of *Cicada* the drums are not visible on the outside of the body, but are covered by convex triangular pieces on each side of the first ring behind the thorax, which must be cut away in order to expose them. On raising the large valves of the belly, however, there is seen, close to each side of the body, a little opening, like a pocket, in which the drum is lodged, and from which the sound issues when the insect opens the valves. The hinder extremity of the body of the female is conical, and the under side has a longitudinal channel for the reception of the piercer, which is furthermore protected by four short grooved pieces fixed in the sides of the channel. The piercer itself consists of three parts in close contact with each other; namely, two outer ones grooved on the inside and enlarged at the tips, which externally are beset with small teeth like a saw, and a central, spear-pointed borer, which plays between the other two. Thus this instrument has the power and does the work both of an awl and of a double-edged saw, or rather of two key-hole saws cutting opposite to each other. No species of *Cicada* possesses the power of leaping. The legs are rather short, and the anterior thighs are armed beneath with two stout spines.

The duration of life in winged insects is comparatively very short, seldom exceeding two or three weeks in extent, and in many is limited to the same number of days or hours.

To increase and multiply is their principal business in this period of their existence, if not the only one, and the natural term of their life ends when this is accomplished. In their previous states, however, they often pass a much longer time, the length of which depends, in great measure, upon the nature and abundance of their food. Thus maggots, which subsist upon decaying animal or vegetable matter, come more quickly to their growth than caterpillars and other insects which devour living plants; the former are appointed to remove an offensive nuisance, and do their work quickly; the latter have a longer time assigned to them, corresponding in some degree to the progress or continuance of vegetation. The facilities afforded for obtaining food influence the duration of life; hence those grubs that live in the solid trunks of perennial trees, which they are obliged to perforate in order to obtain nourishment, are longer lived than those that devour the tender parts of leaves and fruits, which last only for a season, and require no laborious efforts to be prepared for food. The harvest-flies continue only a few weeks after their final transformation, and their only nourishment consists of vegetable juices, which they obtain by piercing the bark and leaves of plants with their beaks; and during this period they lay their eggs, and then perish. They are, however, amply compensated for the shortness of their life in the winged state by the length of their previous existence, during which they are wingless and grub-like in form, and live under ground, where they obtain their food only by much labor in perforating the soil among the roots of plants, the juices of which they imbibe by suction. To meet the difficulties of their situation and the precarious supply of their food, for which they have to grope in the dark in their subterranean retreats, a remarkable longevity is assigned to them; and one species has obtained the name of *Cicada septendecim*, on account of its life being protracted to the period of seventeen years.

This insect has been observed in the southeastern parts of

Massachusetts, and in the valley of the Connecticut River, as far north at least as Hadley; but does not seem to have extended to other parts of the State. The earliest account that we have of it is contained in Morton's "Memorial," wherein it is stated that "there was a numerous company of flies, which were like for bigness unto wasps or bumblebees," which appeared in Plymouth in the spring of 1633. "They came out of little holes in the ground, and did eat up the green things, and made such a constant yelling noise as made the woods ring of them, and ready to deafen the hearers." Judge Davis, in the Appendix to his edition of Secretary Morton's "Memorial," states that these insects appeared in Plymouth, Sandwich, and Falmouth, in the year 1804; but, if the exact period of seventeen years had been observed, they should have returned in 1803. Circumstances may occasionally retard or accelerate their progress to maturity, but the usual interval is certainly seventeen years, according to the observations and testimony of many persons of undoubted veracity. Their occurrence in large swarms at long intervals, like that of the migratory locusts of the East, probably suggested the name of locusts, which has commonly been applied to them in this country. The following extract from a letter * from the late Rev. Ezra Shaw Goodwin, of Sandwich, contains some interesting particulars which this gentleman had the kindness to communicate to me.

"I have not been unmindful of what you said to me respecting the locust insects, nor of the promise I made you with respect to them. They appeared in this town in the year 1821, in the middle of June. Their last previous appearance was in 1804, and their last, previous to that, was in 1787. I ascertained these periods from the statements of individuals, who remembered that it was locust-year when this or that event occurred; as, when this one was married, or that one's eldest son was born; events, the date of which the husband or the parent would not be very likely

* Dated October 19, 1832.

to forget. The remembrance of all, though fixed by different events, concurred in establishing the same years for the appearance of the locusts.

"I first took notice of them in 1821, on the 17th of June, from their noise. They appeared chiefly in the forests, or in thickets of forest-trees, principally oak. Their nearest distance from my dwelling cannot be far from a mile; yet, at a still hour, their music was distinctly heard there. On going to visit them, I found the oak-trees and bushes swarming with them in a winged state. They came up out of the ground a creeping insect. Very soon after they had arrived on the surface of the earth, the skin, or rather the shell of the insect, burst upon the back, and the winged insect came forth, leaving the skin or shell upon the earth, in a perfect form, and uninjured, saving at the rupture on the back; showing an entire withdrawing of the living animal, as much so as does the snake's skin after he has left it. Thus these skins lay in immense numbers under the trees, entirely empty, and perfect in shape. The winged insects did not, so far as I could ascertain, eat anything. Motion and propagation appeared to be the whole object of their existence. They continued about four or five weeks, and then died." Previous to this event "the females laid their eggs in the tender parts of oak branches, near the extremities, making a longitudinal furrow, and depositing rows of eggs therein (Fig. 86). They then sawed the branch partly off below the eggs, so that the wind could twist off the extreme part containing the eggs, and let it fall to the ground. In this way they injured the trees extensively. The forest had a gloomy appearance from the number of these extremities partially twisted off, and hanging, with their dead leaves, ready to fall. In a few weeks they were nearly all separated from the trees, and carried their vital burdens to the earth, which was, certainly, well seeded for a harvest in 1838. I know of no other damage which they did. . . . . . I believe the locusts appear in different places, in different years, and understand that the locust-year, in some places not far distant, is different from their year in this town."

Fig. 86.

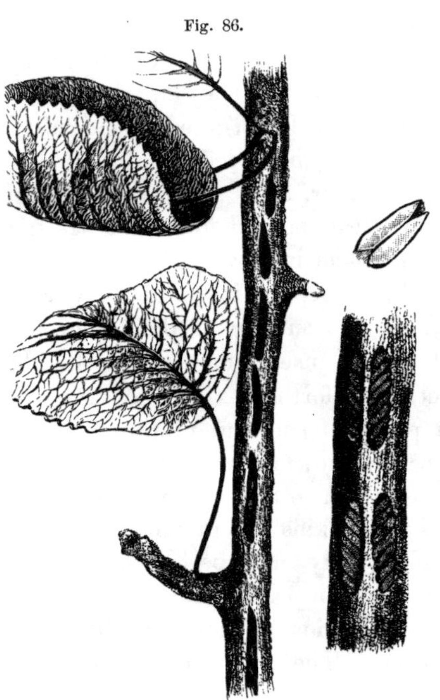

This letter was accompanied by specimens of the insects, in their various states, obtained and preserved by Mr. Goodwin.

The writer of an article in the "Boston Magazine" for November, 1784, observes that Mr. Morton must have been mistaken as to these insects, in saying that they eat up the green things, which from the structure of their mouths we now know could not have been the case. This writer also records the appearance of these insects in 1784, and the place of his residence, in which this occurred, is believed to have been in the County of Bristol; which coincides with the remark made by Mr. Goodwin, that in different places they appear in different years. This remark is furthermore confirmed by the observations of various persons * who have

* Among the authorities which I have consulted upon the history of the seventeen-year Cicada, may be mentioned the Rev. Andrew Sandel, of Philadelphia, an abstract of whose account is given in the 4th vol. of Mitchill and Miller's "Medical Repository," p. 71; the "Columbian Magazine," Vol. I., pages 86 and 108; Mr. Moses Bartram's account in Dodsley's "Annual Register" for 1767, p 103; Dr. McMurtrie, in the 8th vol. of the "Encyclopædia Americana," p. 43; Dr. S. P. Hildreth's interesting account in the 10th vol. of Silliman's "American Journal of Science," p. 327; and a pamphlet entitled "Notes on the Locusta," &c., with which I have been favored by the author, Professor Nathaniel Potter,

published accounts of the occurrence of these insects in the Middle, Southern, and Western States, where, at regular intervals of seventeen years, varying according to the locality, they are seen even in greater abundance than in Massachusetts. The following dates and places of their ascent are given in Professor Potter's "Notes on the Locusta decem Septima" (*Cicada septendecim*): Maryland, 1749, 1766, 1783, 1800, 1817, 1834; South Carolina and Georgia, 1817, 1834; Middlesex County, New Jersey, 1826; Louisiana, 1829; Gallipolis, Ohio, 1821, and Muskingum, 1829; western parts of Pennsylvania, 1832; Fall River, Massachusetts, 1834. To these may be added from other sources, Pennsylvania, 1715, 1766, 1783, 1800, 1817;* Marietta, Ohio, 1795, 1812; Plymouth, 1633, 1804; Sandwich, 1787, 1804, 1821; Hadley, 1818; Westfield, 1835; North Haven, Conn., 1724, 1741, 1758, 1792, 1809, 1826, 1843; Genesee County, New York, 1832; Martha's Vineyard, 1833. From information derived from various sources it appears that this species is widely spread over the country, with the exception only of the northern parts of New England; and that it may be seen in some portion of the United States almost every year; and, although certain disturbing causes may occasionally accelerate or retard the return of individuals, or even of an entire swarm, in any one place, yet the lineal descendants of one particular family or swarm will ordinarily come forth only once in seventeen years, while those of other swarms may appear, after equally regular intervals, in the intervening period, in other places.

of Baltimore. This last work is exclusively devoted to the history of this insect, and has afforded me much valuable information. From these various sources I have selected the principal facts which follow. Mr. Collins's "Observations on the Cicada of North America," published in the "Philosophical Transactions" of London, Vol. LIV. p. 65, with a plate, probably refer to the seventeen-year Cicada, but the insects figured are not the same, and seem to be the *Cicada pruinosa* of Mr. Say.

* A writer in the "United States Gazette" records the appearance of these insects in great numbers in Germantown, Pennsylvania, *on the 25th of May*, at four successive periods.

The seventeen-year Cicada (*Cicada septendecim* of Linnæus), (Plate III. Fig. 7,) in the winged state, is of a black color, with transparent wings and wing-covers, the thick anterior edge and larger veins of which are orange-red, and near the tips of the latter there is a dusky zigzag line in the form of the letter W; the eyes when living are also red; the rings of the body are edged with dull orange; and the legs are of the same color. The wings expand from 2½ to 3¼ inches.

In those parts of Massachusetts which are subject to the visitation of this Cicada, it may be seen in forests of oak about the middle of June. Here such immense numbers are sometimes congregated, as to bend and even break down the limbs of the trees by their weight, and the woods resound with the din of their discordant drums from morning to evening. After pairing, the females proceed to prepare a nest for the reception of their eggs. They select, for this purpose, branches of a moderate size, which they clasp on both sides with their legs, and then, bending down the piercer at an angle of about forty-five degrees, they repeatedly thrust it obliquely into the bark and wood in the direction of the fibres, at the same time putting in motion the lateral saws, and in this way detach little splinters of the wood at one end, so as to form a kind of fibrous lid or cover to the perforation. The hole is bored obliquely to the pith, and is gradually enlarged by a repetition of the same operation, till a longitudinal fissure is formed of sufficient extent to receive from ten to twenty eggs. The side-pieces of the piercer serve as a groove to convey the eggs into the nest, where they are deposited in pairs, side by side, but separated from each other by a portion of woody fibre, and they are implanted into the limb somewhat obliquely, so that one end points upwards. When two eggs have been thus placed, the insect withdraws the piercer for a moment, and then inserts it again and drops two more eggs in a line with the first, and repeats the operation till she has filled the fissure from one end to the other, upon which she removes to a little distance, and begins to make another nest to contain two more rows of eggs. She is about fifteen minutes in preparing a single nest and filling it with eggs; but it is not unusual for her to make fifteen or twenty fissures in the same limb; and one observer counted fifty nests extending along in a line, each containing fifteen or twenty eggs in two rows, and all of them apparently the work of one insect.*. After one limb is thus sufficiently stocked, the Cicada goes to another, and passes from limb to limb and from tree to tree, till her store, which consists of four or five hundred eggs, is exhausted. At length she becomes so weak by her incessant labors to provide for a succession of her kind, as to falter and fall in attempting to fly, and soon dies.

Although the Cicadas abound most upon the oak, they resort occasionally to other forest-trees, and even to shrubs, when impelled by the necessity for depositing their eggs, and not unfrequently commit them to fruit-trees, when the latter are in their vicinity. Indeed there seem to be no trees or shrubs that are exempted from their attacks, except those of the pine and fir tribes, and of these even the white cedar is sometimes invaded by them. The punctured limbs languish and die soon after the eggs which are placed in them are hatched; they are broken by the winds or by their own weight, and either remain hanging by the bark alone, or fall with their withered foliage to the ground. In this way orchards have suffered severely in consequence of the injurious punctures of these insects.

The eggs are one twelfth of an inch long, and one sixteenth of an inch through the middle, but taper at each end to an obtuse point, and are of a pearl-white color. The shell is so thin and delicate that the form of the included insect can be seen before the egg is hatched, which occurs, according to Dr. Potter, in fifty-two days after it is laid, but

* See also my communication in Downing's Horticulturist, Vol. III. p. 278, Dec., 1848.

Miss Morris says in forty-two days, and other persons say in fourteen days.

The young insect when it bursts the shell is one sixteenth of an inch long, and is of a yellowish-white color, except the eyes and the claws of the fore legs, which are reddish; and it is covered with little hairs. In form it is somewhat grub-like, being longer in proportion than the parent insect, and is furnished with six legs, the first pair of which are very large, shaped almost like lobster-claws, and armed with strong spines beneath. On the shoulders are little prominences in the place of wings; and under the breast is a long beak for suction. These little creatures when liberated from the shell are very lively, and their movements are nearly as quick as those of ants. After a few moments their instincts prompt them to get to the ground, but in order to reach it they do not descend the body of the tree, neither do they cast off themselves precipitately; but, running to the side of the limb, they deliberately loosen their hold, and fall to the earth. It seems, then, that they are not borne to the ground in the egg state by the limbs in which their nests are contained, but spontaneously make the perilous descent, immediately after they are hatched, without any clew, like that of the canker-worm, to carry them in safety through the air and break the force of their fall. The instinct which impels them thus fearlessly to precipitate themselves from the trees, from heights of which they can have formed no conception, without any experience or knowledge of the result of their adventurous leap, is still more remarkable than that which carries the gosling to the water as soon as it is hatched. In those actions that are the result of foresight, of memory, or of experience, animals are controlled by their own reason, as in those to which they are led by the use of their ordinary senses, or by the indulgence of their common appetites, they may be said to be governed by the laws of their organization; but in such as arise from special and extraordinary instincts, we see the most striking proofs of that creative wisdom which has implanted in them an unerring guide, where reason, the senses, and the appetites would fail to direct them. The manner of the young Cicadas' descent, so different from that of other insects, and seeming to require a special instinct to this end, would be considered incredible, perhaps, if it had not been ascertained and repeatedly confirmed by persons who have witnessed the proceeding. On reaching the ground the insects immediately bury themselves in the soil, burrowing by means of their broad and strong fore feet, which, like those of the mole, are admirably adapted for digging. In their descent into the earth they seem to follow the

Fig. 87

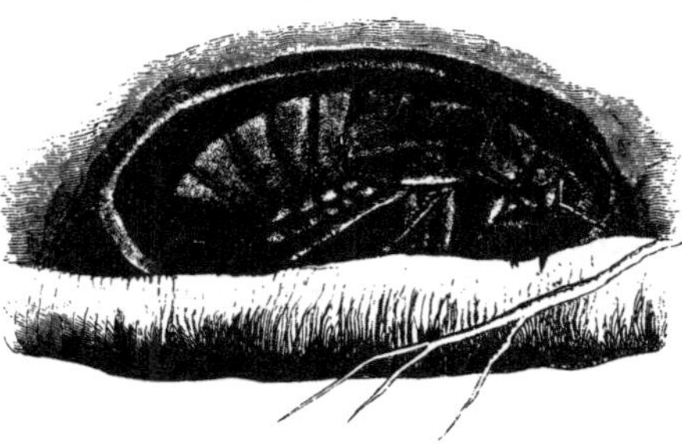

roots of plants, and are subsequently found attached to those which are most tender and succulent, perforating them with their beaks, and thus imbibing the vegetable juices which constitute their sole nourishment. (Fig. 87.)

Miss Margaretta H. Morris, who attributes the decline of the pear-tree and the failure of its fruits to depredations of the young Cicadas on its roots, has given interesting accounts of her observations upon these insects. On removing the earth from "a pear-tree that had been declining for years, without any apparent cause," she "found the larvæ of the Cicada in countless numbers clinging to the roots of the tree, with their suckers piercing the bark, and so deep and firmly placed, that they remained hanging for half an hour after being removed from the earth. From a root a yard long, and about an inch in diameter, she gathered twenty-three larvæ; they were of various sizes, from a quarter of an inch to an inch in length. They were on all the roots that grew deeper than six inches below the surface. The roots were unhealthy, and bore the appearance of external injury from small punctures. On removing the outer coat of bark, this appearance increased, leaving no doubt as to the cause of the disease." *

The grubs do not appear ordinarily to descend very deeply into the ground, but remain where roots are most abundant; and it is probable that the accounts of their having been discovered ten or twelve feet from the top of the ground have been founded on some mistake, or the occurrence of the insects at such a depth may have been the result of accident. The only alteration to which the insects are subject, during the long period of their subterranean confinement, is an increase of size, and the more complete development of the four small scale-like prominences on their backs, which represent and actually contain their future wings.

As the time of their transformation approaches, they gradually ascend towards the surface, making in their progress cylindrical passages, oftentimes very circuitous, and seldom exactly perpendicular, the sides of which, according to Dr. Potter, are firmly cemented and varnished so as to be waterproof. These burrows are about five eighths of an inch in diameter, are filled below with earthy matter removed by the insect in its progress, and can be traced by the color and compactness of their contents to the depth of from one to two feet, according to the nature of the soil; but the upper portion to the extent of six or eight inches is empty, and serves as a habitation for the insect till the period for its exit arrives. Here it remains during several days, ascending to the top of the hole in fine weather for the benefit of the warmth and the air, and occasionally peeping forth, apparently to recon-

* Proceedings of the Academy of Natural Sciences, Philadelphia, Nov. and Dec., 1846; and Downing's Horticulturist, Vol. II. p. 16, July, 1847.

noitre, but descending again on the occurrence of cold or wet weather.

During their temporary residence in these burrows near the surface, the Cicada grubs, or more properly pupæ, for such they are to be considered at this period, though they still retain something of a grub-like form, acquire strength for further efforts by exposure to the light and air, and seem then only to wait for a favorable moment to issue from their subterranean retreats. When at length this arrives, they issue from the ground in great numbers in the night, crawl up the trunks of trees, or upon any other object in their vicinity to which they can fasten themselves securely by their claws. After having rested awhile, they prepare to cast off their skins, which, in the mean time, have become dry and of an amber color. By repeated exertions, a longitudinal rent is made in the skin of the back, and through this the included Cicada pushes its head and body, and withdraws its wings and limbs from their separate cases, and, crawling to a little distance, it leaves its empty pupa-skin, apparently entire, still fastened to the tree. At first the wing-covers and wings are very small and opaque, but, being perfectly soft and flexible, they soon stretch out to their full dimensions, and in the course of a few hours the superfluous moisture of the body evaporates, and the insect becomes strong enough to fly.

During several successive nights the pupæ continue to issue from the earth; above fifteen hundred have been found to arise beneath a single apple-tree, and in some places the whole surface of the soil, by their successive operations, has appeared as full of holes as a honeycomb. In Alabama the species under consideration leaves the ground in February and March, in Maryland and Pennsylvania in May, but in Massachusetts it does not come forth till near the middle of June. Within about a fortnight after their final transformation they begin to lay their eggs, and in the space of six weeks the whole generation becomes extinct.

Fortunately these insects are appointed to return only at

periods so distant that vegetation often has time to recover from the injury inflicted by them; but were they to appear at shorter intervals, our forest and fruit trees would soon be entirely destroyed by them. They are moreover subject to many accidents, and have many enemies, which contribute to diminish their numbers. Their eggs are eaten by birds; the young, when they first issue from the shell, are preyed upon by ants, which mount the trees to feed upon them, or destroy them when they are about to enter the ground. Blackbirds eat them when turned up by the plough in fields, and hogs are excessively fond of them, and, when suffered to go at large in the woods, root them up, and devour immense numbers just before the arrival of the period of their final transformation, when they are lodged immediately under the surface of the soil. It is stated that many perish in the egg state, by the rapid growth of the bark and wood, which closes the perforations and buries the eggs before they have hatched; and many, without doubt, are killed by their perilous descent from the trees.

There are several other harvest-flies in the United States, the males of which are musical; but their drums are concealed within little cavities in the sides of the first abdominal ring. One of these is found in Massachusetts, and, though it never appears in such great numbers as the preceding species, it is more common or more generally met with throughout the State. It may be called the dog-day harvest-fly, or *Cicada canicularis* (Fig. 88), from the circumstance of its invariably appearing with the beginning of dog-days. During many years in succession, with only one or two exceptions, I have heard this insect, on the 25th of July, for the first time in the season, drumming in the trees, on some part of the day between the hours of ten in the morning and two in the afternoon. It is true that all do not muster on the same day; for at first they are few in number, and scattered at great distances from each other; new-comers, however, are added from day to day, till, in a short time, almost every tree seems to have its musician, and the rolling of their drums may be heard in every direction. This circumstance, however, does not render it any the less remarkable that the first of the band should keep their appointed time with such extreme regularity. The dog-day harvest-fly measures about one inch and six tenths from the front to the tips of the wing-covers, which, when spread, expand about three inches.

Fig. 88.

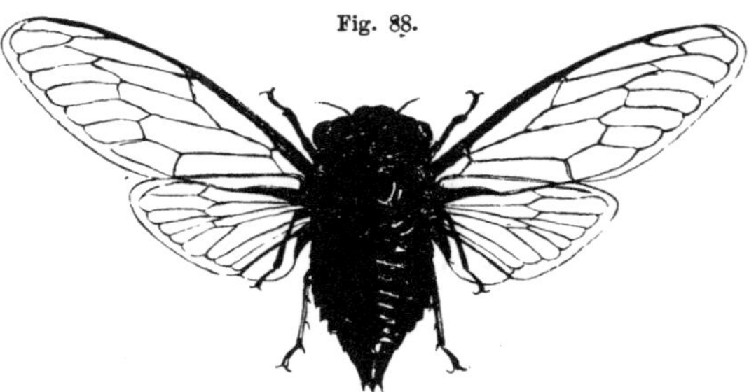

Its body is black on the upper side; the under side of the head, the breast, and the sides of the belly are covered with a white substance resembling flour; the top of the head and the thorax are ornamented with olive-green lines and characters, one of which, in the shape of the letter W, is very conspicuous; the legs, and the front edge and principal veins of the wing-covers and of the wings are also green, and there is a dusky zigzag spot on the little cross-veins near the tip of the wing-covers; and the valves beneath the body of the males are wider than long. This species has heretofore been mistaken for the *Cicada pruinosa*, or frosted harvest-fly, described by Mr. Say, which is found in the Middle States, measures two inches to the tips of the wing-covers, has a white spot each side of the base of the abdomen, a second on the middle of the sides, and a third near to the tip, and has the valves of the males longer than wide.* I am not aware

* The form and proportions of the abdominal valves have decided me to separate the *canicularis* from Mr. Say's *pruinosa*, although, with the exception of their difference in size, they present no other constant characters which will invariably

that the females of the dog-day harvest-fly prefer to lay their eggs in one rather than in another kind of tree; for I have taken the pupæ emerging from the ground beneath cherry, maple, and elm trees, and it is probable that they could not have travelled far from the trees upon which, when young, they were hatched, and upon the trunks of which they finally leave their vacant shells. These have much the same form and appearance as the pupa-shells of the seventeen-year harvest-fly, but are considerably larger. Some individuals of this species continue with us as late as the end of September. As they are not very numerous, the injury sustained by the trees from their punctures is comparatively small.

The other harvest-flies of this country have only two eyelets, and are not furnished with musical instruments; but they enjoy the faculty of leaping, which the Cicadas do not. This faculty does not, as in the grasshoppers and other leaping insects, result from an enlargement of their hindmost thighs, which do not differ much in thickness from the others; but is owing to the length of their hindmost shanks, or to the bristles and spines with which these parts are clothed and tipped. These spines serve to fix the hind legs securely to the surface, and, when the insect suddenly unbends its legs, its body is launched forward in the air. Some of these harvest-flies, when assisted by their wings, will leap to the distance of five or six feet, which is more than two hundred and fifty times their own length; in the

serve to distinguish them from each other.[2] In my collection are four more native species of *Cicada;* namely, the *auletes* of Germar, our largest species, from North Carolina; a second species, apparently undescribed, about equal to this in magnitude, from Long Island, New York; the *tibicen* of Linnæus, also from New York, and quite common even within the city; and the *hieroglyphica* of Say, which, I believe, was captured in Florida, and was presented to me by Mr. Edward Doubleday. A specimen of the *tibicen*, or some other large species, has been taken in Massachusetts, but I have not the individual to refer to at this time.

[[2] This is nothing more than a local variety of *C. pruinosa*, Say; there is no persistency in the form and length of the abdominal valves, and the coloration and extent of *pruinoseness* upon the insect depend upon various contingencies to which it is liable. — UHLER.]

same proportion, "a man of ordinary stature should be able at once to vault through the air to the distance of a quarter of a mile." Some of these leaping harvest-flies have the face nearly vertical, and the thorax very large, tapering to a point behind, covering the whole of the upper side of the body, and overtopping even the head, which is not visible from above. These belong chiefly to the genus *Membracis*, to which allusion has already been made; and, as they are found mostly on the limbs of trees and shrubs, they may receive the name of tree-hoppers.* In others the face slopes downwards towards the breast, the thorax is of moderate size, and does not extend much, if at all, beyond the base of the wing-covers, and does not conceal the head when viewed from above. Some of the insects, with this small-sized thorax, are familiarly called, in English works, cuckoo-spit, and frog-hoppers, and to others may be applied the name of leaf-hoppers, because they live mostly on the leaves of plants.

The thorax differs very much in shape in different kinds of tree-hoppers (MEMBRACIDIDÆ), and the variations of this part are productive of many odd forms among these insects, and particularly in foreign species. Among the species inhabiting Massachusetts, there are some in which the thorax forms a thin and high arched crest over the body, as in *Membracis camelus* of Fabricius, and the *vau* of my Catalogue.[3] To these the name of *Membracis*, which means sharp-edged, is most applicable. In other species (*M. emarginata* and *sinuata* of Fabricius, and *concava* of Say[4]) the crest of the thorax is deeply notched on the top. In others the whole of the thorax is not elevated longitudinally in the middle, but only in some part; thus *M. Ampelopsidis*[5] has an oblong square crest on the middle of the thorax; *M. bi-*

* Mr. Rennie, in the "Library of Entertaining Knowledge," has misapplied this name to the *Cicadas*, which do not leap.

[3 Both belong to the genus *Smilia*, Amyot. — UHLER.]

[4 *M. emarginata*, *sinuata*, and *concava* belong to *Entilia*, Amyot. —UHLER.]

[5 *M. ampelopsidis* belongs to *Telamona*, Fitch. — UHLER.]

maculata of Fabricius and *univittata*[6] of my Catalogue have a thin horn-like projection, blunt, however, at the end, extending obliquely forwards and upwards from the fore part of the thorax; and *M. binotata* and *latipes*[7] of Say have a similarly situated horn, narrower however, and curved, so as to give to the insects, when viewed sidewise, the shape of a bird; and, lastly, in *M. bubalus* of Fabricius, *diceros* of Say, and *taurina*[8] of my Catalogue, the ridge of the thorax, viewed from above, has somewhat the shape of the letter T, becoming broad at the fore part, and extending outwards on each side like a pair of short thick horns, which gave rise to the foregoing specific names, meaning buffalo, two-horned, and kine-like.

The habits of some of the tree-hoppers are presumed to be much the same as those of the musical harvest-flies, for they are found on the limbs of trees, where they deposit their eggs, only during the adult state, and probably pass the early period of their existence in the ground. Others, however, are known to live and undergo all their changes on the stems of plants. Among the former is our largest native species, the two-spotted tree-hopper, or *Membracis bimaculata** of Fabricius (Fig. 89), which may be found in great abundance on the limbs of the locust-tree (*Robinia pseudacacia*) during the months of September and October. These, as well as other tree-hoppers, show but little activity when undisturbed, remaining without motion for hours together on the limbs of the trees; but on the approach of the fingers, they leap vigorously, and, spreading their wings at the same time,

Fig. 89.

* Fabricius describes the male only under this name; the female is his *Membracis acuminata.* This species belongs to Professor Germar's new genus, *Hemiptycha.*[9]

[[6] *M. bimaculata* and *univittata* belong to *Thelia*, Amyot. — UHLER.]

[[7] *M. binotata* and *latipes* belong to *Euchenopa*, Amyot. — UHLER.]

[[8] *M. bubalus, diceros*, and *taurina* belong to *Ceresa*, Amyot. — UHLER.]

[[9] It might be added, that this genus is now restricted to *Membracis punctata*, Fab., and a few allied species. — UHLER.]

fly to another limb and settle there, in the same position as before. They never sit across the limbs, but always in the direction of their length, with the head or forepart of the body towards the extremity of the branches. On account of their peculiar form, which is that of a thick cone with a very oblique direction, their dark color, and their fixed posture while perching, they would readily be mistaken for the thorns of the tree, a circumstance undoubtedly intended for their preservation. Other instances have been mentioned displaying proofs of equal wisdom in the formation of insects. Thus, in the leaf-insects, grasshoppers, and walking-sticks, which live in trees, the latter exactly simulating a little twig in appearance, and the others having the form and color of leaves, their resemblance to the objects among which they have been destined to live has doubtless been given to them with the express design of screening them from their enemies of the feathered race. Many other examples of the same kind might be mentioned, did time and the limits of my subject warrant; but these alone suffice to show that special provision has been wisely made in the construction of certain defenceless animals with a view to secure them from observation. Surely insects, the most despised of God's creation, are not unworthy our study, since they are objects of His care and subjects of a special providence.

But to return to our locust tree-hopper, which remains to be described;—it measures about half an inch from the tip of the horn to the end of the body; the male is blackish above, with a long yellow spot on each side of the back; and the female is ash-colored, and without spots. While on the trees, these insects, though perfectly still, are not unemployed; but puncture the bark with their sharp and slender beaks, and imbibe the sap for nourishment. The female also appears to commit her eggs to the protection of the tree, being furnished with a piercer beneath the extremity of her body, with which to make suitable perforations in the branches. As I have never seen the young on these trees,

I presume that, as soon as they are hatched, they make their way to the ground, and remain under the surface of the soil, sucking the sap from the roots of plants, until they are about to enter upon their last period of existence, when they crawl up the trunks of the trees, throw off their coats, and appear in the perfect or winged state. From the great numbers of these tree-hoppers which exist in certain seasons, the locust-trees undoubtedly suffer much, not only in consequence of the quantity of sap abstracted from their branches, but from the numerous punctures made by the insects in obtaining it and in laying their eggs.

The oak-tree is attacked by another species, the white-lined tree-hopper (*M. univittata*), which may be found upon it during the month of July. It is about four tenths of an inch in length; the thorax is brown, has a short obtuse horn extending obliquely upwards from its fore part, and there is a white line on the back, extending from the top of the horn to the hinder extremity.

The common creeper (*Ampelopsis quinquefolia*) is inhabited by a tree-hopper, which has an oblong square and thin elevation or crest on the middle of the thorax. Its body is usually of a reddish ash-color, and the thorax is ornamented with three reddish-brown bands, one of which is above the head and extends transversely between the lateral projecting angles of the thorax, the second is a short and oblique line on each side of the front part of the crest, and the third is also oblique, and begins on the outer edge of the thorax, and passes obliquely forwards on each side to the top of the hind part of the crest. This species may be called *Membracis Ampelopsidis* * (Figs. 90 and 91), from the plant on which it is found in the perfect state. The young appear to live in the earth till they are fully grown and have acquired the rudiments of wing-covers and wings, or have become pupæ,

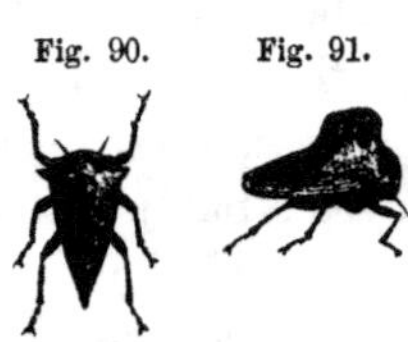
Fig. 90. Fig. 91.

* It is the *Membracis Cissi* of my Catalogue.

after which they are seen ascending the stems of the creeper, on which they change their skins for the last time. This occurs from the middle to the end of June.

There is a little tree-hopper, which is found during the months of July and August on the wax-work, or *Celastrus scandens*, accompanied usually by its young. When fully grown, it is nearly three tenths of an inch in length, including the horn of the thorax; is of a dusky brown color, with two yellowish spots on the ridge of the back; and the first four shanks are exceedingly broad and flat. It is the two-spotted tree-hopper, or *Membracis binotata* of Say. When seen sidewise it presents a profile much like that of a bird, the head and neck of which are represented by the curved projecting horn of the thorax; and a group of these little tree-hoppers, of various sizes, clustered together on a stem of the wax-work, may be likened to a flock of old and young partridges. They appear to pass through all their transformations on the plant, are fond of society, and sit close together, with their heads all in the same direction.

Tree-hoppers are often surrounded by ants, for the sake of their castings, and for the sap which oozes from the punctures made by the former, of which the ants are very fond. Those kinds that live on the stems of plants from the time when they are hatched till they are fully grown, are very closely attended by ants; and as from their constant sucking the young become often wet, their careful attendants, the ants, find regular employment in wiping them clean and dry with their antennæ and tongues.

The remaining Homopterous insects have a thorax of moderate size, not tapering to a point behind, and not covering the whole body as in the preceding species. Their heads are visible from above, and the face slopes downwards towards the breast.

Here may be arranged the singular insects called frog-hoppers (CERCOPIDIDÆ), which pass their whole lives on plants, on the stems of which their eggs are laid in the autumn. The following summer they are hatched, and the young immediately perforate the bark with their beaks, and begin to imbibe the sap. They take in such quantities of this, that it oozes out of their bodies continually, in the form of little bubbles, which soon completely cover up the insects. They thus remain entirely buried and concealed in large masses of foam, until they have completed their final transformation, on which account the names of cuckoo-spittle, frog-spittle, and frog-hoppers have been applied to them. We have several species of these frog-hoppers in Massachusetts, and the spittle, with which they are sheltered from the sun and air, may be seen in great abundance, during the summer, on the stems of our alders and willows. In the perfect state they are not thus protected, but are found on the plants, in the latter part of summer, fully grown and preparing to lay their eggs. In this state they possess the power of leaping in a still more remarkable degree than the tree-hoppers ; and, for this purpose, the tips of their hind shanks are surrounded with little spines, and the first two joints of their feet have a similar coronet of spines at their extremities. Their thorax narrows a little behind, and projects somewhat between the bases of the wing-covers ; their bodies are rather short, and their wing-covers are almost horizontal and quite broad across the middle, which, with the shortness of their legs, gives them a squat appearance.*

The leaf-hoppers (TETTIGONIADÆ) leap almost as well as the spittle-insects just mentioned ; but their hind legs are longer, are not surrounded with coronets of short spines, but are three-sided, and generally fringed on two of their edges

* The following species are found in Massachusetts, namely: *Cercopis ignipecta* of my Catalogue, and the *parallela*, *quadrangularis*, and *obtusa*, of Say. The last three belong to Germar's genus *Aphrophora*,† which means spume-bearer. *Cercopis*, which may be translated impostor, was applied by the Greeks to a small Cicada.

† [*Clastoptera proteus*, an insect of this class which does great injury to the cranberry crop in some parts of Massachusetts, but of whose habits very little has been ascertained, is figured on Plate III. Fig. 6. — ED.]

with numerous long and slender spines, which contribute, like the coronets of the frog-hoppers, to fix their shanks firmly when they are about to leap. The leaf-hoppers have been divided, by Professor Germar and other entomologists, into many genera, according to the structure of their legs, the situation of the eyelets, and the form of the head; but we may retain them, without inconvenience, in the genus *Tettigonia*, proposed for them by Geoffroy, or rather adopted from the ancient Greeks, who gave this name to the small kinds of harvest-flies, calling the larger ones *Tettix*.

The Tettigonians, or leaf-hoppers, have the head and thorax somewhat like those of frog-hoppers, but their bodies are, in general, proportionally longer, not so broad across the middle, and not so much flattened. The head, as seen from above, is broad, and either crescent-shaped, semicircular, or even extended forwards in the form of a triangle; its upper side is more or less flattened, and the face slopes downwards towards the breast at an acute angle with the top of the head. The thorax is wider than long, with the front margin curving forwards, the hind margin transverse, or not extended between the wing-covers, which space is filled by a pretty large triangular scutel or escutcheon. The wing-covers are generally opaque, rather long and narrow, and more or less inclined at the sides of the body, not flat however, but moulded somewhat to the form of the body, and the wings are rather shorter and broader, not netted like those of the tree-hoppers, but strengthened by a few longitudinal veins. The eyes, which are distant from each other, and placed at the sides of the head, are pretty large, but flattish, and not globular as in the Cicadas; and the eyelets, which are rarely wanting, vary in their situation, being sometimes on the top and sometimes below the front edge of the head. Notwithstanding the small size of most of these insects, they are deserving our attention on account of their beauty, delicacy, and surprising agility, as well as for the injury sustained by vegetation from them.

It is stated by the late Mr. Fessenden, in the "New American Gardener," that some persons in this country have entirely "abandoned their grape-vines" in consequence of the depredations of a small insect, which, for many years, was supposed to be the vine-fretter of Europe. It is not, however, the same insect, but is a leaf-hopper, and was first described by me in the year 1831, in the eighth volume of the "Encyclopædia Americana,"* under the name of *Tettigonia Vitis* (Plate III. Fig. 5). In its perfect state it measures one tenth of an inch in length. It is of a pale yellow or straw color; there are two little red lines on the head; the back part of the thorax, the scutel, the base of the wing-covers, and a broad band across their middle, are scarlet; the tips of the wing-covers are blackish, and there are some little red lines between the broad band and the tips. The head is crescent-shaped above, and the eyelets are situated just below the ridge of the front.

The vine-hoppers, as they may be called, inhabit the foreign and the native grape-vines, on the under surface of the leaves of which they may be found during the greater part of the summer; for they pass through all their changes on the vines. They make their first appearance on the leaves in June, when they are very small and not provided with wings, being then in the larva state. During most of the time they remain perfectly quiet, with their beaks thrust into the leaves, from which they derive their nourishment by suction. If disturbed, however, they leap from one leaf to another with great agility. As they increase in size they have occasion frequently to change their skins, and great numbers of their empty cast-skins, of a white color, will be found, throughout the summer, adhering to the under sides of the leaves and upon the ground beneath the vines.

When arrived at maturity, which generally occurs during the month of August, they are still more agile than before, making use of their delicate wings as well as their legs in

* Article *Locust*, p. 43.

their motions from place to place ; and when the leaves are agitated, they leap and fly from them in swarms, but soon alight and begin again their destructive operations. The infested leaves at length become yellow, sickly, and prematurely dry, and give to the vine at midsummer the aspect it naturally assumes on the approach of winter. But this is not the only injury arising from the exhausting punctures of the vine-hoppers. In consequence of the interruption of the important functions of the leaves, the plant itself languishes, the stem does not increase in size, very little new wood is formed, or, in the language of the gardeners, the canes do not ripen well, the fruit is stunted and mildews, and, if the evil be allowed to go on unchecked, in a few years the vines become exhausted, barren, and worthless. In the autumn the vine-hoppers desert the vines, and retire for shelter during the coming winter beneath fallen leaves and among the decayed tufts and roots of grass, where they remain till the following spring, when they emerge from their winter-quarters, and in due time deposit their eggs upon the leaves of the vine, and then perish.

As the vine-hoppers are much more hardy and more vivacious than the European vine-fretters or plant-lice, the applications that have proved destructive to the latter are by no means so efficacious with the former. Fumigations with tobacco, beneath a movable tent placed over the trellises, answer the purpose completely.* They require frequent repetition, and considerable care is necessary to prevent the escape and insure the destruction of the insects ; circumstances which render the discovery of some more expeditious method an object to those whose vineyards are extensive.

There is another little leaf-hopper that has been mistaken for a vine-fretter or Thrips, though never found upon the grape-vine. It lives upon the leaves of rose-bushes, and is

* See Fessenden's "New American Gardener," p. 299, for a description of the tent and of the process of fumigation.

very injurious to them. In its perfect state it is rather less than three twentieths of an inch long. Its body is yellowish white, its wing-covers and wings are white and transparent, and its eyes, claws, and piercer brown. The male has two recurved appendages at the tip of its hind body. It may be called *Tettigonia Rosæ*.* Swarms of these insects may be found, in various stages of growth, on the leaves of the rose-bush, through the greater part of summer, and even in winter upon housed plants. Their numerous cast skins may be seen adhering to the lower side of the leaves. They pair and lay their eggs about the middle of June, and they probably live through the winter in the perfect state, concealed under fallen leaves and rubbish on the surface of the ground. Fumigations with tobacco, and the application of a solution of whale-oil soap in water with a syringe, are the best means for destroying these leaf-hoppers.

I have found that the Windsor bean, a variety of the *Vicia Faba* of Linnæus, is subject to the attacks of a species of leaf-hopper, particularly during dry seasons, and when cultivated in light soils. In the early part of summer the insects are so small and so light-colored that they easily escape observation, and it is not till the beginning of July, when the beans are usually large enough to be gathered for the table, that the ravages of the insects lead to their discovery. A large proportion of the pods will then be found to be rough, and covered with little dark-colored dots or scars, and many of them seem to be unusually spongy and not well filled. On opening these spongy pods, we find that the beans have not grown to their proper size, and if they are left on the plant they cease to enlarge. At the same time the leaves, pods, and stalks are more or less infested with little leaf-hoppers, not fully grown, and unprovided with wings. Usually between the end of July and

* This insect may be the *Cicada Rosæ* of Linnæus, or *Iassus Rosæ* of Fabricius. It belongs to Dr. Fitch's genus *Empoa*, as also does *Tettigonia Fabæ*. The *Tettigonia Vitis* is an *Erythroneura* of the same author.

the middle of August the insects come to their growth and acquire their wings; but the mischief at this time is finished, and the plants have suffered so much that all prospect of a second crop of beans, from new shoots produced after the old stems are cut down, is frustrated.

These leaf-hoppers have the same agility in their motions, and apparently the same habits, as the vine-hoppers; but in the perfect state they are longer, more slender, and much more delicate. They are of a pale green color; the wing-covers and wings are transparent and colorless; and the last joint of the hind feet is bluish. The head, as seen from above, is crescent-shaped, and the two eyelets are situated on its front edge. The male has two long recurved feathery threads at the extremity of the body. The length of this species is rather more than one tenth, but less than three twentieths of an inch. It may be called *Tettigonia Fabæ.* Probably it passes the winter in the same way as the vine-hopper.

2. Plant-lice. (*Aphididæ.*)

The Aphidians, in which group we include the insects commonly known by the name of plant-lice, differ remarkably from all the foregoing in their appearance, their formation, and their manner of increase. Their bodies are very soft, and usually more or less oval. The females are often without wing-covers and wings; and the former, when they exist, do not differ in texture from the wings, but are usually much larger and more useful in flight. We may therefore cease to call these parts wing-covers, in all the remaining insects of this order, and apply to them the name of upper wings.

Some of the Aphidians have the power of leaping, like the leaf-hoppers, from which, however, they differ in having very large and transparent upper wings, which cover the sides of the body like a very steep roof; and their antennæ are pretty long and threadlike, and are tipped with two short bristles at the end. Both sexes, when arrived at maturity, are winged, and some of the females are provided with a kind of awl at the end of the body, very different, however, from the piercers of the foregoing insects. With this they prick the leaves, in which they deposit their eggs, and the wounds thus made sometimes produce little excrescences or swellings on the plant. These leaping plant-lice belong to a genus called *Psylla*, which was the Greek name for a small jumping insect. They are by no means so prolific as the other plant-lice, for ordinarily they produce only one brood in the year. They live in groups, composed of about a dozen individuals each, upon the stems and leaves of plants, the juices of which they imbibe through their tubular beaks. The young are often covered with a substance resembling fine cotton arranged in flakes. This is the case with some which are found on the alder and birch in the spring of the year.

Within a few years, a kind of *Psylla*, before unknown here, has appeared upon pear-trees in the western parts of Connecticut and of Massachusetts, particularly in the valley of the Housatonic, and in the adjoining counties of Dutchess and Columbia in New York. It was first made known to me, in December, 1848, by Dr. Ovid Plumb, of Salisbury, Connecticut, and it is the subject of a communication in the "American Agriculturist," for January, 1849. Since that time, Dr. Plumb has favored me with additional observations, and an account of his experiments, with various remedies, and towards the end of July, 1851, a brief visit to Salisbury gave me an opportunity of seeing the insects in a living condition, and in the midst of their operations upon the trees.

This *Psylla*, or jumping plant-louse, is one of the kinds whose young are naked, or not covered with a coat of cotton. In some of its forms it is found on pear-trees during most of the time from May to October; and probably two if not more broods are produced in the course of the summer.

It was first observed by Dr. Plumb in the spring of 1833, on some imported pear-trees, which had been set the year before. These trees, in the autumn after they were planted, wore an unhealthy aspect, and had patches of a blackish rust upon their branches. During the second summer, these trees died; and other trees, upon which the same rusty matter was found, proved to be infested with the same insects.

Like the aphides, or plant-lice, these insects live by suction. By means of their suckers, which come from the lower side of the head near the breast, they puncture the bark of the twigs and small branches, and imbibe the sap. They soon gorge themselves to such a degree, that the fluid issues constantly from their bodies in drops, is thrown over the surface of the twigs, and, mingled with their more solid castings, defiles the bark, and gives it the blackish color above noticed. Swarms of flies and ants upon the trees are a sure indication of the presence of these sap-suckers, being attracted by the sweetish fluid thrown out by them.

Young trees suffer excessively by the attacks of these insects, nor do old trees escape without injury from them. In consequence apparently of their ravages alone, Dr. Plumb lost several hundred pear-trees from 1834 to 1838 inclusive; his trees have continued to suffer, to some extent, from this cause, since that time; and he informs me that the same destructive depredations have been observed in all the adjacent region. On the 23d of July, I saw these insects on the trees, some already provided with wings, and others advancing towards maturity. The young ones were of a dull orange-yellow color. They were short, and were obtuse behind, and had little wing-scales on the sides of their bodies. The perfect, or winged individuals, were about one tenth of an inch long from the forehead to the tips of the closed wings. The front of the head was notched in the middle. The eyes were large and prominent. The head and thorax were brownish orange, and the hind body greenish. Their four ample wings were colorless and transparent, and were marked with a few dark veins. The body of the female is pointed at the end, and inclines to a reddish hue.

The pear-tree, in Europe, is subject to the attacks of a similar insect, called *Psylla Pyri*, the pear-tree Psylla. The European species is said to vary in color at different ages, and in different seasons of the year, being of a dull crimson color, shaded with black in the spring, when it comes forth to lay its eggs. Not having seen any of our pear-tree *Psyllæ* in their spring dress, I cannot say whether they agree with those of Europe in being of the same crimson color at this season of the year. As, however, they do correspond very nearly in other respects to the descriptions given of the European species, and have precisely the same destructive habits, and as they were first detected upon imported pear-trees, I apprehend that they were introduced from abroad, and that they will prove to be the same species as the European *Psylla Pyri*.

The following particulars, abridged from Köllar's "Treatise," if confirmed by future observations, will serve to complete the history of the American insect. The European pear-tree Psylla comes forth from its winter retreat, provided with wings, as soon as the buds of fruit-trees begin to expand. After pairing, the female lays her eggs in great numbers near each other on the young leaves and blossoms, or on the newly-formed fruit and shoots. The eggs are oblong, yellowish, and look somewhat like grains of pollen. The young insects hatched therefrom resemble wingless plant-lice, and are of a dark yellow color. They change their skins and color repeatedly, and acquire wing-scales, or rudimentary wings. They then fix themselves to the bark in rows, and remain sucking the sap till their last change approaches, at which time they disperse among the leaves, cast off their skins, and appear in the winged form.

When considerable numbers attack a pear-tree, the latter

soon assumes an unhealthy appearance, its growth is checked, its leaves and shoots curl up, and the tree dies by degrees, if not freed from its troublesome guests. Köllar recommends brushing off the insects, when young, with a brush of hog's bristles, and crushing under foot those that fall; and also advises to search for the winged females in the spring, and destroy them by hand. Such a process would be altogether too tedious and uncertain here. I would therefore suggest the expediency of washing the twigs with a brush dipped in a mixture of strong soap-suds and flour of sulphur. If this be done *before the buds expand*, the latter will not be injured thereby, while the application will be likely to deter the insects from laying their eggs on the tree. A weaker application of the same, or the common solution of whale-oil soap, may suffice to kill the young insects after they have fastened themselves upon the bark. If the latter be thrown upon the trees with a syringe, it will destroy the insects on the leaves also.

Others, both sexes of which are also winged, have long and slender bodies, very narrow wings, which are fringed with fine hairs, and lie flatly on the back when not in use. They are exceedingly active in all their motions, and seem to leap rather than fly. They live on leaves, flowers, in buds, and even in the crevices of the bark of plants, but are so small that they readily escape notice, the largest being not more than one tenth of an inch in length. These minute and slender insects belong to the genus *Thrips*. Their punctures appear to poison plants, and often produce deformities in the leaves and blossoms. The peach-tree sometimes suffers severely from their attacks, as well as from those of the true plant-lice; and they are found beneath the leaves, in little hollows caused by their irritating punctures.

The same applications that are employed for the destruction of plant-lice may be used with advantage upon plants infested with the *Thrips*. Mrs. N. G. S. Gage, formerly of

Concord, N. H., to whom I am indebted for much valuable information respecting the wheat-fly, or *Cecidomyia Tritici*, has discovered another pernicious insect in the ears of growing wheat. It seems to agree with the accounts of the *Thrips cerealium*, which sometimes infests wheat, in Europe, to a great extent. This insect, in its larva state, is smaller than the wheat maggot, is orange-colored, and is provided with six legs, two antennæ, and a short beak, and is very nimble in its motions. It is supposed to suck out the juices of the seed, thus causing the latter to shrink, and become what the English farmers call pungled. This little pest may probably be destroyed by giving the grain a thorough coating of slacked lime.

Aphides, or plant-lice, as they are usually called, are among the most extraordinary of insects. They are found upon almost all parts of plants, the roots, stems, young shoots, buds, and leaves, and there is scarcely a plant which does not harbor one or two kinds peculiar to itself. They are, moreover, exceedingly prolific, for Réaumur has proved that one individual, in five generations, may become the progenitor of nearly six thousand millions of descendants.

It often happens, that the succulent extremities and stems of plants will, in an incredibly short space of time, become completely coated with a living mass of these little lice. These are usually wingless, consisting of the young and of the females only; for winged individuals appear only at particular seasons, usually in the autumn, but sometimes in the spring, and these are small males and larger females. After pairing, the latter lay their eggs upon or near the leaf-buds of the plant upon which they live, and, together with the males, soon afterwards perish.

The genus to which plant-lice belong is called *Aphis*, (Plate III. Fig. 4, *Aphis mali*,) from a Greek word which signifies to exhaust. The following are the principal characters by which they may be distinguished from other insects. Their bodies are short, oval, and soft, and are furnished at the hinder extremity with two little tubes, knobs, or pores, from which exude almost constantly minute drops of a fluid as sweet as honey; their heads are small, their beaks are very long and tubular, their eyes are globular, but they have not eyelets, their antennæ are long, and usually taper towards the extremity, and their legs are also long and very slender, and there are only two joints to their feet. Their upper are nearly twice as large as the lower wings, are much longer than the body, are gradually widened towards the extremity, and nearly triangular; they are almost vertical when at rest, and cover the body above like a very sharp-ridged roof.

The winged plant-lice provide for a succession of their race by stocking the plants with eggs in the autumn, as before stated. These are hatched in due time in the spring, and the young lice immediately begin to pump up sap from the tender leaves and shoots, increase rapidly in size, and in a short time come to maturity. In this state, it is found that the brood, without a single exception, consists wholly of females, which are wingless, but are in a condition immediately to continue their kind. Their young, however, are not hatched from eggs, but are produced alive, and each female may be the mother of fifteen or twenty young lice in the course of a single day. The plant-lice of this second generation are also wingless females, which grow up and have their young in due time; and thus brood after brood is produced, even to the seventh generation or more, without the appearance or intervention, throughout the whole season, of a single male. This extraordinary kind of propagation ends in the autumn with the birth of a brood of males and females, which in due time acquire wings and pair; eggs are then laid by these females, and with the death of these winged individuals, which soon follows, the race becomes extinct for the season.

Plant-lice seem to love society, and often herd together in dense masses, each one remaining fixed to the plant by means of its long tubular beak; and they rarely change their places till they have exhausted the part first attacked. The attitudes and manners of these little creatures are exceedingly amusing. When disturbed, like restive horses, they begin to kick and sprawl in the most ludicrous manner. They may be seen, at times, suspended by their beaks alone, and throwing up their legs as if in a high frolic, but too much engaged in sucking to withdraw their beaks. As they take in great quantities of sap, they would soon become gorged if they did not get rid of the superabundant fluid through the two little tubes or pores at the extremity of their bodies. When one of them gets running-over full, it seems to communicate its uneasy sensations, by a kind of animal magnetism, to the whole flock, upon which they all, with one accord, jerk upwards their bodies, and eject a shower of the honeyed fluid. The leaves and bark of plants much infested by these insects are often completely sprinkled over with drops of this sticky fluid, which, on drying, become dark colored, and greatly disfigure the foliage. This appearance has been denominated honey-dew; but there is another somewhat similar production observable on plants, after very dry weather, which has received the same name, and consists of an extravasation or oozing of the sap from the leaves.

We are often apprised of the presence of plant-lice on plants growing in the open air by the ants ascending and descending the stems. By observing the motions of the latter, we soon ascertain that the sweet fluid discharged by the lice is the occasion of these visits. The stems swarm with slim and hungry ants running upwards, and others lazily descending with their bellies swelled almost to bursting. When arrived in the immediate vicinity of the plant-lice, they greedily wipe up the sweet fluid which has distilled from them, and when this fails, they station themselves among the lice, and catch the drops as they fall.

The lice do not seem in the least annoyed by the ants, but live on the best possible terms with them; and, on the other hand, the ants, though unsparing of other insects weaker than themselves, upon which they frequently prey, treat the plant-lice with the utmost gentleness, caressing them with their antennæ, and apparently inviting them to give out the fluid by patting their sides. Nor are the lice inattentive to these solicitations, when in a state to gratify the ants, for whose sake they not only seem to shorten the periods of the discharge, but actually yield the fluid when thus pressed. A single louse has been known to give it drop by drop successively to a number of ants, that were waiting anxiously to receive it. When the plant-lice cast their skins, the ants instantly remove the latter, nor will they allow any dirt or rubbish to remain upon or about them. They even protect them from their enemies, and run about them in the hot sunshine to drive away the little ichneumon flies that are forever hovering near to deposit their eggs in the bodies of the lice.

Plant-lice differ very much in form, color, clothing, and in the length of the honey-tubes. Some have these tubes quite long, as the rose-louse, *Aphis Rosæ*, which is green, and has a little conical projection or stylet, as it is called, at the extremity of the body, between the two honey-tubes. The cabbage-louse, *Aphis Brassicæ*, has also long honey-tubes, but its body is covered with a whitish mealy substance. This species is very abundant on the under side of cabbage-leaves in the month of August.

The largest species known to me is found in clusters beneath the limbs of the pig-nut hickory (*Carya porcina*), in all stages of growth, from the first to the middle of July. It is the *Aphis* * *Caryæ* of my Catalogue. Its body, in the winged state, measures one quarter of an inch to the end of the abdomen, and above four tenths of an inch to the tips of the upper wings, which expand rather more than seven tenths of an inch. It has no terminal stylet, and the honey-tubes are very short. Its body is covered with a bluish-white

* It probably belongs to the genus *Lachnus* of Illiger, or *Cinara* of Curtis.

substance like the bloom of a plum, with four rows of little transverse black spots on the back; the top of the thorax and the veins of the wings are black, as are also the shanks, the feet, and the antennæ, which are clothed with black hairs; the thighs are reddish brown. This species sucks the sap from the limbs, and not from the leaves, of the hickory.

There is another large species, living in the same way on the under side of the branches of various kinds of willows, and clustered together in great numbers. About the first of October they are found in the winged state. The body measures one tenth of an inch in length, and the wings expand about four tenths. The stylet is wanting; the body is black and without spots; the wings are transparent, but their veins, the short honey-tubercles, the third joint of the antennæ, and the legs, are tawny yellow. This species cannot be identical with the willow-louse, *Aphis Salicis* of Linnæus, which has a spotted body; and therefore I propose to call it *Aphis Salicti*,[10] the plant-louse of willow groves. When crushed, it communicates a stain of a reddish or deep orange color.

Some plant-lice live in the ground, and derive their nourishment from the roots of plants. We annually lose many of our herbaceous plants, if cultivated in a light soil, from the exhausting attacks of these subterranean lice. Upon pulling up China asters, which seem to be perishing from no visible cause, I have found hundreds of little lice, of a white color, closely clustered together on the roots. I could never discover any of them that were winged, and therefore conclude from this circumstance, as well as from their peculiar situation, that they never acquire wings. Whether these are of the same species as the *Aphis radicum* of Europe, I cannot ascertain, as no sufficient description of the latter

[[10] The name *Salicti* was long ago appropriated by Schrank to a very different species of *Aphis*, inhabiting Europe. This name must therefore fall as a synonyme to some other which may be applied to it. It might be called *Aphis Salicicola*. — UHLER.]

has ever come to my notice.[11] These little lice are attended by ants, which generally make their nests near the roots of the plants, so as to have their milch kine, as the plant-lice have been called, within their own habitations; and in consequence of the combined operations of the lice and the ants, the plants wither and prematurely perish.

When these subterranean lice are disturbed, the attendant ants are thrown into the greatest confusion and alarm; they carefully take up the lice which have fallen from the roots, and convey them in their jaws into the deep recesses of their nests; and here the lice still contrive to live upon the fragments of the roots left in the soil.

It is stated * that the ants bestow the same care and attention upon the root-lice as upon their own offspring, that they defend them from the attacks of other insects, and carry them about in their mouths to change their pasture; and that they pay particular attention to the eggs of the lice, frequently moistening them with their tongues, and in fine weather bringing them to the surface of the nest to give them the advantage of the sun. On the other hand, the sweet fluid supplied in abundance by these lice forms the chief nutriment both of the ants and their young, which is sufficient to account for their solicitude and care for their valuable herds.

The peach-tree suffers very much from the attacks of plant-lice, which live under the leaves, causing them by their punctures to become thickened, to curl or form hollows beneath, and corresponding crispy and reddish swellings above, and finally to perish and drop off prematurely. Whether our insect is the same as the European Aphis of

[[11] It is very probable that the *Aphis* infesting China asters is the same with the *radicis* of Europe. Many foreign species of plant-lice have become naturalized in this country, and we may thus expect to find most, if not all, of the commoner European species infesting our vegetation. The *Aphis* (*Trama*) *radicis* of Europe corresponds with our own in color, and, as supposed by Dr. Harris, winged specimens have never been discovered. — UHLER.]

* See Kirby and Spence's Introduction to Entomology, Vol. II. pp. 91, 92.

the peach-tree (*Aphis Persicæ* of Sulzer) I cannot determine, for the want of a proper description of the latter.

The injuries occasioned by plant-lice are much greater than would at first be expected from the small size and extreme weakness of the insects; but these make up by their numbers what they want in strength individually, and thus become formidable enemies to vegetation. By their punctures, and the quantity of sap which they draw from the leaves, the functions of these important organs are deranged or interrupted, the food of the plant, which is there elaborated to nourish the stem and mature the fruit, is withdrawn, before it can reach its proper destination, or is contaminated and left in a state unfitted to supply the wants of vegetation.

Plants are differently affected by these insects. Some wither and cease to grow, their leaves and stems put on a sickly appearance, and soon die from exhaustion. Others, though not killed, are greatly impeded in their growth, and their tender parts, which are attacked, become stunted, curled, or warped.

The punctures of these lice seem to poison some plants, and affect others in a most singular manner, producing warts or swellings, which are sometimes solid and sometimes hollow, and contain in their interior a swarm of lice, the descendants of a single individual, whose punctures were the original cause of the tumor. I have seen reddish tumors of this kind, as big as a pigeon's egg, growing upon leaves, to which they were attached by a slender neck, and containing thousands of small lice in their interior. Naturalists call these tumors galls, because they seem to be formed in the same way as the oak-galls which are used in the making of ink. The lice which inhabit or produce them generally differ from the others, in having shorter antennæ, being without honey-tubes, and in frequently being clothed with a kind of white down, which, however, disappears when the insects become winged.

These downy plant-lice are now placed in the genus *Eriosoma*, which means woolly body, and the most destructive species belonging to it was first described, under the name of *Aphis lanigera*, by Mr. Hausmann,* in the year 1801, as infesting the apple-trees in Germany. It seems that it had been noticed in England as early as the year 1787, and has since acquired there the name of American blight, from the erroneous supposition that it had been imported from this country. It was known, however, to the French gardeners † for a long time previous to both of the above dates, and, according to Mr. Rennie, ‡ is found in the orchards about Harfleur in Normandy, and is very destructive to the apple-trees in the department of Calvados.

There is now good reason to believe that the miscalled American blight is not indigenous to this country, and that it has been introduced here with fruit-trees from Europe. Some persons, indeed, have supposed that it was not to be found here at all, but the late Mr. Buel has stated § that it existed on his apple-trees, and I have once or twice seen it on apple-trees in Massachusetts, where, however, it still appears to be rare, and consequently I have not been able to examine the insects sufficiently myself. The best account that I have seen of them is contained in Knapp's "Journal of a Naturalist," from which, and from Hausmann's description, the following observations are chiefly extracted.

The eggs of the woolly apple-tree louse are so small as not to be distinguished without a microscope, and are enveloped in a cotton-like substance furnished by the body of the insect. They are deposited in the crotches of the branches and in the chinks of the bark at or near the surface of the ground, especially if there are suckers springing from the same place. The young, when first hatched, are covered with a very short and fine down, and appear in

* Illiger's Magazin, Vol. I. p. 440.

† Salisbury's Hints on Orchards, p. 39.

‡ Insect Miscellanies, p. 180.

§ New England Farmer, Vol. VII. p. 169; Vol. IX. p. 178.

the spring of the year like little specks of mould on the trees (Fig. 92). As the season advances, and the insect increases in size, its downy coat becomes more distinct, and grows in length daily. This down is very easily removed, adheres to the fingers when it is touched, and seems to issue from all the pores of the skin of the abdomen. When fully grown, the insects of the first brood are one tenth of an inch in length, and, when the down is rubbed off, the head, antennæ, sucker, and shins are found to be of a blackish color, and the abdomen honey-yellow. The young are produced alive during the summer, are buried in masses of the down, and derive their nourishment from the sap of the bark and of the alburnum or young wood immediately under the bark.

Fig. 92.

The adult insects never acquire wings, at least such is the testimony both of Hausmann and Knapp, and are destitute of honey-tubes, but from time to time emit drops of a sticky fluid from the extremity of the body. These insects, though destitute of wings, are conveyed from tree to tree by means of their long down, which is so plentiful and so light, as easily to be wafted by the winds of autumn, and thus the evil will gradually spread throughout an extensive orchard. The numerous punctures of these lice produce on the tender shoots a cellular appearance, and wherever a colony of them is established, warts or excrescences arise on the bark; the limbs thus attacked become sickly, the leaves turn yellow and drop off; and, as the infection spreads from limb to limb, the whole tree becomes diseased, and eventually perishes.

In Gloucestershire, England, so many apple-trees were destroyed by these lice in the year 1810, that it was feared the making of cider must be abandoned. In the North of England the apple-trees are greatly injured, and some annually destroyed by them, and in the year 1826 they abounded there in such incredible luxuriance, that many trees seemed, at a short distance, as if they had been whitewashed.

Mr. Knapp thinks that remedies can prove efficacious in removing this evil only upon a small scale, and that when the injury has existed for some time, and extended its influence over the parts of a large tree, it will take its course, and the tree will die. He says that he has removed this blight from young trees, and from recently attacked places in those more advanced, by painting over every node or infected part of the tree with a composition consisting of three ounces of melted resin mixed with the same quantity of fish-oil, which is to be put on while warm, with a painter's brush. Sir Joseph Banks succeeded in extirpating the insects from his own trees by removing all the old and rugged bark, and scrubbing the trunk and branches with a hard brush. The application of the spirits of tar, of spirits of turpentine, of oil, urine, and of soft soap, has been recommended. Mr. Buel found that oil sufficed to drive the insects from the trunks and branches, but that it could not be applied to the roots, where numbers of the insects harbored.

The following treatment I am inclined to think will prove as successful as any which has heretofore been recommended. Scrape off all the rough bark of the infected trees, and make them perfectly clean and smooth early in the spring; then rub the trunk and limbs with a stiff brush wet with a solution of potash as hereafter recommended for the destruction of bark-lice; after which remove the sods and earth around the bottom of the trunk, and with the scraper, brush, and alkaline liquor, cleanse that part as far as the roots can conveniently be uncovered. The earth and sods should immediately be carried away, fresh loam should be placed around the roots, and all cracks and wounds should be filled with grafting cement or clay mortar. Small limbs and extremities of branches, if infected, and beyond reach of the applications, should be cut off and burned.

There are several other species of *Eriosoma* or downy lice in this State, inhabiting various forest and ornamental trees, some of which may also have been introduced from abroad. The descriptions of foreign plant-lice are mostly so brief and imperfect, that it is impossible to ascertain from them which of our species are identical with those of Europe; I shall therefore omit any further account of these insects, and close this part of the subject with a few remarks on the remedies to be employed for their destruction generally, and some notice of the natural enemies of plant-lice.

Solutions of soap, or a mixture of soapsuds and tobacco-water, used warm and applied with a watering-pot or with a garden engine, may be employed for the destruction of these insects. It is said that hot water may also be employed for the same purpose with safety and success. The water, tobacco-tea, or suds should be thrown upon the plants with considerable force, and if they are of the cabbage or lettuce kind, or other plants whose leaves are to be used as food, they should subsequently be drenched thoroughly with pure water. Professor Lindley recommends syringing plants, as often as necessary to remove the lice, with a solution of half an ounce of strong carbonate of ammonia in one quart of water, which has the merit of being clean as well as effectual. Lice on the extremities of branches may be killed by bending over the branches and holding them for several minutes in warm and strong soapsuds, or in a solution of whale-oil soap.

Against the depredations of the plant-lice that sometimes infest potato-fields, dusting the plants with lime has been found a good remedy. Lice multiply much faster, and are more injurious to plants, in a dry than in a wet atmosphere; hence in green-houses, attention should be paid to keep the air sufficiently moist; and the lice are readily killed by fumigations with tobacco or with sulphur. To destroy subterranean lice on the roots of plants, I have found that watering with salt water was useful, if the plants were hardy; but tender herbaceous plants cannot be treated in this way, but may sometimes be revived, when suffering from these hidden foes, by free and frequent watering with soapsuds.

Plant-lice would undoubtedly be much more abundant and destructive, if they were not kept in check by certain redoubtable enemies of the insect kind, which seem expressly created to diminish their numbers. These lice-destroyers are of three sorts. The first are the young or larvæ of the hemispherical beetles familiarly known by the name of lady-birds, and scientifically by that of *Coccinella.* These little beetles are generally yellow or red, with black spots, or, black, with white, red, or yellow spots; there are many kinds of them, and they are very common and plentiful insects, and are generally diffused among plants. They live, both in the perfect and young state, upon plant-lice, and hence their services are very considerable. Their young are small flattened grubs (Fig. 93) of a bluish or blue-black color, spotted usually with red or yellow, and furnished with six legs near the fore part of the body. They are hatched from little yellow eggs, laid in clusters among the plant-lice, so that they find themselves at once within reach of their prey, which, from their superior strength, they are enabled to seize and slaughter in great numbers.

Fig. 93.

In July, 1848, a friend sent to me a whole brood of lady-bird grubs, which, being found upon potato-vines, were thought by some of his neighbors to be the cause of *the rot.* In a few weeks the grubs were transformed to beetles, about as big as half a pea, and having nine black dots on their dull orange-colored wing-shells. Hence they derive their name of *Coccinella novemnotata*, (Fig. 94, pupa and imago, and Plate II. Fig. 4,) the nine-dotted Coccinella. It need hardly be added, that these little insects were wholly innocent of all offence to the plants, upon which, when infested with

Fig. 94.

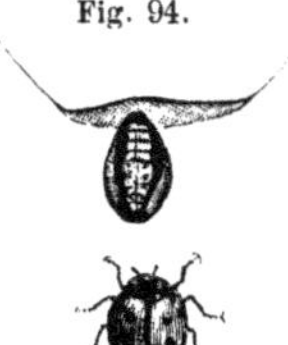

the common potato plant-lice, they may always be found. It is amusing, however, that both of these kinds of insects should have been charged with the same fault, one having no more to do with producing the disease than the other.

There are some lady-birds, of a very small size, and blackish color, sparingly clothed with short hairs, and sometimes with a yellow spot at the end of the wing-covers, whose young are clothed with short tufts or flakes of the most delicate white down. These insects belong to the genus *Scymnus*, which means a lion's whelp, and they well merit such a name, for their young, in proportion to their size, are as sanguinary and ferocious as the most savage beasts of prey. I have often seen one of these little tufted animals preying upon plant-lice, catching and devouring, with the greatest ease, lice nearly as large as its own body, one after another, in rapid succession, without apparently satiating its hunger or diminishing its activity.

The second kind of plant-lice destroyers are the young of the golden-eyed lace-winged fly, *Chrysopa perla*[12] (Plate III. Fig. 8). This fly is of a pale green color, and has four wings, resembling delicate lace, and eyes of the brilliancy of polished gold, as its generical name implies; but notwithstanding its delicacy and beauty, it is extremely disgusting from the offensive odor that it exhales. It suspends its eggs, by threads, in clusters beneath the leaves where plant-lice abound. The young, or larva, (Plate III. Fig. 9; Fig. 10, cocoon,) is a rather long and slender grub, provided with a pair of large curved and sharp teeth (*jaws*), moving laterally, and each perforated with a hole, through which it sucks the juices of its victims. The havoc it makes is astonishing; for one minute is all the time it requires to kill the largest plant-louse, and suck out the fluid contents of its body.

The last of the enemies of plant-lice are the maggots or

[[12] *Chrysopa perla* is not found in this country; probably *C. euryptera*, Burm., or some other species common to New England, will be found destructive of these pernicious plant-lice. — UHLER.]

young of various two-winged flies belonging to the genus *Syrphus*. Many of these flies are black, with yellow bands on their bodies. I have often seen them hovering over small trees and other plants, depositing their eggs, which they do on the wing, like the bot-fly, curving their tails beneath the leaves, and fixing here and there an egg, wherever plant-lice are discovered. Others lay their eggs near the buds of trees, where the young may find their appropriate nourishment as soon as they are hatched.

The young are maggots, which are thick and blunt behind, tapering and pointed before; their mouths are armed with a triple-pointed dart, with which they pierce their prey, elevate it above their heads, and feast upon its juices at leisure. Though these maggots are totally blind, they are enabled to discover their victims without much groping about, in consequence of the provident care of the parent flies, which leave their eggs in the very midst of the sluggish lice. Mr. Kirby says, that, on examining his currant-bushes, which but a week before were infested by myriads of aphides, not one was to be found; but beneath each leaf were three or four full-fed maggots, surrounded by heaps of the slain, the trophies of their successful warfare. He also says that he has found it very easy to clear a plant or small tree of lice, by placing upon it several larvæ of Coccinella or Syrphi.

3. Bark-lice. (*Coccidæ.*)

The celebrated scarlet in grain, which has been employed in Asia and the South of Europe, from the earliest ages, as a coloring material, was known to the Romans by the name of *Coccus*, derived from a similar Greek word, and was, for a long time, supposed to be a vegetable production, or grain, as indeed its name implies. At length it was ascertained that this valuable dye was an insect, and others agreeing with it in habits, and some also in properties, having been discovered, Linnæus retained them all under the same name. Hence in the genus *Coccus* are included, not only the *Thola* of the Phœnicians and Jews, the *Kermes* of the Arabians, or the *Coccus* of the Greeks and Romans, but the scarlet grain of Poland, and the still more valuable Cochenille of Mexico, together with various kinds of bark-lice, agreeing with the former in habits and structure.

These insects vary very much in form; some of them are oval and slightly convex scales, and others have the shape of a muscle; some are quite convex, and either formed like a boat turned bottom upwards, or are kidney-shaped, or globular. They live mostly on the bark of the stems of plants; some, however, are habitually found upon leaves, and some on roots. In the early state, the head is completely withdrawn beneath the shell of the body and concealed, the beak or sucker seems to issue from the breast, and the legs are very short and not visible from above. The females undergo only a partial transformation, or rather scarcely any other change than that of an increase in size, which in some species, indeed, is enormous, compared with the previous condition of the insect; but the males pass through a complete transformation before arriving at the perfect or winged state. In both sexes we find threadlike or tapering antennæ, longer than the head, but much shorter than those of plant-lice, and feet consisting of only one joint, terminated by a single claw. The mature female retains the beak or sucker, but does not acquire wings; the male on the contrary has two wings, but the beak disappears. In both there are two slender threads at the extremity of the body, very short in some females, usually quite long in the males, which moreover are provided with a stylet at the tip of the abdomen, which is recurved beneath the body.

The following account * contains a summary of nearly all that is known respecting the history and habits of these insects. Early in the spring the bark-lice are found apparently torpid, situated longitudinally in regard to the branch,

* It was drawn up by me in the year 1828, and published in the seventh volume of the "New England Farmer," pp. 186, 187.

the head upwards, and sticking by their flattened inferior surface closely to the bark. On attempting to remove them they are generally crushed, and there issues from the body a dark-colored fluid. By pricking them with a pin, they can be made to quit their hold, as I have often seen in the common species, *Coccus Hesperidum*, infesting the myrtle. A little later the body is more swelled, and, on carefully raising it with a knife, numerous oblong eggs will be discovered beneath it, and the insect appears dried up and dead, and only its outer skin remains, which forms a convex cover to its future progeny. Under this protecting shield the young are hatched, and, on the approach of warm weather, make their escape at the lower end of the shield, which is either slightly elevated or notched at this part. They then move with considerable activity, and disperse themselves over the young shoots or leaves.

The shape of the young Coccus is much like that of its parent, but the body is of a paler color and more thin and flattened. Its six short legs and its slender beak are visible under a magnifier. Some are covered with a mealy powder, as the *Coccus Cacti*, or cochenille of commerce, and the *Coccus Adonidum*, or mealy bug of our greenhouses. Others are hairy or woolly; but most of them are naked and dark-colored. These young lice insert their beaks into the bark or leaves, and draw from the cellular substance the sap that nourishes them.

Réaumur observed the ground quite moist under peach-trees infested with bark-lice, which was caused by the dripping of the sap from the numerous punctures made by these insects. While they continue their exhausting suction of sap, they increase in size, and during this time are in what is called the larva state. When this is completed, the insects will be found to be of different magnitudes, some much larger than the others, and they then prepare for a change that is about to ensue in their mode of life, by emitting from the under side of their bodies numerous little white downy threads, which are fastened, in a radiated manner, around their bodies, to the bark, and serve to confine them securely in their places. After becoming thus fixed they remain apparently inanimate; but under these lifeless scales the transformation of the insect is conducted; with this remarkable difference, that in a few days the large ones contrive to break up and throw off, in four or five flakes, their outer scaly coats, and reappear in a very similar form to that which they before had; the smaller ones, on the contrary, continue under their outer skins, which serve instead of cocoons, and from which they seem to shrink and detach themselves, and then become perfect pupæ, the rudiments of wings, antennæ, feet, &c. being discoverable on raising the shells.

If we follow the progress of these small lice, which are to produce the males, we shall see, in process of time, a pair of threads and the tips of the wings protruding beneath the shell at its lower elevated part, and through this little fissure the perfect insect at length backs out. After the larger lice have become fixed, and have thrown off their outer coats, they enter upon the pupa or chrysalis state, which continues for a longer or shorter period, according to the species. But when they have become mature, they do not leave the skins or shells covering their bodies, which continue flexible for a time. These larger insects are the females, and are destined to remain immovable, and never change their place after they have once become stationary. The male is exceedingly small in comparison to the female, and is provided with only two wings, which are usually very large, and lie flatly on the top of the body.

After the insects have paired, the body of the female increases in size, or becomes quite convex, for a time, and ever afterwards remains without alteration; but serves to shelter the eggs which are to give birth to her future offspring. These eggs, when matured, pass under the body of the mother, and the latter by degrees shrinks more and more, till nothing is left but the dry outer convex skin, and the insect perishes on the spot. Sometimes the insect's body is not large enough to cover all her eggs, in which case she beds them in a considerable quantity of the down that issues from the under or hinder part of her body (Fig. 95). There are several broods of some species in the year; of the bark-louse of the apple-tree at least two are produced in one season. It is probable that the insects of the second or last brood pair in the autumn, after which the males die, but the females survive the winter, and lay their eggs in the following spring.

Fig. 95.

Young apple-trees, and the extremities of the limbs of older trees, are very much subject to the attacks of a small species of bark-louse. The limbs and smooth parts of the trunks are sometimes completely covered with these insects, and present a very singularly wrinkled and rough appearance from the bodies which are crowded closely together. In the winter these insects are torpid, and apparently dead. They measure about one tenth of an inch in length, are of an oblong oval shape, gradually decreasing to a point at one end, and are of a brownish color very near to that of the bark of the tree. These insects resemble in shape one which was described by Réaumur* in 1738, who found it on the elm in France, and Geoffroy named the insect *Coccus arborum linearis*, while Gmelin called it *conchiformis* (Fig. 96). This, or one much like it, is very abundant upon apple-trees in England, as we learn from Dr. Shaw† and Mr.

Fig. 96.

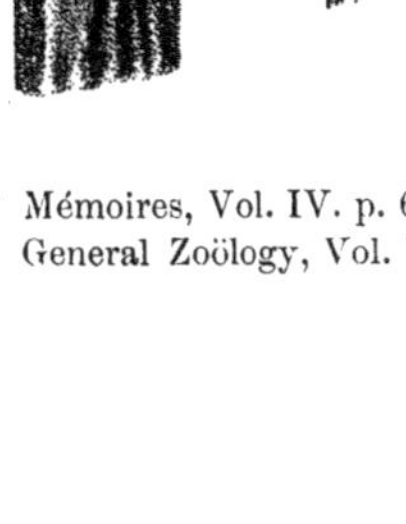

* Mémoires, Vol. IV. p. 69, plate 5, figs. 5, 6, 7.

† General Zoölogy, Vol. VI. Part I. p. 196.

Kirby;* and Mr. Rennie† states that he found it in great plenty on currant-bushes.

It is highly probable that we have received this insect from Europe, but it is somewhat doubtful whether our apple-tree bark-louse be identical with the species found by Réaumur on the elm; and the doubt seems to be justified by the difference in the trees and in the habits of the insects, our species being gregarious, and that of the elm nearly solitary. It is true that on some of our indigenous forest-trees bark-lice of nearly the same form and appearance have been observed; but it is by no means clear that they are of the same species as those on the apple-tree. The first account that we have of the occurrence of bark-lice on apple-trees, in this country, is a communication by Mr. Enoch Perley, of Bridgeton, Maine, written in 1794, and published among the early papers of the Massachusetts Agricultural Society.‡

These insects have now become extremely common, and infest our nurseries and young trees to a very great extent. In the spring the eggs are readily to be seen on raising the little muscle-shaped scales beneath which they are concealed. These eggs are of a white color, and in shape nearly like those of snakes. Every shell contains from thirty to forty of them, imbedded in a small quantity of whitish friable down. They begin to hatch about the 25th of May, and finish about the 10th of June, according to Mr. Perley. The young, on their first appearance, are nearly white, very minute, and nearly oval in form. In about ten days they become stationary, and early in June throw out a quantity of bluish-white down, soon after which their transformations are completed, and the females become fertile, and deposit their eggs. These, it seems, are hatched in the course of the summer, and the young come to their growth and provide for a new brood before the ensuing winter.

Among the natural means which are provided to check the increase of these bark-lice are birds, many of which,

* Introduction to Entomology, Vol. I. p. 201.

† Insect Transformations, p. 92.

‡ See Papers for 1796, p. 32.

especially those of the genera *Parus* and *Regulus*, containing the chickadee and our wrens, devour great quantities of these lice. I have also found that these insects are preyed upon by internal parasites, minute ichneumon-flies, and the holes (which are as small as if made with a fine needle), through which these little insects come forth, may be seen on the backs of a great many of the lice which have been destroyed by their intestine foes.

The best application for the destruction of the lice is a wash made of two parts of soft soap and eight of water, with which is to be mixed lime enough to bring it to the consistence of thick whitewash. This is to be put upon the trunks and limbs of the trees with a brush, and as high as practicable, so as to cover the whole surface, and fill all the cracks in the bark. The proper time for washing over the trees is in the early part of June, when the insects are young and tender. These insects may also be killed by using in the same way a solution of two pounds of potash in seven quarts of water, or a pickle consisting of a quart of common salt in two gallons of water.

There has been found on the apple and pear tree another kind of bark-louse, which differs from the foregoing in many important particulars, and approaches nearest to a species inhabiting the aspen in Sweden, of which a description has been given by Dalman in the "Transactions of the Royal Academy of Sciences of Stockholm," * for the year 1825, under the name of *Coccus cryptogamus*. This species is of the kind in which the body of the female is not large enough to cover her eggs, for the protection whereof another provision is made, consisting, in this species, of a kind of membranous shell, of the color and consistence almost of paper. In the autumn and throughout the winter, these insects are seen in a dormant state, and of two different forms and sizes on the bark of the trees.

The larger ones measure less than a tenth of an inch in length, and have the form of a common oyster-shell,

* Kongl. Vetenskaps Academ. Nya Handlingar.

being broad at the hinder extremity, but tapering towards the other, which is surmounted by a little oval brownish scale. The small ones, which are not much more than half the length of the others, are of a very long oval shape, or almost four-sided, with the ends rounded; and one extremity is covered by a minute oval dark-colored scale. These little shell-like bodies are clustered together in great numbers, are of a white color and membranous texture, and serve as cocoons to shelter the insects while they are undergoing their transformations. The large ones are the pupa-cases or cocoons of the female, beneath which the eggs are laid; and the small ones are the cases of the males, and differ from those of the females not only in size and shape, but also in being of a purer white color, and in having an elevated ridge passing down the middle. The minute oval dark-colored scales on one of the ends of these white cases are the skins of the lice while they were in the young or larva state, and the white shells are probably formed in the same way as the down which exudes from the bodies of other bark-lice, but which in these assumes a regular shape, varying according to the sex, and becoming membranous after it is formed. Not having seen these insects in a living state, I have not been able to trace their progress, and must therefore refer to Dalman's memoir above mentioned, for such particulars as tend to illustrate the remaining history of this species.

The body of the female insect, which is covered and concealed by the outer case above described, is minute, of an oval form, wrinkled at the sides, flattened above, and of a reddish color. By means of her beak, which is constantly thrust into the bark, she imbibes the sap, by which she is nourished; she undergoes no change, and never emerges from her habitation. The male becomes a chrysalis or pupa, and about the middle of July completes its transformations, makes its escape from its case, which it leaves at the hinder extremity, and the wings with which it is provided are reversed over its head during the operation, and are the last to be extricated. The perfect male is nearly as minute as a point, but a powerful magnifier shows its body to be divided into segments, and endued with all the important parts and functions of a living animal.

To the unassisted eye, says Dalman, it appears only as a red atom, but it is furnished with a pair of long whitish wings, long antennæ or horns, six legs with their respective joints, and two bristles terminating the tail. This minute insect perforates the middle of the case covering the female, and thus celebrates its nuptials with its invisible partner. The latter subsequently deposits her eggs and dies. In due time the young are hatched and leave the case, under which they were fostered, by a little crevice at its hinder part. These young lice, which I have seen, are very small, of a pale yellowish brown color, and of an oval shape, very flat, and appearing like minute scales. They move about for a while, at length become stationary, increase in size, and in due time the whitish shells are produced, and the included insects pass from the larva to the pupa state. The means for destroying these insects are the same as those recommended for the extermination of the previous species.

Many years ago, when on a visit from home, I observed on a fine native grape-vine, that was trained against the side of a house, great numbers of reddish-brown bark-lice, of a globular form, and about half as large as a small pea, arranged in lines on the stems. An opportunity for further examination of this species did not occur till the summer of 1839, when I was led to the discovery of a few of these lice on my Isabella grape-vines, by seeing the ants ascending and descending the stems. Upon careful search I discovered the lice, which were nearly of the color of the bark of the vine, partly imbedded in a little crevice of the bark, and arranged one behind another in a line. They drew great quantities of sap, as was apparent by their exudations, by which the ants were attracted. Further observations were arrested by a fire which consumed the house and the vines that were trained to it.

CHAPTER V.

LEPIDOPTERA.

CATERPILLARS. — BUTTERFLIES. — SKIPPERS. — HAWK-MOTHS. — ÆGERIANS OR BORING-CATERPILLARS. — GLAUCOPIDIANS. — MOTHS. — SPINNERS. — LITHOSIANS. — TIGER-MOTHS. — ERMINE-MOTHS. — TUSSOCK-MOTHS. — LACKEY-MOTHS. — LAPPET-MOTHS. — SATURNIANS. — CERATOCAMPIANS. — CARPENTER-MOTHS. — PSYCHIANS. — NOTODONTIANS. — OWL-MOTHS. — CUT-WORMS. — GEOMETERS, OR SPAN-WORMS, AND CANKER-WORMS. — DELTA-MOTHS. — LEAF-ROLLERS. — BUD-MOTHS. — FRUIT-MOTHS. — BEE-MOTHS. — CORN-MOTHS. — CLOTHES-MOTHS. — FEATHER-WINGED MOTHS.

THERE are perhaps no insects which are so commonly and so universally destructive as caterpillars; they are inferior only to locusts in voracity, and equal or exceed them in their powers of increase, and in general are far more widely spread over vegetation. Caterpillars are the young of butterflies and of moths; and of these, five hundred species, which are natives of Massachusetts, are already known to me, and probably there are at least as many more kinds to be discovered within the limits of this Commonwealth.[1] As each female usually lays from two hundred to five hundred eggs, one thousand different kinds of butterflies and moths will produce, on an average, three hundred thousand caterpillars; if one half of this number, when arrived at

[[1] The number of species in the United States may fairly be estimated at 3,500, or even more. My Catalogue, published by the Smithsonian Institute, contains the names of nearly 1,800 already described by various authors, exclusive of Microlepidoptera, which is a numerous family of itself, and comparatively little progress has as yet been made in the discovery of our indigenous species generally. The latest and most complete work on German and Swiss Lepidoptera (*Die Schmetterlinge Deutschlands und der Schweiz*, von H. v. Heinemann, Brunswick, 1859) gives 1,387 species, exclusive of Microlepidoptera, in those two countries alone, and we can confidently reckon on finding over three times that number in the United States. — MORRIS.]

maturity, are females, they will give forty-five millions of caterpillars in the second, and six thousand seven hundred and fifty millions in the third generation. These data suffice to show that the actual number of these insects, existing at any one time, must be far beyond the limits of calculation. The greater part of caterpillars subsist on vegetable food, and especially on the leaves of plants; hence their injuries to vegetation are immense, and are too often forced upon our notice. Some devour the solid wood of trees, some live only in the pith of plants, and some confine themselves to grains and seeds. Certain species attack our woollens and furs, thereby doing us much injury; even leather, meat, wax, flour, and lard afford nourishment to particular kinds of caterpillars.

Caterpillars vary greatly in form and appearance, but, in general, their bodies are more or less cylindrical, and composed of twelve rings or segments, with a shelly head, and from ten to sixteen legs. The first three pairs of legs are covered with a shelly skin, are jointed and tapering, and are armed at the end with a little claw; the other legs are thick and fleshy, without joints, but elastic or contractile, and are generally surrounded at the extremity by numerous minute hooks. There are six very small eyes[2] on each side of the head, two short antennæ, and strong jaws or nippers, placed at the sides of the mouth, so as to open and shut sidewise. In the middle of the lower lip is a little conical tube, from which the insects spin the silken threads that are used by them in making their nests and their cocoons, and in various other purposes of their economy. Two long and slender bags, in the interior of their bodies, and ending in the spinning tube, contain the matter of the silk. This is a sticky fluid, and it flows from the spinner in a fine stream, which hardens into a thread so soon as it comes

[[2] Though Dr. Harris mentions the "eyes" of caterpillars, yet be it understood, he does not *assert* that they *see*. It is very doubtful whether they have the faculty of vision. — MORRIS.]

to the air. Some caterpillars make but very little silk; others, such as the silk-worm and the apple-tree caterpillar, produce it in great abundance.

Some caterpillars herd together in great numbers, and pass the early period of their existence in society; and of these there are species which unite in their labors, and construct tents serving as a common habitation in which they live, or to which they retire occasionally for shelter. Others pass their lives in solitude, either exposed to the light and air, or sheltered in leaves folded over their bodies, or form for themselves silken sheaths, which are either fixed or portable. Some make their abodes in the stems of plants, or mine in the pulpy substance of leaves; and others conceal themselves in the ground, from which they issue only when in search of food.

Caterpillars usually change their skins about four times before they come to their growth. At length they leave off eating entirely, and prepare for their first transformation. Most of them, at this period, spin around their bodies a sort of shroud or cocoon, into which some interweave the hairs of their own bodies, and some employ, in the same way, leaves, bits of wood, or even grains of earth. Other caterpillars suspend themselves, in various ways, by silken threads, without enclosing their bodies in cocoons; and again, there are others which merely enter the earth to undergo their transformations.

When the caterpillar has thus prepared itself for the approaching change, by repeated exertions and struggles it bursts open the skin on the top of its back, withdraws the fore part of its body, and works the skin backwards till the hinder extremity is extricated. It then no longer appears in the caterpillar form, but has become a pupa or chrysalis, shorter than the caterpillar, and at first sight apparently without a head or limbs. On close examination, however, there may be found traces of a head, tongue, antennæ, wings, and legs, closely pressed to the body, to which these parts are cemented by a kind of varnish. Some chrysalids are angular, or furnished with little protuberances; but most of them are smooth, rounded at one end, and tapering at the other extremity. While in the pupa state these insects take no food, and remain perfectly at rest, or only move the hinder extremity of the body when touched. After a while, however, the chrysalis begins to swell and contract, till the skin is rent over the back, and from the fissure there issues the head, antennæ, and body of a butterfly or moth. When it first emerges from its pupa-skin the insect is soft, moist, and weak, and its wings are small and shrivelled; soon, however, the wings stretch out to their full dimensions, the superfluous moisture of the body passes off, and the limbs acquire their proper firmness and elasticity.

The conversion of a caterpillar to a moth or butterfly is a transformation of the most complete kind. The form of the body is altered, some of the legs disappear, the others and the antennæ become much longer than before, and four wings are acquired. Moreover, the mouth and digestive organs undergo a total change; for the insect, after its final transformation, is no longer fitted to subsist upon the same gross aliment as it did in the caterpillar state; its powerful jaws have disappeared, and instead thereof we find a slender tongue, by means of which liquid nourishment is conveyed to the mouth of the insect, and its stomach becomes capable of digesting only water and the honeyed juice of flowers.

Ceasing to increase in size, and destined to live but a short time after their final transformation, butterflies and moths spend this brief period of their existence in flitting from flower to flower and regaling themselves with their sweets, or in slaking their thirst with dew or with the water left standing in puddles after showers, in pairing with their mates, and in laying their eggs; after which they die a natural death, or fall a prey to their numerous enemies.

These insects belong to an order called LEPIDOPTERA, which means scaly wings; for the mealy powder with which their wings are covered, when seen under a powerful microscope, is found to consist of little scales, lapping over each other like the scales of fishes, and implanted into the skin of the wings by short stems. The body of these insects is also more or less covered with the same kind of scales, together with hair or down in some species. The tongue consists of two tubular threads placed side by side, and thus forming an instrument for suction, which, when not in use, is rolled up spirally beneath the head, and is more or less covered and concealed on each side by a little scaly or hairy jointed feeler. The shoulders or wing-joints of the fore wings are covered, on each side, by a small triangular piece, forming a kind of epaulette, or shoulder-cover; and between the head and the thorax is a narrow piece, clothed with scales or hairs sloping backwards, which may be called the collar. The wings have a few branching veins,[3] generally forming one or two large meshes on the middle. The legs are six in number, though only four are used in walking by some butterflies, in which the first pair are very short and are folded like a tippet on the breast; and the feet are five-jointed, and are terminated, each, by a pair of claws.

It would be difficult, and indeed impossible, to arrange the Lepidopterous insects according to their forms, appearance, and habits, in the caterpillar state, because the caterpillars of many of them are as yet unknown; and therefore it is found expedient to classify them mostly according to the characters furnished by them in the winged state.

We may first divide the Lepidoptera into three great sections, called butterflies, hawk-moths, and moths, corre-

[[3] The systematists of the present day determine genera, and even *species*, by the peculiar and various modifications of these veins. The main veins are called *nervures*, the branches *nervules*, and the whole system *Pterology*. The French and the Germans differ as to the names of the distinct veins, so that, unless a student knows to which of the schools a describer belongs, he would be apt to be misled. — MORRIS.]

sponding to the genera *Papilio*, *Sphinx*, and *Phalæna* of Linnæus.[4]

The BUTTERFLIES (*Papiliones*) have threadlike antennæ, which are knobbed at the end; the fore wings in some, and all the wings in the greater number, are elevated perpendicularly, and turned back to back, when at rest; they have generally two little spurs on the hind legs; and they fly by day only.

The HAWK-MOTHS (*Sphinges*) generally have the antennæ thickened in the middle, and tapering at each end, and most often hooked at the tip; the wings are narrow in proportion to their length, and are confined together by a bristle or bunch of stiff hairs on the shoulder of each hind wing, which is retained by a corresponding hook on the under side of each fore wing; all the wings, when at rest, are more or less inclined like a roof, the upper ones covering the lower wings; there are two pairs of spurs on the hind legs. A few fly by day, but the greater number in the morning and evening twilight.

In the MOTHS (*Phalænæ*) the antennæ are neither knobbed at the end nor thickened in the middle, but taper from the base to the extremity, and are either naked, like a bristle, or are feathered on each side; the wings are confined together by bristles and hooks, the first pair covering the hind wings, and are more or less sloping when at rest; and there are two pairs of spurs to the hind legs. These insects fly mostly by night.

I. BUTTERFLIES. (*Papiliones.*)

Besides the characters already given, which distinguish this section of the Lepidoptera, it may be stated that their

[[4] Modern writers divide them into two great divisions: 1st, *Rhopalocera*, with filiform antennæ, terminating in a club or knob, from ῥόπαλον, club, and κέρας, horn; and 2d, *Heterocera*, with antennæ of variable form, sometimes prismatic, linear, pectinated, plumose, &c., &c., from ἕτερος, variable, and κέρας, horn.— MORRIS.]

caterpillars always have sixteen legs; namely, two, which are tapering, jointed, and scaly, to each of the first three segments behind the head, and a pair of thick fleshy legs, without joints, to all the remaining segments, except the fourth, fifth, tenth, and eleventh.

The butterflies are divisible into two tribes; namely, the true butterflies, which carry all their wings upright when at rest; and the skippers, which have only the fore wings upright, the hind wings being nearly horizontal when at rest.

1. Butterflies.

In these insects all the wings are erect when at rest, and the antennæ are knobbed, but never hooked, at the end. Their caterpillars have a head of moderate size, suspend themselves by the tail when about to transform, and are not enclosed in cocoons. Some of these butterflies have the six legs all equally fitted for walking; their caterpillars are more or less cylindrical, and secure themselves by a transverse band, as well as by the tail, previously to their transformation to chrysalids; and the latter are angular. All these characters exist in the following species.

In the month of June there may be found on the leaves of the parsley and carrot certain caterpillars, (Plate IV. Fig. 6,) more commonly called parsley-worms, which are somewhat swelled towards the fore part of the body, but taper a little behind. When first hatched they are less than one tenth of an inch in length, are of a black color, with a broad white band across the middle, and another on the tail; and the back is studded with little black projecting points. After they have increased in size, and have cast their coats, it is found that the white band covers only the sixth and seventh segments, that the black projecting points spring from spots of an orange color, and on the lower part of the sides is a row of white spots, two more spots of the same color on the top of the first segment, and one larger spot on the tail. These caterpillars alter in color and appearance with each successive moulting, and before they are half grown the projecting points and the white band and spots entirely disappear, the skin becomes perfectly smooth, and of a delicate apple-green color, rather paler at the sides of the body and whitish beneath, and on each segment there is a transverse band consisting of black and yellow spots alternately arranged. When touched, they thrust forth, from a slit in the first segment of the body, just behind the head, a pair of soft orange-colored horns, growing together at the bottom, and somewhat like the letter Y in form. The horns are scent-organs, and give out a strong and disagreeable smell, perceptible at some distance, and seem to be designed to defend the caterpillars from the annoying attacks of flies and ichneumons. These caterpillars usually come to their full size between the 10th and 20th of July, and then measure about one inch and a half in length. After this they leave off eating, desert the plants, and each one seeks some sheltered spot, such as the side of a building or fence, or the trunk of a tree, where it prepares for its transformation. It first spins a little web or tuft of silk against the surface whereon it is resting, and entangles the hooks of its hindmost feet in it, so as to fix them securely to the spot; it then proceeds to make a loop or girth of many silken threads bent into the form of the letter U, the ends of which are fastened to the surface on which it rests on each side of the middle of its body; and under this, when finished, it passes its head, and gradually works the loop over its back, so as to support the body, and prevent it from falling downwards. Though it generally prefers a vertical surface on which to fasten itself in an upright posture, it sometimes selects the under side of a limb or of a projecting ledge, where it hangs suspended, nearly horizontally, by its feet and the loop. Within twenty-four hours after it has taken its station, the caterpillar casts off its caterpillar-skin and becomes a chrysalis, or pupa, (Plate IV. Fig. 7,) of a pale green, ochre-yellow, or ash-gray color, with two short ear-like projections above the head, just below which, on the upper part of the back, is a little prominence like a pug-nose. The chrysalis hangs in the same way as the caterpillar, and remains in this state from nine to fifteen days, according to the temperature of the atmosphere, cold and wet weather having a tendency to prolong the period. When this is terminated, the skin of the chrysalis bursts open, and a butterfly issues from it, clings to the empty shell till its crumpled and drooping wings have extended to their full dimensions, and have become dried, upon which it flies away in pursuit of companions and food.

This butterfly is the *Papilio Asterias*[5] of Cramer. (Plate IV. Fig. 4.) It is of a black color, with a double row of yellow dots on the back; a broad band, composed of yellow spots, across the wings, and a row of yellow spots near the hind margin; the hind wings are tailed, and have seven blue spots between the yellow band and the outer row of yellow spots, and, near their hinder angle, an eye-like spot of an orange color with a black centre; and the spots of the under side are tawny orange. The female (Plate IV. Fig. 5) differs from the male, above described, in having only a few small and distinct yellow spots on the upper side of the wings. The wings of this butterfly expand from three and a half to four inches.

During the month of July the Asterias butterflies may be seen in great abundance upon flowers, and particularly on those of the sweet-scented Phlox. They lay their eggs, in this and the following month, on various umbellate plants, placing them singly on different parts of the leaves and stems. I have found the caterpillars on the parsley, carrot, parsnip, celery, anise, dill, caraway, and fennel of our gardens, as well as on the conium, cicuta, sium, and other native plants of the same natural family, which originally

[5 The synonymes of *P. Asterias* are *P. Troilus* Smith Abbot, I. pl. 1; *P. Ajax* Clerck, Icon., t. 83; *P. polyxenes* Fab. — MORRIS.]

constituted the appropriate food of these insects, before the exotic species furnished them with a greater variety and abundance.

Their injury to these cultivated plants is by no means inconsiderable; they not only eat the leaves, but are particularly fond of the blossoms and young seeds. I have taken twenty caterpillars on one plant of parsley, which was going to seed. The eggs laid in July and August are hatched soon afterwards, and the caterpillars come to their growth towards the end of September, or the beginning of October; they then suspend themselves, become chrysalids, in which state they remain during the winter, and are not transformed to butterflies till the last of May or the beginning of June in the following year.

I know of no method so effectual for destroying these caterpillars as gathering them by hand and crushing them. An expert person will readily detect them by their ravages on the plants which they inhabit; and a few minutes devoted, every day or two, to a careful search in the garden, during the season of their depredations, will suffice to remove them entirely.

There is another butterfly which bears a close resemblance to the female of the Asterias butterfly, and is nearly of the same size; but the blue spots on the hind wings are much larger, and cover nearly one third of the surface; the yellow spots around the margin are larger and paler; the eye-like spot near the hind angle has not a black centre, and there is a large orange-colored spot near the middle of the front margin of the same wings. This species is the Troilus butterfly, or *Papilio Troilus* of Linnæus.

The caterpillar is entirely different from that of the Asterias butterfly. It lives on the leaves of the sassafras-tree, upon the upper surface of which it spins a little web, and folds over the sides of the leaf so as to form a furrow or case, in which it resides. The fore part of its body is large and swollen, and it tapers thence to the tail. When first hatched it is slate-colored above, with a black spot like an eye on each side of the third segment, below and behind which is a large and long white spot, and the top of the eleventh segment is white. After changing its skin, it becomes of a pale brownish olive color, the white spots disappear, and on the top of the back we find two rows of minute blue dots. When fourteen or fifteen days old it changes its skin and its colors again, the back becoming pea-green, with blue dots, the sides yellowish, and the head, belly, and legs pink; there is a transverse black line on the top of the first segment, and there are two large orange-colored spots on the fourth segment, and two of the same color, with a black centre, on the third segment. The caterpillar retains these colors from ten to sixteen days, increasing greatly in size during this period, and finally attains to the length of two inches or more. It comes to its full growth when about four weeks old, and then eats no longer, but, deserting its leafy habitation, it seeks a suitable place in which to undergo its transformation, previously to which it casts off its green coat, and appears in one of an ochre-yellow color. It then suspends itself in the same way as the caterpillar of the Asterias butterfly, and within two or three days after its last change of skin it moults again, and becomes a chrysalis.

The chrysalis is generally of a pale wood-color, smoother than that of the preceding species, and with rather longer and sharper ear-like projections. The chrysalids, which are produced from caterpillars hatched in August and September, remain unchanged through the winter, and are not transformed to butterflies till the middle of the following June. It is possible that these butterflies may lay their eggs so early as to produce a brood of caterpillars in the summer, and these may come to their growth, and pass through their transformations, before September; but I have only found the caterpillars towards the end of summer. I once discovered them on the leaves of the lilac, on which they appeared to thrive quite as well as on the sassafras.

One more butterfly is found in Massachusetts, resembling the preceding in its larva state and in its habits. It is our largest species, expanding from four and a half to five inches. The prevailing color of the wings is yellow, with a broad black margin, on which is a row of yellow spots; the fore

wings have four short black bands extending from their front edge, and the hind wings are tailed, and are ornamented with an orange-red spot near the hind angle. It is the *Papilio Turnus* of Linnæus (Fig. 97).*

The caterpillar of the Turnus butterfly (Fig. 98) lives upon the leaves of apple and wild-cherry trees, folding them up in the same way as does that of the Troilus butterfly, which, moreover,

Fig. 98.

[* In this figure, and others which follow, the under side of the wing, detached from the body of the insect, is represented, as well as the upper side, which in this figure is on the left, and connected with the body. — ED.]

it resembles in form. When fully grown, it measures from two to two and a half inches in length; it is of a green color above, with little blue dots in rows, a yellow eye-spot with a black centre on each side of the third segment, a yellow and black band across the fourth segment, and the head, belly, and legs are pink. It suspends itself and becomes a chrysalis about the first of August, and is not changed to a butterfly till the month of June in the following summer. Great numbers of these butterflies are sometimes seen around puddles of water left by rain in New Hampshire, where this species is much more common and abundant than in Massachusetts.

The caterpillars of the three foregoing species are the only ones in Massachusetts which are provided with forked scent-organs, capable of being withdrawn and concealed within the first segment of the body. All which follow are destitute of this means of defence.

In Europe there are several kinds of caterpillars which live exclusively on the cruciferous or oleraceous plants, such as the cabbage, broccoli, cauliflower, kale, radish, turnip, and mustard, and oftentimes do considerable injury to them. The prevailing color of these caterpillars is green, and that of the butterflies produced from them, white.

They belong to a genus called *Pontia;* in which the hind wings are not scalloped nor tailed, but are rounded and entire on the edges, and are grooved on the inner edge to receive the abdomen; the feelers are rather slender, but project beyond the head; and the antennæ have a short flattened knob; their caterpillars are nearly cylindrical, taper a very little towards each end, and are sparingly clothed with short down, which requires a microscope to be distinctly seen; they suspend themselves by the tail and a transverse loop; and their chrysalids are angular at the sides, and pointed at both ends.

In the northern and western parts of Massachusetts there is a white butterfly, which, in all its states, agrees with the foregoing characters. It is the *Pontia oleracea*[6] (Fig. 99), potherb Pontia, or white butterfly, and was first described by me in the year 1829, in the seventh volume of the "New England Farmer."* About the last of May, and the beginning of June, it is seen fluttering over cabbage, radish, and turnip beds, and patches of mustard, for the purpose of depositing its eggs. These are fastened to the under sides of the leaves, and but seldom more than three or four are left upon one leaf. The eggs are yellowish, nearly pear-shaped, longitudinally ribbed, and are one fifteenth of an inch in length. They are hatched in a week or ten days after they are laid, and the caterpillars produced from them attain their full size when three weeks old, and then measure about one inch and a half in length. Being of a pale green color, they are not readily distinguished from the ribs of the leaves beneath which they live. They do not devour the leaf at its edge, but begin indiscriminately upon any part of its under side, through which they eat irregular holes.

Fig. 99.

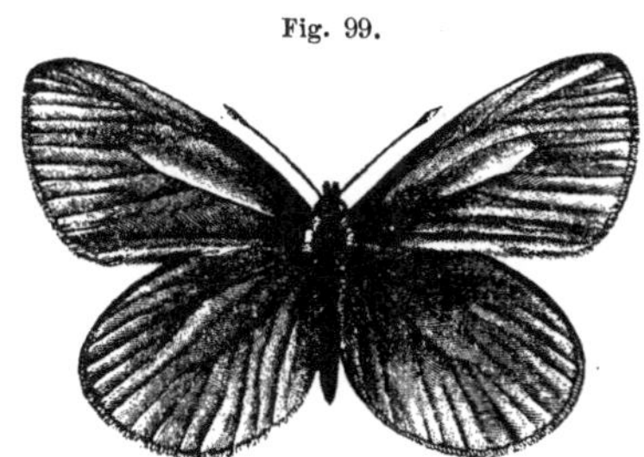

When they have completed the feeding stage, they quit the plants, and retire beneath palings, or the edges of stones, or into the interstices of walls, where they spin a little tuft of silk, entangle the hooks of their hindmost feet in it, and then proceed to form a loop to sustain the fore part of the body in a horizontal or vertical position. Bending its head on one side, the caterpillar fastens to the surface, beneath the middle of its body, a silken thread, which it carries across

[[6] *Pontia oleracea* belongs to the genus *Pieris* Schrk. (Morris's Catalogue). The *P. casta* of Kirby, in Faun. Bor., IV. 288, is only a variety of Harris's *P. oleracea;* and Kirby's *casta* is the *cruciferarum* of Boisd. Spec. Gen., I. 519.—MORRIS.]

* Page 402. For a figure of it, see "Lake Superior," by Agassiz and Cabot, pl. 7, fig. 1.

its back and secures on the other side, and repeats this operation till the united threads have formed a band or loop of sufficient strength. On the next day it casts off the caterpillar skin, and becomes a chrysalis. This is sometimes of a pale green, and sometimes of a white color, regularly and finely dotted with black; the sides of the body are angular, the head is surmounted by a conical tubercle, and over the fore part of the body, corresponding to the thorax of the included butterfly, is a thin projection, having in profile some resemblance to a Roman nose.

The chrysalis state lasts eleven days, at the expiration of which the insect comes forth a butterfly. The wings are white, but dusky next to the body; the tips of the upper ones are yellowish beneath, with dusky veins; the under side of the hinder wings is straw-colored, with broad dusky veins, and the angles next to the body are deep yellow; the back is black, and the antennæ are blackish, with narrow white rings, and ochre-yellow at the tips. The wings expand about two inches.

I have seen these butterflies in great abundance during the latter part of July and the beginning of August, in pairs, or laying their eggs for a second brood of caterpillars. The chrysalids produced from this autumnal brood survive the winter, and the butterflies are not disclosed from them till May or June. In gardens or fields infested by the caterpillars, boards, placed horizontally an inch or two above the surface of the soil, will be resorted to by them when they are about to change to chrysalids, and here it will be easy to find, collect, and destroy them, either in the caterpillar or chrysalis state. The butterflies also may easily be taken by a large and deep bag-net of muslin, attached to a handle of five or six feet in length; for they fly low and lazily, especially when busy in laying their eggs. In Europe the caterpillars of the white butterflies are eaten by the larger titmouse (*Parus major*), and probably our own titmouse or chickadee, with other insect-eating birds, will be found equally useful, if properly protected.

Twice a year our pastures and road-sides are enlivened by great numbers of the small yellow Philodice butterfly (*Colias Philodice* of Godart). (Fig. 100, male; Fig. 101, female.) They begin to appear towards the end of April, are common throughout the month of May, after which no more are seen till near the end of July, when a new brood begins to come forth, and some of them continue till late in the autumn. Their wings are yellow, with a black hind border, which in the females is quite broad on the fore wings, and spotted with yellow; the fringes of the wings, the antennæ, and the shanks are red; the fore wings have a small narrow black spot on both sides near the middle; the hind wings have a round orange-colored spot in the middle of the upper side, which on the under side is replaced by a large and a small silvery spot close together, and surrounded by a rust-colored ring.

Fig. 100.

Fig. 101.

The males are generally smaller than the females. The caterpillars live upon clover, medicago, and lucerne, and I have occasionally found them on pea-vines. They are green, slightly downy, paler or yellowish at the sides, and grow to the length of about one inch and a half. They suspend themselves to the stems of plants by the tail and a transverse loop, in the same way as the preceding species. The chrysalis (Fig. 102) is straw-colored, not angulated at the sides, with a slight prominence over the thorax, and the anterior extremity ends in a short and blunt point. The genus *Colias*, to which the Philodice butterfly belongs, is distinguished by the following characters. Six legs formed for walking; short antennæ, gradually thickened towards the end; wings entire, hinder ones rounded, with a gutter on their inner edge to receive the abdomen, and the central mesh closed behind by an angular vein; caterpillars cylindrical, smooth or downy; not striped on the top of the back; suspending themselves by the tail and a loop round the body; chrysalids somewhat gibbous or bulging, not angulated at the sides, and conical at the upper extremity.

Fig. 102.

We have several kinds of small six-footed butterflies, some of which are found, during the greater part of the summer, in the fields and around the edges of woods, flying low and frequently alighting, and oftentimes collected together in little swarms on the flowers of the clover, mint, and other sweet-scented plants. Their caterpillars secure themselves by the hind feet and a loop, when about to transform; but they are very short and almost oval, flat below and more or less convex above, with a small head, which is concealed under the first ring; and the feet, which are sixteen in number, are so short, that these caterpillars in moving seem to glide rather than creep. The chrysalids (Fig. 103) are short and thick, with the under side flat, the upper side very convex, and both extremities rounded or obtuse. They belong to a little group which may be called Lycenians (Lycænadæ), from the principal genus included in it.

Fig. 103.

The most common of these butterflies has generally been mistaken for the European *Lycæna Phlæas*, but I am convinced that it is distinct, and propose to call it the American copper butterfly, *Lycæna Americana* (Fig. 104). The fore wings on the upper side are coppery red, with about eight small square black spots, and the hind margin broadly bordered with dusky brown; hind wings with a few small black spots on the middle, and a broad coppery-red band on the hind margin. The wings expand from $1\frac{1}{10}$ to $1\frac{1}{5}$ inch. This butterfly is found throughout the summer fluttering on the grass and other low plants. The caterpillar is long, oval, and slightly convex above, and of a greenish color; it probably lives, like the *Phlœas*, on the leaves of dock and sorrel. The chrysalis, which is usually suspended under a stone, is light yellowish-brown, and spotted with black dots.

Fig. 104.

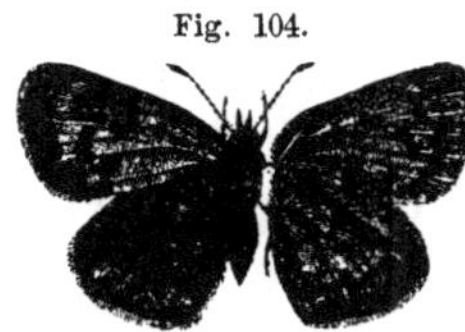

The Epixanthe butterfly, *Lycæna Epixanthe* (Boisduval), resembles the preceding in form and size, but is of a dusky brown color above, with a few black spots on the middle of the wings, and a narrow, wavy band, or a few contiguous spots of an orange color on the hinder margin of the posterior wings. This species is rather rare. The wings in both these butterflies are entire, or not notched or tailed, and the knobs of their antennæ are short, thick, and nearly oval.

There are others with the hind wings also entire and rounded, but the knobs of the antennæ are longer and not near so thick, and their caterpillars are shorter and very convex above. These characters exist in the beautiful azure-blue butterfly, *Polyommatus Pseudargiolus* (Boisd.), (Fig. 105, male, Fig. 106, var. profile,) which measures from $1\frac{1}{10}$ to $1\frac{1}{5}$ inch across the wings. These in the male are light blue on the upper side, with the lustre of satin; the fore wings of the female have a broad blackish outer margin, and on that of the hind wings is a row of small blackish spots; all the wings on the under side are pearl-gray, with little blackish spots; the fringes of the wings are white.

Fig. 105.

Fig. 106.

The blue Lucia butterfly (*Polyommatus Lucia* of Kirby) greatly resembles the preceding, but the black border of the fore wings in the female is not so broad, the fringes of the wings are spotted with black, and all the wings on the under side are dusky gray, with larger blackish spots, and a broad blackish border behind. Mr. Kirby has described only the male of this butterfly, in the fourth volume of the Fauna Boreali-Americana. It is found in April and May.

The Comyntas butterfly (*Polyommatus Comyntas* of Godart) is readily distinguished from the foregoing by having a little thread-like tail on the edge of the hind wings. The wings in the males are violet blue, and in the females blackish glossed with blue on the upper side, with whitish fringes; there are several blackish spots around the hind margins, and on the hind wings near the posterior margin two crescents of a deep orange-color. The under sides of all the wings are gray, with black spots encircled with white, and each of the two orange-colored crescents of the hind wings encloses a deep black spot encircled with silvery blue. The wings expand about one inch. This butterfly is found in dry woods and pastures in July and August, and the caterpillars live on the leaves of the *Lespedeza*, which grows in those places. They are oval, convex, and downy, of a pale green color with three darker green lines, the sides of the body reddish, and the head black. The chrysalis, which is usually fastened to a leaf, is at first pale green, but becomes brownish afterwards; it is sparingly clothed with whitish hairs, and there are three rows of black dots on the back. The chrysalis state lasts from nine to eleven days.

We have several more of these small butterflies with thread-like tails on their hind wings, but they differ from all the preceding species in having the knobs of the antennæ longer and nearly cylindrical, the eyes covered with a very fine down, and an oval opaque spot on the fore wings, near the front margin in the males. They belong to the genus *Thecla*. Their caterpillars are longer and flatter than those in the genus *Polyommatus*, and they usually live on trees. One of our largest kinds is the Falacer butterfly (*Thecla Falacer* of Godart). Its wings expand from $1\frac{1}{10}$ inch to $1\frac{4}{10}$ inch, are dark brown on the upper side, with two slender tails, one of which is very short, on each of the hind wings; and on the hind margin of the same wings is an orange-colored spot, larger and more conspicuous in the females than in the other sex; the under side of the wings is lighter brown; and on each wing near the middle is a dark-brown spot margined within and without with white, and beyond the middle there are two rows of spots of the same color, bordered on one side only with white; besides these spots, there are on the hind wings near the margin three or four orange-colored crescents, the inner one of which is separated from the others by a large blue spot. This insect is found among bushes in July and August. The caterpillar is said to live upon various kinds of hawthorns.

The streaked Thecla (*Thecla strigosa*) has a long and a short tail on each of the hind wings, and is of a dark-brown color without spots on the upper side; the wings beneath are ornamented with wavy transverse white streaks, and near the hind margin of the posterior wings is a row of deep orange-colored crescents, with a large blue spot near the hindmost angle. It measures one inch and one tenth across the wings. I took it on Blue Hill on the 1st of August. In the markings of the under side of the wings it nearly resembles *Thecla Liparops*.

The heads of the common hop are frequently eaten by the little green and downy caterpillars of a very pretty butterfly, which has been mistaken for the *Thecla Favonius*, figured in Mr. Abbot's "Natural History of the Insects of Georgia"; but it differs from it in so many respects, that I do not hesitate to give it another name, and will therefore call it the hop-vine Thecla, *Thecla Humuli* [7] * (Plate IV. Fig. 3).

[[7] *T. Humuli* is the *T. melinus* of Hübner. — Morris.]

* M. Boisduval has figured and described this species under the name of *Thecla Favonius* in his "Histoire des Lépidoptères de l'Amérique Septentrionale."

The wings on the upper side are dusky brown, with a tint of blue-gray, and, in the males, there is an oval darker spot near the front edge; the hind wings have two short, thread-like tails, the inner one the longest, and tipped with white; along the hind margin of these same wings is a row of little pale blue spots, interrupted by a large orange-red crescent enclosing a small black spot; the wings beneath are slate-gray, with two wavy streaks of brown edged on one side with white, and on the hind wings an orange-colored spot near the hind angle, and a larger spot of the same color enclosing a black dot just before the tails. It expands one inch and one tenth.

The last of these butterflies with two tails to each of the hind wings, does not seem to have been described, unless it is to be referred to the *Simaethis* of Drury, the *Damon* of Cramer, or the *Smilacis* of Boisduval, with the descriptions of which it does not fully agree. I propose, therefore, to call it the Auburn Thecla (*Thecla Auburniana*), from a favorite spot near Cambridge, formerly known by the name of Sweet Auburn, where I have repeatedly taken it before the place was converted to a cemetery. As in the preceding species, the outermost of the tails is very short, and often nothing remains of it but a little tooth on the edge of the wing. It varies considerably in color; the females are generally deep brown above, but sometimes the wings are rust-colored or tawny in the middle, as they always are in the males; the oval opaque spot which characterizes the latter sex is ochre-yellow. Upon the under side the wings in both sexes are green, the anterior pair tinged with brown from the middle to the inner edge; externally, next to the fringe, they are all margined by a narrow wavy white line, bordered internally with brown; this line on the fore wings does not reach the inner margin; on the hind wings it consists of six spots arranged in a zigzag manner, and the last spot next to the inner margin is remote from the rest; besides these there are on the same wings three more white spots bordered with brown between the zigzag band and the base; and between the same band and the margin three black spots, behind the middle one of which is a rust-red spot with a black centre. The wings expand from $1\frac{1}{20}$ to $1\frac{1}{10}$ inch. This pretty species is found on the mouse-ear (*Gnaphalium plantagineum*) in May, and on the flowers of the spearmint in August.

Some kinds of *Thecla* have the hind edges of the wings notched, but not tailed. This is the case with the Niphon butterfly (*Thecla Niphon* of Hübner), (Fig. 107,) which has been taken at Sweet Auburn early in May. As in the Auburn butterfly, the wings are deep brown above, with a large rusty space on each; the notches on their edges are white, and the teeth between them are rounded and of a black color; on the under side the wings are light brown, with dark brown wavy and zigzag lines, two of which are bordered on one side with white. The wings expand $1\frac{1}{5}$ inch.

Fig. 107.

The Mopsus butterfly (*Thecla Mopsus* of Hübner) differs from all the foregoing in having the hind wings entire and not tailed; but the inner angle projects a little, as it does in some species of *Lycæna*. In form, and in the color and arrangement of the spots on the under side of the wings, it approaches to the *Phlœas* and *Americana;* but in these species the eyes are not downy, and the males have not the oval opaque spot near the front margin of the anterior wings. The Mopsus butterfly is dark brown above, with a row of seven or eight deep orange-colored spots near the margin of the hind wings, larger and much more conspicuous on the under than on the upper side. The wings beneath are light brown, with a row of deep orange or vermilion-colored spots near the hind margins of all the wings, an inner and more irregular row of small black spots encircled with white on the same, and two more similar spots close together on the middle of the hind wings. It expands $1\frac{4}{10}$ inch. My only specimen of this fine butterfly was taken at Sandwich, by Mr. John Bethune.

Fig. 108.

Thecla Augusta.

Some butterflies have the first pair of legs so much shorter than the others that they cannot be used in walking, and are folded on the breast like a tippet. Their caterpillars, when about to transform, do not make a loop to support the fore part of the body, but suspend themselves vertically by the hindmost feet. As they all secure themselves pretty much in the same way, it may be proper to explain the process. Having finished eating, the caterpillar wanders about till it has discovered a suitable situation in which to pass through its transformations. This may be the under side of a branch or of a leaf, or any other horizontal object beneath which it can find sufficient room for its future operations.

Here it spins a web or tuft of silk, fastening it securely to the surface beneath which it is resting, entangles the hooks of its hindmost feet among the threads, and then contracts its body and lets itself drop so as to hang suspended by the hind feet alone, the head and fore part of the body being curved upwards in the form of a hook. After some hours, the skin over the bent part of the body is rent, the fore part of the chrysalis protrudes from the fissure, and, by a wriggling kind of motion, the caterpillar-skin is slipped backwards till only the extremity of the chrysalis remains attached to it. The chrysalis has now to release itself entirely from the caterpillar-skin, which is gathered in folds around its tail, and to make itself fast to the silken tuft by the minute hooks with which the hinder extremity is provided. Not having the assistance of a transverse loop to support its body while it disengages its tail, the attempt would seem perilous in the extreme, if not impossible. Without having witnessed the operation, we should suppose that the insect would inevitably fall, while endeavoring to accomplish its object. But, although unprovided with ordinary limbs, it is not left without the means to extricate itself from its present difficulty.

The hinder and tapering part of the chrysalis consists of several rings or segments, so joined together as to be capable of moving from side to side upon each other; and these supply to it the place of hands. By bending together two of these rings near the middle of the body, the chrysalis seizes, in the crevice between them, a portion of the empty caterpillar-skin, and clings to it so as to support itself while it withdraws its tail from the remainder of the skin.

It is now wholly out of the skin, to which it hangs suspended by nipping together the rings of its body; but, as the chrysalis is much shorter than the caterpillar, it is yet at some distance from the tuft of silk, to which it must climb before it can fix in it the hooks of its hinder extremity. To do this, it extends the rings of its body as far apart as possible, then, bending together two of them above those by which it is suspended, it catches hold of the skin higher up, at the same time letting go below, and, by repeating this process with different rings in succession, it at length reaches the tuft of silk, entangles its hooks among the threads, and then hangs suspended without further risk of falling.. It next contrives to dislodge the cast caterpillar-skin by whirling itself around repeatedly, till the old skin is finally loosened from its attachment and falls to the ground. The whole of this operation, difficult as it may seem, is performed in the space of a very few minutes, and rarely does the insect fail to accomplish it successfully and safely.

We may see the whole process in the caterpillars of the Archippus butterfly (*Danais Archippus* of Fabricius), which lives on the common silk-weed or milk-weed (*Asclepias Syriaca*) in June and July. This caterpillar is cylindrical, with a pair of thread-like black horns on the top of the second segment, and a shorter pair on the eleventh segment, and its body is marked with alternate transverse bands of yellow, black, and white. It comes to its growth in about fourteen days, during which it changes its skin three times, and finally attains to the length of nearly two inches. The chrysalis is about an inch long, but very thick, nearly cylindrical in the middle, and rounded at each end, with a very slender black point, by which it is suspended. Its skin is exceedingly thin and delicate, of a light green color, and ornamented with golden spots and a transverse stripe of black and gold. The chrysalis state lasts ten or twelve days, at the expiration of which the butterfly comes forth. The Archippus butterfly is very common on flowers, particularly on low lands, from the middle of July to the first of September. The wings on the upper side are tawny orange, on the under side deep nankin-yellow; they are surrounded by a black border spotted with white; the veins are black, and there are several yellow and white spots on the black tips of the fore wings. The males are distinguished by an elevated black spot contiguous to one of the veins near the middle of the hind wings. This butterfly measures across the wings from 3¾ to 4½ inches. The antennæ in the genus Danais have a long and curved knob; the head and thorax are spotted with white; the males have an elevated spot near the middle of the hind wings, which in both sexes are rounded, and never tailed or indented. The caterpillars are furnished with projecting thread-like horns in pairs, and the chrysalids are short and thick, somewhat oval, and are ornamented with golden spots. The other characters of the genus are the same as those of the division to which it belongs.

We have another four-footed butterfly which closely resembles the Archippus in color and markings, but differs from it entirely in the chrysalis and caterpillar state. It is the Disippe butterfly (*Nymphalis Disippe** of Godart). (Fig. 109.) It is of a tawny yellow above, and of a paler yellow beneath, the wings are surrounded by a broad black border spotted with white, the veins are black, there is a triangular patch spotted with white near the tips of the fore wings, and

* This is the *Misippus* of Fabricius, but not of Linnæus.

on the hind wings a curved black band. It expands from three to three and a half inches. The caterpillar lives on the poplar and willow; it is of a pale brown color, more or less variegated with white on the sides, and sometimes with green on the back; the head is notched on the top; there is a hump on the second segment, from which proceed two

Fig. 109.

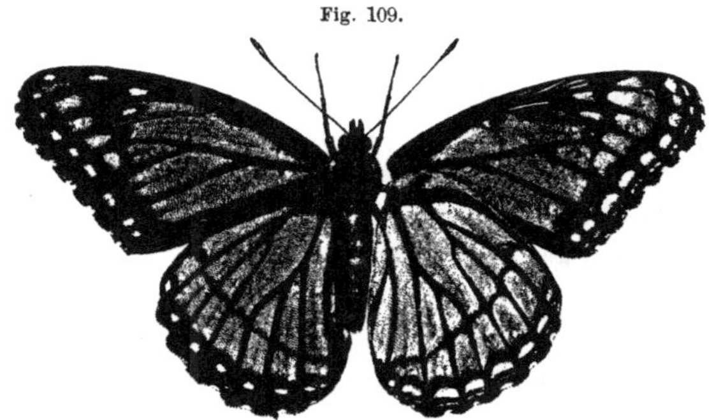

slender blackish horns, barbed on all sides with little points; the third, fourth, and fifth segments are also somewhat humped above, and on the tenth and eleventh are short tubercles. It suspends itself by the hind feet, before changing to a chrysalis. The latter is angular, and tapers towards the tail; it is of a pale brown or ashen-gray color, with the sides of the back and the extremity of the body whitish; and there is a thin almost circular projection standing vertically on its edge on the middle of the back. The butterfly appears in September, and lays its eggs for a second brood of caterpillars, which are transformed to chrysalids in the autumn, and remain without further change till the following spring, when they are changed to butterflies.

The genus *Nymphalis* * is readily distinguished by the following characters. Four-footed butterflies, with a long straight and slender knob to the antennæ, the edges of the

* The name Limenitis, under which I formerly included our species, is now appropriated by Dr. Boisduval to certain butterflies of the eastern continent, such as the *Camilla*, &c.

wings, particularly of the hinder ones, scalloped but not tailed, the inner margin grooved so as to receive and conceal the abdomen below, no closed mesh in the middle of the wings, and no elevated spot on them in the males; caterpillars and chrysalids in form like those of the Disippe, and suspended only by the hindmost extremity.

The caterpillar of the Ephestion butterfly (*Nymphalis Ephestion* of Stoll) is of a brownish color, more or less variegated with white on the sides, and with green above, and, like that of the Disippe, has two long barbed brown horns on the second segment. I have found it on the scrub-oak (*Quercus ilicifolia*) in June, but Mr. Abbot says it lives on the whortleberry-bush and the cherry-tree.

The chrysalis is not to be distinguished from that of the Disippe in form and color, and the butterfly leaves it eleven days after the insect has changed from a caterpillar. This butterfly is found about the middle of June; I have seen it again in September, though rarely, and the caterpillars of the last brood remain in the chrysalis state throughout the winter, and are changed to butterflies in the months of April and May following. This butterfly is of a blue-black color, finely glossed with blue on the hinder part of the wings, the scalloped edges of which are white, and the hind margins bordered with three black lines; near the tips of the fore wings are two or three white spots, and just within the border a row of orange-colored spots; these spots are more distinct on the under side of the fore wings, which are more or less tinged with brown, and have near the body two large orange-colored spots; on the under side of the hind wings is a row of seven orange-colored spots inside of the hind border, and three more of the same color near the shoulders of the wings. It expands from 3 to 3¾ inches.

The Arthemis butterfly (*Nymphalis Arthemis* of Drury) (Plate I. Fig. 7) is very rare in Massachusetts, but more common in the hilly parts of New Hampshire. It is smaller than the preceding, measuring from 2¾ to 3 inches, resembles it a good deal in form and general color, but is readily distinguished from it, and from all the other American butterflies, by the broad white arched band on the wings, which, beginning just beyond the middle of the front edge of the fore wings, curves backwards, crossing both wings, and ends on the inner edge of the hind wings. The male differs from the female in having a row of orange-colored spots on the upper side of the hind wings next to the border, as well as on the under side. The caterpillar and chrysalis of this species are unknown to me.

The caterpillars of many of the four-footed butterflies are spiny, or have their backs armed with numerous projecting points; these, in some, are short and soft, and beset all around with very small stiff hairs, in others they are long, hard, and sharp prickles, which generally are furnished with little stiff branches. The butterflies have the knobs of the antennæ short and broad; the feelers are rather long, and placed close together, at the base at least; the inner margin of the hind wings is folded downwards, and grooved for the reception of the body; the central mesh of these wings is not closed behind; and the nails of the four hind feet are divided so as to appear double. This group may be called Vanessians (VANESSADÆ), and contains the genera *Argynnis*, *Melitæa*, *Cynthia*, and *Vanessa*.

In *Argynnis* the wings are never angulated or toothed, and the hind ones are generally ornamented with silvery or pearly spots beneath; the feelers spread apart at their points; the caterpillars have a round head, and are furnished with branched spines on all their segments, two of those on the first segment being usually longer than the rest, and directed forwards; chrysalids somewhat angular, arched, rather thick at both ends, with the head squared or very slightly notched, without a prominent nose-like projection on the thorax, and on the back are two rows of projecting points, which are usually golden-colored. Most of the caterpillars in this genus are observed to live on various kinds of violets, and on these plants we may expect to find the caterpillars of our native species, which as yet are mostly unknown, in the months of May, June, and July.

Argynnis Idalia, Drury. Idalia Butterfly. (Fig. 110.)

Fore wings deep tawny orange, spotted with black, and with a broad black hind border, around which, in the females, is a row of white spots; hind wings blue-black above,

Fig. 110.

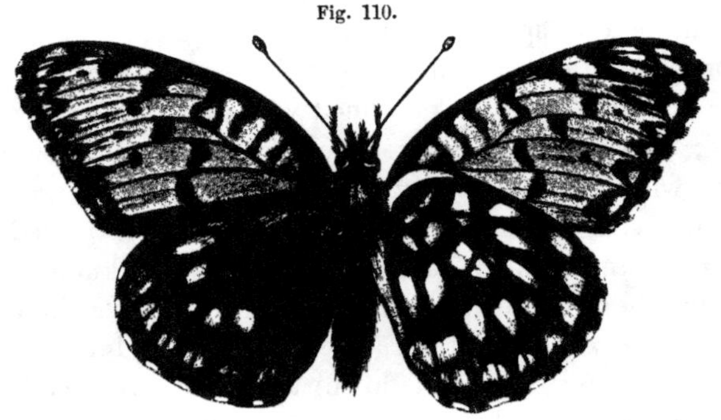

with two rows of spots behind, both of which in the female are cream-colored, but in the males the spots of the outer row are deep tawny orange; all the wings on the under side have a row of pearly-white crescents within the black border; and on the hind wings, which are brown, are seventeen more pearly-white spots; the fringes of all the wings are spotted with white.

Expands from 3½ to 3¾ inches or more.

This large and fine butterfly is found in meadows in the latter part of July and beginning of August.

Argynnis Aphrodite, Fabricius. Aphrodite Butterfly. (Fig. 111.)

Wings tawny-yellow in the males, ochre-yellow in the females, in both brownish next to the body, with a black line near the hinder margins, within which is a row of black crescents, and within the latter is a row of round black spots; the rest of the surface is more or less covered with large irregular black spots; beneath the tips of the fore wings are seven or eight silvery spots, and on the under

Fig. 111.

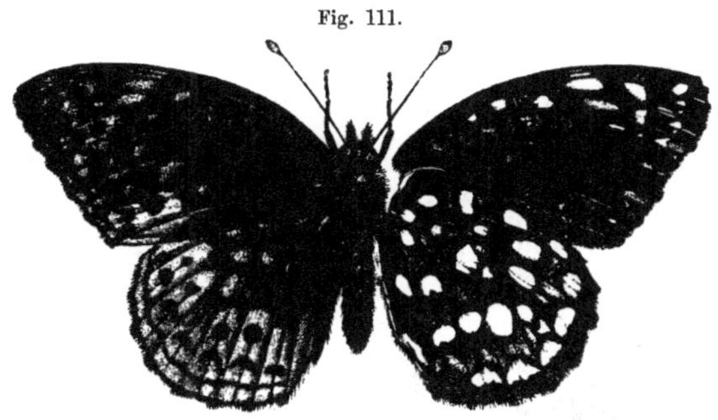

side of the hind wings are above twenty large silvery-white spots, six of which are near the base, and the rest are arranged in three curved rows.

Expands from $2\frac{3}{4}$ to $3\frac{1}{2}$ inches.

Very common on flowers in low grounds in the latter part of July and the beginning of August.

Argynnis Myrina, Cramer. Myrina Butterfly. (Fig. 112.)

Wings tawny, bordered with black above, with a row of black crescents adjoining the border, and another of round black spots at a distance from it; the remainder of the surface from the base to the middle with irregular black spots; under side of the hind wings variegated with brown, with a few ochre-yellow spaces interposed, and above twenty silvery-white spots arranged in four rows; between the two outer rows is a series of black dots, and between the two inner rows a single black dot encircled with silvery white.

Fig. 112.

Expands from $1\frac{3}{4}$ to $1\frac{8}{10}$ inch.

The wings and the feelers of this and the following species are proportionally more elongated than in the Idalia and Aphrodite butterflies. The Myrina begins to appear about the last of May, and may be found till the end of June; it reappears again in August and September.

Argynnis Bellona, Fabricius. Bellona Butterfly. (Figs. 113, 114.)

Wings tawny above, with two rows of black spots around the hind margins, at a distance from which is a row of round spots of the same color; from the base to beyond the middle

Fig. 113. Fig. 114.

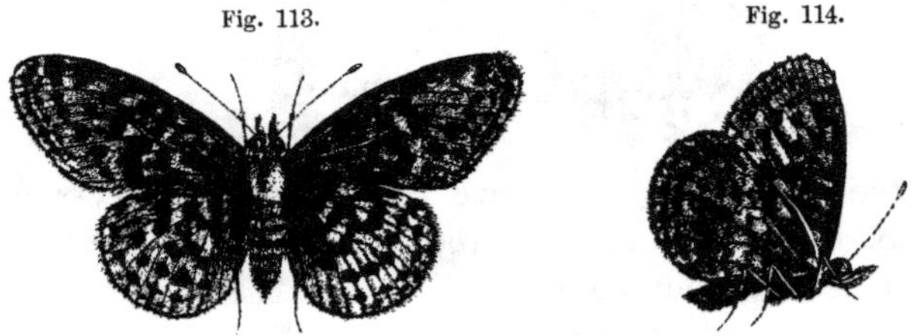

the wings are covered with blackish spots, running together more or less, as in the preceding species; tips of the fore wings beneath, and under side of the hind wings, brownish, and glossed with purplish white on the posterior half of the latter, which are variegated with dark brown lines and spots.

Expands from $1\frac{3}{4}$ to $1\frac{9}{10}$ inch.

Very closely resembles the Myrina in form and color of the upper surface of the wings, but is easily distinguished from it by the want of the silvery spots beneath. It is found on flowers in the latter part of July.

The butterflies of the genus *Melitæa* agree in most respects with those of *Argynnis*, except that the under side of the hind wings is usually checkered with various colors, but not ornamented with silvery or pearly spots. Their caterpillars are very different, being covered with blunt tubercles beset with very short stiff bristles, and most of them live on various kinds of plantain. The chrysalids are of the same form as those of *Argynnis*, and spotted with black or brown, but are not ornamented with golden spots.

Melitæa Phaeton, Drury. Phaeton Butterfly. (Fig. 115.)

Wings black, with a row of orange-red crescents around the hind margin, within which are from two to four rows of cream-colored spots; on the fore wings, behind the middle of the front margin, are two orange-red spots, and sometimes another of the same color on the middle of the hind wings. All the wings are black beneath, and spotted in the same way as on the upper side, with the addition of several large orange-red and pale yellow spots between the middle and the base; the abdomen has three rows of cream-colored dots on the top.

Fig. 115.

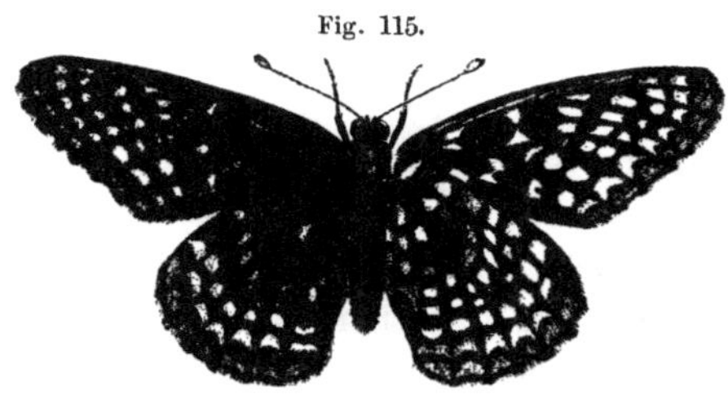

Expands from 2 to 2¼ inches or more.

This species is rare in Massachusetts; it appears in low grounds in June. The wings are elongated, as in *Argynnis Myrina*, but the feelers are short.

Melitæa Ismeria? Boisduval. Ismeria Butterfly.

Wings tawny above, blotched with blackish narrow spots at the base, the fore wings blackish on the hind margins and tips; the hind wings veined and edged with black, with a row of black crescents near the hind border, next to which is a row of round black dots; body covered with white down beneath; under side of the wings ochre-yellow, with a row of pale yellow crescents edged with black near the hind margin; the rest of the surface of the fore wings variegated with small black and large yellowish spots; next to the external row of crescents of the hind wings is a row of yellowish dots encircled with black, across the middle a broad pale yellow band traversed and edged with wavy black lines, which with the black veins divide it into a series of checkers; on the shoulders of these wings a long pale yellow spot surrounded with black, behind which are three square ones of the same colors, contiguous by their sides, and behind these two more joining each other by their angles.

Expands 1¾ inch.

I think it possible that this species may be distinct from the *Ismeria*, which is known to me only by Dr. Boisduval's figure.* The wings are short and broad, and the feelers longer and more slender at their tips than in the *Phaeton*. In the markings of the under side of its hind wings it approaches to the *Maturna*, *Cynthia*, and *Ossianus* of Europe. The only specimen which I have seen was sent to me by Dr. D. S. C. H. Smith of Sutton.

Melitæa Pharos, Drury. Pharos Butterfly. (Fig. 116, male. Fig. 117, female.)

Wings short and broad, tawny-orange above, with a broad black hind border, on which is a row of narrow tawny crescents, and before these a row of round black spots, much

Fig. 116. Fig. 117.

more distinct on the hind than on the fore wings; the rest of the wings, from the middle to the base, is marked with narrow black spots, running together like network; and on the fore wings is a large black spot, extending nearly half across the wing; the under side of the fore wings is tawny, variegated with black and brown, with a buff-colored

* Hist. des Lépidopt. de l'Amérique Septent., pl. 46.

spot at tip, and a crescent-shaped one of the same color on the middle of the hind margin; under side of the hind wings pale ochre-yellow or buff, variegated with brown lines and spots, with a very large brown spot on the hinder margin, on the middle of which is a whitish crescent, and before this a row of blackish dots.

Expands from $1\frac{3}{10}$ to $1\frac{1}{2}$ inch.

The chrysalis is about half an inch long, brown and sprinkled with white dots before, and reddish brown with black dots behind, and three rows of minute points on the back; the anterior extremity is square and the top of the thorax arched, with three little points disposed in a triangle. The butterfly comes out about the first of June. This little and very common butterfly varies considerably in the depth and quantity of its dark markings. It is found on flowers in June, July, and August.

The genus Cynthia was proposed by Fabricius to contain certain butterflies which some entomologists now place in Vanessa. Taken, however, in a more limited sense than was originally intended, it may be retained for some of the species which differ from the others in the form and coloring of the wings, in the habits of the caterpillars, and in the shape of the chrysalids. As thus restricted, the genus Cynthia is distinguished by the wings of the butterflies included in it being more or less scalloped on the edges, but not indented or tailed, and not marked with metallic characters beneath; their feelers are much longer than the head, are tapering, curve upwards and are contiguous to their extremity, giving the head of the insect, when viewed sideways, somewhat the form of the bows of a ship. The caterpillars are armed with branched spines, about equal in length on all the segments except the first and last, on which they are often wanting, and the head is heart-shaped, with little elevated points or short spines on the top. They are solitary, and conceal themselves under a web, or within a folded leaf, and suspend themselves by the hind feet alone when about to transform. The chrysalids are angular on the sides, with two or three rows of sharp tubercles on the back, the anterior extremity is nearly square, or hardly notched, and there is a short and thick prominence on the top of the thorax. The tubercles, and oftentimes the greater part of the surface of the chrysalis, have the color and lustre of burnished gold; from which originated the name chrysalis, derived from the Greek name for gold, now, however, applied to other insects in their second stage of transformation, which are not golden-colored.

Cynthia Cardui. Thistle Butterfly. (Fig. 118.)

Wings tawny above, with a tinge of rose-red, spotted with black and white; hind wings marbled beneath, with a

Fig. 118.

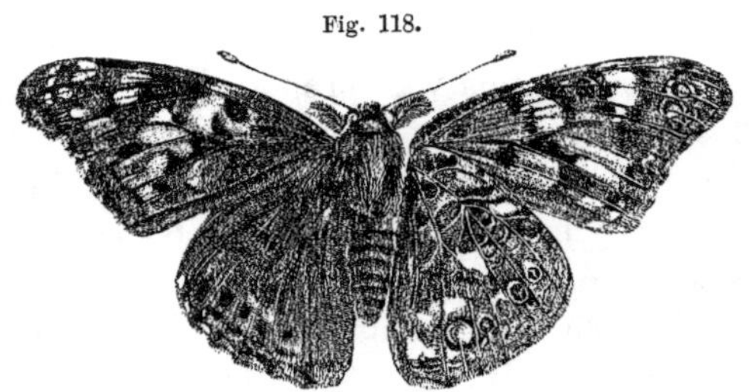

triangular white spot in the middle, and a row of five eye-like spots near the hind margin.

Expands $2\frac{1}{2}$ to $2\frac{3}{4}$ inches or more.

The caterpillars of this butterfly are found on thistles, particularly the spear-thistle (*Cnicus lanceolatus*) and cotton-thistle (*Onopordon acanthium*), on the leaves of the sunflower, hollyhock, burdock, and other rough-leaved plants, in June and July. Though there may be several on the same plant, they keep at some distance from each other. Each one spins for itself a thin web on the surface of the leaf, usually near the edge, to which it is also fastened, so as to draw over a part of the leaf, and thus form a little tent beneath which the caterpillar lives. It devours the skin and pulpy substance of the leaf, without touching the under skin; and when it has exhausted the part under its tent, it removes to another place, and makes a larger habitation as before. Very young caterpillars, which are distinguished by their darker color as well as their inferior size from the older ones, cover themselves with a very small portion of the leaf, and are principally protected by means of the silken tent. The full-grown caterpillar is about one inch and a half long. Its head is black, its feet reddish, its body striped with black and yellow interrupted lines, with about seven branched spines, of a white color tipped with black, on each segment except the first, those on the fore part of the body being more obscure than the rest. These caterpillars frequently suspend themselves to the plants on which they live, and they seldom wander far in search of a place wherein to prepare for transformation. The chrysalis varies in color, being most often brown, with golden or brassy spots on the sides and back, sometimes entirely golden, and sometimes white with a silvery lustre. The chrysalis state lasts from eleven to fourteen days. The butterflies appear from the middle to the end of July, and are found on the flowers of thistles and other plants. I have also found them early in May, and as late as the month of August.

Cynthia Huntera, Fab. Hunter's Butterfly. (Fig. 119.)

Wings tawny above, variegated and spotted with black and white; hind wings marbled and streaked beneath, with two large eye-like spots near the hind margin.

Expands from 2¼ to 2½ inches.

The caterpillars are found on the same plants as those of the thistle butterfly, and particularly on the burdock and cotton-thistle in June and July. Mr. Abbot says that they live on a species of everlasting (*Gnaphalium polycephalum*) also. They, as well as the chrysalids, are very much like those of the preceding species. The butterflies appear in August and September.

Fig. 119

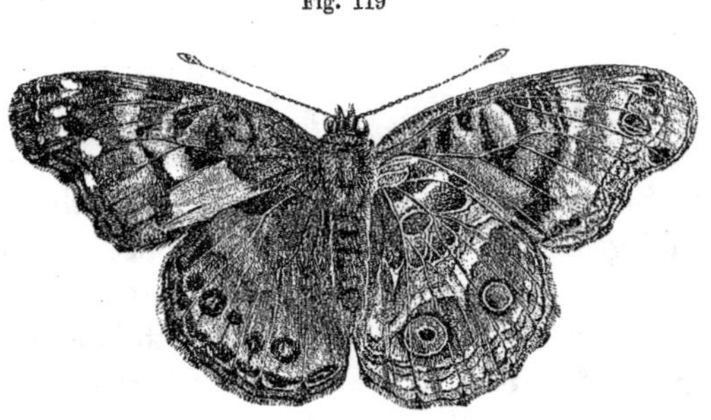

Cynthia Lavinia,* Fab. Lavinia Butterfly.

Wings dark brown above, each with a large and a small eye-like spot on both sides; the fore wings with two orange-red spots near the middle of the front margin, and a large whitish band enclosing the hinder eye-like spots; hind wings with a reddish band near the hind margin.

Expands from 2 to 2½ inches.

The caterpillar is said to be blackish and dotted with white, with the belly and legs tawny, and two white lines on each side, the uppermost one of which is spotted with tawny orange; the spines (of which there are two short ones on the head, besides those on the body) are black and branched. According to Mr. Abbot, it lives on the Canada snap-dragon (*Antirrhinum Canadense*), and remains in the chrysalis state sixteen days. The chrysalis resembles in form that of the two preceding species, but is said to be destitute of metallic spots. I took one of these butterflies in a meadow in Milton, on the 19th of August, 1827, and have never met with it since in this State. It is very common in the Southern States throughout the whole of the summer.

* Dr. Boisduval has described this insect under the specific name of *Cœnia*.

Cynthia Atalanta, L. Atalanta Butterfly. (Fig. 120.)

Wings black above, spotted with white near the tips of the first pair, on which is also an orange-red band across the middle; hind wings with a marginal orange-red band,

Fig. 120.

on which is a row of black dots, the two nearest to the hind angle having a pale blue centre.

Expands from 2¼ to 3 inches.

The Atalanta butterfly was probably introduced into America from Europe with the common nettle, which it inhabits. It deposits its eggs in May upon the youngest and smallest leaves of this plant, being cautious to drop only one upon a single leaf. The young caterpillar is guarded against injury from the poisonous prickles of the leaf by the numerous branching spines with which it is covered, and which, being longer than the prickles, prevent its body from coming in contact with the latter. The head is covered with a tough shell, which sufficiently protects this part, while its strong and horny jaws are adapted for cutting and chewing the leaves and their prickles with impunity. As soon as the caterpillar is hatched, it spins a little web to cover itself, securing the threads all around to the edges of the leaf, so as to bend upwards the sides and form a kind of trough, in which it remains concealed. One end of the cavity is open, and through this the caterpillar thrusts its head while eating. It begins with the extremity of the folded leaf, and eats downwards, and, as it gradually consumes its habitation, it retreats backwards, till at last, having, as it were, eaten itself out of house and home, it is forced to abandon its imperfect shelter, and construct a new one. This is better than the first; for the insect has become larger and stronger, and withal more skilful from experience. The sides of the larger leaf selected for its new habitation are drawn together by silken threads, so that the edges of the leaf meet closely and form a light and commodious cavity, which securely shelters and completely conceals the included caterpillar. This in time is eaten like the first, and another is formed in like manner. At length the caterpillar, having eaten up and constructed several dwellings in succession, and changed its skin three or four times, comes to its full size, leaves off eating, and seeks a suitable place in which to undergo its transformations. The young caterpillars are almost black; the full-grown ones measure about one inch and a half, are generally of a brown color more or less dotted with white, with a black head, rough with elevated white points, with white branching spines on the back, and on each side there is a row of yellow crescents. The chrysalis is gray, with a whitish bloom upon it like that on a plum, and the little pointed tubercles on its back are gold-colored. The chrysalis state continues about ten days, or longer if the weather be cool and wet. The butterflies from the first brood appear in July, and from the second in September.

In the butterflies belonging to the genus *Vanessa*, the wings are jagged or tailed on the hind edges. The under side of the hind wings, in many, is marked with a golden or silvery character in the middle; the feelers are long, curving, and contiguous, and form a kind of projecting beak. The head of the chrysalis is deeply notched or furnished with two ear-like prominences; the sides are very angular; on the middle of the thorax there is a thin projection, in profile somewhat like a Roman nose; and on the back are two rows of very sharp tubercles of a golden color. The caterpillars are cylindrical, and armed with branching spines; they live in company, at least during the early period of their existence, and do not conceal themselves under a web or within a folded leaf.

Vanessa Antiopa, L. Antiopa Butterfly.[8] (Fig. 121.)

Wings purplish brown above, with a broad buff-yellow margin, near the inner edge of which there is a row of pale blue spots.

Expands from 3 to 3½ inches.

This butterfly passes the winter in some sheltered place in a partially torpid state. I have found it in mid-winter

Fig. 121.

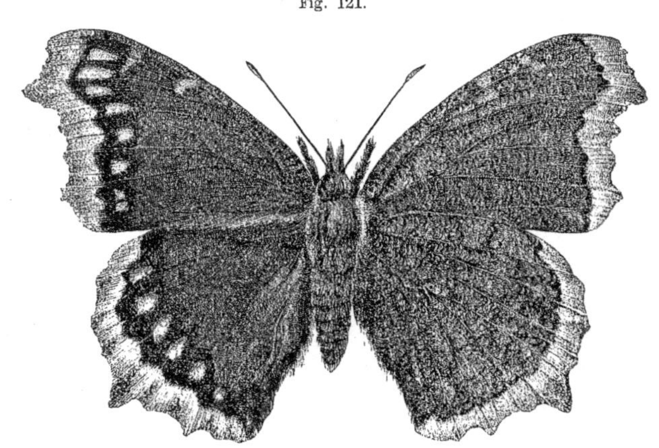

sticking to the rafters of a barn, and in the crevices of walls and stone-heaps, huddled together in great numbers, with the wings doubled together above the back, and apparently benumbed and lifeless; but it soon recovers its activity on being exposed to warmth. It comes out of its winter quar-

[8 This is one of the few butterflies common to this country and Europe, and has probably been introduced here. — MORRIS.]

ters very early in spring, often before the snow has entirely left the ground, but with ragged and faded wings; and may be seen sporting in warm and sheltered spots in the beginning of March, and through the months of April and May. Wilson, in his beautiful lines on the blue-bird, alludes to its early coming in the spring,

"When first the lone butterfly flits on the wing."

The caterpillars (Fig. 122) of the Antiopa butterfly live together in great numbers on the poplar, willow, and elm, on which the first broods may be found early in June. They are black, minutely dotted with white, with a row of eight dark brick-red spots on the top of the back. The head is black and rough with projecting points; the spines, of which there are six or seven on each segment except the first, are black, stiff, and branched, and the intermediate legs are reddish. When fully grown they measure an inch and three quarters in length, and appear very formidable with their thorny armature, which is doubtless intended to defend them from their enemies. It was formerly supposed that they were venomous, and capable of inflicting dangerous wounds; and within my remembrance many persons were so much alarmed on this account as to cut down all the poplar-trees around their dwellings. This alarm was unfounded; for, although there are some caterpillars that have the power of inflicting venomous wounds with their spines and hairs, this is not the case with those of the Antiopa butterfly. The only injury which can be laid to their charge is that of despoiling of their foliage some of our most ornamental trees, and this is enough to induce us to take all proper measures for exterminating the insects, short of destroying the trees that they infest. I have sometimes seen them in such profusion on the willow and elm, that the limbs bent under their weight, and the long leafless branches,

Fig. 122.

which they had stripped and deserted, gave sufficient proof of the voracity of these caterpillars. The chrysalis (Fig. 123) is of a dark brown color, with large tawny spots around the pointed tubercles on the back. The butterflies come forth in eleven or twelve days after the insects have entered upon the chrysalis state, and this occurs in the beginning of July. A second brood of caterpillars is produced in August, and they pass through all their changes before winter.

Fig. 123.

Vanessa J Album. The White J Butterfly.

Wings pale tawny red above, each with a white spot between two black ones near the outer angle on the front margin; the fore wings with a larger black spot on the middle of the front edge, and five smaller roundish black spots near the middle of the wings; hind wings with a silvery-white character somewhat in the shape of the letter J in the middle of the under side.

Expands from 2½ to 3 inches.

The caterpillar and chrysalis of this butterfly are unknown to me. The butterfly probably survives the winter like the Antiopa, for it has been observed late in the autumn, and again early in the ensuing spring, sometimes in great numbers; but it is very inconstant in its appearance. It is more common in New Hampshire than in Massachusetts.

Vanessa Interrogationis, F. Semicolon Butterfly.[9] (Fig. 124.)

Wings on the upper side tawny orange, with brown spots running together on the hinder part, and with black spots in the middle; hind wings in the male most often black above, except at the base, and sometimes of this color in the other sex also; the edges and the tails glossed with reddish white; under side of the wings in some rust-red, in others marbled with light and dark brown, glossed with reddish white, and

[9 *Vanessa Interrogationis* belongs to the genus *Grapta*, Kirby. — MORRIS.]

with a pale gold-colored semicolon on the middle of the hinder pair.

Expands from $2\frac{1}{2}$ to $2\frac{3}{4}$ inches, or more.

The paly-gold character beneath the hind wings has much more nearly the shape of a semicolon than of a note of interrogation;* for which reason I have called this the semicolon butterfly, instead of translating the specific name. It first appears in May, and again in August and September, and is frequently seen on the wing, in warm and sunny places, till the middle of October. The caterpillars live on the American elm and lime trees, and also on the hop-vine,

Fig. 124.

and on the latter they sometimes abound to such a degree as totally to destroy the produce of the plant. In the latter part of August the hop-vine caterpillars come to their full growth, and suspend themselves beneath the leaves and stems of the plant, and change to chrysalids. This fact affords a favorable opportunity for destroying the insects in this their stationary and helpless stage, at some loss, however, of the produce of the vines, which, when the insects have become chrysalids, should be cut down, stripped of the fruit that is sufficiently ripened, and then burnt. There is prob-

[* This butterfly received its name from the Greek note of interrogation, which is identical with our semicolon. — Ed.]

ably an early brood of caterpillars in June or July, but I have not seen any on the hop-vine before August; the former are therefore confined to the elm and other plants, in all probability. The caterpillar is brownish, variegated with pale yellow, or pale yellow variegated with brown, with a yellowish line on each side of the body; the head is rust-red, with two blackish branched spines on the top; and the spines of the body are pale yellow or brownish and tipped with black. The chrysalis is ashen brown, with the head deeply notched, and surmounted by two conical ears, a long and thin nose-like prominence on the thorax, and eight silvery spots on the back. The chrysalis state usually lasts from eleven to fourteen days; but the later broods are more tardy in their transformations, the butterfly sometimes not appearing in less than twenty-six days after the change to the chrysalis. Great numbers of the chrysalids are annually destroyed by little maggots within them, which, in due time, are transformed to tiny four-winged flies (*Pteromalus Vanessæ*), which make their escape by eating little holes through the sides of the chrysalis. They are ever on the watch to lay their eggs on the caterpillars of this butterfly, and are so small as easily to avoid being wounded by the branching spines of their victims.

Vanessa Comma. Comma Butterfly.[10] (Plate IV. Fig. 1.)

Upper side tawny orange; fore wings bordered behind and spotted with black; hind wings shaded behind with dark brown, with two black spots on the middle, and three more in a transverse line from the front edge, and a row of bright orange-colored spots before the hind margin; hind edges of the wings powdered with reddish white; under side marbled with light and dark brown, the hinder wings with a silvery comma in the middle.

Expands from 2⅛ to 2⅜ inches.

This butterfly very closely resembles the white C (*C*

[10 *V. Comma* belongs to the genus *Grapta*, Kirby. — MORRIS.]

album) of Europe, for which it has probably been mistaken. On a close and careful comparison of several specimens of both together, I am satisfied that the American Comma is a distinct species, and the hinder edges of the wings, which are not so deeply indented, will at once serve to distinguish it. I have therefore now named and described it for the first time. The caterpillar lives upon the hop, and, as nearly as I can recollect, has a general resemblance to that of the semicolon butterfly. The chrysalis (Plate IV. Fig. 2, chrysalis from which the butterfly has escaped) is brownish gray, or white variegated with pale brown, and ornamented with golden spots; there are two conical ear-like projections on the top of the head, and the prominence on the thorax is shorter and thicker than that of the semicolon butterfly, and more like a parrot's beak in shape. The butterflies appear first in the beginning of May; I have obtained them from the chrysalids in the middle of July, and on the first of September.

*Vanessa Progne,** Fab. Progne Butterfly.

Upper side tawny orange; fore wings bordered and spotted with black; hind wings blackish on the posterior half, with two black spots before the middle, and a row of small orange-colored spots before the hind margin; tails and posterior edges of the wings powdered with reddish white; under side gray, with fine blackish streaks, and an angular silvery character somewhat in the form of the letter L on the middle of the hind wings.

Expands from $1\frac{7}{8}$ to $2\frac{3}{8}$ inches.

This butterfly appears in August, and probably also at other times. Though very much like the preceding in general appearance, it is readily distinguished from it by the darker color of the hind wings and the angular shape of the silvery character on their under side. This character is very

* Mr. Kirby, whose work on the insects of North America abounds in mistakes, has redescribed this old and well-known species under the name of *Vanessa C. argenteum.*

slender, and is sometimes entirely wanting. I have raised the Progne and Comma butterflies from caterpillars which were so much alike, that I am not certain to which of them the following description belongs. These caterpillars were found on the American elm in August; they were pale yellow, with a reddish-colored head, white branching spines tipped with black, and a row of four rusty spots on each side of the body. They were suspended on the 21st and 22d of August, changed to chrysalids within twenty-four hours, and were transformed to butterflies sixteen days afterwards. At another time, a Progne butterfly was obtained from a caterpillar, which I neglected to describe, on the 18th of August, the chrysalis state having continued only eleven days. The chrysalis is brownish gray, with silvery spots on the back, a short, thick, and rounded nose-like prominence on the thorax, and two conical double-pointed horns or ears on the head, the outer points very short, and the inner ones longer and curving inwards.

Vanessa Milberti,* Godart. Milbert's Butterfly. (Fig. 125.)

Fig. 125.

Black above, with a broad orange-red band near the hinder margin of all the wings, behind which on the hind wings is a row of pale blue crescents; fore wings with a small white spot near the tips, and two orange-red spots near the middle of the front edge; under side deep brown, with a pale band near the extremity of the wings, and no metallic characters on the hinder pair.

Expands from $2\frac{1}{8}$ to $2\frac{3}{8}$ inches.

This showy butterfly is rare in the vicinity of Boston, but

* This is the *Vanessa furcillata* of Mr. Say; but Godart's name has the priority in point of time.

abundant in the northwestern part of the State and in New Hampshire. It appears in May, and again in July and August. The caterpillars live together on the common nettle. They vary in color, some being much darker than others; generally, however, they are pale brown, minutely dotted with yellowish white, with a dark brown longitudinal line on the top of the back, a whitish one on each side just above the feet, and above this a row of brown spots; the head is small, black, and rough, with little black and white tubercles; the spines are blackish, short, and with very small branches or lateral bristles. It measures when fully grown an inch and a quarter or more in length, the chrysalis is pale brown with golden spots, the top of the head widely but not deeply notched, and the nose-like prominence very small.

The last of the four-footed butterflies remaining to be described may be called Hipparchians (*Hipparchiadæ*). The wings of the butterflies belonging to this group are entire, with the veins of the first pair swelled at their origin, and the central mesh of the second pair closed behind. Their caterpillars are not spiny, and are of a green color, spindle-shaped, or cylindrical, tapering at both ends, with the hinder extremity notched or terminating in two conical points, and the head is either rounded or notched above. They live exclusively on various kinds of grasses, for the most part concealing themselves during the day among the stubble, and suspend themselves by the hindmost feet alone when about to transform.

The chrysalis is either oblong and somewhat angular at the sides, with the head notched and two rows of pointed tubercles on the back, or short and rounded, with the head obtuse; but never ornamented with metallic spots. The small size and uniformly green color of the caterpillars of our native species, and the obscurity in which they generally live, render it very difficult to discover them; and hence they rarely pass under our observation. This being the case, and not having much to communicate respecting the habits of individual species, I shall confine my further remarks to a description of the insects in their final state, when they are exposed to view, and attract our notice by their neat and modest coloring, and their graceful and gentle motions. They are mostly found in thickets and woods, and more rarely in places more open and exposed.

Hipparchia semidea, Say. The Mountain Butterfly. (Fig. 126.)

Wings dusky brown above, thin, delicate, and almost transparent, in the male paler, and with more of an ochre-yellow tint; fringes black, barred with ochre-yellow, and a row of faint ochre-yellow spots near the hind margin of the second pair; the under side of these wings and of the tips of the fore wings is marbled with black and white, a portion of the white forming an irregular band beyond the middle of the hind wings.

Fig. 126.

Expands $1\frac{8}{10}$ inch to 2 inches.

This butterfly has hitherto been taken only on the summit of the White Mountains of New Hampshire in June and July. It was observed in great abundance flying about on the top of Mount Washington on the 29th of July last. It has also been seen on the Monadnoc Mountain, and will probably be discovered on the tops of the high mountains in our own State, if looked for at the proper season. It closely resembles the *Fortunatus* of Lapland, with which I have compared it, and find it to be specifically distinct. Mr. Say was the first describer of it, and it is well figured in his American Entomology. Dr. Boisduval has since re-described and figured it under the name of *Chionabas Also*.*

* Icones Lépidopt. Nouv.,. I. p. 197, Pl. 40, fig. 1, 2, and Lépidopt. Amer., I. p. 222.

Hipparchia Alope, Fab. Alope Butterfly. (Fig. 127.)

Dark brown; fore wings with a broad ochre-yellow band beyond the middle, enclosing two round black spots, with a sky-blue centre; hind wings notched behind, with from one to three eye-like spots of a black color, with a blue centre on the upper side, and four or five of the same kind, but of unequal size, beneath; the under side of the wings is pale brown, with numerous dark brown streaks. The eye-spots on the hind wings are sometimes wanting in the males.

Fig. 127.

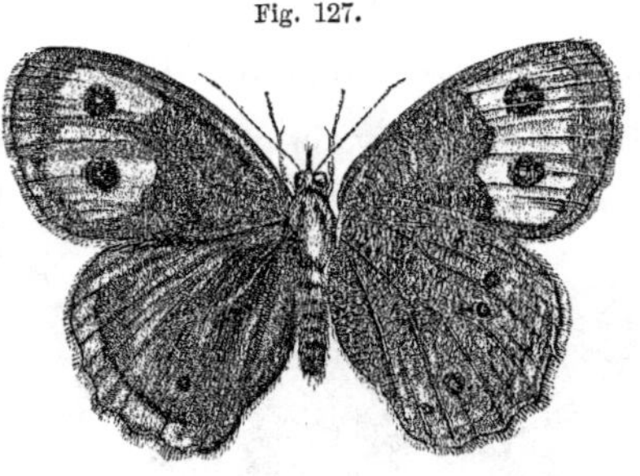

Expands from 2 to 2½ inches. In the Southern States individuals are found measuring three inches.

The Alope butterfly is found from the first of July to the middle of September in open woods and in orchards. The caterpillar is pale green with dark green stripes; the head is round, and the tail ends in a short fork. The chrysalis is elongated, roundish at the sides, with the head notched.

Hipparchia Boisduvallii. Boisduval's Butterfly. (Fig. 128.)

Pale yellowish-brown; the fore wings upon both sides have four eye-like, blackish spots, with a white centre, and the hind wings have six, the external spot remote from the others, and the two next to the hind angle very small and close together. In some individuals the white centre is wanting in some of the eye-spots on the upper side of the wings.

Fig. 128.

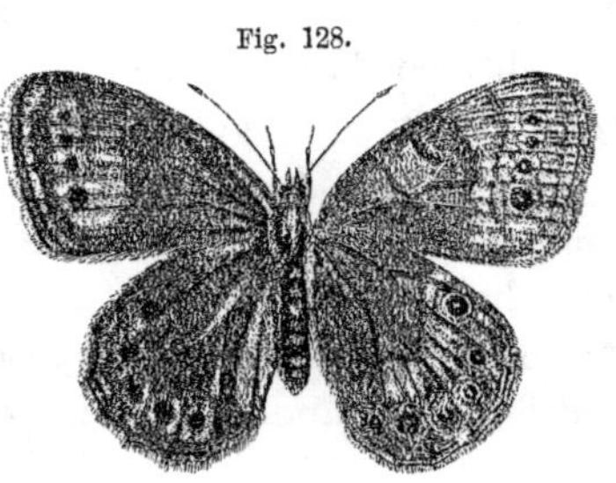

Expands 2 inches or more.

This butterfly is figured in Dr. Boisduval's *Histoire des Lépidoptères de l'Amérique*, under the name of *Satyrus Canthus*; but as it does not agree with the descriptions of the *Canthus* of Linnæus and of Fabricius, in both of which there are no eye-spots on the upper side of the wings, I have thought it entitled to a new name, and am happy to dedicate it to one of the most accomplished entomologists now living. This delicate butterfly delights in open and elevated situations, and is found in July on the sides of the highest hills, and in the mountain meadows of the northwestern parts of this State.

Hipparchia Eurytris, Fab. Eurytris Butterfly. (Fig. 130.)

Fig. 129.

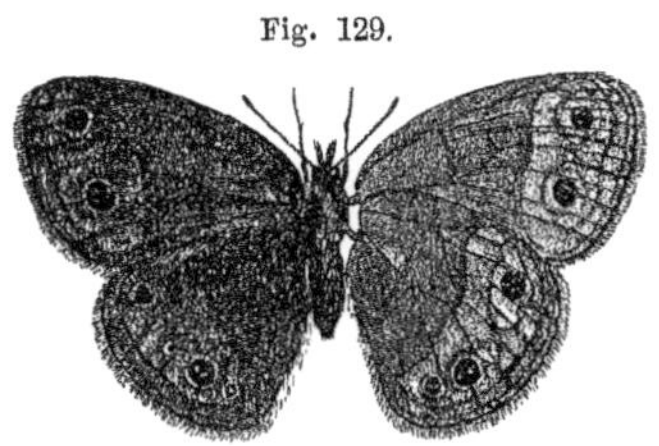

Dark brown above, paler beneath, with two longitudinal dusky stripes; on the upper side of the wings are two black eye-spots, enclosed in an ochre-yellow ring, with two lead-colored dots in the centre of each spot; on the hind wings there is another smaller spot, with a lead-colored centre, near the hinder angle; all these spots are found on the under side of the wings, and between them are interposed the same number of small lead-colored spots.

Expands 1 inch and 6 or 7 tenths.

Fig. 130.

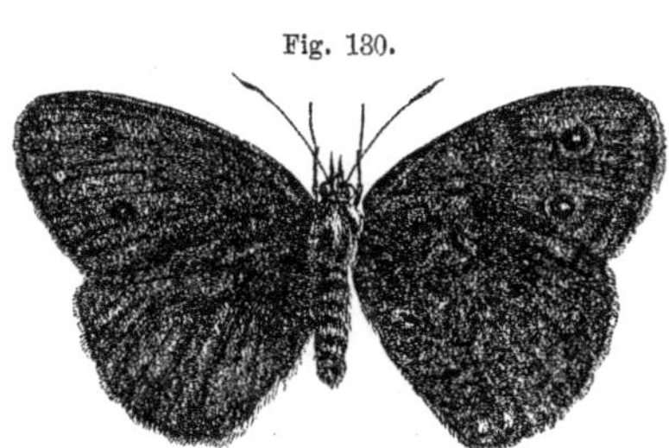

Hipparchia Nephele.

This butterfly is found in June and July among bushes and in the paths of woods, seeking the shade rather than the sunshine. The caterpillar resembles that of the Alope butterfly, but the chrysalis is shorter, with the head obtuse.

2. Skippers. (*Hesperiadæ.*)

The butterflies of this tribe frequent grassy places, and low bushes and thickets, flying but a short distance at a time, with a jerking motion, whence they are called skippers by English writers. When they alight, they usually keep the hind wings extended horizontally, and the fore wings somewhat raised, but spreading a little, and not entirely closed, as in other butterflies; some of them, however, have all the wings spread open when at rest, and there are others in which they are all elevated. Notwithstanding this difference in the position of the wings, the Hesperians all have certain characters in common, by which they are readily distinguished from other butterflies. Their bodies are short and thick, with a large head, and very prominent eyes; the feelers are short, almost square at the end, and thickly clothed with hairs, which give them a clumsy appearance; the antennæ are short, situated at a considerable distance from each other, and in most of these insects with the knob at the end either curved like a hook, or ending with a little point bent to one side; the legs are six in number, and the four hinder shanks are armed with two pairs of spurs.

Their caterpillars are somewhat spindle-shaped, cylindrical in the middle, and tapering at each extremity, without spines, and generally naked or merely downy, with a very large head and a small neck. They are solitary in their habits, and many of them conceal themselves within folded leaves, like the caterpillars of the thistle and nettle butterflies (*Cynthia Cardui* and *Atalanta*), and undergo their transformations within an envelope of leaves or of fragments of stubble gathered together with silken threads. Their chrysalids are generally conical or tapering at one end, and rounded, or more rarely pointed, at the other, never angular or ornamented with golden spots, but most often covered with a bluish-white powder or bloom. They are mostly fastened by the tail and a few transverse threads, within some folded leaves, which are connected together by a loose internal web of threads, forming a kind of imperfect cocoon.

Heteropterus marginatus. Bordered Skipper. (Fig. 131.)

Fore wings tawny yellow above, shaded with brown behind, and with an indistinct brownish streak in the middle; beneath, brown, with the front and hind margin broadly bordered with tawny yellow; hind wings tawny yellow, with a broad brownish outer margin above, and without a border beneath; antennæ and legs ringed with black and white; body slender, longer than the hind wings, which are horizontal in repose, and the fore wings raised and spread a little.

Fig. 131.

Expands about $\frac{7}{8}$ of an inch.

This pretty species does not appear to have been described before. The chrysalis from which it was obtained, on the 20th of July, is rather long, nearly cylindrical, but tapering at the hinder extremity, and with an obtusely rounded head. It is reddish ash-colored, minutely sprinkled with brown dots. I am not sure that this skipper belongs to the genus *Heteropterus*, but have placed it in this genus on account of the antennæ, which are not hooked at the end, but terminate much like those of the genus *Polyommatus.*

In the greater number of our skippers the antennæ are curved or hooked at the end. This is the case in the kinds belonging to the genus *Thanaos*, which have the knobs of the antennæ long, tapering, and curved, the body thick, and shorter than the wings; the latter are generally spread in repose, and the fringes are of one uniform color, or not spotted. The males are distinguished by having the middle of the front edge of the fore wings doubled back on the upper surface.

Thanaos Juvenalis, Fab. Juvenal's Skipper.

Smoky brown on both sides ; fore wings variegated above with gray, with transverse rows of dusky spots, and six or seven small semi-transparent white spots near the tips ; six of these spots are disposed in a transverse row, but the two hindmost are separated from the others by a considerable interval, and the seventh spot, which is sometimes wanting, is placed nearer the middle of the wing ; hind wings with a row of blackish spots near the hind margin.

Expands $1\frac{6}{10}$ inch.

There is a local variety of this skipper, that is much more common in Massachusetts than the preceding, of inferior size, seldom expanding more than $1\frac{4}{10}$ inch, in which the white spots are smaller, and the seventh is wanting near the middle of the fore wing. This skipper is found in meadows in May, and again in August. The caterpillar lives on various pea-blossomed plants, such as the *Glycine*, or groundnut, the *Lathyrus*, or vetchling, &c. It is green, with pale stripes, and a heart-shaped brown head. The chrysalis is rather long and tapering, according to Mr. Abbot of a green color, and is enclosed in a cocoon of leaves and threads ; in my specimens pale yellowish brown, with a few minute hairs on the body, and with the tongue-case prominent and projecting beyond the middle of the breast ; and the cocoon was composed of stubble. Mr. Abbot informs us that in summer the skipper leaves the chrysalis in nine days ; but the autumnal brood continues in the chrysalis state throughout the winter.

Thanaos Brizo. Brizo Skipper. (Fig. 132.)

Dark brown ; fore wings almost black on the upper side, and variegated with gray externally ; near their hind margin is a row of gray dots, within which is a transverse band, composed of another row of oval gray spots, between two slender black zigzag lines, and across the middle is another band of the same kind ; on the hind wings are two wavy rows of ochre-yellow dots near the hind margin; all the wings beneath have two rows of dots of the same color behind.

Expands from $1\frac{4}{10}$ to $1\frac{7}{10}$ inch.

Fig. 132.

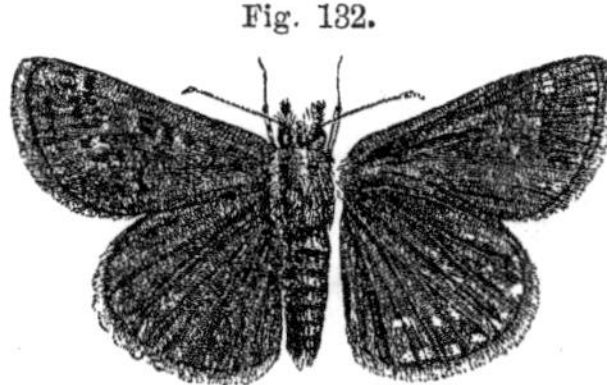

This skipper* has not been described before, but is figured in Dr. Boisduval's work under the name above given. It is found in the same places and at the same times as the preceding species, to which also it bears a close resemblance in the caterpillar and chrysalis states, and lives on the same kind of plants.

In the skippers which Dr. Boisduval arranges under the name of *Eudamus*, the knobs of the antennæ are very long, gradually taper to a point, and are suddenly bent like a hook in the middle; the front edge of the fore wings, in the males, is doubled over; the hind wings are often tailed, or are furnished with a little projection on the hinder angle; the fringes are spotted; and all the wings are raised when at rest.

Eudamus Tityrus, Fab. Tityrus Skipper.[11] (Plate V. Fig. 1.)

Wings brown; first pair with a transverse semi-transparent band across the middle, and a few spots towards the tip, of a honey-yellow color; hind wings with a short rounded tail on the hind angles, and a broad silvery band across the middle of the under side.

Expands from 2 to $2\frac{1}{2}$ inches.

This large and beautiful insect makes its appearance, from the middle of June till after the beginning of July, upon sweet-scented flowers, which it visits during the middle of the day. Its flight is vigorous and rapid, and its strength is

* It is figured in Abbot's Insects of Georgia as one of the sexes, or a variety, of the *Juvenalis*; but the sexes of both of these species are known to me.

[11 *Eudamus Tityrus* belongs to the genus *Goniloba* Doubleday. — MORRIS.]

so great that it cannot be captured without danger of its being greatly defaced in its struggles to escape. The females lay their eggs, singly, on the leaves of the common locust-tree (*Robinia pseudacacia*), and on those of the viscid locust (*Robinia viscosa*), which is much cultivated here as an ornamental tree. The caterpillars are hatched in July, and when quite small conceal themselves under a fold of the edge of a leaf, which is bent over their bodies and secured by means of silken threads. When they become larger they attach two or more leaves together, so as to form a kind of cocoon or leafy case to shelter them from the weather, and to screen them from the prying eyes of birds. The full-grown caterpillar (Fig. 133), which attains to the length of about two inches, is of a pale green color, transversely streaked with darker green, with a red neck, a very large head roughened with minute tubercles, slightly indented or furrowed above, and of a dull red color, with a large yellow spot on each side of the mouth. Although there may be and often are many of these caterpillars on the same tree and branch, yet they all live separately within their own cases. One end of the leafy case is left open, and from this the insect comes forth to feed. They eat only, or mostly, in the night, and keep themselves closely concealed by day. These caterpillars are very cleanly in their habits, and make no dirt in their habitations, but throw it out with a sudden jerk, so that it shall fall at a considerable distance. They frequently transform to chrysalids within the same leaves which have served them for a habitation, but more often quit the trees and construct in some secure place a cocoon (Fig. 134) of leaves or fragments of stubble, the interior of which is lined with a loose web

Fig. 133.

Fig. 134.

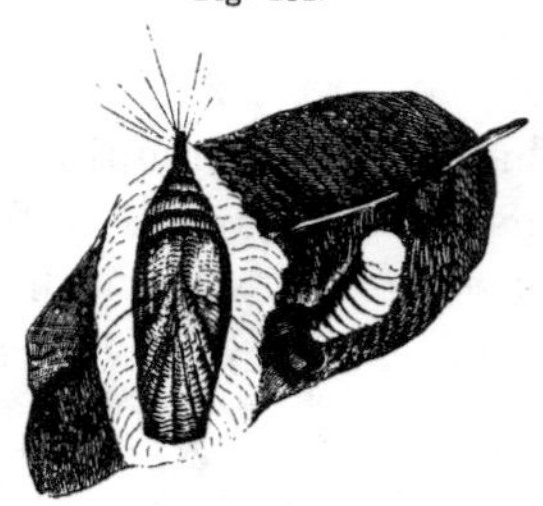

of silk. They remain in their cocoons without further change throughout the winter, and are transformed to butterflies in the following summer. The viscid locust-tree is sometimes almost completely stripped of its leaves by these insects, or presents only here and there the brown and withered remains of foliage, which has served as a temporary shelter to the caterpillars.

Eudamus Bathyllus, Smith. Bathyllus Skipper. (Fig. 135.)

In Massachusetts we have what I suppose to be only a local variety of the Bathyllus skipper, differing from Southern specimens in the inferior size of the white spots on the fore wings, the less prominent hind angle of the hind wings, and the darker color of the fringes. It is of a dark brown color; on the fore wings is a row of small white spots across the middle, and another shorter row of only three or four contiguous spots between the first and the tip; the wings beneath are light brown, shaded at the base with dark brown; the hinder pair with a slightly prominent posterior angle, and two dark brown transverse bands.

Fig. 135.

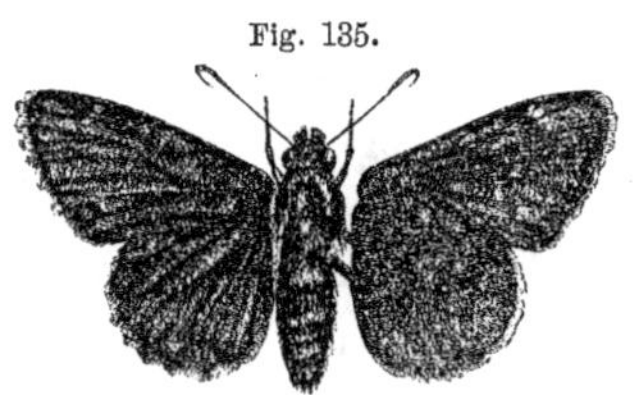

Expands from $1\frac{1}{2}$ to $1\frac{7}{10}$ inch.

This species is found on flowers in June and July; in the Southern States it appears also in March and April. The caterpillar is very similar to that of the Tityrus skipper, and is found on various kinds of *Glycine*, *Hedysarum*, &c., in May and June.

The rest of our skippers belong to the old genus *Hesperia* of Fabricius, which, as now restricted by the French entomologists, very nearly coincides with *Pamphila* of the English writers. The American species are quite numerous, and moreover vary a good deal; which, with the difference existing between the sexes, renders it quite difficult to determine and characterize them. In the distribution of the Hesperians, by far the largest portion of the family or group seems to have been assigned to the Western Continent; and it is probable that New England, or perhaps Massachusetts alone, contains a larger number of species than the whole of Europe. The insects of this group recede in many striking characters, and in their general habits, from the true butterflies, and seem to form the connecting link between the latter and the sphinges or hawk-moths. Those belonging to the genus *Hesperia* delight in cool and shady places, and most commonly appear on the wing towards the evening, which led Fabricius to give them a generic name indicative of this circumstance. Their antennæ are considerably shorter than in those included in *Thanaos* and *Eudamus*, and the knob at the end, which is thick and oblong oval, terminates suddenly in a little point directed to one side. The upper wings are raised and the lower are expanded when at rest; and the fringes are not spotted. The body is thick, and about as long as the hind wings. Most of the males are distinguished by an oblique black dash near the middle of the fore wings. The caterpillar lives chiefly on low herbaceous plants. The chrysalis (Fig. 136) is described as being conical, with a pointed head, and a long tongue-case, folded on the breast, but not confined at the point. The transformation takes place in a slight cocoon of stubble or grass, connected by a few threads within. These skippers frequent meadows, and other grassy and somewhat shady places, during the middle and latter part of summer. They are of smaller size than the preceding Hesperians, and are much more common and abundant. Their flight, though short and intermitting, is exceedingly swift, and they possess a great deal of muscular strength.

Fig. 136.

Hesperia Hobomok. Hobomok Skipper. (Fig. 137.)

Dark brown above; on each of the wings a large tawny-yellow spot occupying the greater part of the middle, four or five minute spots of the same color near the tips of the fore wings, on which is also a short brownish line at the outer extremity of the central mesh; under side of the fore wings similar to the upper, but paler; hind wings brown beneath, with a yellow spot near the shoulder, and a very broad deep yellow band, which does not attain the inner margin, and has a tooth-like projection extending towards the hinder edge. The male has not the usual distinguishing oblique dash on the fore wings, which differ from those of the female only in the greater size of the tawny portion, which extends to the front margin.

Fig. 137.

Expands from $1\frac{7}{20}$ to $1\frac{4}{10}$ inch.

This skipper comes very near to the *Otho* of Smith and Abbot (which is not the same as the *Otho* of Boisduval), and also approaches closely to a species that is figured in Dr. Boisduval's work under the name of *Zabulon;* but does not sufficiently agree with either of them, and, in the belief that it has not been described before, I have given it the name of one of our celebrated Indian chiefs. It is found in June and July.

Hesperia Leonardus. Leonard's Skipper. (Fig. 138.)

Dark brown above; fore wings of the male tawny yellow on the front margin from the base to beyond the middle; behind this tawny portion is a short black line, and behind the latter a row of contiguous tawny spots, extending from the middle of the inner edge towards the tip; the spots at this extremity small and separated from the oth-

Fig. 138.

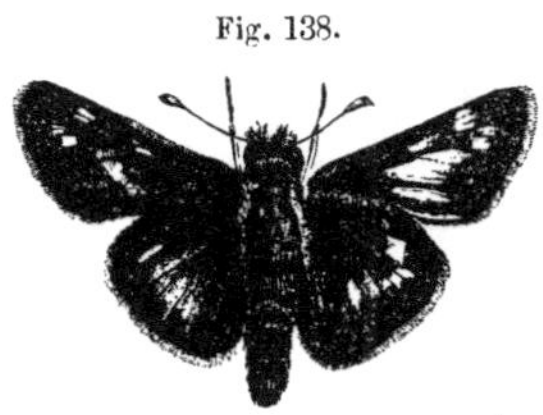

ers ; fore wings in the female without the tawny front edge and black line ; hind wings, in both sexes, with a central, curved, tawny-yellow band ; wings beneath bright red-brown ; the first pair blackish from the middle to the inner edge, and spotted as on the upper side ; hind wings with a yellow dot in the middle, and a curved row of seven bright yellow spots behind it.

Expands from $1\frac{7}{20}$ to $1\frac{1}{2}$ inch.

This very distinct and strongly marked skipper does not seem to have been described before. For a specimen of the male I am indebted to the Rev. L. W. Leonard, to whom I have dedicated the species. The females I have taken in the beginning of September.

Hesperia Sassacus. Sassacus Skipper.

Dark brown above; all the wings with a tawny-yellow spot occupying the greater part of the middle of each, and with two or three little detached spots of the same color near the extremity of the first pair; beneath ochre-yellow, with small pale yellow spots near the tip, corresponding to those on the upper side of the fore wings ; and on the hind wings seven small, square, pale yellow spots, namely, one before the middle and the others in pairs behind it.

Expands $1\frac{1}{4}$ inch.

Of this skipper I have seen only the female, which was taken in Cambridge in the month of June. Its upper side is very much like that of the Hobomok skipper, but it differs from it in the color and markings of the under side, and seems not to have been described before. I have therefore given it, as a new species, the name of an Indian warrior.

Hesperia Peckius, Kirby. Peck's Skipper. (Fig. 139.)

Dark brown above ; fore wings with a row of contiguous tawny-yellow spots, extending from the middle of the inner margin towards the tip, where the spots are more distant, and a tawny line from the base to the middle, behind which, in the male, is a short, curved, deep black line; hind wings with an indented tawny band, or row of unequal spots, behind the middle, which, in the male, are very indistinct; beneath, light brown; fore wings marked with bright yellow spots; hind wings with a very large, irregular, bright yellow spot, covering nearly the whole under surface, and almost divided in two near the middle.

Fig 139.

Expands from $1\frac{1}{10}$ to $1\frac{2}{10}$ inch.

This skipper was named by Mr. Kirby in honor of the late Professor Peck of Cambridge, and is figured and described in the fourth volume of the "Fauna Boreali Americana." The upper surface of the female resembles that of the same sex of the *Phylæus* of Drury or *Vitellius* of Fabricius; but the under side is different. It is found on flowers in meadows in the latter part of July and in August.

Hesperia Cernes? Boisduval. Cernes? Skipper.

Dark brown above, fore wings of the male with a large brassy-yellow spot, extending from the front edge beyond the middle, and an oblique wavy black line; hind wings with a brassy gloss; under side of the fore wings tawny yellow before, dusky behind, with a pale yellow oblique spot near the middle, and two or three minute spots of the same color near the front margin; hind wings dusky ochre-yellow beneath, with a transverse row of four small paler yellow almost obsolete spots; head and body glossed with green above, yellowish white beneath.

Expands $1\frac{3}{10}$ inch.

In one individual from the Southern States there are two or three minute yellow dots on the fore wings between the oblique line and the tip. I think it probable that this may be the species figured, but not described, by Dr. Boisduval, under the above name. It is found in the latter part of July, but seems to be rare, and the female is unknown to me.

Hesperia Metacomet. Metacomet Skipper.

Dark brown, slightly glossed with greenish yellow above, the male with a short, oblique black line on the middle of the fore wings, on both sides of which, in the female, are two yellowish dots on the middle, and two more near the front margin and tip; hind wings beneath with a transverse row of four very faint yellowish dots, which, however, are often wanting.

Expands $1\frac{3}{10}$ inch.

It resembles the preceding in some respects, but is of a uniform dark color above, and is probably a distinct species. It appears in July. Metacomet was the Indian name of the celebrated King Philip.

Hesperia Ahaton. Ahaton Skipper. (Fig. 140.)

Dark brown above; fore wings in the male tawny before the middle from the base nearly to the tip, the tawny portion ending externally in three minute wedge-shaped spots; on the middle an oblique velvet-black line, near the outer extremity of which are two or three small tawny spots; under side spotted as above; hind wings without spot above; of a greenish or dusky yellow tinge below, with a transverse curved row of four minute yellowish dots, which are often very faint or entirely wanting. In the female there is a tawny dash along the front margin of the fore wings, and the oblique black line is wanting, but the other spots are larger and more distinct.

Fig. 140.

Expands from 1 inch to $1\frac{1}{10}$.

The markings on the fore wings somewhat resemble those of H. Leonardus, but in other respects it is different, and is much inferior in size. It was captured many years ago in Milton, and I have given it the name of an Indian from that vicinity.

Hesperia Wamsutta. Wamsutta Skipper. (Fig. 141.)

Dark brown above; fore wings with a broken row of small tawny spots towards the tip, and in the males a large tawny patch covering the whole of the fore part of the wings from the base to the middle, and an oblique curved black line behind it; hind wings with a small tawny dot before the middle, and an indented tawny band, or row of contiguous unequal spots; under side of the fore wings light brown, and with larger yellow spots than on the other side, hind wings light brown, with two large irregular bright yellow spots connected in the middle and covering nearly the whole surface.

Fig. 141.

Expands from $\frac{9}{10}$ of an inch to nearly an inch.

This species hardly differs from Peck's skipper, except in being uniformly smaller. It is a very common kind, and is found in meadows in the latter part of summer, particularly through the month of August. Wamsutta, whose name I have given it, was the oldest son of the Sachem Massasoit.

There are a few more skippers in my collection, which were taken in Massachusetts, but some of them are not sufficiently perfect to be described, and of the others I have only one sex.

II. HAWK-MOTHS. (*Sphinges.**)

Linnæus was led to give the name of Sphinx to the insects in his second group of the Lepidoptera, from a fancied resemblance that some of their caterpillars, when at rest, have to the Sphinx of the Egyptians. The attitude of these caterpillars is indeed very remarkable. Supporting themselves by their four or six hind legs, they elevate the

* See page 262.

fore part of the body, and remain immovably fixed in this posture for hours together. In the winged state, the true *Sphinges* are known by the name of humming-bird moths, from the sound which they make in flying, and hawk-moths, from their habit of hovering in the air while taking their food. These humming-bird or hawk moths may be seen during the morning and evening twilight, flying with great swiftness from flower to flower. Their wings are long, narrow, and pointed, and are moved by powerful muscles, to accommodate which their bodies are very thick and robust. Their tongues, when uncoiled, are, for the most part, excessively long, and with them they extract the honey from the blossoms of the honeysuckle and other tubular flowers, while on the wing. Other Sphinges fly during the daytime only, and in the brightest sunshine. Then it is that our large clear-winged Sesiæ make their appearance among the flowers, and regale themselves with their sweets. The fragrant Phlox is their especial favorite. From their size and form and fan-like tails, from their brilliant colors, and the manner in which they take their food, poised upon rapidly vibrating wings above the blossoms, they might readily be mistaken for humming-birds. The Ægerians are also diurnal in their habits. Their flight is swift, but not prolonged, and they usually alight while feeding. In form and color they so much resemble bees and wasps as hardly to be distinguished from them. The Smerinthi are heavy and sluggish in their motions. They fly only during the night, and apparently, in the winged state, take no food, for their tongues are very short, and indeed almost invisible. The Glaucopidians, or Sphinges with feathered antennæ, fly mostly by day, and alight to take their food, like many moths, which some of them resemble in form, and in their transformations. The caterpillars of the Sphinges have sixteen legs, placed in pairs beneath the first, second, third, sixth, seventh, eighth, ninth, and last segments of the body; all of them, except the Ægerians and Glaucopidians, have either a kind of horn or a tubercle on the top of the last segment, and, when at rest, sit with the fore part of the body elevated.

Having devoted a large portion of this treatise to a description of the spinning-moths, my observations on the other insects of this order must be brief, and confined to a few species, which are more particularly obnoxious on account of their devastations in the caterpillar state. Those persons who are curious to know more about the Sphinges than can be included in this essay, are referred to my descriptive catalogue of these insects, contained in the thirty-sixth volume of Professor Silliman's "Journal of Science."[12]

Every farmer's boy knows the potato-worm, as it is commonly called; a large green caterpillar (Fig. 142), with a kind of thorn upon the tail, and oblique whitish stripes on the sides of the body. This insect, which devours the leaves of the potato, often to the great injury of the plant, grows to the thickness of the fore-finger, and the length of three inches or more. It attains its full size from the middle of August to the first of September, then crawls down the stem of the plant and buries itself in the ground. Here, in a few days, it throws off its caterpillar-skin, and becomes a chrysalis (Fig. 143), of a bright brown color, with a long and slender tongue-case, bent over from the head so as to touch the breast only at the end, and somewhat resembling the handle of a pitcher. It remains in the ground through the winter, below the reach of frost, and in the following summer the chrysalis-skin bursts open, a large moth crawls out of it, comes to the surface of the ground, and, mounting upon some neighboring plant, waits till the approach of evening invites it to expand its untried wings and fly in search of food. This large insect has generally been con-

[[12] A more complete monograph of the Sphinges has been lately published in the Journal of the Academy of Natural Sciences of Philadelphia, 1859, Art. V., p. 97, by Dr. Brackinridge Clemens, of Easton, Penn. —MORRIS.]

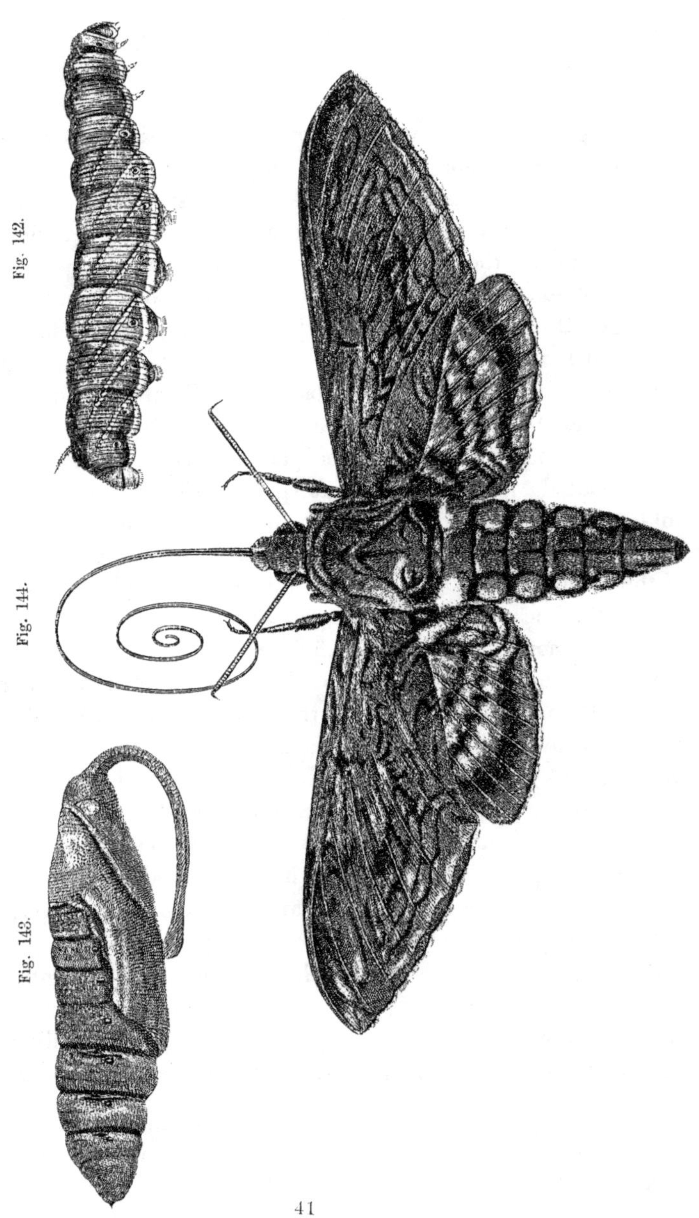
Fig. 142.
Fig. 144.
Fig. 143.

Fig. 145.

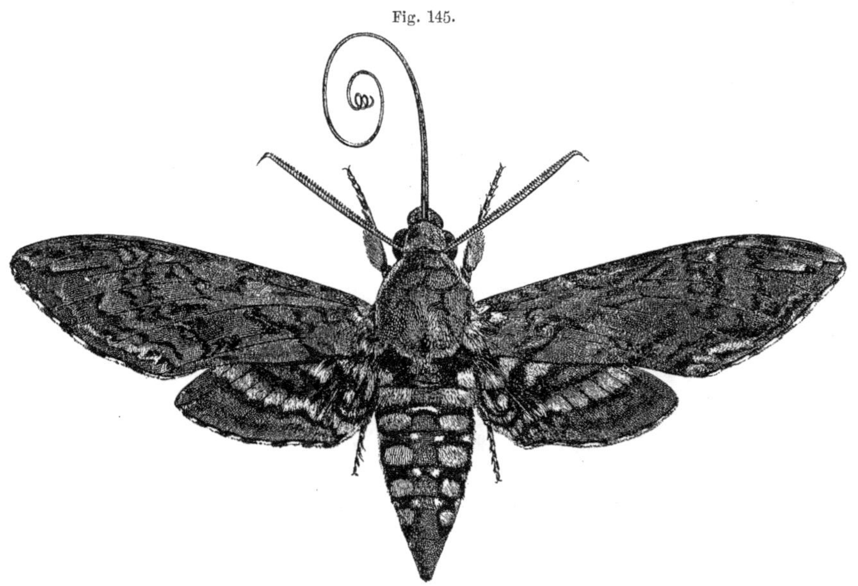

founded with the Carolina Sphinx (*Sphinx Carolina* of Linnæus, Fig. 145, Fig. 146, larva, Fig. 147, pupa), which it

Fig. 146.

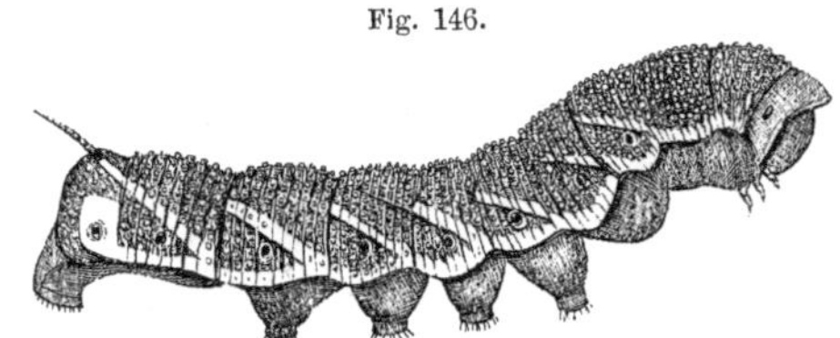

closely resembles. It measures across the wings about five inches; is of a gray color, variegated with blackish lines and bands; and on each side of the body there are five round, orange-colored spots encircled with black. Hence it is called by English entomologists *Sphinx quinquemaculatus* (Fig. 144), the five-spotted Sphinx. Its tongue can be unrolled to the

Fig. 147.

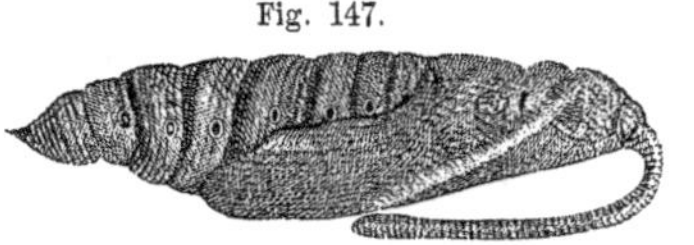

length of five or six inches, but, when not in use, is coiled like a watch-spring, and is almost entirely concealed between two large and thick feelers, under the head.

Among the numerous insects that infest our noble elms, the largest is a kind of Sphinx, which, from the four short horns on the fore part of the back, I have named *Ceratomia* * *quadricornis* (Fig. 148), or four-horned Ceratomia. On

Fig. 148.

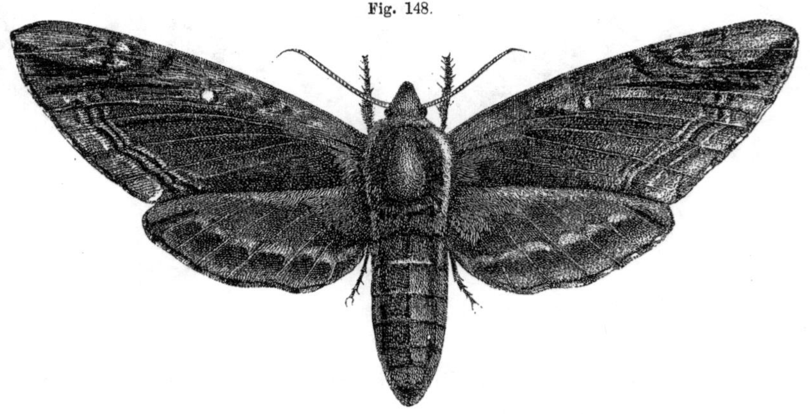

some trees these Sphinges exist in great numbers, and their ravages then become very obvious; while a few, though capable of doing considerable injury, may escape notice among the thick foliage which constitutes their food, or will only be betrayed by the copious and regularly formed pellets of excrement beneath the trees. They are very abundant during the months of July and August on the large elms which surround the northern and eastern sides of the Common in Boston; and towards the end of August, when they descend from the trees for the purpose of going into the ground, they may often be seen crawling in the Mall in considerable numbers. These caterpillars (Fig. 149), at this period of their existence, are about three inches and a

* Ceratomia, derived from the Greek, means *having horns on the shoulders*, a peculiarity which I have not observed in any other Sphinx.

half in length, are of a pale green color, with seven oblique white lines on each side of the body, and a row of little notches, like saw-teeth, on the back. The four short horns

Fig. 149

on their shoulders are also notched, and, like most other Sphinges, they have a long and stiff spine on the hinder extremity of the body. They enter the earth to become chrysalids, and pass the winter, and come forth in the winged state in the month of June following, at which time the moths may often be found on the trunks of trees, or on fences in the vicinity. In this state their wings expand nearly five inches, are of a light brown color, variegated with dark brown and white, and the hinder part of the body is marked with five longitudinal dark brown lines. A young friend of mine, in Boston, once captured on the trunks of the trees a large number of these moths during a morning's walk in the Mall, although obliged to be on the alert to escape from the guardians of the Common, whose duty it was to prevent the grass from being trodden down. Nearly all of these specimens were females, ready to deposit their eggs, with which their large bodies were completely filled. On being taken they made scarcely any efforts to escape, and were safely carried away. It would not be difficult, by such means, very considerably to reduce the number of these destructive insects; in addition to which it might be expedient, during the proper season, for our city authorities to employ persons to gather and kill every morning the caterpillars which may be found in those public walks where they abound.

From the genus Sphinx I have separated another group

to which I have given the name of *Philampelus*,* from the circumstance that the larvæ or caterpillars live upon the grape-vine. When young they have a long and slender tail recurved over the back like that of a dog; but this, after one or two changes of the skin, disappears, and nothing remains of it but a smooth, eye-like, raised spot on the top of the last segment of the body. Some of these caterpillars are pale green and others are brown, and the sides of their body are ornamented by six cream-colored spots, of a broad oval shape, in the species which produces the *Satellitia* of Linnæus; narrow oval and scalloped, in that which is transformed to the species called *Achemon* (Fig. 150) by Drury.[13]

Fig. 150.

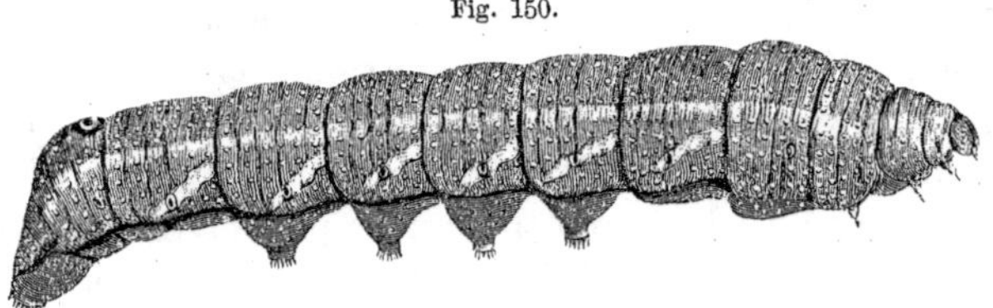

They have the power of withdrawing the head and the first three segments of the body within the fourth segment, which gives them a short and blunt appearance when at rest. As they attain to the length of three inches or more, and are thick in proportion, they consume great quantities of leaves; and the long leafless branches of the vine too often afford evidence of their voracity. They also devour the leaves of the common creeper (*Ampelopsis quinquefolia*), which, with those of our indigenous vines, were their only food till the introduction and increased cultivation of foreign vines afforded them an additional supply. They come to their growth during the month of August, enter the earth to transform, and appear in the winged or moth state the following summer, in June and July. The *Satellitia* Hawk-moth (Plate V. Fig. 2) expands from four to five inches,

* The literal signification of this word is, *I love the vine.*

[[13] *P. achemon* is *Sphinx crantor* Cramer and Hübner. — Morris.]

and is of a light olive color, variegated with patches of darker olive. The *Achemon* (Plate V. Fig. 3; Fig. 151, pupa) expands from three to four inches, is of a reddish ash-color, with two triangular patches of deep brown on the thorax, and two square ones on each fore wing; the hind wings are pink, with a deeper red spot near the middle, and a broad ash-colored border behind.

Fig. 151.

The grape-vine suffers still more severely from the ravages of another kind of Sphinx caterpillars, smaller in size than the preceding, and like them solitary in their habits, but more numerous, and, not content with eating the leaves alone, in their progress from leaf to leaf down the stem, they stop at every cluster of fruit, and, either from stupidity or disappointment, nip off the stalks of the half-grown grapes, and allow them to fall to the ground untasted. I have gathered under a single vine above a quart of unripe grapes thus detached during one night by these caterpillars.

They are naked and fleshy, like those of the *Achemon* and *Satellitia*, and are generally of a pale green color (sometimes, however, brown), with a row of orange-colored spots on the top of the back, six or seven oblique darker green or brown lines on each side, and a short spine or horn on the hinder extremity. The head is very small, and, with the fore part of the body, is somewhat retractile, but not so completely as in the two preceding species. The fourth and fifth segments being very large and swollen, while the three anterior segments taper abruptly to the head, the fore part of the body presents a resemblance to the head and snout of a hog. This suggested the generical name of *Chœrocampa*, or hog-caterpillar, which has been applied to some of these insects. (Fig. 152, caterpillar covered with cocoons of a parasitic Hymenopterous insect; Fig. 153, the parasite, natural size and magnified.)

The species under consideration is found on the vine and the creeper in July and August; when fully grown, it descends to the ground, conceals itself under fallen leaves,

Fig. 152.

Fig. 153.

which it draws together by a few threads so as to form a kind of cocoon, or covers itself with grains of earth and rubbish in the same way, and under this imperfect cover it changes to a pupa or chrysalis (Fig. 154), and finally appears in the winged state in the month of July of the following year. The moth, to which Sir James Edward Smith gave the name of *Pampinatrix* [14] (Plate V. Fig. 4), from its living on the shoots of the vine, expands from two and a half to three inches, is of an olive-gray color, except the hind wings, which are rust-colored, and the fore wings and shoulder-covers are traversed with olive-green bands.

Fig. 154.

Among the Sphinges of Massachusetts may be mentioned those belonging to the genus *Smerinthus*, whose tongue is very short and scarcely visible, and whose fore wings are generally scalloped on the outer edge. Their caterpillars are rough or granulated, with a stout thorn on the tail, and a triangular head, the apex of the triangle corresponding to the crown. The blind-eyed Smerinthus (*S. excœcata*, Fig. 155) is fawn-colored, clouded with brown, except the hind wings, which are rose-colored in the middle, and ornamented with an eye-like black spot having a pale blue centre. The caterpillar lives on the apple-tree, but is not

[[14] *C. pampinatrix* is *Sphinx myron* Cramer, and *Sphinx cnotus* Hübner. — Morris.]

common enough to prove seriously injurious. The same observation will apply to that of the chocolate brown-eyed Sphinx (*Smerinthus myops*), which lives on the wild-cherry-

Fig. 155.

tree, and to the walnut Sphinx (*Smerinthus Juglandis*), which lives on the black walnut and butternut. The latter species is destitute of eye-like spots on the hind wings.

Of those belonging to the genus Sphinx proper, that which bears the specific name *drupiferarum* inhabits the hackberry (*Celtis occidentalis*) and the plum-tree ; *Sphinx Kalmiæ* inhabits the broad-leaved laurel (*Kalmia latifolia*) ; the caterpillar of the *Gordius* is found on the apple-tree ; that of the great ash-colored Sphinx (*S. cinerea*) on the lilac ; *Hylæus* on the black alder (*Prinos glaber*, &c.) and whortleberry ; and the curiously checkered caterpillar of *Sphinx coniferarum* on pines. Of the hog-caterpillars, those of *Chærocampa chœrilus* and *versicolor* may be found on swamp pinks (*Azalea viscosa* and *nudiflora*). The caterpillar of the white-lined morning Sphinx (*Deilephila lineata*) feeds upon purslane and turnip leaves ; and that of *Deilephila Chamœnerii* on the willow-herb (*Epilobium angustifolium*). The clear-winged Sphinges, *Sesia pelasgus*[15] (Fig. 156) and *diffinis*, are distinguished by their transparent wings and their fan-shaped tails. They hover over flowers,

[[15] *S. pelasgus* is *S. thisbe* Fab. = *S. cimbiciformis* Stephens = *S. ruficaudis* Kirby. — MORRIS.]

Fig. 156.

like humming-birds, during the daytime, in the months of July and August. Their caterpillars bear a general resemblance to those of the genus Sphinx, and, as far as they are known, seem to possess the same habits.

The Ægerians (ÆGERIADÆ) constitute a very distinct group among Sphinges. They are easily recognized, in the perfected or winged state, by their resemblance to bees, hornets, or wasps, by their narrow wings, which are mostly transparent, and by the tufts or brush at the end of the body, which they have the power of spreading out like a fan at pleasure. They fly only in the daytime, and frequently alight to bask in the sunshine. Their habits, in the caterpillar state, are entirely different from those of the other Sphinges; the latter living exposed upon plants whose leaves they devour, while the caterpillars of the Ægerians are concealed within the stems or roots of plants, and derive their nourishment from the wood and pith. Hence they are commonly called borers, a name, however, which is equally applicable to the larvæ or young of many insects of other orders.

The caterpillars of the Ægerians are whitish, soft, and slightly downy. Like those of other Sphinges they have sixteen feet, but they are destitute of a thorn or prominence on the last segment of the body. When they have come to their full size, they enclose themselves in oblong oval cocoons (Fig. 157), made of fragments of wood or bark cemented by a gummy matter, and within these are transformed to chrysalids. The latter are of a shining bay color,

Fig. 157.

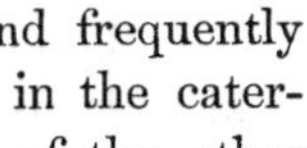

and the edges of the abdominal segments are armed with transverse rows of short teeth. By means of these little teeth, the chrysalis, just before it is about to be transformed to a winged insect, works its way out of the cocoon, and partly through the hole, in the stem or root, which the caterpillar had previously made; and the shell of the chrysalis (Fig. 158) is left half emerging from the orifice, after the moth has escaped from it.

Fig. 158.

The ash-tree suffers very much from the attacks of borers of this kind, which perforate the bark and sap-wood of the trunk from the roots upwards, and are also found in all the branches of any considerable size. The trees thus infested soon show symptoms of disease, in the death of branches near the summit; and, when the insects become numerous, the trees no longer increase in size and height, and premature decay and death ensue. These borers assume the chrysalis form in the month of June, and the chrysalids may be seen projecting half-way from the round holes in the bark of the tree in this and the following month, during which time their final transformation is effected, and they burst open and escape from the shells of the chrysalis in the winged or moth state. Under this form this insect was described, in my paper in Professor Silliman's "Journal of Science," by the name of *Trochilium* * *denudatum;* as the habits of the larva are now ascertained, we may call it the ash-tree *Trochilium.* Its general color is brown; the edges of the collar and of the abdominal rings, the shins, the feet, and the under side of the antennæ are yellowish. The hind wings are transparent; the fore wings are opaque and brown, variegated with rust-red; they have a transparent space near the tips, and expand about an inch and a half.

* The word *Trochilium* is derived from *Trochilus*, the scientific name of the humming-bird genus; and these insects are sometimes called humming-bird moths.

During the month of August, the squash and other cucurbitaceous vines are frequently found to die suddenly down to the root. The cause of this premature death is a little borer (Fig. 159, larva), which begins its operations near the ground, perforates the stem, and devours the interior. It afterwards enters the soil, forms a cocoon (Fig. 160, cocoon containing chrysalis) of a gummy substance covered with particles of earth, changes to a chrysalis, and comes forth the next summer a winged insect. This is conspicuous for its orange-colored body, spotted with black, and its hind legs fringed with long orange-colored and black hairs. The hind wings only are transparent, and the fore wings expand from one inch to one inch and a half. It deposits its eggs on the vines close to the roots, and may be seen flying about the plants from the 10th of July till the middle of August. This insect, which may be called the squash-vine Ægeria, was first described by me in the year 1828, under the name of *Ægeria*[16] *Cucurbitæ* (Plate V. Fig. 8), the trivial name indicating the tribe of plants on which the caterpillar feeds.*

Fig. 159.

Fig. 160.

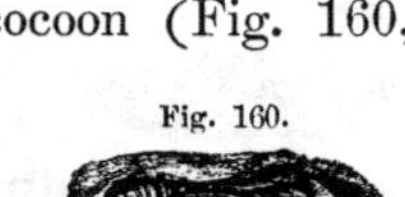

The pernicious borer (Fig. 161, larva) which, during many years past, has proved very destructive to peach-trees throughout the United States, is a species of *Ægeria*, named *exitiosa* (Plate V. Fig. 6, male), or the destructive, by Mr. Say, who first scientifically described it in the third volume of the "Journal of the Academy of Natural Sciences of Philadelphia," and subsequently gave a representation and account of it in his "American En-

Fig. 161.

[16 The genus *Ægeria* Fab. is now rejected by the best authorities, and all the species are put under *Trochilium* Scopoli, which has the priority by thirty years. — Morris.]

* See New England Farmer, Vol. VIII. p. 33; my Discourse before the Massachusetts Horticultural Society, in 1832, p. 26; and Silliman's Journal, Vol. XXXVI. p. 310.

tomology." In the fifth volume of the "New England Farmer" I have given the history of this insect, have mentioned the principal authors who have noticed it, and recommended preventive measures, which have been found effectual in protecting the peach-tree from its most serious attacks.

The eggs, from which these borers are hatched, are deposited, in the course of the summer, upon the trunk of the tree near the root; the borers penetrate the bark, and devour the inner bark and sap-wood. The seat of their operations is known by the castings and gum which issue from the holes in the tree. When these borers are nearly one year old, they make their cocoons either under the bark of the trunk or of the root, or in the earth and gum contiguous to the base of the trees; soon afterwards they are transformed to chrysalids (Fig. 162), (Plate V. Fig. 7, chrysalis from which the moth has escaped,) and finally come forth in the winged state, and lay the eggs for another generation of borers. The last transformation takes place from June to October, most frequently, however, during the month of July, in the State of Massachusetts. Here, although there are several broods produced by a succession of hatches, there is but one rotation of metamorphoses consummated within a year. Hence borers, of all sizes, will be found in the trees throughout the year, although it seems to be necessary that all of them, whether more or less advanced, should pass through one winter before they appear in the winged state.

Fig. 162.

Under its last form, this insect is a slender, dark-blue, four-winged moth, having a slight resemblance to a wasp or ichneumon-fly, to which it is sometimes likened. The two sexes differ greatly from each other, so much so as to have caused them to be mistaken for two distinct species. The male (Plate V. Fig. 6), which is much smaller than the female, has all the wings transparent, but bordered and veined with steel-blue, which is the general color of the body in both sexes; the palpi or feelers, the edges of the collar, of the shoulder-covers, of the rings of the abdomen, and of the brush on the tail, are pale yellow, and there are two rings of the same yellow color on the shins. It expands about one inch. The fore wings of the female are blue, and opaque, the hind wings transparent, and bordered and veined like those of the male, and the middle of the abdomen is encircled by a broad orange-colored belt. It expands an inch and a half, or more.

This insect does not confine its attacks to the peach-tree. I have repeatedly obtained both sexes from borers inhabiting the excrescences which are found on the trunks and limbs of the cherry-tree; and, moreover, I have frequently taken them in connection on the trunks of cherry and of peach-trees. They sometimes deposit their eggs in the crotches of the branches of the peach-tree, where the borers will subsequently be found; but the injury sustained by their operations in such parts bears no comparison to that resulting from their attacks at the base of the tree, which they too often completely girdle, and thus cause its premature decay and death.

The following plan, which was recommended by me in the year 1826, and has been tried with complete success by several persons in this vicinity, will effectually protect the neck, or most vital part of the tree, from injury. Remove the earth around the base of the tree, crush and destroy the cocoons and borers which may be found in it, and under the bark, cover the wounded parts with the common clay composition, and surround the trunk with a strip of sheathing-paper eight or nine inches wide, which should extend two inches below the level of the soil, and be secured with strings of matting above. Fresh mortar should then be placed around the root, so as to confine the paper and prevent access beneath it, and the remaining cavity may be filled with new or unexhausted loam. This operation should be performed in the spring, or during the month of June. In the winter the strings may be removed, and in the following spring the trees should again be examined for any borers that may have escaped search before, and the protecting applications should be renewed.

In Europe there is a species of *Ægeria*, named by Linnæus *tipuliformis*, which has long been known to inhabit the stems of the currant-bush. This, or an insect closely resembling it, is far too common in America, in the cultivated currant, with which it may have been introduced from Europe. The caterpillars are produced from eggs laid singly, near the buds; when hatched, they penetrate the stem to the pith, which they devour, and thus form a burrow of several inches in length in the interior of the stem. As the borer increases in size, it enlarges the hole communicating with its burrow, to admit of the more ready passage of its castings, and to afford it the means of escape when it is transformed to a moth. The inferior size of the fruit affords an indication of the operations of the borers; and the perforated stems frequently break off at the part affected, or, if of sufficient size still to support the weight of the foliage and fruit, they soon become sickly, and finally die.

In some gardens, nearly every currant-bush has been attacked by these borers; and instances are known to me wherein all attempts to raise currant-bushes from cuttings have been baffled, during the second or third year of the growth of the plants, by the ravages of these insects. They complete their transformations, and appear in the moth state, about the middle of June.

The moth is of a blue-black color; its wings are transparent, but veined and fringed with black, and across the tips of the anterior pair there is a broad band, which is more or less tinged with copper-color; the under side of the feelers, the collar, the edges of the shoulder-covers, and three very narrow rings on the abdomen, are golden yellow. The wings expand three quarters of an inch, or a little more.

Some years ago, it was ascertained that a species of *Ægeria* inhabited the pear-tree in this State; and it is said that considerable injury has resulted from it. An infested tree may be known by the castings thrown out of the small perforations made by the borers, which live under the bark of the trunk, and subsist chiefly upon the inner bark. They make their cocoons under the bark, and change to chrysalids in the latter part of summer. The winged insects appear in the autumn, having, like others of this kind, left their chrysalis-skins projecting from the orifice of the holes which they had previously made. In its winged form, this Ægeria is very much like that which inhabits the currant-bush; but it is a smaller species. It was described by me in the year 1830, under the name of *Ægeria Pyri* (Plate V. Fig. 5), the pear-tree Ægeria; and my account of it will be found on the second page of the ninth volume of the "New England Farmer."

Its wings expand rather more than half an inch; are transparent, but veined, bordered, and fringed with purplish black, and across the tips of the fore wings is a broad dark band glossed with coppery tints; the prevailing color of the upper side of the body is purple-black; but most of the under side is golden yellow, as are the edges of the collar, of the shoulder-covers, and of the fan-shaped brush on the tail, and there is a broad yellow band across the middle of the abdomen, preceded by two narrow bands of the same color.

There are several more insects * belonging to this group in Massachusetts, one of which lives in the stems of the lilac, and another inhabits those of the wild currant, *Ribes floridum*. The winged male of the latter species is remarkable for the very long, slender, and cylindrical tuft or pencil at the extremity of the body. Of the rest, there is nothing particularly worthy of note.

* See Silliman's Journal, Vol. XXXVI. pp. 309 to 313.

The Glaucopidians,* so named from the glaucous or bluish-green color of some of the species, are distinguished from the other Sphinges by their antennæ, which, in the males at least, and sometimes in both sexes, are feathered, or furnished on each side with little slender branches, parallel to each other like the teeth of a comb. In scientific works such antennæ are called pectinated, from *pecten*, the Latin for comb.

The caterpillars of the Glaucopidians have sixteen feet, are slender, and cylindrical, with a few hairs scattered generally over the surface of the body, or arranged in little tufts arising from minute warts, and are without a horn on the hinder extremity. They devour the leaves of plants, and make for themselves cocoons of coarse silk, in which they undergo their transformations. The chrysalids are oblong oval, rounded at one end, tapering at the other, and are not provided with transverse rows of teeth on the surface of the body. In the caterpillar and winged states, in the nature of their transformations, and in their habits, these insects approach very closely to the *Phalænæ*, or moths, forming the third division of Lepidopterous insects, among which they are arranged by some naturalists. There are not many of them in Massachusetts, and only one species requires to be noticed here.†

This is the *Procris Americana* (Fig. 163), a small moth of a blue-black color, with a saffron-colored collar, and a notched tuft on the extremity of the body. The wings, which are very narrow, expand nearly one inch. This little insect is the American representative of the *Procris vitis* or *ampelophaga* of Europe, which, in the caterpillar state, sometimes proves very injurious to the grape-vine. The habits of our species are exactly the same; but have been overlooked, or

Fig. 163.

* See additional observations on page 319.

† For the other species see Silliman's Journal, Vol. XXXVI. pp. 315 to 319.

very rarely observed, in this vicinity. The caterpillars are gregarious, that is, considerable numbers of them live and feed together, collected side by side on the same leaf, and only disperse when they are about to make their cocoons. They are of a yellow color, with a transverse row of black velvety tufts on each ring, and a few conspicuous hairs on each extremity of the body. They are hatched from eggs, which are laid in clusters of twenty or more together on the lower sides of the leaves of the grapevine and creeper; and they come to their growth from the middle to the end of August. They then measure six tenths or rather more than one half of an inch in length. Their feet are sixteen in number, and rather short, and their motions are sluggish. When touched, they curl their bodies sidewise and fall to the ground, or, more rarely, hang suspended from the leaves by a silken thread. When young, they eat only portions of the surface of the leaf; but as they grow older, they devour all but the stalk and principal veins, and, passing from leaf to leaf, thus strip whole branches of their foliage. When numerous, they do much damage to the vines and fruit, by stripping off the leaves in midsummer, when most needed. I have found them in Massachusetts on the grape-vine and on the common creeper, or *Ampelopsis quinquefolia*, and conjecture that the latter constitutes their natural food.

About the year 1830, Professor Hentz found them in swarms upon cultivated grape-vines at Chapel Hill, in North Carolina; and constant care was required to check their ravages there, during several successive years. Several broods appeared there in the course of the summer; but hitherto, only one annual brood has been observed in Massachusetts, although two or more broods may occasionally be produced. When about to make their cocoons, the caterpillars leave the vines, and retire to some sheltered spot. They then enclose themselves, each in a very thin, but tough, oblong oval cocoon, and soon afterwards are transformed to shining brown chrysalids. Early in July, and in the middle of the day, I have seen the moths flying about grape-vines and creepers, at which time, also, they pair and lay their eggs. A more full account of this insect, illustrated by figures, will be found in Hovey's Magazine, for June, 1844.

III. MOTHS. (*Phalænæ.*)*

The third great section of the Lepidoptera, which Linnæus named *Phalæna*, includes a vast number of insects, sometimes called millers, or night-butterflies, but more frequently moths. The latter term, thus applied, comprehends not only those domestic moths which, in the young or caterpillar state, devour cloth, but all other insects belonging to the order Lepidoptera which cannot be arranged among the butterflies and hawk-moths.

These insects vary greatly in size, color, and structure. Some of them, particularly those with gilded wings, are very minute; while the Atlas-moth of China (*Attacus Atlas*), when its wings are expanded, covers a space measuring nearly nine inches by five and a half; and the owl-moth (*Erebus Strix*) has wings which, though not so broad, expand eleven inches. Some female moths are destitute of wings, or have but very small ones, wholly unfitted for flight; and there are species whose wings are longitudinally cleft into several narrow rays, resembling feathers. The stalk of the antennæ of moths generally tapers from the base to the end. These parts sometimes resemble simple or naked bristles, and sometimes they are plumed on each side of the stalk, like feathers. There is often a good deal of difference in the antennæ, according to the sex; feathered or pectinated antennæ being generally narrower in the females than in the males; and there are some moths the males of which have feathered antennæ,

* See page 320.

while those of the other sex are not feathered at all, or only furnished with very short projections, like teeth, at the sides. Most moths have a sucking-tube, commonly called the tongue, consisting of two hollow and tapering threads, united side by side, and when not in use rolled up in a spiral form; but in many this member is very short, and its two threads are not united; and in some it is entirely wanting, or is reduced to a mere point. Two *palpi* or feelers are found in most moths. They grow from the lower lip, generally curve upwards, and cover the face on each side of the tongue. Some have, besides these, another pair, which adhere to the roots of the tongue. Many moths are said to have no feelers; these parts being in them very small, and invisible to the naked eye.

The caterpillars of these insects differ more from each other than the moths. In general they are of a cylindrical shape, and are provided with sixteen legs; there are many, however, which have only ten, twelve, or fourteen legs; and in a few the legs are so very short as hardly to be visible, so that these caterpillars seem to glide along in the manner of slugs. Some caterpillars are naked, and others are clothed with hairs or bristles, and the hairs are either uniformly distributed, or grow in tufts. Sometimes the surface of the body is even and smooth; sometimes it is covered with little warts or tubercles; or it is beset with prickles and spines, which not unfrequently are compound or branched.

Many caterpillars, previous to their transformation, enclose themselves in cocoons, composed entirely of silk, or of silk interwoven with hairs stripped from their own bodies, or with fragments of other substances within their reach. Some go into the ground, where they are transformed without the additional protection of a cocoon; others change to chrysalids in the interior of the stems, roots, leaves, or fruits of plants. The chrysalids of moths are generally of an elongated oval shape, rounded at one end, and tapering almost to a point at the other; and they are destitute of the angular elevations which are found on the chrysalids of butterflies.

These brief remarks, which are necessarily of a very general nature, and comprise but a few of the principal differences observable in these insects, must suffice for the present occasion.

Linnæus divided the Moths into eight groups; namely, *Attaci*, *Bombyces*, *Noctuæ*, *Geometræ*, *Tortrices*, *Pyralides*, *Tineæ*, and *Alucitæ;* and these (with the exception of the *Attaci*, which are to be divided between the *Bombyces* and *Noctuæ*) have been recognized as well-marked groups, and have been adopted by some of the best entomologists * who succeeded him.

1. Spinners. (*Bombyces.*)

The Bombyces, so called from *Bombyx*, the ancient name of the silk-worm, are mostly thick-bodied moths, with antennæ in the greater number feathered or pectinated, at least in the males, the tongue and feelers very short or entirely wanting, the thorax woolly, but not crested, or very rarely, and the fore legs often very hairy. Their caterpillars have sixteen legs, are generally spinners, and, with few exceptions, make cocoons within which they are transformed.

This tribe has been subdivided into a number of lesser groups or families; but naturalists are not at all agreed upon the manner in which these should be arranged. We might place at the head of the tribe those large moths, whose Sphinx-like caterpillars are naked and warty, and which, in the winged state, are ornamented with eye-like spots like the *Smerinthi;* or we might place first in the series the moths whose caterpillars are wood-eaters, with the habits

* It is hardly necessary to say that among these are Denis and Schiffermüller, the authors of the celebrated Vienna Catalogue, besides Latreille, Leach, Stephens, and others, whose classifications of the Moths, how much soever varied, enlarged, or improved, are essentially based on the arrangement proposed by Linnæus.

and transformations of the *Ægerians;* or we may begin with the smaller species, with hairy caterpillars, whose habits and transformations are like those of the *Glaucopidians*, and which resemble the latter closely in the winged state; and thus the series, from *Procris* and other moth-like *Sphinges* to the true *Moths*, will be uninterrupted. The latter, on the whole, seems to be the most natural course, and it agrees with the arrangement of Dr. Boisduval, which I shall follow, with some slight changes only.

Agreeably to this arrangement the first family of the Bombyces will be the Lithosians (LITHOSIADÆ), so named from two Greek words,* meaning a stone, and to live; for the caterpillars of many of these insects live in stony places, and devour the lichens growing on rocks. (Such also are the habits of *Glaucopis Pholus* (Fig. 164), one of the Glaucopidians.) On this account they are not properly subjects for notice in this essay; but as some of the larger species are grass-eaters, are conspicuous for their beauty, and naturally conduct to another family particularly obnoxious to the cultivators of the soil, it may be interesting to point out their distinguishing traits.

Fig. 164.

The Lithosians are slender-bodied moths, mostly of small size, whose rather narrow upper or fore wings, when at rest, generally lie flatly on the top of the back, crossing or overlapping each other on their inner margins, and entirely covering the under wings, which are folded longitudinally, and, as it were, moulded around the body; more rarely the wings slope a little at the sides, and cover the back like a low roof. The antennæ are rather long, and bristle-formed; sometimes naked in both sexes, more often slightly feathered with a double row of short hairs beneath,

* This is the derivation given by M. Godart, Hist. Nat. Lépidopt. de France Vol. V. p. 10.

in the males. The tongue and one pair of feelers are very distinct and of moderate length. The back is smooth, neither woolly nor crested, but thickly covered with short and close feather-like scales. The wings of many of the Lithosians are prettily spotted, and they frequently fly in the daytime like the Glaucopidians. Their caterpillars are sparingly clothed with hairs, growing in little clusters from minute warts on the surface of the body. They enclose themselves in thin oblong cocoons of silk interwoven with their own hairs. The rings of their chrysalids are generally so closely joined as not to admit of motion.

Of about a dozen kinds inhabiting Massachusetts, I shall describe only two. The first of these may be called *Gnophria vittata*,* [17] the striped Gnophria. It is of a deep scarlet color; its fore wings, which expand one inch and one eighth, have two broad stripes, and a short stripe between them at the tip, of a lead-color, and the hind wings have a very broad lead-colored border behind; the middle of the abdomen and the joints of the legs are also lead-colored. The caterpillar lives upon lichens, and may be found under loose stones in the fields in the Spring. It is dusky, and thinly covered with stiff, sharp, and barbed black bristles, which grow singly from small warts. Early in May it makes its cocoon, which is very thin and silky; and twenty days afterwards is transformed to a moth.

By far the most elegant species is the *Deïopeia bella* (Plate VI. Fig. 3), the beautiful Deïopeia. This moth has naked bristle-formed antennæ; its fore wings are deep yellow, crossed by about six white bands, on each of which is a row of black dots; the hind wings are scarlet red, with an irregular border of black behind; the body is

* This moth has all the essential characters of the European *Gnophria rubricollis*, an insect closely resembling in its colors the *Procris Americana*. The name of the genus is derived from a Greek word signifying dusky, in allusion to the dark colors of the insects.

[[17] *Gnophria vittata* is *Lithosia miniata* Kirby. — MORRIS.]

white, and the thorax is dotted with black. It expands from one inch and a half to one and three quarters. Its time of appearance here is from the middle of July till the beginning of September. The caterpillar is unknown to me; but Drury states that he was informed it was of the same color as the fore wings of the moth, (that is, yellow and white dotted with black,) and that it feeds upon the blue lupines.* The European *Deïopeia pulchella*, which is very much like our species, feeds, in the caterpillar state, on the leaves of the mouse-ear, *Myosotis arvensis* and *palustris;* and it is probable that ours may be found on plants of the same kind here.

Some of the large and richly colored Lithosians resemble, in many respects, the insects in the next family, called, by the English, tiger and ermine moths. The caterpillars of most of these tiger-moths are thickly covered with hairs, whence they have received the name of woolly bears, and the family, including them, that of ARCTIADÆ, or Arctians, from the Greek word for bear. The Arctians, or tiger-moths, have shorter and thicker feelers than the Lithosians; their tongue is also for the most part very short, not extending, when unrolled, much beyond the head; their antennæ, with few exceptions, are doubly feathered on the under side; but the feathering is rather narrow, and is hardly visible in the females; their wings are not crossed on the top of the back,† but are roofed or slope downwards on each side of the body, when at rest; the thorax is thick, and the abdomen is short and plump, and generally ornamented with rows of black spots. Their fore wings are often variegated with dark-colored spots on a light ground, or light-colored veins on a dark ground; and the hind wings are frequently red, orange, or yellow, spotted with black or blue. They fly only in the night. Their caterpil-

* Drury's Illustrations, Vol. I. p. 52, pl. 24, fig. 3.

† To this character there is an exception in the *Lophocampa tessellaris*, the wings of which are closed like those of *Lithosia quadra*.

lars are covered with coarse hairs, spreading out on all sides like the bristles of a bottle-brush, and growing in clusters or tufts from little warts regularly arranged in transverse rows on the surface of the body. They run very fast, and when handled roll themselves up almost into the shape of a ball. Many of them are very destructive to vegetation, as, for example, the salt-marsh caterpillar, the yellow bear-caterpillar of our gardens, and the fall web-caterpillar. When about to transform, they creep into the chinks of walls and fences, or hide themselves under stones and fallen leaves, where they enclose themselves in rough oval cocoons, made of hairs plucked from their own bodies, interwoven with a few silken threads. The chrysalis is smooth, and not hairy, and its joints are movable.

Some of the slender-bodied Arctians, with bristle-formed antennæ, which are not distinctly feathered in either sex, and having the feelers slender, and the tongue longer than the others, come so near to the Lithosians that naturalists arrange them sometimes among the latter, and sometimes among the Arctians. They belong to Latreille's genus *Callimorpha** (meaning beautiful form), one species of which inhabits Massachusetts, and is called *Callimorpha militaris* (Fig. 165), the soldier-moth, in my Catalogue. Its fore wings expand about two inches, are white, almost entirely bordered with brown, with an oblique band of the same color from the inner margin to the tip; and the

Fig. 165.

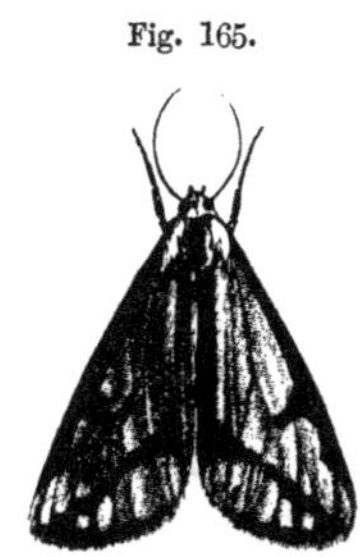

* The French naturalists, whom I have followed, include in this genus the European moths called *Hera*, *Dominula*, *Donna*, *Jacobææ*, &c. Closely allied to the *Hera*, and still more so to the *militaris*, is a large and fine species, which inhabits the Southern States, and which I have named *Callimorpha Carolina*. It differs from the *militaris* in being larger, measuring across the wings two inches and a quarter, or more, and in having the hind wings of a deep Indian-yellow or ochre color, with one or two black spots near the hind margin; the abdomen also is ochre-yellow. It is possible that this may be the *Clymene* of Esper and Ochsenheimer, or the *Colona* of Hübner, whose works I have not seen.

brown border on the front margin generally has two short angular projections extending backwards on the surface of the wing. The hind wings are white, and without spots. The body is white; the head, collar, and thighs, buff-yellow; and a longitudinal brown stripe runs along the top of the back from the collar to the tail. This is a very variable moth; the brown markings on the fore wings being sometimes very much reduced in extent, and sometimes, on the contrary, they run together so much that the wings appear to be brown, with five large white spots. This latter variety is named *Callimorpha Lecontei* by Dr. Boisduval. The caterpillar is unknown to me. The caterpillars of the Callimorphas are more sparingly clothed with hairs than the other Arctians; and they are generally dark-colored, with longitudinal yellow stripes. They feed on various herbaceous and shrubby plants, and conceal themselves in the daytime under leaves or stones.

Most of the other tiger and ermine moths of Massachusetts may be arranged under the general name of *Arctia.** The first of them would probably be placed by Mr. Kirby in *Callimorpha,*† from which, however, they differ in their shorter and more robust antennæ, always very distinctly feathered, at least in the males. They are distinguished from the rest by having two black spots on the collar, and three short black stripes on the thorax. The largest and most rare of these moths is the *Arctia virgo*, or virgin tiger-moth. On account of the peculiarly strong and disagreeable odor which it gives out, it might with greater propriety have been named the stinking tiger-moth. It is a very beautiful insect. Its

* *Chelonia* of the French, *Euprepia* of the Germans (from a Greek word signifying pre-eminent beauty), and subdivided, by the English entomologists, into many genera, founded on minute differences in the length of the joints of the feelers, &c., which it is unnecessary to regard in this treatise.

† Mr. Kirby's *Callimorpha parthenice* and *virguncula* closely resemble the first two or three species which follow. The European *pudica*, and probably also the *Nemeophila plantaginis* belong to the same group. See Fauna Boreali Americana, Vol. IV. pp. 304, 305, pl. 4, fig. 6.

fore wings expand from two inches to two and a half, are flesh-red, fading to reddish buff, and covered with many stripes and lance-shaped spots of black; the hind wings are vermilion-red, with seven or eight large black blotches; the under side of the body is black, the upper side of the abdomen vermilion-red, with a row of black spots close together along the top of the back. The caterpillar is brown, and pretty thickly covered with tufts of brown hairs. The moth appears here in the latter part of July and August.

The *Arge* tiger-moth resembles the preceding, but is smaller, and not so highly colored, and the black markings on the fore wings are smaller, and separated from each other by wider spaces. Its general tint is a light flesh-color, fading to nankin; the fore wings are marked with streaks and small triangular spots of black; the hind wings are generally deeper-colored than the fore wings, and have from five to seven or eight black spots of different sizes upon them; there are two black spots on the collar, and three on the thorax, as in the preceding species; the abdomen is of the color of the hind wings, with a longitudinal row of black dots on the top, another on each side, and two rows of larger size beneath. The wings expand from one inch and three quarters to two inches. I have taken this moth from the 20th of May till the middle of July. The caterpillar appears here sometimes in large swarms in the month of October, having then become fully grown, measuring about one inch and a half in length, and being at this time in search of proper winter quarters wherein to make their cocoons. They are of a dark greenish-gray color, but appear almost black from the black spots with which they are thickly covered; there are three longitudinal stripes of flesh-white on the back, and a row of kidney-shaped spots of the same color on each side of the body. The warts are dark gray, and each one produces a thin cluster of spreading blackish hairs. They eat the leaves of plantain and of other herbaceous plants, and it is stated*

* Abbot's Insects of Georgia, p. 125, pl. 63.

that they sometimes make great devastation among young Indian corn in the Southern States.

A much more abundant species in Massachusetts is that which has been called the harnessed moth, *Arctia phalerata* (Fig. 166) of my Catalogue. It makes its appearance from the end of May to the middle of August, and probably breeds throughout the whole summer. It is of a pale buff or nankin color; the hind wings next to the body, and the sides of the body, are reddish; on the fore wings are two longitudinal black stripes and four triangular black spots, the latter placed near the tip; and these stripes and spots are arranged so that the buff-colored spaces between them somewhat resemble horse-harness; the hind wings have several black spots near the margin; there are two dots on the collar, three stripes on the thorax, and a stripe along the top of the back, of a black color; the under side of the body and the legs are also black. The wings expand from one inch and a half to one inch and three quarters. The caterpillar is not yet known to me. This moth, in many respects, resembles one called *Phyllira** by Drury, rarely found here, but abundant in the Southern States; the fore wings of which are black, with one longitudinal line, two transverse lines, and near the tip two zigzag lines forming a W, of a buff color.

Fig. 166.

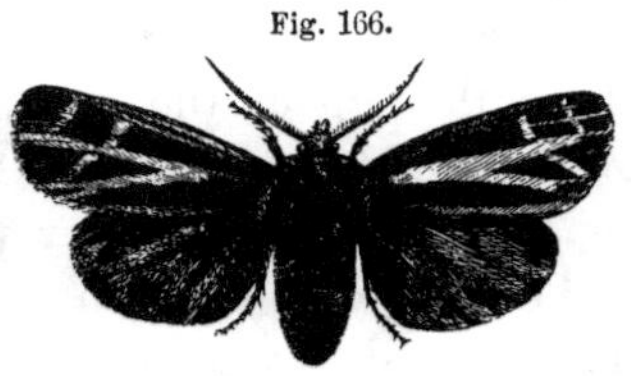

The feelers and tongue of the foregoing moths, though short, are longer than in the following species, which have these parts, as well as the head, smaller and more covered with hairs. Some of the latter may be said to occupy the centre or chief place among the Arctians, exceeding all the rest in the breadth of their wings, the thickness of their bodies, and the richness of their colors. Among these is the great American tiger-moth, *Arctia Americana*, an unde-

* More properly *Philyra*.

scribed species, which some of the French entomologists* have supposed to be the same as the great tiger, *Arctia Caja*, of Europe. Of this fine insect I have a specimen, which was presented to me by Mr. Edward Doubleday, who obtained it, with several others, near Trenton Falls in New York. It has not yet been discovered in Massachusetts, but will probably be found in the western part of the State. The fore wings of the *Arctia Americana* expand two inches and a half or more; they are of a brown color, with several spots and broad winding lines of white, dividing the brown surface into a number of large irregular blotches; the hind wings are ochre-yellow, with five or six round blue-black spots, three of them larger than the rest; the thorax is brown and woolly; the collar edged with white before, and with crimson behind; the outer edges of the shoulder-covers are white; the abdomen is ochre-yellow, with four black spots on the middle of the back; the thighs and fore legs are red, and the feet dark brown. This moth closely resembles the European *Caja*, and especially some of its varieties, from all of which, however, it is essentially distinguished by the white edging of the collar and shoulder-covers, and the absence of black lines on the sides of the body. It is highly probable that specimens may occur with orange-colored or red hind wings like the *Caja*, but I have not seen any such. The caterpillar of our species probably resembles that of the *Caja*, which is dark chestnut-brown or black, clothed with spreading bunches of hairs, of a foxy-red color on the fore part and sides of the body, and black on the back; but the clusters of hairs, though thick, are not so close as to conceal the breathing holes, which form a distinct row of pearly-white spots on each side of the body. These caterpillars eat the leaves of various kinds of garden plants without much discrimination, feeding together in considerable numbers on the same plant when young, but scattering as they grow older.

* Godart, Lépidopt. de France, Tom. IV. p. 303. It is figured in the "Lake Superior" of Agassiz and Cabot, pl. 7, fig. 5.

The largest of the American Arctians is the *Scribonia*, or great white leopard-moth, which varies in expansion from two and a half to three and a half inches, the females being invariably much larger than the males. It is of a white color; the fore wings and thorax are ornamented with many small oval black rings, the hind wings are more or less spotted with black; and the abdomen is yellow, with rows of large blue-black spots on the back and sides.

The caterpillar, as represented by Mr. Abbot,* is the counterpart of that of the *Hebe* of Europe, being chestnut-brown with transverse red bands between the rings, and is clothed with clusters of dark brown hairs. It is said to eat the leaves of the wild sunflower and of various other plants. It has been confidently reported to me that the great leopard-moth has been seen in Brookline; but it must be very rare here, for I have never heard of its being taken in any part of New England. Specimens of this fine insect would be a very acceptable addition to any collection of such objects.

Of all the hairy caterpillars frequenting our gardens, there are none so common and troublesome as that which I have called the yellow-bear (Fig. 167). Like most of its genus, it is a very general feeder, devouring almost all kinds of herbaceous plants with equal relish, from the broad-leaved plantain at the door-side, the peas, beans, and even the flowers of the garden, and the corn and coarse grasses of the fields, to the leaves of the vine, the currant, and the gooseberry, which it does not refuse when pressed by hunger. This kind of caterpillar varies very much in its colors; it is perhaps most often of a pale yellow or straw color, with a black line along each side of the body, and a transverse line

Fig 167.

* Insects of Georgia, p. 137, pl. 69.

of the same color between each of the segments or rings, and it is covered with long pale yellow hairs. Others are often seen of a rusty or brownish yellow color, with the same black lines on the sides and between the rings, and they are clothed with foxy-red or light brown hairs. The head and ends of the feet are ochre-yellow, and the under side of the body is blackish in all the varieties. They are to be found of different ages and sizes from the first of June till October. When fully grown they are about two inches long, and then creep into some convenient place of shelter, make their cocoons, in which they remain in the chrysalis state during the winter, and are changed to moths in the months of May or June following. Some of the first broods of these caterpillars appear to come to their growth early in summer, and are transformed to moths by the end of July or the beginning of August, at which time I have repeatedly taken them in the winged state; but the greater part pass through their last change in June. The moth (Fig. 168) is familiarly known by the name of the white miller, and is often seen about houses. Its scientific name is *Arctia Virginica*,[18] and, as it nearly resembles the insects commonly called ermine-moths* in England, we may give to it the name of the Virginia ermine-moth. It is white, with a black point on the middle of the fore wings, and two black dots on the hind wings, one on the middle and the other near the posterior angle, much more distinct on the under than on the upper side; there is a row of black dots on the top of the back, another on each side, and between these a longitudinal deep yellow stripe; the hips and thighs of the fore legs are also ochre-yellow.

Fig. 168.

[18 *Arctia Virginica* belongs to the genus *Spilosoma.* — MORRIS.]

* It is most like the *Arctia Urticæ*, but is of a much purer white color.

It expands from one inch and a half to two inches. Its eggs are of a golden-yellow color, and are laid in patches upon the leaves of plants. In some parts of France, and in Belgium, the people have been required by law to *écheniller*, or uncaterpillar, their gardens and orchards, and have been punished by fine for the neglect of the duty. Although we have not yet become so prudent and public-spirited as to enact similar regulations, we might find it for our advantage to offer a bounty for the destruction of caterpillars; and though we should pay for them by the quart, as we do for berries, we should be gainers in the end, while the children whose idle hours were occupied in the picking of them would find this a profitable employment.

The salt-marsh caterpillar (Fig. 169), an insect by far too well known on our seaboard, and now getting to be common in the interior of the State, whither it has probably been introduced, while under the chrysalis form, with the salt hay annually carried from the coast by our inland farmers, closely resembles the yellow bear in some of its varieties. The history of this insect forms the subject of a communication made by me to the Agricultural Society of Massachusetts, in the year 1823, and printed in the seventh volume of the "Massachusetts Agricultural Repository and Journal," with figures representing the insect in its different stages. At various times and intervals since the beginning of the present century, and probably before it also, the salt marshes about Boston have been overrun and laid waste by swarms of caterpillars. These appear towards the end of June, and grow rapidly from that time till the first of August. During this month they come to their full size, and begin to run, as the phrase is, or retreat from the marshes, and disperse through the adjacent uplands, often committing very extensive ravages

Fig. 169.

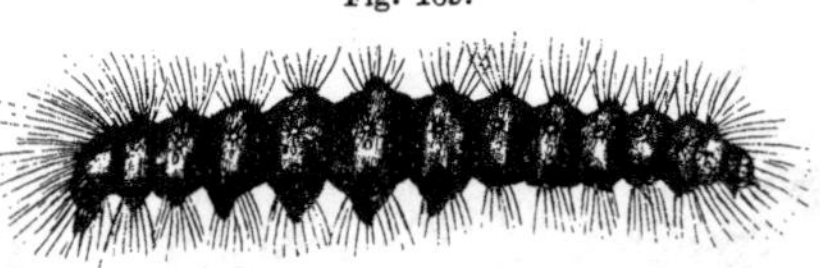

in their progress. Corn-fields, gardens, and even the rank weeds by the way-side, afford them temporary nourishment while wandering in search of a place of security from the tide and weather. They conceal themselves in walls, under stones, in hay-stacks and mows, in wood-piles, and in any other places in their way, which will afford them the proper degree of shelter during the winter. Here they make their coarse hairy cocoons, and change to chrysalids, in which form they remain till the following summer, and are transformed to moths in the month of June.

In those cases where, from any cause, the caterpillars, when arrived at maturity, have been unable to leave the marshes, they conceal themselves beneath the stubble, and there make their cocoons. Such, for the most part, is the course and duration of the lives of these insects in Massachusetts; but in the Middle and Southern States two broods are brought to perfection annually, and even here some of them run through their course sooner, and produce a second brood of caterpillars in the same season; for I have obtained the moths between the 15th and 20th of May, and again between the 1st and the 10th of August. Those which were disclosed in May passed the winter in the chrysalis form, while the moths which appeared in August must have been produced from caterpillars that had come to their growth and gone through all their transformations during the same summer. This, however, in Massachusetts, is not a common occurrence; for by far the greater part of these insects appear at one time, and require a year to complete their several changes.

The full-grown caterpillar measures one inch and three quarters or more in length. It is clothed with long hairs, which are sometimes black and sometimes brown on the back and fore part of the body, and of a lighter brown color on the sides. The hairs, like those of the other Arctians, grow in spreading clusters from warts, which are of a yellowish color in this species. The body, when stripped of the hairs, is yellow, shaded at the sides with black, and there is a blackish line extending along the top of the back. The breathing-holes are white, and very distinct even through the hairs. These caterpillars, when feeding on the marshes, are sometimes overtaken by the tide, and when escape becomes impossible they roll themselves up in a circular form, as is common with others of the tribe, and abandon themselves to their fate. The hairs on their bodies seem to have a repelling power, and prevent the water from wetting their skins, so that they float on the surface, and are often carried by the waves to distant places, where they are thrown on shore and left in winrows with the wash of the sea. After a little time, most of them recover from their half-drowned condition, and begin their depredations anew. In this way, these insects seem to have spread from the places where they first appeared to others at a considerable distance.

From the marshes about Cambridge they were once, it is said, driven in great numbers by a high tide and strong wind upon Boston Neck, near to Roxbury line. Thence they seem to have migrated to the eastern side of the Neck, and, following the marshes to South Boston and Dorchester, they have spread in the course of time to those which border upon Neponset River and Quincy. How far they have extended north of Boston I have not been able to ascertain ; but I believe that they are occasionally found on all the marshes of Chelsea, Saugus, and Lynn. Although these insects do not seem ever entirely to have disappeared from places where they have once established themselves, they do not prevail every year in the same overwhelming swarms ; but their numbers are increased or lessened at irregular periods from causes which are not well understood.

These caterpillars are produced from eggs, which are laid by the moths on the grass of the marshes about the middle of June, and are hatched in seven or eight days afterwards ; and the number of eggs deposited by a single female is, on an average, about eight hundred. The moths themselves vary in color. In the males (Plate VI. Fig. 9), the thorax and upper side of the fore wings are generally white, the latter spotted with black ; the hind wings and abdomen, except the tail, deep ochre-yellow, the former with a few black spots near the hind margin, and the abdomen with a row of six black spots on the top of the back, two rows on the sides, and one on the belly; the under side of all the wings and the thighs are deep yellow. It expands from one inch and seven eighths to two inches and a quarter. The female (Plate VI. Fig. 10) differs from the male either in having the hind wings white, instead of ochre-yellow, or in having all the wings ashen-gray with the usual black spots. It expands two inches and three eighths or more. Sometimes, though rarely, male moths occur with the fore wings ash-colored or dusky. Professor Peck called this moth *pseuderminea*, that is, false ermine, and this name was adopted by me in my communication to the Agricultural Society. Professor Peck's name, however, cannot be retained, inasmuch as the insect had been previously named and described. Drury, the first describer of the moth, called the male *Caprotina*, and the female *Acrea*,* supposing them to be different species ; but the latter name alone has been retained for this species by most naturalists.

In order to lessen the ravages of the salt-marsh caterpillars, and to secure a fair crop of hay when these insects abound, the marshes should be mowed early in July, at which time the caterpillars are small and feeble, and, being unable to wander far, will die before the crop is gathered in. In defence of early mowing, it may be said that it is the only way by which the grass may be saved in those meadows where the caterpillars have multiplied to any exent ; and if the practice is followed generally, and continued during several years in succession, it will do much towards exterminating these destructive insects.

By the practice of late mowing, where the caterpillars abound, a great loss in the crop will be sustained, immense

* The proper orthography is *Acræa*.

numbers of caterpillars and grasshoppers will be left to grow to maturity and disperse upon the uplands, by which means the evil will go on increasing from year to year; or they will be brought in with the hay to perish in our barns and stacks, where their dead bodies will prove offensive to the cattle, and occasion a waste of fodder. To get rid of "the old fog" or stubble, which becomes much thicker and longer in consequence of early mowing, the marshes should be burnt over in March. The roots of the grass will not be injured by burning the stubble, on the contrary they will be fertilized by the ashes; while great numbers of young grasshoppers, cocoons of caterpillars, and various kinds of destructive insects, with their eggs, concealed in the stubble, will be destroyed by the fire. In the Province of New Brunswick, the benefit arising from burning the stubble has long been proved; and this practice is getting into favor here.

During the autumn there may be seen in our gardens and fields, and even by the way-side, a kind of caterpillar (Fig. 170) whose peculiar appearance must frequently have excited attention. It is very thickly clothed with hairs, which are stiff, short, and perfectly even at the ends, like the bristles of a brush, as if they had all been shorn off with the shears to the same length.

Fig. 170.

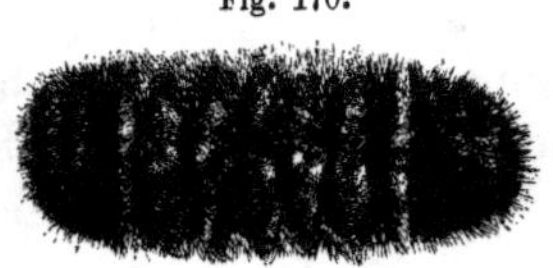

The hairs on the first four and last two rings are black; and those on the six intermediate rings of the body are tan-red. The head and body of the caterpillar are also black. When one of these insects is taken up, it immediately rolls itself into a ball, like a hedge-hog, and, owing to its form and to the elasticity of the diverging hairs with which it is covered, it readily slides from the fingers and hand of its captor. It eats the leaves of the clover, dandelion, narrow-leaved plantain, and of various other herbaceous plants, and on the approach of winter creeps under stones, rails, or boards on the ground, where it remains in a half-torpid state till spring. In April or May it makes an oval blackish cocoon, composed chiefly of the hairs of its body, and comes forth in the moth state in June or July.

My specimens remained in the chrysalis form five weeks; but Mr. Abbot* states that a caterpillar of this kind, which made its cocoon in Georgia on the 24th of June, was transformed to a moth on the 5th of July, having remained only eleven days in the chrysalis state. The moth is the *Arctia Isabella*, or Isabella tiger-moth, and it differs essentially from those which have been described in the antennæ, which are not feathered, but are merely covered on the under side with a few fine and short hairs, and even these are found only in the males. Its color is a dull grayish tawny-yellow; there are a few black dots on the wings, and the hinder pair are frequently tinged with orange-red; on the top of the back is a row of about six black dots, and on each side of the body a similar row of dots. The wings expand from two inches to two inches and three eighths. The specific name, which was first given to this moth by Sir James Edward Smith, is expressive of its peculiar shade of yellow.

We have a much smaller tiger-moth, with naked antennæ like those of the *Isabella*. Its wings are so thinly covered with scales as to be almost transparent. It has not yet been described, and it may be called the ruddle tiger-moth, *Arctia rubricosa* (Fig. 171). Its fore wings are reddish-brown, with a small black spot near the middle of each; its hind wings are dusky, becoming blacker behind (more rarely red, with a broad blackish border behind), with two black dots near the middle, the inner margin next to the body, and the fringe, of a red color; the thorax is reddish-brown; and the abdomen is cinnabar-red, with a row of black dots on the top, and another row on each side. It expands about one inch and one quarter. This moth is

Fig. 171.

* Insects of Georgia, p. 131, pl. 66.

rare; and it appears here in July and August. It closely resembles the ruby tiger-moth, *Arctia fuliginosa*, of Europe, the wings of which are not so transparent, and have two black dots on each of them, with a distinct row of larger black spots around the outer margin of the hind pair. The caterpillar of our moth is unknown to me; it will probably be found to resemble that of the ruby tiger, which is blackish, and thickly covered with reddish-brown or reddish-gray hairs. It eats the leaves of plantain, dock, and of various other herbaceous plants, grows to the length of one inch and three eighths, passes the winter concealed beneath stones, or in the crevices of walls, and makes its cocoon in the spring.

The caterpillars of all the foregoing Arctians live almost entirely upon herbaceous plants; those which follow (with one exception only) devour the leaves of trees. Of the latter, the most common and destructive are the little caterpillars known by the name of fall web-worms, whose large webs, sometimes extending over entire branches with their leaves, may be seen on our native elms, and also on apple and other fruit trees, in the latter part of summer. The eggs, from which these caterpillars proceed, are laid by the parent moth in a cluster upon a leaf near the extremity of a branch; they are hatched from the last of June till the middle of August, some broods being early and others late, and the young caterpillars immediately begin to provide a shelter for themselves by covering the upper side of the leaf with a web, which is the result of the united labors of the whole brood. They feed in company beneath this web, devouring only the upper skin and pulpy portion of the leaf, leaving the veins and lower skin of the leaf untouched. As they increase in size they enlarge their web, carrying it over the next lower leaves, all the upper and pulpy parts of which are eaten in the same way, and thus they continue to work downwards, till finally the web covers a large portion of the branch with its dry, brown, and filmy foliage, reduced to this unseemly condition by these little spoilers. These caterpillars (Plate

VII. Fig. 12, young caterpillar), when fully grown, measure rather more than one inch in length; their bodies are more slender than those of the other Arctians, and are very thinly clothed with hairs of a grayish color, intermingled with a few which are black. The general color of the body is greenish yellow dotted with black; there is a broad blackish stripe along the top of the back, and a bright yellow stripe on each side. The warts, from which the thin bundles of spreading, silky hairs proceed, are black on the back, and rust-yellow or orange on the sides. The head and feet are black.

I have not observed the exact length of time required by these insects to come to maturity; but towards the end of August and during the month of September they leave the trees, disperse, and wander about, eating such plants as happen to lie in their course, till they have found suitable places of shelter and concealment, where they make their thin and almost transparent cocoons (Plate VII. Fig. 10; Fig. 11, pupa), composed of a slight web of silk intermingled with a few hairs. They remain in the cocoons in the chrysalis state through the winter, and are transformed to moths in the months of June and July. These moths are white and without spots; the fore thighs are tawny yellow, and the feet blackish. Their wings expand from one inch and a quarter to one inch and three eighths. Their antennæ and feelers do not differ essentially from those of the majority of the Arctians, the former in the males being doubly feathered beneath, and those of the females having two rows of minute teeth on the under side. This species was first described by me in the seventh volume of the New England Farmer, page 33, where I gave it the name of *Arctia textor*, the weaver, from the well-known habits of its caterpillar. Should it be found expedient to remove it from the genus *Arctia*, I propose to call the genus which shall include it *Hyphantria*, a Greek name for weaver, and place in the same genus the many-spotted ermine-moth, *Arctia punctatissima*[19] of Sir J.

[[19] *Arctia punctatissima* is *Spilosoma cunea* Drury. — MORRIS.]

E. Smith, which is found in the Southern States, and agrees with our weaver in habits. From the foregoing account of the habits and transformations of the fall web-worm, or *Hyphantria textor*,[20] it is evident that the only time in which we can attempt to exterminate these destructive insects with any prospect of success is when they are young and just beginning to make their webs on the trees. So soon, then, as the webs begin to appear on the extremities of the branches, they should be stripped off, with the few leaves which they cover, and the caterpillars contained therein, at one grasp, and should be crushed under foot.

There are many kinds of hairy caterpillars in Massachusetts, differing remarkably from those of the other *Arctians*, and resembling in some respects those belonging to the next tribe, with which they appear to connect the true Arctians. The first of these are little party-colored tufted caterpillars (Fig. 172), which may be found in great plenty on the common milk-weed, *Asclepias Syriaca*, during the latter part of July and the whole of August. Although the plants on which these insects live are generally looked upon as weeds and cumberers of the soil, yet the insects themselves are deserving of notice, on account of their singularity, and the place that they fill in the order to which they belong. They keep together in companies, side by side, beneath the leaves, their heads all turned towards the edge of the leaf while they are eating, and when at rest they arch up the fore part of the body and bend down the head, which is then completely concealed by long overhanging tufts of hairs, and if disturbed they jerk their heads and bodies in a very odd way. These harlequin caterpillars have sixteen legs, which, with the head, are black. Their bodies are black also, with a whitish line on each side, and are thickly covered with short tufts of hairs

Fig. 172.

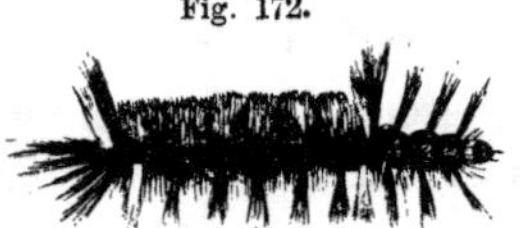

[20 *Hyphantria textor* is *Spilosoma textor*. — MORRIS.]

proceeding from little warts. Along the top of the back is a row of short black tufts, and on each side, from the fifth to the tenth ring inclusive, are alternate tufts of orange and of yellow hairs, curving upwards so as nearly to conceal the black tufts between them; below these, along the sides of the body, is a row of horizontal black tufts; on the first and second rings are four long pencil-like black tufts extending over the head, on each side of the third ring is a similar black pencil, and two, which are white, placed in the same manner on the sides of the fourth and of the tenth rings. About the last of August, and during the month of September, these caterpillars leave the milk-weed, disperse, conceal themselves, and make their cocoons (Fig. 173), which mostly consist of hairs. The chrysalis (Fig. 174) is short, almost egg-shaped, being quite blunt and rounded at the hind end, and is covered with little punctures like those on the head of a thimble, only much smaller. The chrysalids are transformed to moths between the middle of June and the beginning of July. These moths, though not so slender as the Callimorphas, are not so thick and robust as the Arctias, their antennæ resemble those of the latter, but are rather longer, the feelers are also longer, and spread apart from each other, and the tongue is but little longer than the head, when unrolled. The wings are rather long, thin, and delicate, of a bluish-gray color, paler on the front edge, and without spots; the head, thorax, under side of the body, and the legs are also gray; the neck is cream-colored; the top of the abdomen bright Indian-yellow, with a row of black spots, and two rows on each side. It expands from one inch and three quarters to nearly two inches. This moth was figured and described many years ago by Drury, who named it *Egle*. Though marked and colored like some of the Arctias (for example, the *luctifera* of Europe), it cannot with propriety be included in the same genus, and therefore I have proposed to call it *Euchætes Egle;* the first

Fig. 173.

Fig. 174.

name, signifying fine-haired, or having a flowing mane, is given to it on account of the long tuft of hairs overhanging the fore part of the caterpillar like a mane. This moth, in some of its characters, approaches to the Lithosians, but seems, in others, too near to the Arctians to be removed from the latter tribe, and it is evidently, in the caterpillar state, nearly allied to the following insects, which are undoubtedly Arctians, but lead apparently to the Liparians. If our Arctians are grouped in a circle, with the larger kinds, such as the great American tiger and leopard moths in the middle, and the others arranged around them, then will these species, which are here described last, be brought round to the Callimorphas, with which the series began, and thus a natural order of succession will be preserved.

During the months of August and September there may be seen on the hickory, and frequently also on the elm and ash, troops of caterpillars (Plate VI. Fig. 1), covered with short spreading tufts of white hairs, with a row of eight black tufts on the back, and two long, slender, black pencils on the fourth and on the tenth ring. The tufts along the top of the back converge on each side, so as to form a kind of ridge or crest; and the warts, from which these tufts proceed, are oblong-oval and transverse, while the other warts on the body are round. The hairs on the fore part of the body are much longer than the rest, and hang over the head; the others are short, as if sheared off, and spreading. The head, feet, and belly are black; the upper side of the body is white, sprinkled with black dots, and with black transverse lines between the rings. These neat and pretty caterpillars, when young, feed in company on the leaves; while not engaged in eating, they bend down the head and bring over it the long hairs on the fore part of the body; and, if disturbed or handled, they readily roll up like the other Arctians. When fully grown, they are nearly one inch and a half long. They leave the trees in the latter part of September, secrete themselves under stones and in the chinks of walls, and make their cocoons (Plate VI. Fig. 2), which are oval, thin, and hairy, like those of the other Arctians. The chrysalis is short, thick, and rather blunt, but not rounded at the hinder end, and not downy. The moths, which come out of the cocoons during the month of June, are of a very light ochre-yellow color; the fore wings are long, rather narrow, and almost pointed, are thickly and finely sprinkled with little brown dots, and have two oblique brownish streaks passing backwards from the front edge, with three rows of white semi-transparent spots parallel to the outer hind margin; the hind wings are very thin, semi-transparent, and without spots; and the shoulder-covers are edged within with light brown. They expand from one inch and seven eighths to two inches and a quarter or more. The wings are roofed when at rest; the antennæ are long, with a double, narrow, feathery edging, in the males, and a double row of short, slender teeth on the under side, in the females; the feelers are longer than in the other Arctians, and not at all hairy; and the tongue is short, but spirally curled. This kind of moth does not appear to have been described before, and it cannot be placed in any of the modern genera belonging to the Arctians; for this reason I propose to call it *Lophocampa*[21] *Caryæ* (Fig. 175); the first name meaning crested caterpillar, and the second being the scientific name of the hickory, on which it lives. In England, the moths that come from caterpillars having long pencils and tufts on their backs are called tussock-moths; we may name the one under consideration the hickory tussock-moth.

Fig. 175.

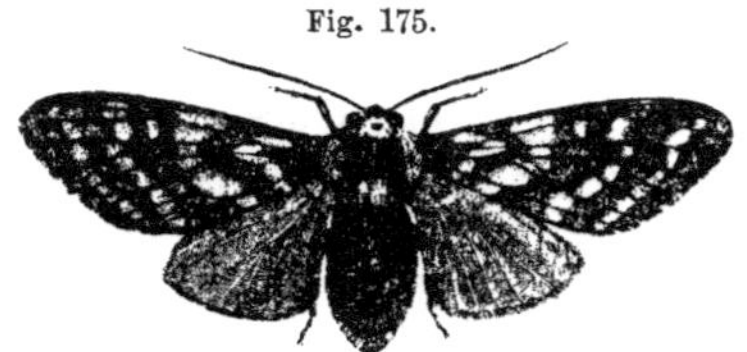

In August and September I have seen on the black walnut, the butternut, the ash, and even on the oak, caterpillars exactly resembling the foregoing in shape, but differing in

[[21] *Lophocampa* is *Halesidota* Walker. — MORRIS.]

color, being covered, when young, with brownish-yellow tufts, of a darker color on the ridge of the back, and having four long white and two black pencils extending over the head from the second ring, and two black pencils on the eleventh ring; when they are fully grown they are covered with ash-colored tufts, those on the ridge blackish; the head is black, the body black or greenish black above, and whitish beneath, and the legs are rust-yellow. This is evidently a different species or kind from the hickory tussock, being differently colored, and having the two hindmost pencils placed on the eleventh, and not on the tenth ring. I have not yet succeeded in keeping these caterpillars alive until they had finished their transformations.

In my collection are specimens of a moth closely resembling the hickory tussock in everything except size and color. It may be named *Lophocampa maculata*, the spotted tussock-moth. It is of a light ochre-yellow color, with large irregular light brown spots on the fore wings, arranged almost in transverse bands. It expands nearly one inch and three quarters. The caterpillar, as far as I can judge from a shrivelled specimen, was covered with whitish tufts forming a crest on the back, in which were situated eight black tufts; there was a black pencil on each side of the fourth and of the tenth ring, and a quantity of long white hairs overhanging the head and the hinder extremity; the head was black; but the color of the body cannot be ascertained.

A fourth kind of *Lophocampa*, or crested caterpillar, remains to be described. It is very common, throughout the United States, on the buttonwood or sycamore, upon which it may be seen in great numbers in July and August. The tufts on these caterpillars are light yellow or straw-colored, the crest being very little darker; on the second and third rings are two orange-colored pencils, which are stretched over the head when the insect is at rest, and before these are several long tufts of white hairs; on each side of the third ring is a white pencil, and there are two pencils, of the same color, directed backwards, on the eleventh ring. The body is yellowish white, with dusky warts, and the head is brownish yellow. These caterpillars leave the trees towards the end of August, and conceal themselves in crevices of fences, and under stones, and make their cocoons, which resemble those of the hickory tussock; and from the middle of June to the end of July the moths come forth. These moths are faintly tinged with ochre-yellow; their long, narrow, delicate, and semi-transparent wings lie almost flatly on the top of the back; the upper pair are checkered with dusky spots, arranged so as to form five irregular transverse bands; the hind edge of the collar, and the inner edges of the shoulder-covers, are greenish blue, and between the latter are two short and narrow deep yellow stripes; the upper side of the abdomen and of the legs are deep ochre-yellow. The wings expand about two inches. The name of this beautiful and delicate moth is *Lophocampa tessellaris*, the checkered tussock-moth. It is figured and described in Smith and Abbot's "Insects of Georgia," where, however, the caterpillar is not correctly represented. Mr. Abbot's figure of the caterpillar has been copied in the illustrations accompanying Cuvier's last edition of the "Règne Animal," and is there referred to Latreille's genus *Sericaria*. This includes, besides various other insects having no resemblance to the foregoing, the true tussock caterpillars belonging to the next group; but from these the caterpillars of all the kinds of *Lophocampa* differ essentially, in being much more hairy, in not having the warts on the sides of the first ring longer than the rest, and in being destitute of the little retractile vesicles on the top of the ninth and tenth rings; moreover, their chrysalids are not covered with short hairs in clusters or ridges. On the other hand, they agree with the Arctians in being covered with warts and spreading bunches of hairs, in rolling up like a ball when handled, and in the form and structure of their cocoons. The position of the wings of the checkered tussock-moth, when at rest, is almost exactly like that of some of the Lithosians ; but the other kinds of Lophocampa do not cross the inner edges of the wings ; and the bodies of all of them are much thicker and more robust than those of the Lithosians.

The third group or family of Bombyces may be called Liparians (LIPARIDÆ*). Of the moths bearing this name, the females have remarkably thick bodies, and are sometimes destitute of wings, while the males are generally slender, and have rather broad wings. Their feelers are very hairy, and for the most part are rather longer than those of the Arctians. Their tongues are very short, and invisible or concealed. Their antennæ are short, and bent like a bow, and doubly feathered on the under side, the feathering of those of the males being very wide, and of the females mostly narrow. When at rest, these moths stretch out their hairy fore legs before their bodies, and keep their upper and lower wings together over their backs, sloping a very little at the sides, and covering the abdomen like a low or flattened roof. The females, even of those kinds that are provided with wings, are very sluggish and heavy in their motions, and seldom go far from their cocoons; the males frequently fly by day in search of their mates. The caterpillars of most of the Liparians are half naked, their thin hairs growing chiefly on the sides of their bodies; the warts which furnish them being only six or eight† in number on each ring; and they have two little soft and reddish warts (one on the top of the ninth, and the other on the tenth ring), which can be drawn in and out at pleasure. Some of them have four or five short and thick tufts, cut off square at the ends, on the top of the back, two long and slender pencils of hairs extending forwards, like antennæ, from the first ring, sometimes two

* From *Liparis*, more properly *Liparus*, the name of a genus of moths belonging to this group. This name means fat or gross, and was probably assigned to the genus on account of the thickness of the bodies of some of these moths.

† The Arctians have ten or more warts on each ring.

more pencils on the fifth ring, and a single pencil on the top of the eleventh ring. The warts which produce these pencils are more prominent or longer than the rest. These caterpillars are called tussocks in England, from the tufts on their backs. They live upon trees and shrubs, and, when at rest, they bend down the head, and bring over it the long plume-like pencils of the first ring. Their cocoons are large, thin, and flattened, and consist of a soft kind of silk, intermixed with which are a few hairs. The chrysalids are covered with down or short hairs, and end at the tail with a long projecting point. In Europe there are many kinds of Liparians, some of them at times exceedingly injurious to vegetation, their caterpillars devouring the leaves of fruit-trees, and not unfrequently extending their devastations to the hedges, and even to the corn and grass.* There do not appear to be many kinds in the United States, and they never swarm to the same extent as in Europe.

During the months of July and August, there may be found on apple-trees and rose-bushes, and sometimes on other trees and shrubs, little slender caterpillars (Plate VII. Fig. 1), of a bright yellow color, sparingly clothed with long and fine yellow hairs on the sides of the body, and having four short and thick brush-like yellowish tufts on the back, that is on the fourth and three following rings, two long black plumes or pencils extending forwards from the first ring, and a single plume on the top of the eleventh ring. The head, and the two little retractile warts on the ninth and tenth rings, are coral-red; there is a narrow black or brownish stripe along the top of the back, and a wider dusky stripe on each side of the body. These pretty caterpillars do not ordinarily herd together, but sometimes our

* These destructive kinds are the caterpillars of the brown-tailed moth (*Porthesia auriflua*), of the golden-tailed moth (*Porthesia chrysorrhœa*), of the gypsy-moth (*Hypogymna dispar*), and of the black arches-moth (*Psilura monacha*). The first of these abounded to such an extent in England, in the year 1782, that prayers were ordered to be read in all the churches, to avert the destruction which was anticipated from them.

apple-trees are much infested by them, as was the case in the summer of 1828. In the summers of 1848, 1849, and 1850, they were very numerous on trees in Boston, both in private yards and on the common, where the horse-chestnuts, which seem ordinarily to escape the attacks of insects, were almost entirely stripped of their leaves by these insects. When they have done eating, they spin their cocoons on the leaves, or on the branches or trunks of the trees, or on fences in the vicinity. The chrysalis is not only beset with little hairs or down, but has three oval clusters of branny scales on the back. In about eleven days after the change to the chrysalis is effected, the last transformation follows, and the insects come forth in the adult state, the females wingless, and the males with large ashen-gray wings, crossed by wavy darker bands on the upper pair, on which, moreover, is a small black spot near the tip, and a minute white crescent near the outer hind angle. The body of the male is small and slender, with a row of little tufts along the back, and the wings expand one inch and three eighths. The females (Plate VII. Figs. 2 and 3) are of a lighter gray color than the males, their bodies are very thick, and of an oblong oval shape, and, though seemingly wingless, upon close examination two little scales, or stinted winglets, can be discovered on each shoulder. These females lay their eggs upon the top of their cocoons (Plate VII. Fig. 5), and cover them with a large quantity of frothy matter, which on drying becomes white and brittle. Different broods of these insects appear at various times in the course of the summer, but the greater number come to maturity and lay their eggs in the latter part of August and the beginning of September, and these eggs are not hatched till the following summer. The name of this moth is *Orgyia* * *leucostigma* (Plate VII.

* This name is derived from a word which signifies to stretch out the hands, and it is applied to this kind of moth on account of its resting with the fore legs extended. The Germans call these moths *streckfüssige Spinner;* the French, *pattes étendues;* and the English, vaporer-moths; the latter probably because the males are seen flying about ostentatiously, or vaporing, by day, when most other moths keep concealed.

Fig. 4, male), the white-marked Orgyia or tussock-moth. It is to the eggs of this insect that the late Mr. B. H. Ives, of Salem, alludes, in an article on "insects which infest trees and plants," published in Hovey's "Gardener's Magazine." * Mr. Ives states, that, on passing through an apple orchard in February, he "perceived nearly all the trees speckled with occasional dead leaves, adhering so firmly to the branches as to require considerable force to dislodge them. Each leaf covered a small patch of from one to two hundred eggs, united together, as well as to the leaf, by a gummy and silken fibre, peculiar to the moth." In March, he "visited the same orchard, and, as an experiment, cleared three trees, from which he took twenty-one bunches of eggs. The remainder of the trees he left untouched until the 10th of May, when he found the caterpillars were hatched from the egg, and had commenced their slow but sure ravages. He watched them from time to time, until many branches had been spoiled of their leaves, and in the autumn were entirely destitute of fruit, while the three trees which had been stripped of the eggs were flush with foliage, each limb, without exception, ripening its fruit." These pertinent remarks point out the nature and extent of the evil, and suggest the proper remedy to be used against the ravages of these insects.

In the New England States there is found a tussock or vaporer moth, seemingly the same as the *Orgyia antiqua*, the antique or rusty vaporer-moth of Europe, from whence possibly its eggs may have been brought with imported fruit-trees. The male moth is of a rust-brown color, the fore wings are crossed by two deeper brown wavy streaks, and have a white crescent near the hind angle. They expand about one inch and one eighth. The female is gray, and wingless, or with only two minute scales on each side in the place of wings, and exactly resembles in shape the female of the foregoing species. The caterpillar is yellow on the back, on which

* Vol. I. p. 52.

are four short square brush-like yellow tufts; the sides are dusky and spotted with red; there are two long black pencils or plumes on the first ring, one on each side of the fifth ring, and one on the top of the eleventh ring; the head is black; and the retractile warts on the top of the ninth and tenth rings are red. These caterpillars live on various trees and shrubs, and are stated by Miss Dix, in Professor Silliman's "Journal of Science,"* to have been "very destructive to the thorn hedges in Rhode Island," "appearing very early in summer, and not disappearing till late in November." The cocoons resemble those of the white-marked vaporer (*Orgyia leucostigma*), and the females, after they have come forth, never leave the outside of their cocoons, but lay their eggs upon them and die there.

The next group may be called Lasiocampians (LASIOCAMPADÆ), after the principal genus† included in it, the name of which signifies hairy caterpillar. The Lasiocampians are woolly and very thick-bodied moths, distinguished by the want of the bristles and hooks that hold together the fore and hind wings of other moths, by the wide and turned-up fore edge of the hind wings, which projects beyond that of the fore wings when at rest, and by their caterpillars, which (with few exceptions) are not warty on the back, and are sparingly clothed with short, soft hairs, mostly placed along the sides of the body, and seldom distinctly arranged in spreading clusters or tufts. These moths fly only by night, and both sexes are winged. Their antennæ generally bend downwards near the middle, and upwards at the points, are longer than those of the Liparians, but not so widely feathered in the males, and very narrowly feathered beneath in the females. The feelers of some are rather longer than common, and are thrust forward like a beak; but more

* Vol. XIX. p. 62.

† To *Lasiocampa* belong the European moths called *Rubi*, *Trifolii*, *Quercus*, *Roboris*, *Dumeti*, &c. I have not seen any insects like these in Massachusetts, and believe that such are seldom if ever to be found in the United States.

often they are very short and small. The tongue, for the most part, is invisible. Their wings cover the back like a steep roof; the under pair, being wider than common, are not entirely covered by the upper wings, but project beyond them at the sides of the body when closed. Their caterpillars live on trees and shrubs, and some kinds herd together in considerable numbers or swarms; they make their cocoons mostly or entirely of silk. The winged insect is assisted in its attempts to come forth, after its last change, by a reddish-colored liquid, which softens the end of its cocoon, and which, as some say, is discharged from its own mouth, or, as others with greater probability assert, escapes from the inside of the chrysalis the moment that the included moth bursts the shell.

To this group belong the caterpillars that swarm in the unpruned nurseries and neglected orchards of the slovenly and improvident husbandman, and hang their many-coated webs upon the wild cherry-trees that are suffered to spring up unchecked by the wayside and encroach upon the borders of our pastures and fields. The eggs, from which they are hatched, are placed around the ends of the branches, forming a wide kind of ring or bracelet, consisting of three or four hundred eggs, in the form of short cylinders standing on their ends close together, and covered with a thick coat of brownish water-proof varnish (Plate VII. Fig. 16).* The caterpillars come forth with the unfolding of the leaves of the apple and cherry tree, during the latter part of April or the beginning of May. The first signs of their activity appear in the formation of a little angular web or tent, somewhat resembling a spider's web, stretched between the forks of the branches a little below the cluster of eggs. Under the shelter of these tents, in making which they all work together, the caterpillars remain concealed at all times when not engaged in eating. In crawling from twig to twig and from

* A good figure of a cluster of these eggs may be seen in the Boston Cultivator, Vol. X. No. 10, for March 4, 1848.

leaf to leaf, they spin from their mouths a slender silken thread, which is a clew to conduct them back to their tents; and as they go forth and return in files, one after another, their pathways in time become well carpeted with silk, which serves to render their footing secure during their frequent and periodical journeys, in various directions, to and from their common habitation. As they increase in age and size, they enlarge their tent, surrounding it, from time to time, with new layers or webs, till at length it acquires a diameter of eight or ten inches. They come out together at certain stated hours to eat, and all retire at once when their regular meals are finished; during bad weather, however, they fast, and do not venture from their shelter. These caterpillars (Plate VII. Fig. 13) are of a kind called lackeys in England, and *livrées* in France, from the party-colored livery in which they appear. When fully grown, they measure about two inches in length. Their heads are black; extending along the top of the back, from one end to the other, is a whitish line, on each side of which, on a yellow ground, are numerous short and fine crinkled black lines, that, lower down, become mingled together, and form a broad longitudinal black stripe, or rather a row of long black spots, one on each ring, in the middle of each of which is a small blue spot; below this is a narrow wavy yellow line, and lower still the sides are variegated with fine intermingled black and yellow lines, which are lost at last in the general dusky color of the under side of the body; on the top of the eleventh ring is a small blackish and hairy wart, and the whole body is very sparingly clothed with short and soft hairs, rather thicker and longer upon the sides than elsewhere. The foregoing description will serve to show that these insects are not the same as either the Neustria *

* *Neustria* was the ancient name of Normandy, from whence this European species was first introduced into England. The Neustria caterpillar has a bluish head, on which, as also on the first ring, are two black dots; the back is tawny-red, with a central white and two black lines from one end to the other; the sides

or the camp * lackey-caterpillars of Europe, for which they have been mistaken. From the first to the middle of June they begin to leave the trees upon which they have hitherto lived in company, separate from each other, wander about awhile, and finally get into some crevice or other place of shelter, and make their cocoons (Plate VII. Fig. 15). These are of a regular long oval form, composed of a thin and very loosely woven web of silk, the meshes of which are filled with a thin paste, that on drying is changed to a yellow powder, like flour of sulphur in appearance. Some of the caterpillars, either from weakness or some other cause, do not leave their nests with the rest of the swarm, but make their cocoons there, and when the webs are opened these cocoons may be seen intermixed with a mass of blackish grains, like gunpowder, excreted by the caterpillars during their stay. From fourteen to seventeen days after the insect has made its cocoon and changed to a chrysalis, it bursts its chrysalis-skin, forces its way through the wet and softened end of its cocoon, and appears in the winged or miller form. Many of them, however, are unable to finish their transformations by reason of weakness, especially those remaining in the webs. Most of these will be found to have been preyed upon by little maggots living upon the fat within their bodies, and finally changing to small four-winged ichneumon wasps, which in due time pierce a hole in the cocoons of their victims, and escape into the air.

The moth (Plate VII. Fig. 14 male, Fig. 17 female) of our American lackey-caterpillar is of a rusty or reddish-brown color, more or less mingled with gray on the middle and base of the fore wings, which, besides, are crossed by

are blue, with a narrow red stripe; on the top of the eleventh ring is a little blackish wart; and the belly is dusky.

* The *castrensis*, or camp-caterpillar, has a narrow broken white line on the top of the back, separating two broad red stripes, which are dotted with black; the sides are blue, with two or three narrow red stripes; the head and first ring are not marked with black dots; there is no wart on the top of the eleventh ring; and the belly is white, marbled with black.

two oblique, straight, dirty white lines. It expands from one inch and a quarter to one inch and a half, or a little more. This moth* closely resembles the *castrensis*, and still more the *Neustria* of Europe, from both of which, however, it is easily distinguished by the oblique lines on the fore wings, which are not wavy as in the foreign species. Moreover, the caterpillar is very different from both of the European lackeys; and it does not seem probable that either of them, if introduced into this country, could have so wholly lost their original characters. Our insect belongs to the same genus, or kind, now called *Clisiocampa*, or tent-caterpillar, from its habits; and I propose to distinguish it furthermore from its near allies by the name of *Americana*, the American tent-caterpillar or lackey. The moths appear in great numbers in July, flying about and often entering houses by night. At this time they lay their eggs, selecting the wild cherry, in preference to all other trees, for this purpose, and, next to these, apple-trees, the extensive introduction and great increase of which, in this country, afford an abundant and tempting supply of food to the caterpillars, in the place of the native cherry-trees that formerly, it would seem, sufficed for their nourishment. These insects, because they are the most common and most abundant in all parts of our country, and have obtained such notoriety that in common language they are almost exclusively known among us by the name of *the caterpillars*, are the worst enemies of the orchard. Where proper attention has not been paid to the destruction of them, they prevail to such an extent as almost entirely to strip the apple and cherry trees of their foliage, by their attacks

* A short but very accurate account of this insect may be found in the late Professor Peck's "Natural History of the Canker-Worm," printed at Boston, among the papers of the Massachusetts Society for promoting Agriculture, in the year 1796. Professor Peck seems to have been aware that it was not identical with the *Neustria*, but he forbore to give it another scientific name. It is figured, in its different forms, in Mr. Abbot's "Natural History of the Insects of Georgia," where it is named *castrensis* by Sir J. E. Smith, the editor of the work.

continued during the seven weeks of their life in the caterpillar form. The trees, in those orchards and gardens where they have been suffered to breed for a succession of years, become prematurely old, in consequence of the efforts they are obliged to make to repair, at an unseasonable time, the loss of their foliage, and are rendered unfruitful, and consequently unprofitable. But this is not all; these pernicious insects spread in every direction, from the trees of the careless and indolent to those of their more careful and industrious neighbors, whose labors are thereby greatly increased, and have to be followed up year after year, without any prospect of permanent relief.

Many methods and receipts for the destruction of these insects have been published and recommended, but have failed to exterminate them, and indeed have done but little to lessen their numbers, as, indeed, might be expected from the tenor of the foregoing remarks. In order to be completely successful, *they must be universally adopted.* These means comprehend both the destruction of the eggs and of the caterpillars. The eggs are to be sought for in the winter and the early part of spring, when there are no leaves on the trees. They are easily discovered at this time, and may be removed with the thumb-nail and forefinger. Nurseries and the lower limbs of large trees may thus be entirely cleared of the clusters of eggs during a few visits made at the proper season. It is well known that the caterpillars come out to feed twice during the daytime, namely, in the forenoon and afternoon, and that they rarely leave their nests before nine in the morning, and return to them again at noon. During the early part of the season, while the nests are small, and the caterpillars young and tender, and at those hours when the insects are gathered together within their common habitation, they may be effectually destroyed by crushing them by hand in the nests. A brush, somewhat like a bottle-brush, fixed to a long handle, as recommended by the late Colonel Pickering, or, for the want thereof, a dried mullein head and its stalk fastened to a pole, will be useful to remove the nests, with the caterpillers contained therein, from those branches which are too high to be reached by hand. Instead of the brush, we may use, with nearly equal success, a small mop or sponge, dipped as often as necessary into a pailful of refuse soapsuds, strong whitewash, or cheap oil. The mop should be thrust into the nest and turned round a little, so as to wet the caterpillars with the liquid, which will kill every one that it touches. These means, to be effectual, should be employed during the proper hours, that is, early in the morning, at midday, or at night, and as soon in the spring as the caterpillars begin to make their nests; and they should be repeated as often, at least, as once a week, till the insects leave the trees. Early attention and perseverance in the use of these remedies will, in time, save the farmer hundreds of dollars, and abundance of mortification and disappointment, besides rewarding him with the grateful sight of the verdant foliage, snowy blossoms, and rich fruits of his orchard in their proper seasons.

Another caterpillar, whose habits are similar to those of the preceding, is now and then met with in Massachusetts, upon oak and walnut trees, and more rarely still upon apple-trees and cherry-trees. According to Mr. Abbot, "it is sometimes so plentiful in Virginia as to strip the oak-trees bare"; and I may add, that it occasionally proves very injurious to orchards in Maine. It may be called *Clisiocampa silvatica*, the tent-caterpillar of the forest (Plate VII. Fig. 19). With us it comes to its full size from the 10th to the 20th of June, and then measures about two inches in length. There are a few short yellow hairs scattered over its body, particularly on the sides, where they are thickest. The general color of the whole body is light blue, clear on the back, and greenish at the sides; the head is blue, and without spots; there are two yellow spots, and four black dots on the top of the first ring; along the top of the back is a row of eleven oval white spots, beginning on the second ring, and two small elevated black and hairy dots on each ring, except the eleventh, which has only one of larger size; on each side of the back is a reddish stripe bordered by slender black lines; and lower down on each side is another stripe of a yellow color between two black lines; the under side of the body is blue-black. This kind of caterpillar lives in communities of three or four hundred individuals, under a common web or tent, which is made against the trunk or beneath some of the principal branches of the trees. When fully grown they leave the trees, get into places sheltered from rain, and make their cocoons, which exactly resemble those of the apple-tree tent-caterpillars in form, size, and materials. The moths (Plate VII. Fig. 18) appear in sixteen or twenty days afterwards. They are of a brownish yellow or nankin color; the hind wings, except at base, are light rusty-brown; and on the fore wings are two oblique rust-brown and nearly straight parallel lines. A variety is sometimes found with a broad red-brown band across the fore wings, occupying the whole space which in other individuals intervenes between the oblique lines. The wings expand from one inch and one quarter to one inch and three quarters. The great difference in the caterpillar will not permit us to refer this species to the *Neustria* of Europe, for which Sir J. E. Smith* mistook it, or to the *castrensis*, which it more closely resembles in its winged form.

Most caterpillars are round, that is, cylindrical, or nearly so; but there are some belonging to this group that are very broad, slightly convex above, and perfectly flat beneath. They seem indeed to be much broader and more flattened than they really are, by reason of the hairs on their sides, which spread out so as nearly to conceal the feet, and form a kind of fringe along each side of the body. These hairs grow mostly from horizontal fleshy appendages or long warts, somewhat like legs, hanging from the sides of every ring; those on the first ring being much longer than the others,

* See Abbot's "Insects of Georgia," where it is figured.

which progressively decrease in size to the last. On the ore part of the body one or two velvet-like and highly colored bands may be seen when the caterpillar is in motion; and on the top of the eleventh ring there is generally a long naked wart. When these singular caterpillars are not eating, they remain at rest, stretched out on the limbs of trees, and they often so nearly resemble the bark in color as to escape observation. From the lappets, or leg-like appendages, hanging to their sides, they are called lappet-caterpillars by English writers.

Twice I have found, on the apple-tree, in the month of September, caterpillars of this kind, measuring, when fully grown, two inches and a half in length, and above half an inch in breadth. The upper side was gray, variegated with irregular white spots, and sprinkled all over with fine black dots; on the fore part of the body there were two transverse velvet-like bands of a rich scarlet color, one on the hind part of the second, and the other on the third ring, and on each of these bands were three black dots; the under side of the body was orange-colored, with a row of diamond-shaped black spots; the hairs on the sides were gray, and many of them were tipped with a white knob. The caterpillar eats the leaves of the apple-tree, feeding only in the night, and remaining perfectly quiet during the day. The moth produced from it was supposed by Sir J. E. Smith * to be the same as the European *Ilicifolia*, or holly-leaved lappet-moth, from which, however, it differs in so many respects that I shall venture to give it another name. It belongs to the genus *Gastropacha*,[22] so called from the very thick bodies of the moths; and the present species may be named *Americana*, the American lappet-moth (Fig. 176).

Fig. 176.

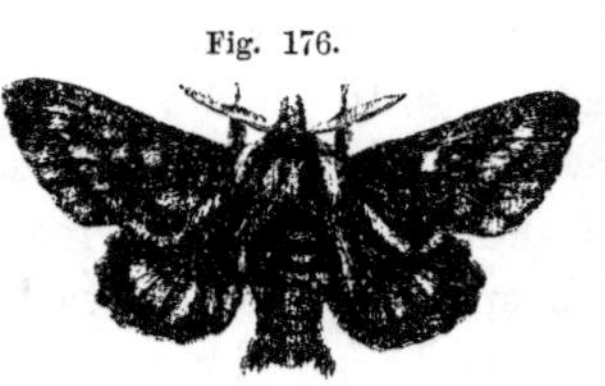

* See Abbot's "Insects of Georgia," p. 101, pl. 51.

[22 *Gastropacha Americana* is *G. occidentalis* Walker. — MORRIS.]

Were it not for its regular shape, it might, when at rest, very easily be mistaken for a dry, brown, and crumpled leaf. The feelers are somewhat prominent, like a short beak; the edges of the under wings are very much notched, as are the hinder and inner edges of the fore wings, and these notches are white; its general color is a red-brown; behind the middle of each of the wings is a pale band, edged with zigzag dark brown lines, and there are also two or three short irregular brown lines running backwards from the front edge of the fore wings, besides a minute pale crescent, edged with dark brown, near the middle of the same. In the females the pale bands and dark lines are sometimes wanting, the wings being almost entirely of a red-brown color. It expands from one inch and a half to nearly two inches. Mr. Abbot, who has figured it, states that the caterpillar lives on the oak and the ash, that it spun itself up in May among the leaves in a gray-brown cocoon, in which the chrysalis was enveloped with a pale brown powder, and that the moth came out in February. My specimens, on the contrary, as above stated, were found on apple-trees, made their cocoons in the autumn, and appeared in the winged form in the early part of the following summer.

The foregoing is the only American lappet-moth, with notched wings, which is known to me; but we have another much larger one, with entire wings. It is the *Velleda* (Fig. 177) of Stoll, so named after a celebrated German female, commemorated by the ancient historian Tacitus. This moth has a very large, thick, and woolly body, and is of a white color, variegated or clouded with blue-gray. On the fore wings are two broad dark gray bands, inter-

Fig. 177.

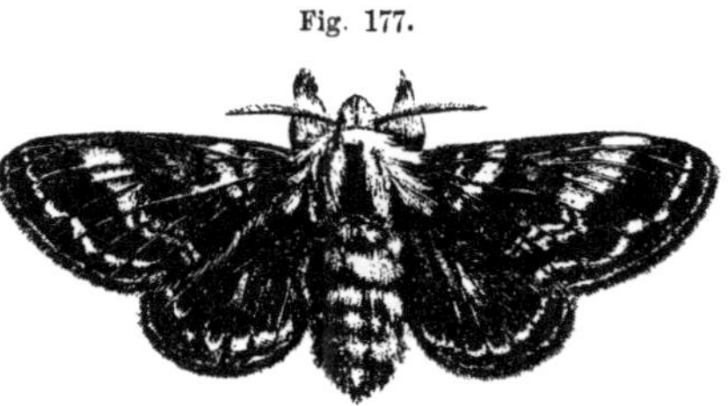

vening between three narrow wavy white bands, the latter being marked by an irregular gray line; the veins are white, prominent, and very distinct; the hind wings are gray, with a white hind border, on which are two interrupted gray lines, and across the middle there is a broad, faint, whitish band; on the top of the thorax is an oblong blackish spot, widening behind, and consisting of long black and pearl-colored erect scales, shaped somewhat like the handle of a spoon. There is a great disparity in the size of the sexes, the males measuring only from one inch and a half to one inch and three quarters across the wings, while the females expand from two and a quarter to two inches and three quarters or more.

Fig. 178.

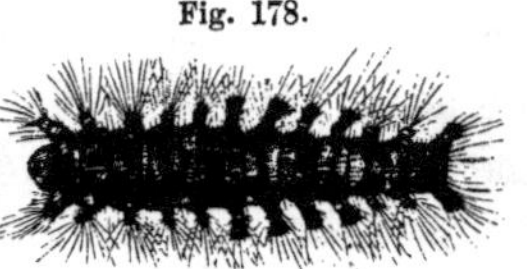

The caterpillar (Fig. 178, young caterpillar) of this fine moth I have never seen alive; but one was sent to me, in the autumn of 1828, by the late T. G. Fessenden, Esq., who received it from Newburyport, from a correspondent, by whom it was found on the 5th of August, sticking so fast to the limb of an apple-tree, that at first it was mistaken for a cankered spot on the bark.* It was said to have measured two inches and a half in length, but when it came into my hands it had spun itself up in its cocoon. A caterpillar of the same kind, found also on an apple-tree, has been described by Miss Dix in Professor Silliman's "Journal of Science." † This observing lady states, that "when at rest the resemblance of its upper surface was so exact with the young bark of the branch on which it was fixed, that its presence might have escaped the most accurate investigation; and this deception was the more complete from the unusual shape of the caterpillar, which might be likened to the external third of a cylinder. The sides of the body were cloaked and fringed with hairs.

* See "New England Farmer," Vol. VII. p. 33.

† Vol. XIX. pp. 62 and 63.

It was of a pale sea-green color above, marked with ash, blended into white; and beneath of a brilliant orange, spotted with vivid black. When in motion its whole appearance was changed, it extended to the length of two inches, and two thirds of an inch in breadth, its colors brightened, and a transverse opening was disclosed on the back, two thirds of an inch from the head, of a most rich velvet-black color. It was sluggish and motionless during the day, and active only at night." Mr. Abbot found the caterpillar of the Velleda lappet-moth on the willow-oak and on the persimmon; and in his figure it is represented of a dark ashen-gray color, with a velvet-like black band across the upper part of the third ring.* The cocoon of the specimen sent to me by Mr. Fessenden resembled grocers' soft brownish-gray paper in color and texture, with a very few blackish hairs interwoven with the silk of which it was made. It was an inch and a half long, and half an inch wide, bordered on all sides by a loose web, which made it seem of larger dimensions; its shape was oval, convex above, and perfectly flat and very thin on the under side. The moth came forth from this cocoon on the 15th of September, or about forty days after the cocoon was spun.

The Chinese silk-worm and its moth, *Bombyx mori*, the Bombyx of the mulberry, should follow these insects in a natural arrangement; for the former is slightly hairy when first hatched from the egg, and, though naked afterwards, it has, like the lappet-caterpillars, a long fleshy wart on the top of the eleventh ring. The history of the silk-worm, however, does not belong to the subject of this treatise.

There are several kinds of caterpillars in the United States whose cocoons are wholly made of a very strong and durable silk, fully equal to that obtained in India from the tusseh and arrindy silk-worms. These insects, together with some others, whose cocoons are much thinner, and consist more of gummy matter than of silk, belong to a family called

* Insects of Georgia, p. 103, pl. 52.

Saturnians (SATURNIADÆ), from *Saturnia*, the name of a genus included in this group. The caterpillars are naked, are generally short, thick, and clumsy, cylindrical, but frequently hunched on the back of each ring, especially when at rest, and are furnished with a few warts, which are either bristled with little points or very short hairs, or are crowned with sharp and branching prickles. They live on trees or shrubby plants, the leaves of which they devour; some of them, when young, keep and feed together in swarms, but separate as they become older. When fully grown and ready to make their cocoons, some of them draw together a few leaves so as to form a hollow, within which they spin their cocoons; others fasten their cocoons to the stems or branches of plants, often in the most artful and ingenious manner; and a very few transform upon or just under the surface of the ground, where they cover themselves with leaves or grains of earth stuck together with a little gummy matter. The escape of the moth from its cocoon is rendered easy by the fluid which is thrown out and softens the threads. The chrysalis offers no striking peculiarities, being smooth, not hairy, and not provided with transverse notched ridges. This group contains some of the largest insects of the order; moths distinguished by great extent and breadth of wings, thick and woolly bodies, and antennæ which are widely feathered on both sides, from one end to the other, in the males at least, and often in both sexes. The tongue and feelers are extremely short and rarely visible. The wings are generally spread out when at rest, so as to display both pairs, and they are held either horizontally, or more or less elevated above the body; a very few, however, turn the fore wings back, so as to cover the hind wings and the body in repose. There are no bristles and hooks to keep the fore hind wings together. In the middle of each wing there is generally a conspicuous spot of a different color from the rest of the surface, often like the eye-spot on peacocks' feathers, sometimes with a transparent space like talc or isinglass in the middle, and sometimes kidney-shaped and opaque. These moths commonly fly towards the close of the day, and in the evening twilight. Their eggs are very numerous, amounting to several hundreds from a single individual.

Although the injuries committed by the caterpillars of the Saturnians are by no means very great, the magnitude and beauty of the moths render them very conspicuous and worthy of notice. The largest kinds belong to that division of the Bombyces called *Attacus* by Linnæus. They are distinguished from the rest of the Saturnians by having wide and flat antennæ, like short oval feathers, in both sexes, and by the fleshy warts on the backs of their caterpillars, which are richly colored, and tipped with minute bristles. Preeminent above all our moths in queenly beauty is the *Attacus Luna* (Fig. 179), or Luna moth, its specific name being the same as that given by the Romans to the moon, poetically styled " fair empress of the night." The wings of this fine insect are of a delicate light-green color, and the hinder angle of the posterior wings is prolonged, so as to form a tail to each, of an inch and a half or more in length; there is a broad purple-brown stripe along the front edge of the fore wings, extending also across the thorax, and sending backwards a little branch to an eye-like spot near the middle of the wing; these eye-spots, of which there is one on each of the wings, are transparent in the centre, and are encircled by rings of white, red, yellow, and black; the hinder borders of the wings are more or less edged or scalloped with purple-brown; the body is covered with a white kind of wool; the antennæ are ochre-yellow; and the legs are purple-brown. The wings expand from four inches and three quarters to five inches and a half. The caterpillar of this moth lives on the walnut and hickory, on which it may be found, fully grown, towards the end of July and during the month of August. It is of a pale and very clear bluish-green color; there is a yellow stripe on each side of the body, and the back is crossed, between the rings, by transverse lines of

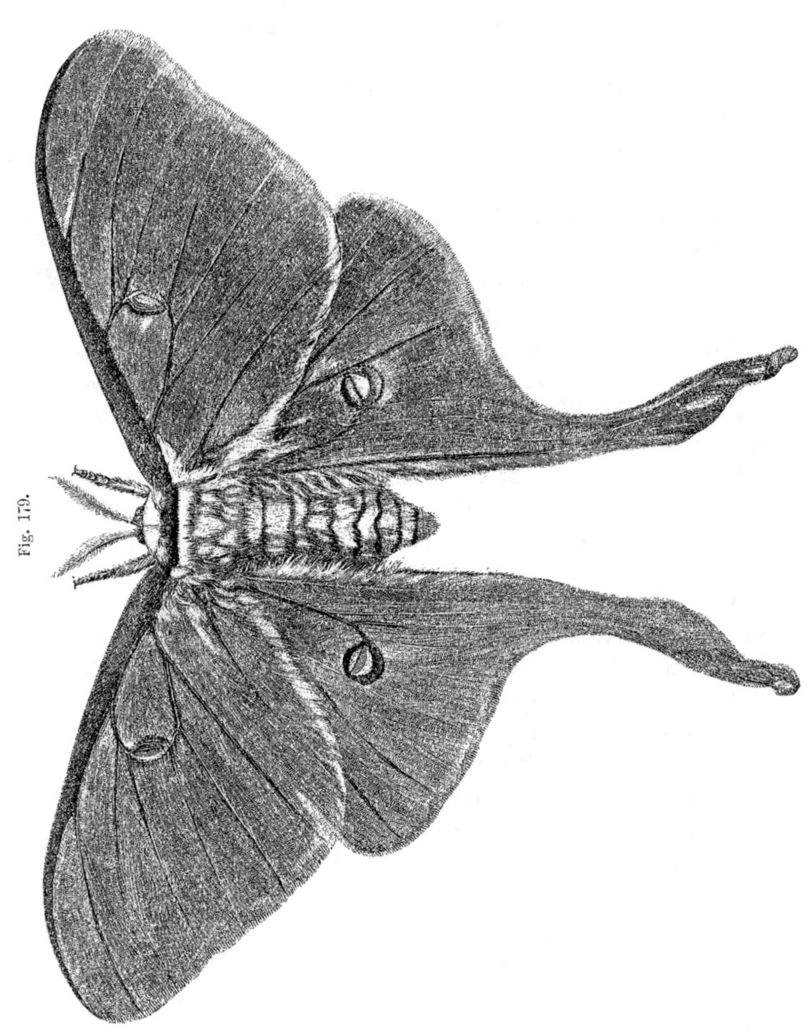

Fig. 179.

the same yellow color; on each of the rings are about six minute pearl-colored warts, tinged with purple or rose-red, and furnishing a few little hairs; and at the extremity of the body are three brown spots, edged above with yellow. When this insect is at rest it is nearly as thick as a man's thumb, its rings are hunched, and its body is shortened, not measuring, even when fully grown, above two inches in length; but, in motion, it extends to the length of three inches or more. When about to make its cocoon, it draws together, with silken threads, two or three leaves of the tree, and within the hollow thus formed spins an oval and very close and strong cocoon (Fig. 180), about one inch and three quarters long, and immediately afterwards changes to a chrysalis. The cocoons fall from the trees in the autumn with the leaves in which they are enveloped; and the moths make their escape from them in June.

Fig. 180.

A caterpillar, closely resembling that of the Luna moth, may be found on oaks, and sometimes also on elm and lime trees, in August and September. Its sides are not striped with yellow, and there are no transverse yellow bands on the back; the warts have a pearly lustre, more or less tinted with orange, rose-red, or purple, and between the two lowermost on the side of each ring is an oblique white line; the head and the feet are brown; and the tail is bordered by a brown V-shaped line. These caterpillars, in repose, cling to the twigs of the trees, with their backs downwards, contract their bodies in length, and hunch up the rings even more than those of the Luna moth, which, when fully grown, they somewhat exceed in size. They make their cocoons upon the trees in the same manner, with an outer covering of leaves, which fall off in the autumn, bearing the enclosed tough oval cocoons to the ground, where they remain through the winter, and the moths come out in the month of June following. Notwithstanding the great similarity of the caterpillar and its cocoon to those of the Luna, the moth is entirely different. Its hind wings are not tailed, but are cut off almost square at the corners. It is of a dull ochre-yellow color, more or less clouded with black in the middle of the wings, on each of which there is a transparent eye-like spot, divided transversely by a slender line, and encircled by yellow and black rings; before and adjoining to the eye-spot of the hind wings is a large blue spot shading into black; near the hinder margin of the wings is a dusky band, edged with reddish white behind; on the front margin of the fore wings is a gray stripe, which also crosses the fore part of the thorax; and near the base of the same wings are two short red lines, edged with white. It expands from five and a quarter to six inches. This moth, on account of its great size, is called *Polyphemus* (Fig. 181), the name of one of the giants in mythology.

*Attacus Cecropia** (Fig. 182) is a still larger insect, expanding from five inches and three quarters to six inches and a half. The hind wings are rounded, and not tailed. The ground-color of the wings is a grizzled dusky brown, with the hinder margins clay-colored; near the middle of each of the wings there is an opaque kidney-shaped dull red spot, having a white centre and a narrow black edging; and beyond the spot a wavy dull red band, bordered internally with white; the fore wings, next to the shoulders, are dull red, with a curved white band; and near the tips of the same is an eye-like black spot, within a bluish-white crescent; the upper side of the body and the legs are dull red; the fore part of the thorax and the hinder edges of the rings of the abdomen are white; and the belly is checkered with red and white. This moth makes its appearance during the month of June. The caterpillar (Fig. 183) is

* *Cecropia* was the ancient name of the city of Athens; its application, by Linnæus, to this moth is inexplicable.

Fig. 181.

Fig. 182.

found on apple, cherry, and plum trees, and on currant and barberry bushes in July and August. When young it is of a deep yellow color, with rows of minute black warts on its back. It comes to its full size by the first of September,

Fig. 183.

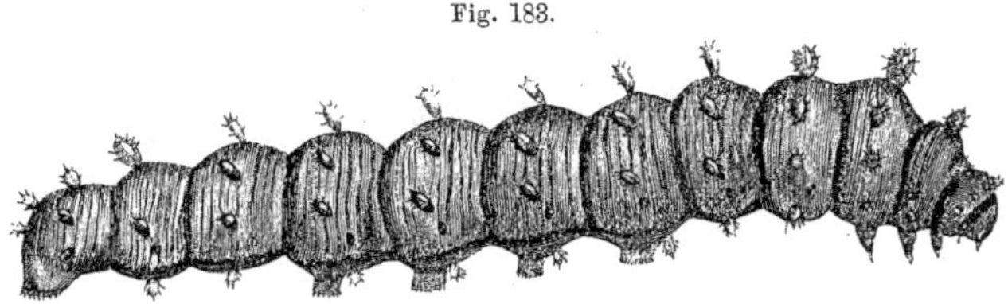

and then measures three inches or more in length, and is thicker than a man's thumb. It is then entirely of a fine, clear, light green color; on the top of the second ring are two large globular coral-red warts, beset with about fourteen very short black bristles; the two warts on the top of the third ring are like those on the second, but rather larger; on the top of the seven following rings there are two very long egg-shaped yellow warts, bristled at the end, and a single wart of larger size on the eleventh ring; on each side of the body there are two longitudinal rows of long light blue warts, bristled at the end, and an additional short row, below them, along the first five rings. This caterpillar does not bear confinement well; but it may be seen spinning its cocoon, early in September, on the twigs of the trees or bushes on which it lives. The cocoon (Fig. 184,

Fig. 184.

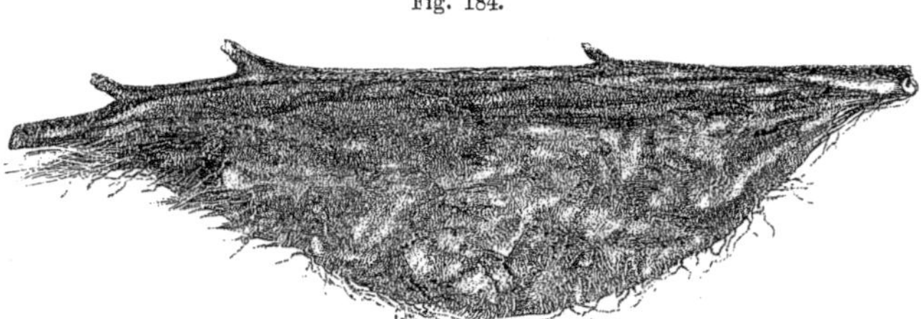

Fig. 185, pupa) is fastened longitudinally to the side of a twig. It is, on an average, three inches long, and one inch in diameter at the widest part. Its shape is an oblong oval, pointed at the upper end. It is double, the outer coat being wrinkled, and resembling strong brown paper in color and thickness; when this tough outer coat is cut open, the inside will be seen to be lined with a quantity of loose, yellow-brown, strong silk, surrounding an inner oval cocoon, composed of the same kind of silk, and closely woven like that of the silkworm. The insect remains in the chrysalis form through the winter. The moth, which comes forth in the following summer, would not be able to pierce the inner cocoon, were it not for the fluid provided for the purpose of softening the threads; but it easily forces its way through the outer cocoon at the small end, which is more loosely woven than elsewhere, and the threads of which converge again, by their own elasticity, so as almost entirely to close the opening after the insect has escaped.

Fig. 185.

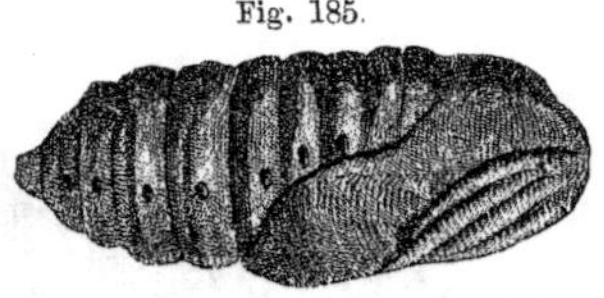

A few brown and curled leaves may frequently be seen hanging upon sassafras-trees during the winter, when all the other leaves have fallen off. If one of these leaves is examined, it will be found to be retained by a quantity of silken thread, which is wound or woolded round the twig to the distance of half an inch or more on each side of the leaf-stalk, and is thence carried downwards around the stalk to an oval cocoon, that is wrapped up by the sides of the leaf. The cocoon itself is about an inch long, of a regular oval shape, and is double, like that of the Cecropia caterpillar; but the outer coat is not loose and wrinkled, and the space between the outer and inner coats is small, and does not contain much floss silk. So strong is the coating of silk that surrounds the leaf-stalk, and connects the cocoon with the branch, that it cannot be severed without great force; and consequently the chrysalis swings securely within its leaf-covered hammock through all the storms of winter.

Cocoons of the same kind are sometimes found suspended to the twigs of the wild cherry-tree, the Azalea, or swamp-pink, and the Cephalanthus, or button-bush, but not so often as on the sassafras-tree. Two of them, hanging close together on one twig, were once brought to me, and a male and a female moth were produced from these twin cocoons in July, the usual time for these insects to leave their winter quarters. Drury called this kind of moth *Promethea*, a mistake probably for *Prometheus*,* the name of one of the Titans, all of whom were fabled to be of gigantic size. The color of *Attacus Promethea* differs according to the sex. The male (Fig. 186) is of a deep smoky brown color on the

Fig. 186.

upper side, and the female (Fig. 187) light reddish brown; in both, the wings are crossed by a wavy whitish line near the middle, and have a wide clay-colored border, which is marked by a wavy reddish line; near the tips of the fore wings there is an eye-like black spot within a bluish-white crescent; near the middle of each of the wings of the female there is an angular reddish-white spot, edged with black; these angular spots are visible on the under side of the wings

* *Atlas* was the brother of Prometheus, and this name, it will be recollected, has been given to another of the Bombyces, an immensely large moth from China.

of the male, but are rarely seen on their upper side; the hind wings in both are rounded and not tailed. These moths expand from three inches and three quarters to four inches and a quarter. The female deposits her eggs on the twigs of the trees, in little clusters of five or six together, and these are hatched towards the end of July or early in August. The caterpillars usually come to their full size by the beginning of September, and then measure two inches or more in length, when extended, and about half an inch in diameter. The body of the caterpillar is very plump, and but very little contracted on the back between the rings. It is of a clear and pale bluish-green color; the head, the

Fig. 187.

feet, and the tail are yellow; there are about eight warts on each of the rings; the two uppermost warts on the top of the second and of the third rings are almost cylindrical, much longer than the rest, and of a rich coral-red color; there is a long yellow wart on the top of the eleventh ring; all the rest of the warts are very small, and of a deep blue color. Before making its cocoon the caterpillar instinctively fastens to the branch the leaf that is to serve for a cover to its cocoon, so that it shall not fall off in the autumn, and then proceeds to spin on the upper side of the leaf, bending over the edges to form a hollow, within which its cocoon is concealed.

The Luna, Polyphemus, Cecropia, and Promethea moths are the only native insects belonging to the genus *Attacus* which are known to me. Their large cocoons, consisting entirely of silk, the fibres of which far surpass those of the silk-worm in strength, might perhaps be employed in the formation of fabrics similar to those manufactured in India from the cocoons of the tusseh and arrindy silk-worms, the durability of which is such, that a garment of tusseh silk "is scarcely worn out in the lifetime of one person, but often descends from mother to daughter; and even the covers of palanquins made of it, though exposed to the influence of the weather, last many years." The method employed by the inhabitants of India for unwinding the cocoons of their native silk-worms would probably apply equally well to those of our country, which have not yet, that I am aware of, been submitted to the same process. It is true that experiments, upon a very limited scale, have been made with the silk of the Cecropia, which has been carded and spun and woven into stockings, that are said to wash like linen. The Rev. Samuel Pullein was among the first to attempt to unwind the cocoons of the Cecropia moth, an account of which is contained in the "Philosophical Transactions of the Royal Society of London," for the year 1759.* Mr. Pullein ascertained that twenty threads of this silk twisted together would sustain nearly an ounce more in weight than the same number of common silk. Mr. Moses Bartram, of Philadelphia, in the year 1767, succeeded in bringing up the caterpillars from the eggs of the Cecropia moth, and obtained several cocoons from them.† In the Paris "Journal des Debats," of the 23d of July, 1840, is an account of the complete success of Mr. Audouin in

* Vol. LI. p. 54.

† See "Transactions of the American Philosophical Society of Philadelphia," Vol. I. p. 294.

rearing the caterpillars of this or of some other American species of Attacus, the cocoons of which were sent to him from New Orleans. The Cecropia does not bear confinement well, and is not so good a subject for experiment as the Luna and Polyphemus, which are easily reared, and make their cocoons quite as well in the house as in the open air. The following circumstances seem particularly to recommend these indigenous silk-worms to the attention of persons interested in the silk culture. Our native oak and nut trees afford an abundance of food for the caterpillars; their cocoons are much heavier than those of the silk-worm, and will yield a greater quantity of silk; and, as the insects remain unchanged in the chrysalis state from September to June, the cocoons may be kept for unwinding at any leisure time during the winter. By a careful search, after the falling of the leaves in the autumn, a sufficient number of cocoons may be found, under the oak and nut trees, with which to begin a course of experiments in breeding the insects, and in the manufacture of their silk.

Two more moths, belonging to the family under consideration, are found in Massachusetts. They may be referred to the genus *Saturnia*,* and are distinguished from the foregoing by their antennæ, which are widely feathered only in the males, the feathering being very narrow in the other sex; their caterpillars, moreover, are furnished with small warts crowned with long prickles or branching spines. None of the caterpillars described in the preceding pages are venomous; all of them may be handled with impunity. This is not the case with the two following kinds, the prickles of which sting severely. The first of these begin to appear by the middle of June, and other broods continue to be hatched till the middle of July. These caterpillars (Fig. 188) live on the balsam poplar and

Fig. 188.

* The surname of Juno, the daughter of Saturn.

the elm, and, according to Mr. Abbot, on the dogwood or cornel, and the sassafras ; they feed well also on the leaves of clover and Indian corn. They are of a pea-green color, with a broad brown stripe edged below with white on each side of the body, beginning on the fourth ring and ending at the tail ; they are covered with spreading clusters of green prickles, tipped with black, and of a uniform length ; each of these clusters consists of about thirty prickles branching from a common centre, and there are six clusters on each of the rings except the last two, on which there are only five, and on the first four rings, on each of which there is an additional cluster low down on each side ; the feet are brown, and there is a triangular brown spot on the under side of each ring, beginning with the fourth. The prickles are exceedingly sharp, sting very severely when the insect is handled, and produce the same kind of irritation as those of the nettle. When young these caterpillars keep together in little swarms. They do not spin a common web, but, when not eating, they creep under a leaf, where they cluster side by side. In going from or returning to their place of shelter they move in regular files, like the processionary caterpillars (*Lasiocampa processionea*) of Europe, a single caterpillar taking the lead, and followed closely by perhaps one or two in single file, after which come two, side by side, close upon the heels of these creep three more, the next rank consists of four, and so on, the ranks continually widening behind, like a flock of wild geese on the wing, but in perfectly regular order. When about half grown they disperse, and each one shirks for himself. At the age of eight weeks they get to their full size, in the meanwhile moulting their skins four times, and finally measure two inches and a half or more in length. At this age they leave off eating, crawl to the ground, and get under leaves or rubbish, which they draw round their bodies to form an outer covering, within which they make an irregular and thin cocoon (Fig. 189), of very gummy brown silk, that has almost the texture of thin parchment. As soon as their cocoons are finished, the insects are changed to chrysalids (Fig. 190), in which form they remain throughout the winter, and in the following summer, during the month of June, or beginning of July, they come out in the winged or moth state. The scientific name of these moths is *Saturnia Io.** Unlike those of the genus *Attacus*, they sit with their wings closed, and covering the body like a low roof, the front edge of the under wings extending a little beyond that of the upper wings, and curving upwards.

Fig. 189.

Fig. 190.

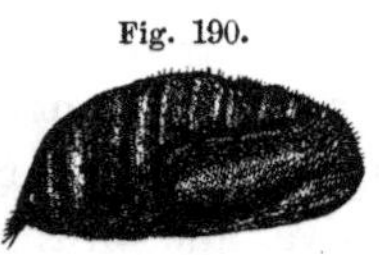

The two sexes differ both in color and size. The male (Fig. 191), which is the smallest, is of a deep or Indian

Fig. 191.

yellow color; on its fore wings there are two oblique wavy lines towards the hind margin, a zigzag line near the base, and several spots so arranged on the middle as to form the letters A H, all of a purplish red color; the hind wings are broadly bordered with purplish red next to the body, and near the hinder margin there is a narrow curved band of the same color; within this band there is a curved black

* *Io*, a priestess of Juno, in Greece, afterwards became the wife of Osiris, the king of Egypt, and received divine honors under the name of Isis.

line, and on the middle of the wing a large round blue spot, having a broad black border and a central white dash. The fore wings of the female (Fig. 192) are purple-brown, min-

Fig. 192.

gled with gray; the zigzag and wavy lines across them are gray, and the lettered space in the middle is replaced by a brown spot surrounded by an irregular gray line; the hind wings resemble those of the male in color and markings; the thorax and legs are purple-brown; and the abdomen is ochre-yellow, with a narrow purple-red band on the edge of each ring. These moths expand from two inches and three quarters to three inches and a half.

The other *Saturnia*, inhabiting Massachusetts, is the *Maia* * (Fig. 193) of Drury, or *Proserpina* † of Fabricius. The

Fig. 193.

moth probably rests with its wings closed, like the *Io* moth,

* *Maia*, in mythology, was one of the seven daughters of *Atlas;* they were placed in the heavens after death, and formed the constellation called *Pleiades.*

† *Proserpina* was the wife of Pluto, the god of the infernal regions.

the fore wings covering the other pair, the front edge of which seems formed to extend a little beyond that of the fore wings in this position. The wings are thin and almost transparent like crape; they are black, and both pairs are crossed by a broad yellow-white band, near the middle of which, on each wing, there is a kidney-shaped black spot having a central yellow-white crescent or curved line on it; the thorax is covered with black hairs on the top, pale yellow hairs on the fore part, and has two tufts of rust-red hairs behind; the abdomen is black, with a few yellowish hairs along the sides, and a patch of a rust-red color at the extremity, in the males. The wings expand from two inches and a half to three inches and one eighth.

Saturnia Maia seems to be a very rare moth in Massachusetts; I have never met with it alive, but have seen several specimens which were taken in this State. The time of its appearance here is not known to me with certainty; but, if I am rightly informed, it has been found in July and the beginning of August, flying by day on the borders of oak woods, or resting on the shrub oaks which cover the sides of some of our high hills. Of the caterpillar I have seen only one specimen, which was found, fully grown, on an oak, towards the end of September; it was destroyed, however, before I had an opportunity of making a description of it. Mr. Abbot * has figured two of the caterpillars, which differ from each other in color and markings. They are nearly three inches long; the head and all the feet are red; and on each of the rings there are six long branched prickles. One of these caterpillars is represented of a dusky brown color mingled with yellow, with yellow warts from which the prickles arise. The other is yellow, with red warts, and two black stripes along the back. Mr. Abbot states that these caterpillars, while small, feed together in company, but disperse as they grow large; they eat the leaves of various kinds of oaks; sting very sharply when

* Insects of Georgia, p. 99, pl. 50.

handled; and that they go into the ground to transform; but he does not inform us whether they make cocoons. Probably their cocoons are like those of the Io moth, composed of a gummy membranaceous substance, covered either with leaves or with grains of earth.

As far as I can ascertain, these six moths are the only Saturnians which have been discovered east of the Mississippi, and they are commonly met with throughout the United States.* The last of them, together with some foreign species, such as the Tau moth of Europe, seem naturally to conduct to the next family, which I call Ceratocampians (CERATOCAMPADÆ), after the name of the chief genus contained in it. This name, moreover, signifying horned caterpillar, serves to point out the principal peculiarity of the caterpillars in this group; they being armed

* Mr. Audubon has figured two more, apparently sexes or varieties of one species, in the fourth volume of his magnificent "Birds of America," pl. 359; but has not named or described them. He informs me that they were taken by Mr. Nuttall near the Rocky Mountains. Through the kindness of Mr. Edward Doubleday, of Epping, England, the present possessor of one of the very specimens from which Mr. Audubon's drawing was made, an opportunity of examining and describing this fine insect has been granted to me.

Though differing somewhat from the other species of *Saturnia*, it approaches so near to the *Maia* that I shall not venture to separate it from this genus, especially as the caterpillar and its habits are unknown. It may be called *Saturnia Hera:* the latter (a generical name proposed for it by Mr. Doubleday) is the name given by the Greeks to Juno. The specimen before me is a male. It resembles the *Maia* in form and size, but the wings are not quite so thin, and are more opaque. The fore wings when the insect is resting probably cover the hind wings, the front edge of which appears to be formed to project a little beyond that of the fore wings. It is of a pale yellow color; on each of the wings there is a kidney-shaped black spot between two transverse wavy black bands; the outer margins are black; the veins from the external black band to the edge are marked with broad black lines; and there is a short black line at the base of the fore wings; the head, fore part of the thorax, and upper sides of the legs, are deep ochre-yellow; and the rings of the abdomen are transversely banded with black at the base, and with ochre-yellow on their hinder edges. The kidney-shaped spots on the fore wings have a very slender central yellow crescent, and those on the hind wings touch the external black band. The wings expand three inches. The other moth, figured on the same plate in Mr. Audubon's work, which is probably the female of the preceding, apparently differs from it only in being of a deep Indian-yellow color, and in having the crescent in the middle of the kidney-shaped spots very distinct, whereas in the male it is almost obsolete.

with thorny points, of which those on the second ring, and sometimes also those on the third, are long, curved, and resemble horns. These caterpillars eat the leaves of forest-trees, and go into the ground to undergo their transformations without making cocoons. The rings of the chrysalis are surrounded by little notched ridges, the teeth of which, together with the strong prickles at the hinder end of the body, assist it in forcing its way upwards out of the earth, just as the moth is about to burst the skin of the chrysalis. The moths are very easily distinguished from all the foregoing by their antennæ, which are short, and in the males are feathered on both sides for a little more than half the length of the stalk, and are naked from thence to the tip; while those of the females are threadlike, and neither feathered nor toothed. The feelers (except in *Ceratocampa*, in which they are very distinct) and the tongue are very small, and not ordinarily visible. There are no bristles and hooks to fasten together the wings, which, when at rest, are not spread, but are closed, the fore wings covering the hinder pair, and the front edge of the latter, in most cases, extends a little beyond that of the fore wings. These are some of the principal characters on which I have ventured to establish this family, which is now, for the first time, pointed out as a peculiar group. I believe that it is exclusively American.

One of the largest and most rare, and withal the most magnificent of our moths, is the *Ceratocampa regalis* (Fig. 194), or regal walnut-moth. Its fore wings are olive-colored, adorned with several yellow spots, and veined with broad red lines; the hind wings are orange-red, with two large irregular yellow patches before, and a row of wedge-shaped olive-colored spots between the veins behind; the head is orange-red; the thorax is yellow, with the edge of the collar, the shoulder-covers, and an angular spot on the top, orange-red; the upper side of the abdomen, and the legs, are also orange-red. Unlike the other moths of the same family, the feelers in this are distinct, cylindrical, and prominent, and the front edge of the hind wings does not seem to be formed to extend beyond that of the other pair when the wings are closed. It expands from five to six

Fig. 194.

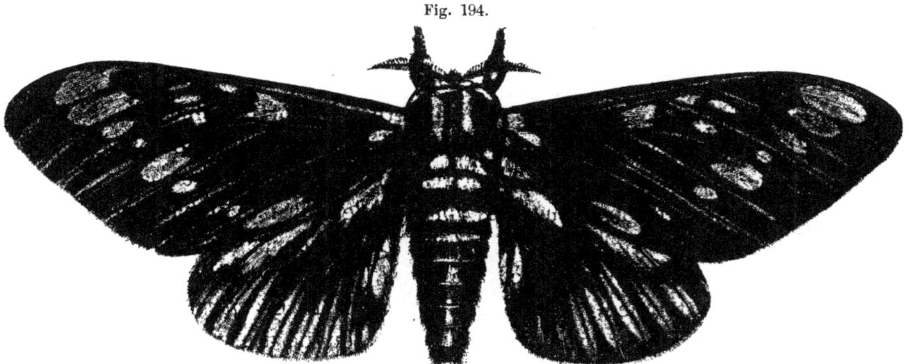

inches. In the year 1828, I found three of the eggs of this fine insect on the black walnut on the 20th of July and the 4th of August. They were just hatched at the time, and the caterpillars were near to them resting on a leaf. The position of these young insects was so peculiar as to attract attention, independently of the long branching spines with which the fore part of their body was armed. They were not stretched out in a straight line, neither were they hunched up like the caterpillars of the Luna and Polyphemus moths; but, when at rest, they bent the fore part of the body sideways, so that the head nearly touched the middle of the side, and their long horn-like spines were stretched forwards, in a slanting direction, over the head. When disturbed, they raised their heads and horns, and shook them from side to side in a menacing manner. These little caterpillars were nearly black; on each of the rings, except the last two, there were six straight yellow thorns or spines, which were furnished on all sides with little sharp points like short branches. Of these branched spines, two on the top of the first ring, and four on the second and the third rings, or ten in all, were very much longer than the rest, and were tipped with little knobs, ending in two points; they were also movable, the insect having the power of dropping them almost horizontally over the head, and of raising them up again perpendicularly. On the eleventh ring there were seven spines, the middle one being long and knobbed like those on the fore part of the body; on the last ring there were eleven short and branched spines. After casting its skin two or three times, the caterpillar becomes lighter-colored, and gradually changes to green; the knobs on the long spines disappear, their little points or branches do not increase in size, and finally these spines become curved, turning backwards at their points, and resemble horns. When fully grown, the caterpillar (Fig. 195) measures from four to

Fig. 195.

five inches in length, and about three quarters of an inch in diameter. It is of a green color, and transversely banded across each of the rings with pale blue; there is a large blue-black spot on each side of the third ring; the head and legs are orange-colored; the ten long horn-like spines on the fore part of the body are orange-colored, with the tips and the points surrounding them black; the other spines are short and black. Notwithstanding the great size, formidable appearance, and menacing motions of this insect, when handled it is perfectly harmless, and unable to sting or wound with its frightful horns. It lives solitary on walnut and hickory trees, the leaves of which it eats; crawls down and goes into the ground towards the end of summer, and changes to a chrysalis without previously making a cocoon. Unfortunately my caterpillars died before the time for their transformation arrived. The chrysalis is short and thick; obtuse behind, but terminated by two minute points; and the transverse notched ridges or little teeth that are found on the chrysalids of the other insects belonging to the same family, are very small and hardly visible on this one. The insect remains in the ground through the winter, and the moth comes out in the following summer, during the month of June, if I am rightly informed. I have not been able to obtain one myself, and my description of the moth was made from a very fine specimen belonging to a friend, who received it from New Bedford.

Between the regal Ceratocampa and the smaller insects of this family belonging to the new genus *Dryocampa* should be placed a noble moth, which partakes, in some respects, of the characters of both; its horned caterpillar, particularly while young, when its horns are proportionally longer and more formidable in appearance than afterwards, resembles somewhat that of the Ceratocampa; its chrysalis is exactly like that of a *Dryocampa*, and like the latter also, in the winged state, its feelers are minute, its hind wings project beyond the front edges of the fore wings when at rest, and its style of coloring is the same. In my Catalogue of the Insects of Massachusetts, I placed this moth, the *imperialis* of Drury, in the genus *Ceratocampa*, from which, however, it must be removed, on account of its very small feelers, and the position of its wings; and I now refer it, with some hesitation, to the genus *Dryocampa*, with which it agrees so well in the moth state, although its caterpillar differs a good deal from those of the other insects of the same genus. The imperial moth, *Dryocampa imperialis* (Fig. 196), has wings of a fine yellow color, thickly sprinkled with purple-brown dots, with a large patch at the base, a small round spot near the middle, and a wavy band towards the hinder margin of each wing, of a light purple-

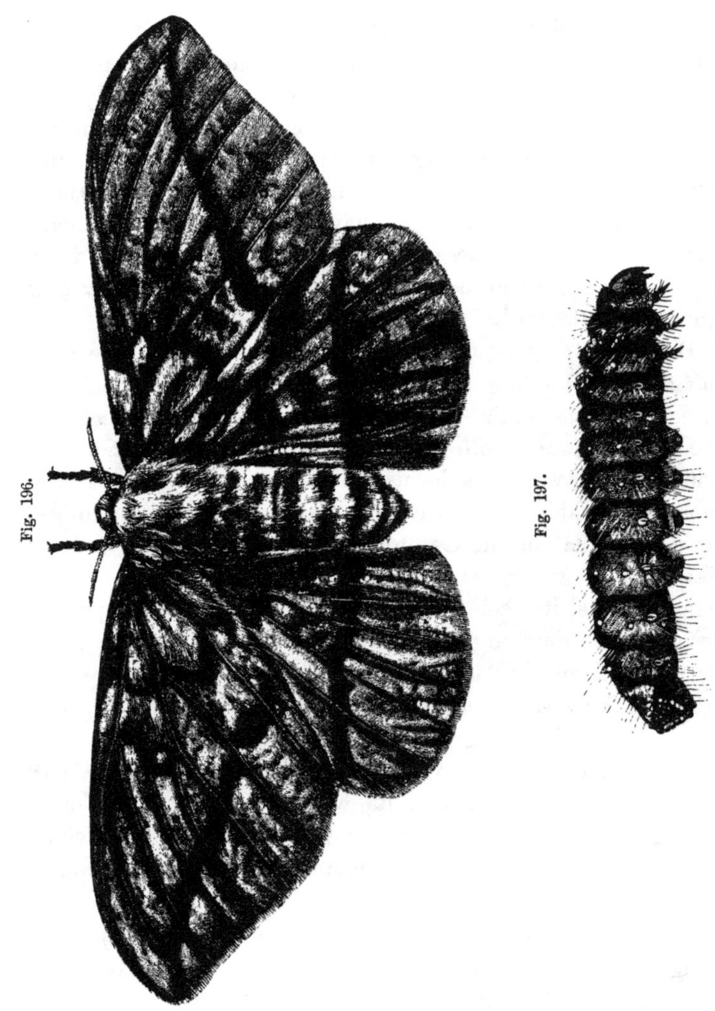

Fig. 196.

Fig. 197.

brown color; in the males there is another purple-brown spot, covering nearly the whole of the outer hind margin of the fore wings, and united to the band near that part; the body is yellow, shaded with purple-brown on the back, and with three spots of the same color on the thorax. It expands from four inches and a half to more than five inches. In a variety of this moth, of which I have a colored drawing done by Mr. Abbot, the purple-brown color prevails so much as to cover the wings, with the exception only of a large triangular yellow spot contiguous to the front margin of each wing. This moth appears here from the 12th of June to the beginning of July, and then lays its eggs on the buttonwood tree.

The caterpillars (Fig. 197) may be found upon this tree, grown to their full size, between the 20th of August and the end of September, during which time they descend from the trees to go into the ground. They are then from three to four inches in length, and more than half an inch in diameter, and, for the most part, of a green color, slightly tinged with red on the back; but many of them become more or less tanned or swarthy, and are sometimes found entirely brown. There are a few very short hairs thinly scattered over the body; the head and the legs are pale orange-colored; the oval spiracles, or breathing-holes, on the sides, are large and white, encircled with green; on each of the rings, except the first, there are six thorny knobs or hard and pointed warts of a yellow color, covered with short black prickles; the two uppermost of these warts on the top of the second and of the third rings are a quarter of an inch or more in length, curved backwards like horns, and are of a deeper yellow color than the rest; the three triangular pieces on the posterior extremity of the body are brown, with yellow margins, and are covered with raised orange-colored dots. The chrysalis, which is not contained in a cocoon, is about two inches long, of a dark chestnut-brown color, rough with little elevated points, particularly on the anterior extremity, ends behind with a long forked spine, and is surrounded, on each ring, with a notched ridge, the little teeth of which point towards the tail. Three of the grooves or incisions between the rings are very deep, thus allowing a great extent of motion to the joints, and these, with the notched ridges, and the long spine at the end of the body, enable the chrysalis to work its way upwards in the earth, above the surface of which it pushes the fore part of its body just before the moth makes its escape.

Dryocampa, oak or forest caterpillar, is a name originally applied by me to certain insects, found sometimes in great numbers on oak-trees, which then suffer very severely from their ravages. Of these caterpillars there are several kinds, resembling each other in shape, and in the form and situation of the thorns with which they are armed, but differing in color, and in the moths produced from them. They live together in swarms, but do not make webs; their bodies are cylindrical, remarkably hard and stiff, naked or not hairy, and have, on each ring, about six short thorns, or sharp points, besides two on the top of the second ring, which are long, slender, and threadlike, but not flexible, and project in the manner of horns. The most common of these caterpillars (Fig. 198) in Massachusetts is black, with four narrow ochre-yellow stripes along the back, and two on each side. It is found in swarms of several hundreds together, on the limbs of the white and red oaks, during the month of August. The eggs from which they proceed are laid in large clusters on the under side of a leaf near the end of a branch. The caterpillars are hatched towards the end of July, but sometimes earlier, and at other times later. At first they eat only the youngest leaves at the end of the branches and twigs, and, as they grow larger and stronger, proceed downwards, devouring every leaf, to the midrib and foot-stalk, from one end of the branch to the other. They

Fig. 198.

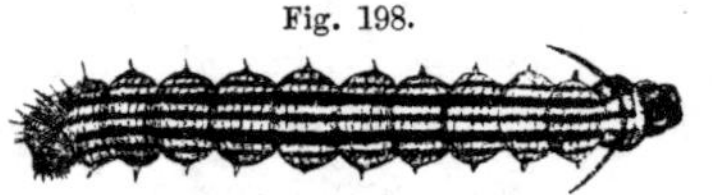

have their regular times for eating and for rest, and when they have finished their meals, they cluster closely together along the twigs and branches. If disturbed, they raise the fore part of their bodies, and shake their heads to signify their displeasure. When fully grown they measure about two inches in length. Commonly in the early part of September, they crawl down the trees and go into the ground, to the depth of four or five inches, where they are changed to chrysalids (Fig. 199). These resemble the chrysalids of the imperial Dryocampa, but are much smaller, and like them they remain in the ground throughout the winter, and work their way up to the surface in the following summer. These chrysalids may often be seen sticking half-way out of the ground under oak-trees in the latter part of June and the beginning of July, at which time the moths burst them open and make their escape. *Dryocampa senatoria* (Fig. 200), the senatorial Dryocampa, which is the name of this kind of moth, is of an ochre-yellow color; the wings are faintly tinged with purplish red, especially on the front and hind margins, and are crossed by a narrow purple-brown band behind the middle; the fore wings are sprinkled with blackish dots, and have a small round white spot near the middle. The male is much smaller than the female, its wings are thinner, and more tinged with dull purple-red. It expands about an inch and three quarters; the female, two inches and a half, or more.

Fig. 199.

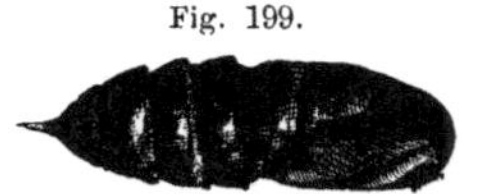

Fig. 200.

Three more kinds of Dryocampa are found in Massachusetts, but they are all rare in this State. The largest of them is the *stigma* of Fabricius, or spotted-wing Dryocampa. It is of a reddish ochre or deep tawny yellow color; the fore wings are tinged with purplish red behind, are thickly sprinkled with blackish dots, have a small round white spot near the middle, and a narrow oblique purple-red band behind; the hind wings have a narrow transverse purple band, behind which the border is sprinkled with a few black dots. It expands from one inch and three quarters to two inches and three quarters. The caterpillar, which I have not seen, is figured in Mr Abbot's work,* where it is colored yellow, with black thorns on its back. It is said to live on the oak, in swarms, while young, but these disperse as the insects grow large.

The following resembles the senatorial Dryocampa; but is rather smaller, and is a more delicate moth. The color of its body is ochre-yellow; the fore wings of the male are purple-brown, with a large colorless transparent space on the middle, near which is a small round white spot, and towards the hinder margin a narrow oblique very faint dusky stripe; the hind wings are purple-brown, almost transparent in the middle, and with a very faint transverse dusky stripe; the wings of the female are purplish red, blended with ochre-yellow, are almost transparent in the middle, and have the same white spots and faint bands as those of the male. It expands from one inch and three quarters to two inches and a quarter, or more, in some females. The distinguishing name, given by Sir J. E. Smith,† to this moth, is *pellucida*, and we may call it the pellucid or clear-wing Dryocampa. I have only once seen the caterpillar, which was found on an oak on the 25th of September. It was about the size of that of the senatorial Dryocampa, and resembled it in everything but color. Its head was rust-yellow, its body pea-green, shaded on the back and sides with red, longitudinally striped with very pale yellowish green, and armed with black thorns.

* Insects of Georgia, p. 111, pl. 56.

† Ibid., p. 115, pl. 58.

The last of these insects is the *rubicunda* (Fig. 201) of Fabricius, or rosy Dryocampa. This delicate and very rare moth is found in Massachusetts in July. Its fore wings are rose-colored, crossed by a broad pale-yellow band; the hind wings are pale yellow, with a short rosy band behind the middle; the body is yellow; the belly and legs are rose-colored. It expands rather more than one inch and three quarters. The caterpillar is unknown to me.*

Fig. 201.

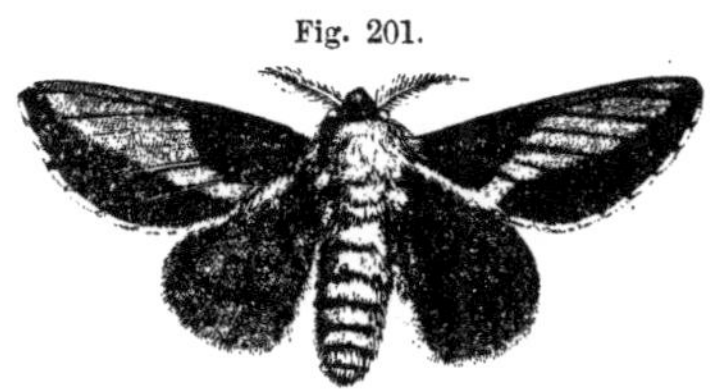

All the Moth caterpillars thus far described in this work live more or less exposed to view, and devour the leaves of plants; but there are others that are concealed from observation in stems and roots, which they pierce in various directions, and devour only the wood and pith; their habits, in this respect, being exactly like those of the Ægerians among the Sphinges. These insects belong to a family of Bombyces, by some naturalists called ZEUZERADÆ, and by others HEPIALIDÆ, both names derived from insects included in the same group. The caterpillars of the Zeuzerians are white or reddish white, soft and naked, or slightly downy, with brown horny heads, a spot on the top of the fore part of the body which is also brown and hard, and sixteen legs. They make imperfect cocoons, sometimes of silk, and sometimes of morsels of wood or grains of earth fastened together by gummy silk. Their chrysalids, like those of the Cerato-

* Only one more North American Dryocampa is known to me. This moth was taken in North Carolina, and does not appear to have been described. It may be called *Dryocampa bicolor*, the two-colored, or gray and red, Dryocampa. The upper side of the fore wings and the under side of the hind wings are brownish gray, sprinkled with black dots, and with a small round white spot near the middle, and a narrow oblique dusky band behind it on the fore wings; the upper side of the hind wings and the under side of the fore wings, except the front edge and hinder margin of the latter, are crimson-red, and the body is brownish gray. The male expands two inches and a quarter. The female and the caterpillar of this insect I have not seen.

campians, are provided with notched transverse ridges on the rings, by means of which they push themselves out of their holes when ready to be transformed. The moths differ a good deal from each other, although the appearance and habits of the caterpillars are so much alike. The antennæ in some are thread-like, or made up of nearly cylindrical joints put together like a string of beads; in others they are more tapering, and doubly pectinated or toothed on the under side, at least in the males; and in *Zeuzera*, a kind of moth not hitherto found in this country, the antennæ resemble those of the Ceratocampians, being half-feathered in the males, and not feathered in the females. The wings are rather long and narrow, and are strengthened by very numerous veins. The female is provided with a kind of tube at the end of the body, that can be drawn in and out, by means of which she thrusts her eggs into the chinks of the bark or into the earth at the roots of plants.

Of the root-eaters there is one kind which is very injurious to the hop-vine in Europe. It is called *Hepiolus humuli*, the hop-vine Hepiolus. The caterpillar is yellowish white; the head, a spot on the top of the first and second rings, and the six fore legs are shining brown, and it is nearly naked, or has only a few short hairs scattered over its body. It lives in the roots of the hop, and, when about to transform, buries itself in the ground, and makes a long, cylindrical cocoon or case, composed of grains of earth held together by a loose silken web. The chrysalis has transverse rows of little teeth on the backs of the abdominal rings, and by means of them it finally works its way out of the cocoon and rises to the surface of the earth; this being done, the included moth bursts its chrysalis shell, and comes forth into the open air. In moths of this kind (genus *Hepiolus*) the antennæ are very short, slender, almost thread-like, and not feathered or pectinated; the tongue is wanting or invisible; and the feelers are excessively small, and concealed in a tuft of hairs.

The hop-vine Hepiolus has not yet been detected in Massachusetts; but we have a much larger species, known to me only in the moth state, which is the reason of my having given the foregoing account of the preparatory stages of a European species. This moth does not appear to have been described. It is named in my Catalogue of the Insects of Massachusetts, *Hepiolus argenteo-maculatus* (Fig. 202), the silver-spotted Hepiolus. Its body and wings are

Fig. 202.

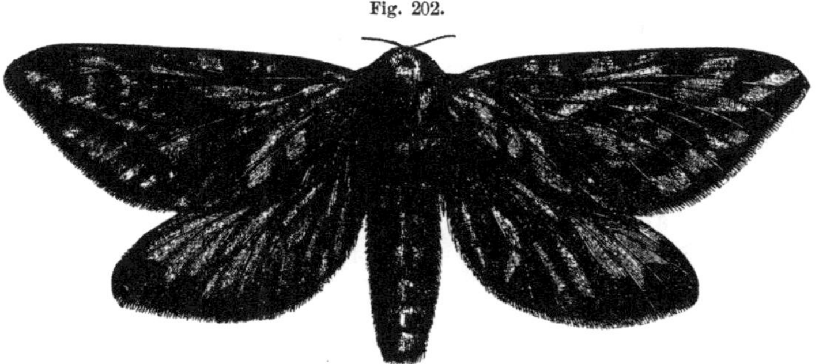

rather long. It is of an ashen-gray color; the fore wings are variegated with dusky clouds and bands, and have a small triangular spot and a round dot of a silvery white color near their base; the hind wings are tinged with ochre-yellow towards the tip. It expands two inches and three quarters. A much larger specimen was found by Professor Agassiz near Lake Superior.*

The locust-tree, *Robinia pseudacacia*, is preyed upon by three different kinds of wood-eaters or borers, whose unchecked ravages seem to threaten the entire destruction and extermination of this valuable tree within this part of the United States. One of these borers is a little reddish caterpillar, whose operations are confined to the small branches and to very young trees, in the pith of which it lives; and by its irritation it causes the twig to swell around the part attacked. These swellings being spongy, and also perforated

* See a figure of it in his "Lake Superior," pl. 7, fig. 6.

by the caterpillar, are weaker than the rest of the stem, which therefore easily breaks off at these places. My attempts to complete the history of this insect have not been successful hitherto.

The second kind of borer of the locust-tree is larger than the foregoing, is a grub, and not a caterpillar, which finally turns to the beetle named *Clytus pictus*, the painted Clytus, already described on a preceding page of this work.

The third of the wood-eaters to which the locust-tree is exposed, though less common than the others, and not so universally destructive to the tree as the painted Clytus, is a very much larger borer, and is occasionally productive of great injury, especially to full-grown and old trees, for which it appears to have a preference. It is a true caterpillar (Fig. 203), belonging to the tribe of moths under consideration,

Fig. 203.

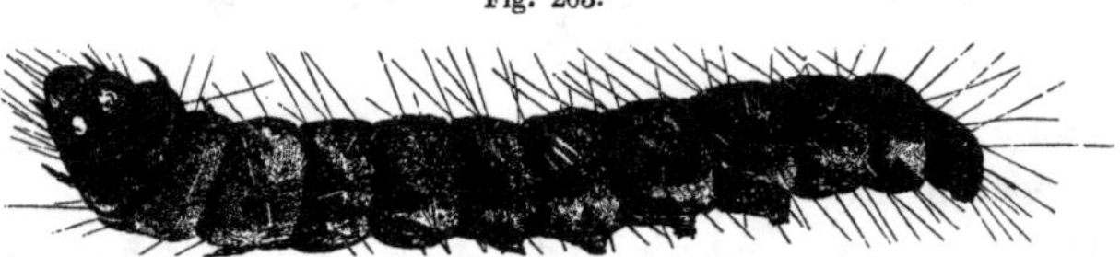

is reddish above, and white beneath, with the head and top of the first ring brown and shelly, and there are a few short hairs arising from minute warts thinly scattered over the surface of the body. When fully grown, it measures two inches and a half, or more, in length, and is nearly as thick as the end of the little finger. These caterpillars bore the tree in various directions, but for the most part obliquely upwards and downwards through the solid wood, enlarging the holes as they increase in size, and continuing them through the bark to the outside of the trunk. Before transforming, they line these passages with a web of silk, and, retiring to some distance from the orifice, they spin around their bodies a closer web, or cocoon, within which they assume the chrysalis form. The chrysalis (Fig. 204) measures one inch and a half or two inches in length, is of an amber color, changing to brown on the fore part of the body; and on the upper side of each abdominal ring are two transverse rows of tooth-like projections. By the help of these, the insect, when ready for its last transformation, works its way to the mouth of its burrow, where it remains while the chrysalis skin is rent, upon which it comes forth on the trunk of the tree a winged moth. In this its perfected state, it is of a gray color; the fore wings are thickly covered with dusky netted lines and irregular spots, the hind wings are more uniformly dusky, and the shoulder-covers are edged with black on the inside. It expands about three inches. The male, which is much smaller, and has been mistaken for another species, is much darker than the female, from which it differs also in having a large ochre-yellow spot on the hind wings, contiguous to their posterior margin. Professor Peck, who first made public the history of this insect,* named it *Cossus Robiniæ*, the Cossus of the locust-tree, scientifically called *Robinia*. It is supposed by Professor Peck to remain three years in the caterpillar state. The moth comes forth about the middle of July. The same insect, or one not to be distinguished from it while a caterpillar, perforates the trunks of the red oak. Mr. Newman † has recently given the name of *Xyleutes*, the carpenter, to the genus including this insect, instead of *Cossus*, which it formerly bore, because the latter, being the name of a species, ought not to have been applied to a genus. The European carpenter-moth, called *Bombyx Cossus* ‡ by Linnæus, will now be the *Xyleutes Cossus;* and our indigenous species will be the *Xyleutes Robiniæ*

Fig. 204.

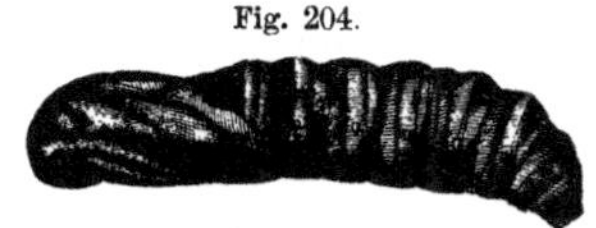

* See "Massachusetts Agricultural Repository and Journal," Vol. V., p. 67, with a plate.

† See "Entomological Magazine," Vol. V., p. 129.

‡ Subsequently named *Cossus ligniperda* by Fabricius.

(Fig. 205), or locust-tree carpenter-moth. The moths of this genus have thick and robust bodies, broad and thickly-veined wings, two very distinct feelers, and antennæ, which

Fig. 205.

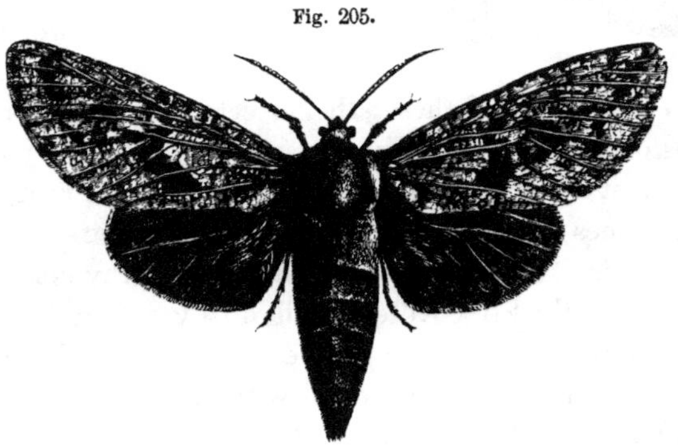

are furnished on the under side, in both sexes, with a double set of short teeth, rather longer in the male than in the female. Their tongue is invisible. They give out a strong and peculiar smell, whence they are sometimes called goat-moths by English writers.

Some caterpillars, which eat the leaves of plants, live in cases or long oval cocoons, open at both ends, and large enough for the insects to turn around within them, so as to go out of either end. They do not entirely leave these cases, even when moving from place to place, but cling to them on the inside with the legs of the hinder part of their bodies, while their heads and fore legs are thrust out. Thus in moving they creep with their six fore legs only, and drag along their cases after them as they go. These cases are made of silk within, and are covered on the outside with leaves, bits of straw, or little sticks. The caterpillars are nearly cylindrical, generally soft and whitish, except the head and upper part of the first three rings, which are brown and hard; they have sixteen legs; the first three pairs are long, strong, and armed with stout claws; the others are very short, consisting merely of slight wart-like elevations provided with numerous minute clinging hooks. When they are about to change their forms, their cases serve them instead of cocoons; they fasten them by silken threads to the plant on which they live, stop up the holes in them, and then throw off their caterpillar-skins. The chrysalids are remarkably blunt at the hinder extremity, and are provided with transverse rows of minute teeth on the back of the abdominal rings. The moths, of which there are several kinds produced by these case-bearing caterpillars, differ very much from each other; but, as they all agree in their habits and general appearance while in the caterpillar form, they are brought together in one family called PSYCHADÆ, the Psychians, from *Psyche*, a genus belonging to it. The Germans give these insects a more characteristic name, that of *Sackträger*,* that is, sack-bearers, and Hübner called them *Canephoræ*, or basket-carriers, because the cases of some of them are made of little sticks somewhat like a wicker basket. The cases of the insects belonging to the European genus *Psyche* are covered with small leaves, bits of grass or of sticks, placed lengthwise on them. The chrysalis of the male *Psyche* pushes itself half-way out of the case when about to set free the moth; the female, on the contrary, never leaves its cocoon, is not provided with wings, and its antennæ and legs are very short. The male *Psyche* resembles somewhat the same sex of *Orgyia*, having pretty broad wings, and antennæ that are doubly feathered on the under side; it has also a bristle and hook to hold the wings together. The cases of *Oiketicus*,† another and much larger kind of sack-bearer, inhabiting the West Indies and South America, are covered with pieces of leaves and of sticks arranged either longitudinally or transversely. The cases of some of the females measure four or five inches in length. Some which I received from Cuba were covered with little

* See Germar's "Magazin der Entomologie," Vol. I. p. 19.

† This name ought to be *Œceticus*. See Mr. Guilding's description of the insect in the "Transactions of the Linnæan Society," Vol. XV.

bits of sticks, about a quarter of an inch long, arranged transversely, and the cases were hung by a thick silken loop or ring to a twig; the lower end of these cases was filled with a large quantity of loose and very soft brownish floss-silk, which completely closed the orifice within. The male *Oiketicus* resembles a *Zeuzera* in the form and great length of its body, in the shape of its wings, and in its antennæ, and in both the latter it resembles also the same sex of a *Dryocampa*, particularly in its antennæ, which are feathered on both sides on the lower part of the stalk, and are bare at the other end. The female has neither wings, antennæ, nor legs, and is said to remain always within its cocoon. Some years ago, a case or cocoon of an *Oiketicus*, which was found on Long Island, was presented to me. It was smaller than the West Indian specimens, measuring only an inch and a half without its loop, and was covered with a few little sticks longitudinally arranged. It contained a female chrysalis, with the remains of the caterpillar. In Philadelphia and the vicinity, cases of a similar kind are very common on many of the trees, particularly on the arbor-vitæ, larch, and hemlock, which are often very much injured by the insects inhabiting them. These are there popularly called *drop-worms* and *basket-worms*.

We have in Massachusetts another sack-bearer, which does not appear to have been described, and differs so much both from *Psyche* and *Oiketicus*, when arrived at maturity, as to induce me to give it another generical name. I therefore call it *Perophora Melsheimerii*,* Melsheimer's sack-bearer (Plate VI. Fig. 5). A case of this insect, containing a living caterpillar, was brought to me towards the end of September, by a student of Harvard College, Mr. H. O. White, who found it on an oak-tree in Cambridge. This case (Plate VI. Fig. 4) was nearly an inch and a half long,

* Named in honor of Dr. F. E. Melsheimer (the son of the Rev. F. V. Melsheimer, the father of American Entomology, as he has been called), from whom I have received specimens of this insect, and its curious case.

and about half an inch in diameter. It was not regularly oval, but somewhat flattened on its lower side. It consisted externally of two oblong oval pieces of a leaf, fastened together in the neatest manner by their edges, but the seams made a little ridge on each side of the case ; this had become dry and faded, and was lined within with a thick and tough layer of brownish silk, in which there was left, at each end, a circular opening just big enough for the caterpillar to pass through. The caterpillar (Fig. 206) was cylindrical, about as thick as a common pipe-stem, of a light reddish-brown color with a paler line along the back; it was rough with little elevated points; its head and the top of the first ring were black, hard, and rough also. The head was provided with a pair of jointed feelers, which the insect extended and drew in at pleasure, and which, when they were out, were kept in continual motion. On each side of the middle of the head, there was a black and flexible kind of antenna, very slender where it joined the head, and broader towards the end, like the handle of a spoon. The first three pairs of legs were equal in length, and armed with stout horny claws. The other legs, if such they could be called, were ten in number, and so short that only the oval soles of the feet were visible, and these were surrounded by numerous minute hooks. The tail end of the body was as blunt as if it had been cut off with a knife; it sloped a little backwards, and consisted of a circular horny plate, of a dark gray color, which, when the caterpillar retired within its case, exactly shut up one of the holes in it. This caterpillar eat the leaves of the oak, and fed mostly by night; while eating, it came half-way, or more, out of its cocoon; and in moving laid hold of the leaf with its fore legs, and then shortened its body suddenly, so as to bring its cocoon after it with a jerk ; and, in this way, it went by jerks from place to place. When it had done eating, it moored its case to a leaf by a few silken threads

Fig. 206.

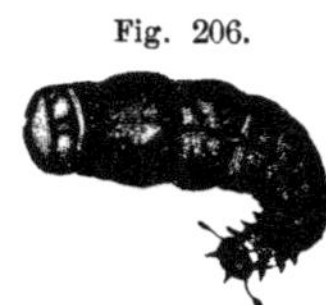

fastened to one, and sometimes to both ends; and before moving again, it came out and bit off these threads close to the case. It could turn round easily within its case, and go out of either end, as occasion required. So tenaciously did it cling to the inside of its case with the little hooks of its hinder feet, that all attempts to make it come wholly out, except by a force which would have been fatal to the insect, were without effect. This kind of caterpillar prepares for transformation by fastening both ends of its cocoon to a branch, and then stops up each of the holes in it with a little circular silken lid, exactly fitting the orifice, and made about the thickness of common brown paper. There is no great difference in the size or form of the chrysalids which produce the male and female moths; they are about three quarters of an inch in length; on both of them the sheaths for the wings, antennæ, and legs are alike, and are as plainly to be seen as on the chrysalids of other winged moths. The chrysalis tapers very little, and does not end with a point, but is blunt behind; and on the edge of each of the rings of the back, there is a transverse row of little pointed teeth which shut into corresponding notches in the ring immediately behind them. These teeth are evidently designed to enable the chrysalis to move towards the mouth of its case, and to hold with, when it is engaged in forcing off the lid in order to allow of the escape of the moth. I do not know at what time the moths come out in Massachusetts; they have been taken in July in Virginia. Both sexes leave their cocoons when arrived at maturity, and both are provided with wings. Their feelers are of moderate size, cylindrical, blunt-pointed, and thickly covered with scales. The tongue is not visible. Their antennæ are curved, and are recurved or bent upwards at the point; the stalk is feathered, in a double row, on the under side, very widely in the males, for more than half its length, and beyond the middle the feathery fringe is suddenly narrowed, and tapers thence to the tip; in the females (Plate VI. Fig. 5) the antennæ are also doubly feathered, but the fringe is narrower throughout than in the other sex. The body and the wings almost exactly resemble those of the foreign silk-worm moth in shape; but the fore wings are rather more pointed and hooked at the tip. There are no bristles and hooks to hold together the wings, which, when at rest, cover the sides like a sloping roof, and the front edge of the hind wings does not project beyond that of the fore wings. These moths are of a reddish-gray color, finely sprinkled all over with minute black dots; the posterior margin of the hind wings above, and the under side of the fore wings, especially behind the tip, are tinged with tawny red; there is a small black dot near the middle of the fore wings; and both the fore and hind wings are crossed by a narrow blackish band, beginning with an angle on the front edge of the former, and passing obliquely backwards to the inner edge of the hind wings. They expand from one inch and three eighths to two inches, or a little more.

The last family of the Bombyces remaining to be noticed may be called Notodontians (NOTODONTADÆ). Many of the caterpillars belonging to it have hunched backs, or tooth-like prominences on the back; and hence the origin of the name of this family, which comes from a word signifying toothed back. Most of these caterpillars are entirely naked; some of them are downy or slightly hairy, but the hairs generally grow immediately from the skin, and not in spreading clusters from little warts on the rings. They have sixteen legs; some raise the last pair when at rest, and some keep these always elevated and do not use them in creeping, in which case these terminal legs are lengthened, and form a forked appendage or tail to the hinder part of the body. Hence such caterpillars are often described as having only fourteen legs, although the wanting members really exist in a modified form. Moreover, the caterpillars of some of the Notodontians seem to be without legs, and even on close examination only the soles of the feet can be perceived. The Notodontians are found chiefly on trees and shrubs, the leaves of which they eat. When about to be transformed, the most of them enclose themselves in cocoons, which are often very hard and thick, made either of silk, or of silk mixed with fragments of wood and bark; some make thin, semi-transparent, and filmy cocoons under a covering of leaves; some merely cover themselves with grains of earth, held together by silken threads; and a very few go into the ground to transform, without making cocoons. The chrysalids taper behind, and are not provided with transverse notched ridges on the back. The moths close their wings over the sides of the body like a sloping roof, when at rest; but the front edges of the hind wings never extend beyond those of the fore wings, and the bristles and hooks for holding the wings together are never wanting. The antennæ are rather long; those of the males are generally doubly feathered on the under side; but the feathery fringe is often very narrow towards the tips, and in the females is always narrower than in the other sex; in a few of both sexes the antennæ are not feathered at all. The feelers and tongue, though short, are generally visible. The body is rather long, and not very thick. In what follows, a few only of the most remarkable species will be described.

Among the many odd-shaped caterpillars belonging to this family, not the least remarkable are those which are called LIMACODES, that is, slug-like, on account of their seeming want of feet, their very slow gliding motions, and the slug-like form of some of them. In these caterpillars the body is very short and thick, and approaches more or less to an oval form; it is naked, or, in some kinds, covered only with short down; the head is small, and can be drawn in and concealed under the first ring; the six fore legs are also small and retractile; and the other legs consist only of little fleshy elevations, without claws or hooks. The under side of the body is smeared with a sticky fluid, which seems designed to render their footing more secure, and leaves a slimy track wherever the insects go. Their co-

53 *

coons are very small, almost round, tough, and parchment-like, and are fastened to the twigs of the plants on which the insects live. The moths of some, if not of all, of the *Limacodes* make their escape by pushing off one end of the cocoon, which separates like a little circular lid.

The most common of these slug-caterpillars, in Massa chusetts, live on walnut-trees. They come to their full size in September and October, and then measure five eighths of an inch in length, and rather more than three eighths across the middle. The body is thick, and its outline nearly diamond-shaped ; the back is a little hollowed, and the middle of each side rises to an obtuse angle ; it is of a green color, with the elevated edges brown. The boat-like form of this caterpillar induced me to name it *Limacodes Scapha*, the skiff Limacodes, in my "Catalogue of the Insects of Massachusetts." My specimens generally died after they had made their cocoons, and consequently the moth is unknown to me.

Fig. 207.

The moth of a *Limacodes*, called *Cippus** (Fig. 207) by Sir J. E. Smith, is sometimes found in Massachusetts, from the middle of July till the 10th of August. It is of a reddish-brown color ; on each of the fore wings there is a small dark brown dot near the middle, and a broad wavy green band beginning at the base, and bending round till it touches the front margin near the tip ; behind a deep notch of this band, near the base of the wing, there is a triangular tawny spot, and another smaller one near the tip. The green band is sometimes broken into three triangular green spots, the middle one of which is wanting in some specimens. One half of the stalk of the antennæ of the male is doubly feathered beneath ; the remainder to

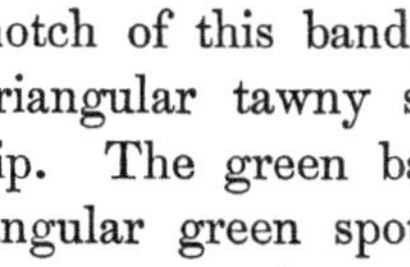

* Probably not the true *Cippus* of Fabricius, which is found in Surinam. There is a figure of our species in Guérin's "Iconographie du Règne Animal," where it is named *Limacodes Delphinii*, but for what reason I know not, for it does not live on the *Delphinium* or larkspur.

the tip is bare. The antennæ of the female are thread-like and not fringed. The wings expand from one inch to one inch and one eighth. The caterpillar figured by Mr. Abbot* is oblong oval, striped with purple and yellow, with twelve fleshy horns, of an orange color, on the sides of its back, namely, six on the fore part, two on the middle, and four on the hind part of the body. Mr. Abbot says that it eats the leaves of the dogwood (*Cornus Florida*), oak, and of other trees; that it makes its cocoon in September, and that the moth comes out in July.

A still more extraordinary slug-caterpillar (Fig. 208), having a very remote resemblance to the last, has been found here on forest-trees, and occasionally in considerable numbers on cherry-trees and apple-trees, from July to September. It is of a dark brown color, and is covered with a short velvet-like down; its body is almost oblong square, but the sides of the rings extend horizontally in the form of flattened teeth; three of these teeth on each side, that is, one on the fore part, the middle, and the hind part of the body, are much longer than the others, and are curved backwards at the end. When fully grown, the caterpillar measures nearly an inch in length. It does not bear confinement well, and often dies before completing its transformations. Dr. Melsheimer, to whom I am indebted for one of the moths, informs me that the caterpillar eats the leaves of the wild cherry, as well as those of the white and red oak, that it makes its cocoon (Fig. 209) about the middle of September, changes to a chrysalis the following April, and that the moth appears in about eight weeks afterwards. The name given to this insect by Sir J. E. Smith † is *pithecium*, the meaning of which is a shrivelled and monkey-faced old woman, bestowed upon it prob-

Fig. 208.

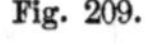

Fig. 209.

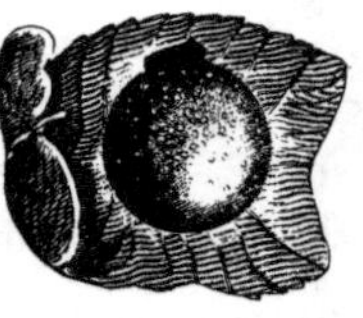

* Insects of Georgia, p. 145, pl. 73.

† Ibid., p. 147, pl. 74.

ably on account of the shrivelled appearance and dark color of the caterpillar. In its winged state, *Limacodes pithecium*, or the hag-moth, as it may be called, is of a dusky brown color; its fore wings are variegated with light yellowish brown, and with a narrow curved and wavy band, of the same light color, edged externally with dark brown near the outer margin, and a light brown spot near the middle; the fringes of all the wings are spotted with light brown; the legs are covered with long hairs; the antennæ, in both sexes, are slender, almost thread-like, and not feathered. It expands from nearly one inch to one inch and a quarter.

There is a kind of caterpillar, found in July and August on the balsam poplar, and sometimes on other poplars and willows, whose form, posture, and motions are so odd as at once to arrest attention. Its body is naked, short, and thick, tapers behind, and ends with a forked kind of tail, which is held upwards at an obtuse angle with the rest of the body. This forked tail, which takes the place of the hindmost pair of legs, the others being only fourteen in number, is not used with the latter in creeping, and consists of two movable hollow tubes, within each of which is concealed a long orange-colored thread, that the insect can push out and draw in at pleasure. The feet are short and small; the head is small, of a purple color, and can be drawn under the front part of the first ring; the body is green, with a triangular purple spot on the top of the fore part, and a large diamond-shaped patch, of the same color, covering the back and middle of the sides like a mantle, and prolonged behind to the tail. When young, these caterpillars have, on the top of the first ring, two little prickly warts, which disappear after one or two changes of the skin. When teased by being touched, or irritated by flies, the caterpillar runs out the threads from its forked tail, which it jerks forwards so as to lash the sides of its body and whip off the intruder. When fully grown, it measures sometimes an inch and a half in length, without including the terminal fork. Caterpillars of this kind are called *Cerura*, horned-tail, by some, and *Dicranura*, forked-tail, by other naturalists. Early in August the one above described makes a tough cocoon of bits of wood and bark glued together with a sticky matter, and fastened to the side of a branch, the lower side being flat and the upper convex. The last transformation occurs about the middle of June, when, after the end of the cocoon has been softened by a liquid thrown out by the insect within, the moth forces its way through. This insect has been figured in Mr. Abbot's work,* where it is called *furcula*, a name, however, which belongs to an European insect. It is also represented in Guérin's "Iconographie," and in Griffith's translation of Cuvier's "Animal Kingdom"; and I have adopted the specific name given to it by Dr. Boisduval in these works. CERURA *borealis*, the northern Cerura, or fork-tail moth, like others of the genus, has the antennæ feathered in both sexes, but narrow, and tapering and bent upwards at the point; the legs, especially the first pair, which are stretched out before the body when at rest, are, like those of our native Limacodes, very hairy; and the wings are thin and almost transparent. The ground-color of our moth is a dirty white; the fore wings are crossed by two broad blackish bands, the outer one of which is traversed and interrupted by an irregular wavy whitish line; the hinder margins of all the wings are dotted with black, and there are several black dots at the base, and a single one near the middle of the fore wings; the top of the thorax is blackish, and the collar is edged with black. In some individuals the dusky bands of the fore wings are edged or dotted with tawny yellow; in others, these wings are dusky, and the bands are indistinct. They expand from one inch and three eighths to one inch and three quarters.

The following insects, for the sake of convenience, may be included in the old genus *Notodonta*. The first of them

* Insects of Georgia, p. 141, pl. 71.

is found in August and September on plum and apple trees, and, according to Mr. Abbot,* on the red-berried alder, *Prinos verticillatus*. The top of the fourth ring of this caterpillar rises in the form of a long horn, sloping forwards a little; the tail, with the hindmost feet, which are rather longer than the others, is always raised when the insect is at rest, but it generally uses these legs in walking; its head is large, and of a brown color; the sides of the second and third rings are green; the rest of the body is brown, variegated with white on the back, and on it there are a very few short hairs, hardly visible to the naked eye. When fully grown, it measures an inch or more in length. Though mostly solitary in their habits, sometimes three or four of these caterpillars are found near together, and eating the leaves of the same twig. Towards the end of September they descend from the trees, and make their cocoons, which are thin and almost transparent, resembling parchment in texture, and are covered generally with bits of leaves on the outside. The caterpillars remain in their cocoons a long time before changing to chrysalids, and the moth does not come out till the following summer. There are probably two broods in the course of one season, for I have taken the moths early in August. In Georgia the caterpillar made its cocoon on the 30th of May, and was transformed to a moth fourteen days afterwards. This moth is the *Notodonta unicornis*, or unicorn moth, so called from the horn on the back of the caterpillar. The fore wings are light brown, variegated with patches of greenish white and with wavy dark brown lines, two of which enclose a small whitish space near the shoulders; there is a short blackish mark near the middle; the tip and the outer hind margin are whitish, tinged with red in the males; and near the outer hind angle there are one small white and two black dashes; the hind wings of the male are dirty white, with a dusky spot on the inner hind angle; those of the female are sometimes

* Insects of Georgia, p. 171, pl. 86.

entirely dusky; the body is brownish, and there are two narrow black bands across the fore part of the thorax. The wings expand from one inch and a quarter to one inch and a half, or nearly.

Our fruit-trees seem to be peculiarly subject to the ravages of insects, probably because the native trees of the forest, which originally yielded the insects an abundance of food, have been destroyed to a great extent, and their places supplied only partially by orchards, gardens, and nurseries. Numerous as are the kinds of caterpillars now found on cultivated trees, some are far more abundant than others, and therefore more often fall under our observation, and come to be better known. Such, for instance, are certain gregarious caterpillars that swarm on the apple, cherry, and plum trees towards the end of summer, stripping whole branches of their leaves, and not unfrequently despoiling our rose-bushes and thorn hedges also. These caterpillars are of two kinds, very different in appearance, but alike in habits and destructive propensities. The first of these may be called the red-humped (Fig. 210), a name that will probably bring these insects to the remembrance of those persons who have ever observed them. Different broods make their appearance at various times during August and September. The eggs from which they proceed are laid, in the course of the month of July, in clusters on the under side of a leaf, generally near the end of a branch. When first hatched they eat only the substance of the under side of the leaf, leaving the skin of the upper side and all the veins untouched; but as they grow larger and stronger, they devour whole leaves from the point to the stalk, and go from leaf to leaf down the twigs and branches. The young caterpillars are lighter-colored than the old ones, which are yellowish brown, paler on the sides, and longitudinally striped with slender black lines; the head is red; on the top of the fourth ring there

Fig. 210.

is a bunch or hump, also of a red color; along the back are several short black prickles; and the hinder extremity tapers somewhat, and is always elevated at an angle with the rest of the body, when the insect is not crawling. The full-grown caterpillars measure one inch and a quarter, or rather more, in length. They rest close together on the twigs, when not eating, and sometimes entirely cover the small twigs and ends of the branches. The early broods come to their growth and leave the trees by the middle of August, and the others between this time and the latter part of September. All the caterpillars of the same brood descend at one time, and disappear in the night. They conceal themselves under leaves, or just beneath the surface of the soil, and make their cocoons, which resemble those of the unicorn Notodonta. They remain a long time in their cocoons before changing to chrysalids, and are transformed to moths towards the end of June or the beginning of July. Mr. Abbot * states that in Georgia these insects breed twice a year, the first broods making their cocoons towards the end of May, and appearing in the winged form fifteen days afterwards. This Notodonta is a neat and trim looking moth, and is hence called *concinna* (Plate VI. Fig. 11) by Sir J. E. Smith. It is of a light brown color; the fore wings are dark brown along the inner margin, and more or less tinged with gray before; there is a dark-brown dot near the middle, a spot of the same color near each angle, a very small triangular whitish spot near the shoulders, and several dark-brown longitudinal streaks on the outer hind margin; the hind wings of the male are brownish or dirty white, with a brown spot on the inner hind angle; those of the other sex are dusky brown; the body is light brown, with the thorax rather darker. The wings expand from one inch to one inch and three eighths.

Every person who has paid any attention to the cultivation of the grape-vine in this country must have observed

* Insects of Georgia, p. 169, pl. 85

upon it, besides the large sphinx caterpillars that devour its leaves, a small blue caterpillar (Fig. 211, and Plate VI. Fig. 7), transversely banded with deep orange across the middle of each ring, the bands being dotted with black, with the head and feet also orange, the top of the eleventh ring somewhat bulging, and the fore part of the body hunched up when the creature is at rest. These caterpillars begin to appear about the middle of July, and others are hatched afterwards, as late, perhaps, as the middle of August. When not eating, they generally rest upon the under sides of the leaves, and, though many may be found on one vine, they do not associate with each other. They live on the common creeper, as well as on the grape-vine. They eat all parts of the leaves, even to the midrib and stalks. When fully grown, and at rest, they measure an inch and a quarter, but stretch out, in creeping, to the length of an inch and a half, or more. Towards the end of August they begin to disappear, and no more will be found on the vines after September. They creep down the vines in the night, and go into the ground, burying themselves three or four inches deep, and turn to chrysalids without making cocoons. The chrysalis is dark brown, and rough with elevated points. The moths begin to come out of the ground as soon as the 25th of June, and others continue to appear till the 20th of July. Though of small size, they are very beautiful, and far surpass all others of the family in delicacy of coloring and design. The name of this moth is *Eudryas grata** (Plate VI. Fig. 8), the first word signifying beautiful wood-nymph, and the second agreeable or pleasing. The antennæ are rather long, almost thread-like, tapering to the end, and not feathered in either sex. The fore wings are pure white, with a broad stripe along the front edge, extending from the shoulder a little beyond the middle of the edge, and a broad band around the

Fig. 211.

* This insect is the *Bombyx grata* of Fabricius.

outer hind margin, of a deep purple-brown color; the band is edged internally with olive-green, and marked towards the edge with a slender wavy white line; near the middle of the wing, and touching the brown stripe, are two brown spots, one of them round and the other kidney-shaped; and on the middle of the inner margin there is a large triangular olive-colored spot; the under side of the same wings is yellow, and near the middle there are a round and a kidney-shaped black spot. The hind wings are yellow above and beneath; on the upper side with a broad purple-brown hind border, on which there is a wavy white line, and on the under side with only a central black dot. The head is black. Along the middle of the thorax there is a broad crest-like stripe of black and pearl-colored glittering scales. The shoulder-covers are white. The upper side of the abdomen is yellow, with a row of black spots on the top, and another on each side; the under side of the body, and the large muff-like tufts on the fore legs, are white; and the other legs are black. This moth rests with its wings closed like a steep roof over its back, and its fore legs stretched forward, like a Cerura. It expands from one inch and a half to one inch and three quarters.

Eudryas unio, of Hübner, the pearl Eudryas, as its name implies, is a somewhat smaller moth, closely resembling the preceding, from which it differs in having the stripe and band on its fore wings of a brighter purple-brown color, the round and kidney-shaped spots contiguous to the former also brown, the olive-colored edging of the band wavy, with a powdered blue spot between it and the triangular olive-colored spot on the inner margin, and a distinct brown spot on the inner hind angle of the posterior wings; all the wings beneath are broadly bordered behind with light brown, and the spots upon them are also light brown. It expands from one inch and three eighths to one inch and a half. This species has been taken in Massachusetts, but it is rare, and the caterpillar is unknown to me.

In the remarks preceding the description of *Notodonta concinna*, mention was made of two kinds of caterpillars, living in great numbers on fruit-trees in the latter part of summer. The second kind (Fig. 212) are now to be described. They grow to a greater size, are longer in coming to their growth, their swarms are more numerous, and consequently they do much more injury, than the red-humped kind. Entire branches of the apple-trees are frequently stripped of their leaves by them, and are loaded with these caterpillars in thickly crowded swarms. The eggs from which they are hatched will be found in patches, of about a hundred together, fastened to the under side of leaves near the ends of the twigs. Some of them begin to be hatched about the 20th of July, and new broods make their appearance in succession for the space of a month or more. At first they eat only the under side and pulpy part of the leaves, leaving the upper side and veins untouched; but afterwards they consume the whole of the leaves except their stems.

Fig. 212.

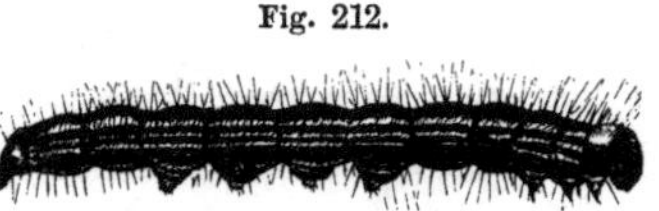

These caterpillars are sparingly covered with soft whitish hairs; the young ones are brown, and striped with white; but, as they grow older, their colors become darker every time they cast their skins. They come to their full size in about five weeks or a little more, and then measure from an inch and three quarters to two inches and a quarter in extent. The head is large and of a black color; the body is nearly cylindrical, with a spot on the top of the first ring, and the legs dull orange-yellow, a black stripe along the top of the back, and three of the same color alternating with four yellow stripes on each side. The posture of these caterpillars, when at rest, is very odd; both extremities are raised, the body being bent, and resting only on the four intermediate pairs of legs. If touched or otherwise disturbed, they throw up their heads and tails with a jerk, at the same time bending the body semicircularly till the two extremities almost meet over the back. They all eat together, and, after they have done, arrange themselves side by side along the twigs and branches which they have stripped. Beginning at the ends of the branches, they eat all the leaves successively from thence towards the trunk, and if one branch does not afford food enough they betake themselves to another. When ready to transform, all the individuals of the same brood quit the tree at once, descending by night, and burrow into the ground to the depth of three or four inches, and, within twenty-four hours afterwards, cast their caterpillar-skins, and become chrysalids without making cocoons. They remain in the ground in this state all winter, and are changed to moths and come out between the middle and end of July.

These moths belong to the genus *Pygœra*, so named because the caterpillar sits with its tail raised up. The antennæ are rather long, those of the males fringed beneath, in a double row, with very short hairs nearly to the tips, which, however, as well as the whole of the stalk of the antennæ in the other sex, are bare; the thorax is generally marked with a large dark-colored spot, the hairs of which can be raised up so as to form a ridge or kind of crest; the hinder margin of the fore wings is slightly notched; and the fore legs are stretched out before the body in repose. Our *Pygœra* was named, by Drury, *ministra*, the attendant or servant. (Plate VI. Fig. 6.) It is of a light brown color; the head and a large square spot on the thorax are dark chestnut-brown; on the fore wings are four or five transverse lines, one or two spots near the middle, and a short oblique line near the tip, all of which, with the outer hind margin, are dark chestnut-brown. One and sometimes both of the dark brown spots are wanting on the fore wings in the males, and the females, which are larger than the other sex, frequently have five instead of four transverse brown lines. It expands from one inch and three quarters to two inches and a half.

I have seen on the oak, the birch, the black walnut, and the hickory trees, swarms of caterpillars slightly differing in color from each other, and from those above described, that live on the apple and cherry trees; they were more hairy than the latter, but their postures and habits appeared to be the same. Whether they were all different species, or only varieties of the *ministra*, arising from difference of food, I have not been able to ascertain.

The cultivation of the balsam and our other large-leaved native poplars seems to have been neglected of late years. It is true that these trees are not so durable and so valuable as many others; but we sometimes meet with noble specimens of them; and the rapidity of their growth, the great size they attain in favorable situations, and the fine shade they afford, are qualities not to be overlooked or despised; nor is the wood entirely worthless, either as fuel or in the arts. If these trees are planted alternately with other more slow-growing trees, we shall have the benefit of the shade and shelter of the former till the others have become large enough to fill their places. They are not subject to be attacked by canker-worms, oak-caterpillars, web-worms, and many other kinds of insects that infest our ornamental and shade trees of hard wood; but, unfortunately, they suffer too often from insect depredators of their own, such as the grubs of two or three kinds of beetles, which bore into their trunks; the spiny caterpillars of the Antiopa butterfly and of the Io moth, the fork-tailed Cerura, the caterpillar of the herald-moth, and another kind of caterpillar now to be described, all which devour the leaves of these trees. This last kind of caterpillar (Fig. 213) is found in little swarms on the trees from the last of July to the beginning of October. It does not raise the hinder part of its body when at rest. It is nearly cylindrical, with two little black warts close together on the top of the fourth and of the eleventh rings.

Fig. 213.

There are a few short, whitish hairs thinly scattered over the body, which is pale yellow, with three slender black lines on the back, and a broad dusky stripe, also marked with three black lines, on each side; and the head, fore legs, and spiracles are black. When fully grown, these caterpillars measure about an inch and a half in length. They live together, in swarms of twenty or more individuals, in a nest (Fig. 214) made of a single leaf folded or curled at the sides, and lined with a thin web of silk. An opening is left at each end of the nest; through the lower one the dirt made by the insects falls, and through the upper one, which is next to the leaf-stalk, the caterpillars go out to feed upon the leaves near to their nests. When young they sometimes fold up one side of a leaf for a nest, and eat the other half. The stalks of the leaves, to which their nests are hung, become covered with silk from the threads carried along by the caterpillars in going over them; and these threads help to secure the nests to the branches. They eat all parts of the leaves except the stalks and larger veins, and frequently strip long shoots of their foliage in a very few days. Towards the end of September or early in October, according to the age of the different broods, they descend from the trees, disperse, and seek a shelter in crevices or under leaves and rubbish on the ground, where they make their cocoons. These are thin, irregular, silken webs, so loosely spun that the in-

Fig. 214.

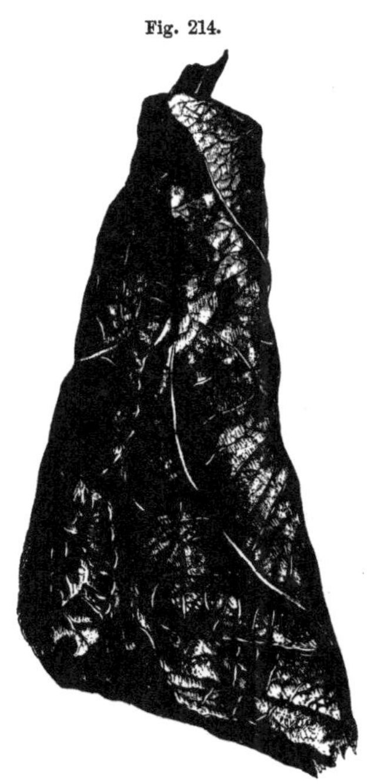

sects can be seen through them; but they are protected by their situation, or by the dead leaves and other matters under which they are made. As soon as the cocoons are finished, the insects become chrysalids (Fig. 215), and remain quiet through the winter; and about the middle of June, or somewhat later, they are transformed to moths. They belong to the genus CLOSTERA, or spinner, so named on account of the spinning habits of the caterpillars. The antennæ are narrowly feathered or pectinated in both sexes; the thorax has an elevated crest in the middle; the tail is tufted and turned up at the end, in the males; the fore legs are thickly covered with hairs to the end, and are stretched out before the body when the insect is at rest.

Fig. 215.

Our poplar spinner may be called *Clostera Americana*,[23] the American Clostera. (Plate VI. Fig. 12.) It closely resembles the European *anastomosis*, from which, however, it differs essentially in its caterpillar state, and the moth presents certain characters, which, on close comparison with the European insect, will enable us to distinguish it from the latter. It is of a brownish-gray color; the fore wings are faintly tinged with pale lilac, and more or less clouded with rust-red; they have an irregular row of blackish dots near the outer hind margin, and are crossed by three whitish lines, of which the first nearest the shoulders is broken and widely separated in the middle, the second divides into two branches, one of which goes straight across the wing to the inner margin, and the other passes obliquely till it meets the end of the third line, with which it forms an angle or letter V; across the middle of the hind wings there is a narrow brownish band, much more distinct beneath than above; on the top of the thorax there is an oblong chestnut-

[[23] This name cannot stand. It is the *C. inclusa,* Hübner, Zutr. Dr. Harris has somewhere said that he had no opportunity of consulting Hübner's works, and hence is not to be blamed for naming what he conceived to be a new species. — MORRIS.]

colored spot, the hairs of which rise upwards behind and form a crest. All the whitish lines on the fore wings are more or less bounded externally with rust-red. It expands from one inch and one quarter to one inch and five eighths. In Georgia this insect breeds twice a year; and the caterpillars eat the leaves of the willow as well as those of the poplar.*

2. Owlet-Moths. (*Noctuæ.*)

Our second tribe of moths, the Noctuæ of Linnæus, appears to have been thus named from *Noctua*, an owl, because they fly chiefly by night, and are hence called *Eulen*, or owl-moths, by the Germans. This tribe contains a very large number of thick-bodied and swift-flying moths, most of which may be distinguished by the following characters. The antennæ are long and tapering, and seldom pectinated even in the males; the tongue is long; the feelers are very distinct, and project more or less beyond the face, the two lower joints being compressed or flattened at the sides, and the last joint is slender and small; the thorax is thick, with rather prominent collar and shoulders, and is often crested on the top; the body tapers behind; the wings are always fastened together by bristles and hooks, are generally roofed, when at rest, and each of the fore wings is marked behind the middle of the front margin with two spots, one of them round and small, and the other larger and kidney-shaped. A few of them fly by day, the others only at night. Their colors are generally dull, and of some shade of gray or brown, and so extremely alike are they in their markings, that it is very difficult to describe them without the aid of figures, which cannot be expected in this treatise. The caterpillars are nearly cylindrical, for the most part naked, though some are hairy, slow in their motions, and generally provided with sixteen legs; those with fewer legs never want

* See *Phalæna anastomosis* of Smith, in Abbot's "Insects of Georgia," p. 143, pl. 72.

the hindmost pair, and never raise the end of the body when at rest. Some of them make cocoons, but the rest go into the ground to transform. Many of the Noctuas vary more or less from the characters above given, and the tribe seems to admit of being divided into several smaller groups or families, under which their peculiarities might be more distinctly pointed out. Unfortunately the history of most of our moths is still imperfectly known; and for this reason, as well as on account of the length to which the foregoing part of this treatise has already extended, I have concluded to suppress a considerable portion of my observations on the owlet-moths and the rest of the *Lepidoptera*, and shall confine my remarks to a few of the most injurious species in each of the remaining tribes.

The injury done to vegetation by the caterpillars of the Noctuas, or owlet-moths, is by no means inconsiderable, and sometimes becomes very great and apparent; but most of these insects are concealed from our observation during the day-time, and come out from their retreats to feed only at night. To turn them out of their hiding-places becomes sometimes absolutely necessary, and it is only by dear-bought experience that we learn how to discover them. This is not the case with all; those of the first family, which I would call Acronyctians (ACRONYCTADÆ*), live exposed on the leaves of trees and shrubs. They have sixteen legs, are cylindrical, and more or less hairy, some of them closely resembling those of the genus *Clostera*, having a wart or prominence on the top of the fourth and the eleventh rings, and some of them have the hair in tufts like Arctians and Liparians. They make tough silken cocoons, in texture almost like stiff brown paper, into which they weave the hairs of their bodies. Their moths have bristle-formed antennæ, and the thorax is not crested. Their fore wings are generally light gray with dark spots, and in many are marked with a character resembling the Greek letter ψ near

* From *Acronycta*, a genus of moths appearing at nightfall, as the name implies.

the inner hind angle. Of those that want this character on the fore wings, the largest American species, known to me, may be called *Apatela Americana*[24] (Fig. 216), which has been mistaken * for *Apatela Aceris*, the maple-moth of Europe. Its body and fore wings are light gray; on the latter

Fig. 216.

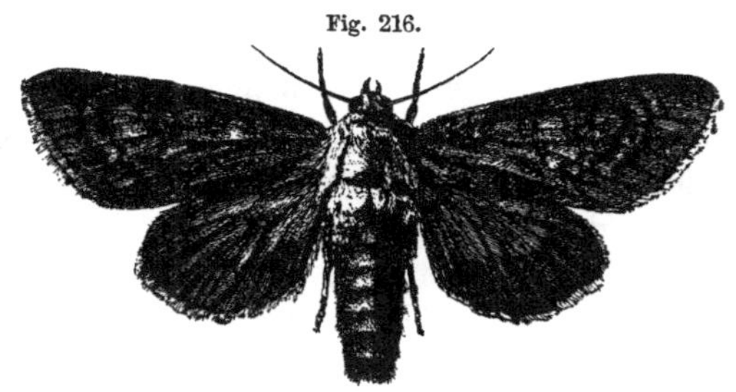

there is a wavy, scalloped white line edged externally with black near the outer hind margin, and the usual round and kidney-shaped spots are also edged with black; the hind wings are dark gray in the male, blackish in the female, with a faintly marked black curved band and central semicircular spot; all the wings are whitish and shining beneath, with a black wavy and curved band and central semicircular spot on each; the fringes are white, scalloped, and spotted with black. It expands from two inches and a quarter to two inches and a half, or more. This kind of moth flies only at night, and makes its appearance between the middle and the end of July. The caterpillar (Fig. 217) eats the leaves of the various kinds of maple, and sometimes also those of the elm, linden, and chestnut. It is one of the largest kinds; and, early in October, when it arrives at maturity,

Fig. 217.

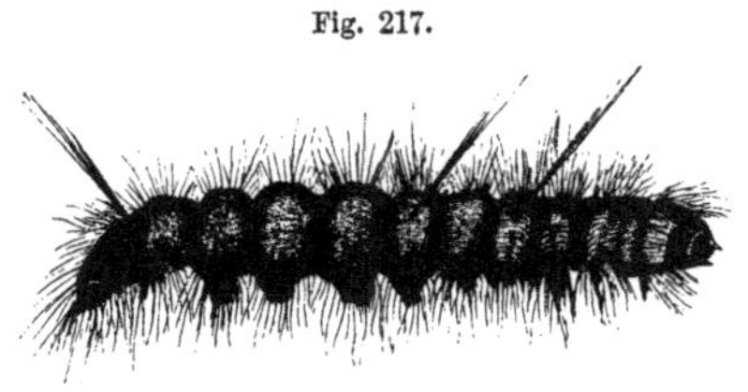

[24 *A. Americana* is synonymous with *Acronycta acericola* Guenée. — MORRIS.]

* See *Phalæna Aceris*, Smith, in Abbot's "Insects of Georgia," p. 185, pl. 93.

measures from one inch and three quarters to two inches or more in length. It is of a greenish-yellow color above, with the head, tail, belly, and feet black; its body is covered with long and soft yellow hairs, growing immediately from the skin; on the top of the fourth ring there are two long, slender, and erect tufts of black hairs, two more on the sixth ring, and a single pencil on the eleventh ring.* While at rest, it remains curled sidewise on a leaf. When about to make its cocoon, it creeps into chinks of the bark, or into cracks in fences, and spins a loose, half-oval web of silk, intermixed with the hairs of its body; under this it then makes another and tougher pod of silk, thickened with fragments of bark and wood, and, when its work is done, changes to a chrysalis (Fig. 218), in which state it remains till the following summer.

Fig. 218.

The caterpillars of the Nonagrians (NONAGRIADÆ †) are naked, long, slender, and tapering at each end, smooth, and generally of a faint reddish or greenish tint, with an oval, dark-colored, horny spot ‡ on the top of the first and last ring. Most of them live within the stems of reeds, flags, and other water-plants; some in the stems and even in the roots of plants remote from the water. They devour the pith and the inside of the roots, and transform in the same situations, having previously gnawed a hole from the inside of their retreat, through the side of the stem or root, to the outside skin, which is left untouched, and which the moth

* Those naturalists who are familiar with the appearance of the European caterpillar of *Apatela Aceris* will perceive the great and essential difference between it and that of our American *Apatela*, which bears about as much resemblance to the former as does that of *Astasia torrefacta* of Sir J. E. Smith, an insect apparently belonging to the Notodontians, and near to *Clostera* and *Pygæra*. *Apatela* signifies deceptive; and this name was probably given to the genus because the caterpillars appear in the dress of Arctians and Liparians, but produce true owlet-moths or Noctuas.

† From *Nonagria*, the meaning of which is uncertain.

‡ These dark horny spots are found on the first ring of most of the caterpillars that burrow in the stems of plants, or in the ground.

can easily break through afterwards. The chrysalids are generally very long and cylindrical, and are blunt at the extremities. Most of the moths have very long bodies, a smooth thorax, and are of a yellowish clay or drab color; the fore wings want the usual spots, are faintly streaked and dotted with black, and have a scalloped hind margin. Those that do not live in water-plants are distinguished by brighter colors of orange-yellow and brown, with the usual spots more or less distinct on the fore wings, the margin of which is wavy; the collar is prominent, and the thorax crested. In all of them the antennæ of the males are slightly thickened with short hairs beneath.

These insects are fatal to the plants attacked, the greater part of which, however, are without value to the farmer. Indian corn must be excepted; for it often suffers severely from the depredations of one of these Nonagrians, known to our farmers by the name of the spindle-worm. The Rev. L. W. Leonard has favored me with a specimen of this insect, its chrysalis, and its moth, together with some remarks upon its habits; and the latter have also been described to me by an intelligent friend, conversant with agriculture. This insect receives its common name from its destroying the spindle of the Indian corn; but its ravages generally begin while the corn-stalk is young, and before the spindle rises much above the tuft of leaves in which it is embosomed. The mischief is discovered by the withering of the leaves, and, when these are taken hold of, they may often be drawn out with the included spindle. On examining the corn, a small hole may be seen in the side of the leafy stalk, near the ground, penetrating into the soft centre of the stalk, which, when cut open, will be found to be perforated, both upwards and downwards, by a slender worm-like caterpillar, whose excrementitious castings surround the orifice of the hole. This caterpillar grows to the length of an inch, or more, and to the thickness of a goose-quill. It is smooth, and apparently naked, yellowish, with the head, the top of the first and of the last rings black, and with a double row, across each of the other rings, of small, smooth, slightly elevated, shining black dots. With a magnifying-glass a few short hairs can be seen on its body, arising singly from the black dots. This mischievous caterpillar is not confined to Indian corn; it attacks also the stems of the Dahlia, as I am informed both by Mr. Leonard and by the Rev. J. L. Russell, both of whom have observed its ravages in the stems of this favorite flower. It has also been found in the pith of the elder, and the same species of moth was produced from it, early in August, as from the spindle-worm of corn. The chrysalis, which is lodged in the burrow formed by the caterpillar, is slender, but not quite so long in proportion to its thickness as are those of most of the Nonagrians. It is shining mahogany-brown, with the anterior edges of four of the rings of the back roughened with little points, and four short spines or hooks, turned upwards, on the hinder extremity of the body. The moth produced from this insect differs from the other Nonagrians somewhat in form, its fore wings being shorter and more rounded at the tip. It may be called *Gortyna* * *Zeæ* (Plate VII. Fig. 9), the corn Gortyna; *Zea* being the botanical name of Indian corn. The fore wings are rust-red; they are mottled with gray, almost in bands, uniting with the ordinary spots, which are also gray and indistinct; there is an irregular tawny spot near the tip, and on the veins there are a few black dots. The hind wings are yellowish gray, with a central dusky spot, behind which are two faint, dusky bands. The head and thorax are rust-red, with an elevated tawny tuft on each. The abdomen is pale brown, with a row of tawny tufts on the back. The wings expand nearly one inch and a half.

In order to check the ravages of these insects they must be destroyed while in the caterpillar state. As soon as our cornfields begin to show, by the withering of the leaves, the

* Gortyna, in ancient geography, was the name of a city in Crete, so called from its founder.

usual signs that the enemy is at work in the stalks, the spindle-worms should be sought for and killed; for, if allowed to remain undisturbed until they turn to moths, they will make their escape, and we shall not be able to prevent them from laying their eggs for another brood of these pestilent insects.

A worm, or caterpillar, something like the spindle-worm, has often been found by farmers in potato-stalks; and the *potato-rot* has sometimes been ascribed to its depredations. On the 9th of July, 1848, one of these caterpillars was brought to me in a potato-stalk from Watertown; and on the 5th of July, 1851, I found another within the stem of the pig-weed, or *Chenopodium*. These caterpillars (Fig. 219) were of a livid hue, faintly striped with three whitish lines along the back. Their transformations have not yet been observed.

Fig. 219.

The roots of the Columbine are attacked by another caterpillar belonging to this family. It burrows into the bottom of the stalk and devours the inside of the roots, which it injures so much that the plant soon dies. One of these caterpillars, which was found in July in the roots of a fine double Columbine in my garden, was of a whitish color, with a few black dots on each of the rings, a brownish head, and the top of the first and of the last rings blackish. It grew to the length of about one inch and a quarter, turned to a chrysalis on the 19th of August, and came out a moth on the 24th of September. The moth closely resembles the *Gortyna flavago* of Europe, but is sufficiently distinct from it. It may be called *Gortyna leucostigma*, the white-spot Gortyna. The fore wings are tawny yellow, sprinkled with purple-brown dots, and with two broad bands and the outer hind margin purple-brown; there is a distinct tawny yellow spot on the tip, followed by a row of faint yellowish crescents between the brown band and margin; the ordinary spots are yellow, margined with brown, and there is a third oval spot of a white color near the round spot. The hind wings are pale buff or yellowish white, with a central spot, and a band behind it, of a brownish color. The head is brown; the thorax is tawny yellow, with a brown tuft; and the edges of the collar and of the shoulder-covers are brown. The wings expand rather more than one inch and a half. I have what appear to be varieties of this moth, expanding one inch and three eighths, with three or four white dots around the kidney-spot, and the ordinary round spot wholly white.

Numerous complaints have been made of the ravages of cut-worms among corn, wheat, grass, and other vegetables, in various parts of the country. After a tiresome search through many of our agricultural publications, I have become convinced that these insects and their history are not yet known to some of the very persons who are said to have suffered from their depredations. Various cut-worms, or more properly subterranean caterpillars, wire-worms, or *Iuli*, and grub-worms, or the young of May-beetles, are often confounded together or mistaken for each other; sometimes their names are interchanged, and sometimes the same name is given to each and all of these different animals. Hence the remedies that are successful in some instances are entirely useless in others. The name of cut-worm seems originally to have been given to certain caterpillars that live in the ground about the roots of plants, but come up in the night, and cut off and devour the tender stems and lower leaves of young cabbages, beans, corn, and other herbaceous plants. These subterranean caterpillars are finally transformed to moths belonging to a group which may be called Agrotidians (AGROTIDIDÆ), from a word signifying rustic, or pertaining to the fields. Some of these rustic moths fly by day, and may be found in the fields, especially in the autumn, sucking the honey of flowers; others are on the wing only at night, and during the day lie concealed in chinks of walls and other dark places. Their wings are nearly horizontal when closed, the upper pair completely covering the lower wings, and often overlapping a little on their inner edges, thus favoring these insects in their attempts to obtain shelter and concealment. The thorax is slightly convex, but smooth or not crested. The antennæ of the males are generally beset with two rows of short points, like fine teeth, on the under side, nearly to the tips. The fore legs are often quite spiny.

Most of these moths come forth in July and August, and soon afterwards lay their eggs in the ground, in ploughed fields, gardens, and meadows. In Europe it is found that the eggs are hatched early in the autumn, at which time the little subterranean caterpillars live chiefly on the roots and tender sprouts of herbaceous plants. On the approach of winter they descend deeper into the ground, and, curling themselves up, remain in a torpid state till the following spring, when they ascend towards the surface, and renew their devastations. The caterpillars of the Agrotidians are smooth, shining, naked, and dark-colored, with longitudinal pale and blackish stripes, and a few black dots on each ring; some of them also have a shining, horny, black spot on the top of the first ring. They are of a cylindrical form, tapering a little at each end, rather thick in proportion to their length, and are provided with sixteen legs. They are changed to chrysalids in the ground, without previously making silken cocoons. The most destructive kinds in Europe are the caterpillars of the corn rustic or winter dart-moth (*Agrotis segetum*), the wheat dart-moth (*Agrotis tritici*), the eagle-moth (*Agrotis aquilina*), and the turf rustic or antler-moth (*Charæas graminis* *). The first two attack both the roots and leaves of winter wheat; the second also destroys buckwheat, and it is stated that sixty bushels of mould, taken from a field where they prevailed, contained twenty-three bushels of the caterpillars; those of the eagle-moth occasionally prove very destructive in vineyards; and the caterpillars of the antler-moth are notorious for their devastations in meadows, and particularly in mountain pastures.

* See Kollar's Treatise, pp. 94, 102, 166, and 136.

The habits of our cut-worms appear to be exactly the same as those of the European Agrotidians. It is chiefly during the months of June and July that they are found to be most destructive. Whole corn-fields are sometimes laid waste by them. Cabbage-plants, till they are grown to a considerable size, are very apt to be cut off and destroyed by them. Potato-vines, beans, beets, and various other culinary plants, suffer in the same way. The products of our flower-gardens are not spared; asters, balsams, pinks, and many other kinds of flowers, are often shorn of their leaves and of their central buds, by these concealed spoilers. Several years ago I procured a considerable number of cut-worms (Fig. 220) in the months of June and July. Some of them were dug up among cabbage-plants, some from potato-hills, and others from the corn-field and the flower-garden. Though varying in length from one inch and a quarter to two inches, they were fully grown, and buried themselves immediately in the earth with which they were supplied. They were all thick, greasy-looking caterpillars, of a dark ashen-gray color, with a brown head, a blackish horny spot on the top of the first and last rings, a pale stripe along the back, and several minute black dots on each ring. They were soon changed to chrysalids, of a shining mahogany-brown color; and between the 20th of July and the 15th of August they came out of the ground in the moth state. Much to my surprise, however, these cut-worms produced five different species of moths; and, when it was too late, I regretted that they had not been more carefully examined, and compared together before their transformation.

Fig. 220.

The largest of these moths may be called *Agrotis telifera*, the lance-rustic. It closely resembles *Agrotis suffusa*, the dark sword-rustic of Europe. The fore wings are light brown, shaded with dark brown along the outer thick edge, and in the middle also in the female; these wings are divided into three nearly equal parts by two transverse bands, each composed of two wavy dark brown lines; in the middle space are situated the two ordinary spots, together with a third oval spot, which touches the anterior band; these spots are encircled with dark brown, and the kidney-spot bears a dark brown lance-shaped mark on its hinder part; the hindmost third of the wing is crossed by a broad pale band, and is ornamented by a narrow wavy or festooned line, and several small blackish spots near the margin. The hind wings are pearly white, and semitransparent, shaded behind, and veined with dusky brown. The thorax is brown or gray-brown, with the edge of the collar blackish. The abdomen is gray. The wings expand two inches or more.

Another of these moths is the counterpart of the *æqua* and *agricola* of Europe. It also resembles the *telifera* in form, but is destitute of the lance-shaped spot on the fore wings; and hence I have named it *Agrotis inermis*, the unarmed rustic-moth. The fore wings are light brown, shaded in the middle and towards the hinder margin with dusky brown; they are crossed by four more or less distinct, wavy bands, each formed of two blackish lines; the kidney-spot is dusky; and there are several blackish spots on the outer thick edge of the wing. The hind wings are pearly white in the middle, shaded behind and veined with dusky brown. The thorax is reddish brown, with the collar and shoulder-covers doubly edged with black. The abdomen is gray. It expands two inches.

The reaping rustic (*Agrotis messoria*), as it may be called, is the representative of the corn-rustic (*Agrotis segetum*) of Europe. The fore wings are reddish gray, crossed by five wavy blackish bands, the first two of which, and generally the fourth also, are double; the two ordinary spots, and a third oval spot near the middle of the wing, are bordered with black. The hind wings are whitish, becoming dusky brown behind, and have a small central crescent and the veins dusky. The head and thorax are chinchilla-gray; the collar is edged with black; and the abdomen is light brownish gray. It expands one inch and four tenths.

The smallest of these rustic moths may be called *Agrotis tessellata* (Fig. 221), the checkered rustic. It probably comes near to the *ocellina* and *aquilina* of Europe, which, however, I have not seen. The fore wings are dark ash-colored, and exhibit only a faint trace of the transverse double wavy bands; the two ordinary spots are large and pale, and alternate with a triangular and a square deep black spot; there is a smaller black spot near the base of the wing. The hind wings are brownish gray in the middle, and blackish behind. It expands one inch and one quarter.

Fig. 221.

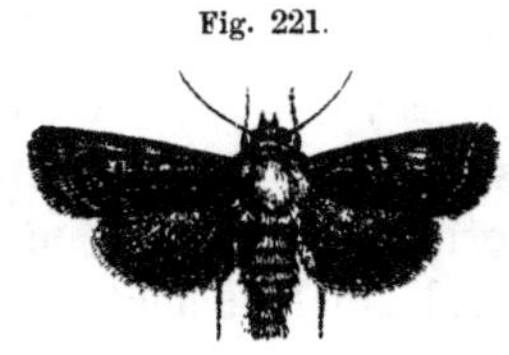

The fifth species I am assured by one of my friends is the moth of the cabbage cut-worm. It agrees, in the main, with the description given of the *Phalæna Noctua devastator*, by Mr. John P. Brace, in the first volume of Professor Silliman's "American Journal of Science"; and may therefore be called *Agrotis devastator*. It somewhat resembles Dr. Boisduval's figures of the *Agrotis latens* of Europe. The fore wings are of a dark ashen-gray color, with a lustre like satin; they are crossed by four narrow wavy whitish bands, which are edged on each side with black; there is a transverse row of white dots followed by a row of black, arrow-shaped spots, between the third and fourth bands, and three white dots on the outer edge near the tip; the ordinary spots are edged with black and white, and there is a third spot, of an oval shape and blackish color, near the middle of the wing, and touching the second band. The hind wings are light brownish gray, almost of a dirty white in the middle, and dusky behind. The head and thorax are chinchilla-gray; and the abdomen is colored like the hind wings. It expands from one inch and five eighths to one inch and three quarters. This kind of moth is very common between the

10th of July and the middle of August. Like all the foregoing species, it flies only at night. According to Mr. Brace, this moth lays its eggs in the beginning of autumn, at the roots of trees, and near the ground; the eggs are hatched early in May; the cut-worms continue their depredations about four weeks, then cast their skin and become pupæ or chrysalids in the earth, a few inches below the surface of the ground; the pupa state lasts four weeks, and the moth comes out about the middle of July; it conceals itself in the crevices of buildings and beneath the bark of trees, and is never seen during the day; about sunset it leaves its hiding-place, is constantly on the wing, is very troublesome about the candles in houses, flies rapidly, and is not easily taken.* From what is known respecting the history of the other kinds of *Agrotis*, and from the size that the cabbage cut-worms are found to have attained in May, I am led to infer that they must generally be hatched in the previous autumn, and that, after feeding awhile on such food as they can find immediately under the surface of the soil, they descend deeper into the ground and remain curled up, in little cavities which each one makes for itself in the earth, till the following spring.

Dr. F. E. Melsheimer, of Dover, Pennsylvania, has favored me with the wing of a moth, which he states is produced from the corn cut-worm. The following remarks on this inseet are extracted from his letters. "There are several species of *Agrotis*, the larvæ of which are injurious to culinary plants; but the chief culprit with us is the same as that which is destructive to young maize." "The corn cut-worms make their appearance in great numbers at irregular periods, and confine themselves in their devastations to no particular vegetables, all that are succulent being relished by these indiscriminate devourers; but, if their choice is not limited, they prefer maize plants when not more than a few inches above the earth, early sown buck-

* American Journal of Science, Vol. I. p. 154.

wheat, young pumpkin-plants, young beans, cabbage-plants, and many other field and garden vegetables." "When first disclosed from the eggs, they subsist on the various grasses. They descend in the ground on the approach of severe frosts, and reappear in the spring about half grown. They seek their food in the night or in cloudy weather, and retire before sunrise into the ground, or beneath stones or any substance which can shelter them from the rays of the sun; here they remain coiled up during the day, except while devouring the food which they generally drag into their places of concealment. Their transformation to pupæ occurs at different periods, sometimes earlier, sometimes later, according to the forwardness of the season, but usually not much later than the middle of July." "The moths, as well as the larvæ, vary much in the depth of their color, from a pale ash to a deep or obscure brown. The ordinary spots of the upper wings of the moth are always connected by a blackish line; where the color is of the deepest shade those spots are scarcely visible, but when the color is lighter they are very obvious."

Since the foregoing was written, I have repeatedly obtained the same moths from cut-worms here. The latter seem, indeed, to be the most common kind; but they differ very little from the cut-worms already described. They vary somewhat in color, as remarked by Dr. Melsheimer. Young ones are always more or less distinctly marked above with pale and dark stripes, and are uniformly paler below. The moth is very abundant in the New England States, from the middle of June till the middle or end of August. The fore wings are generally of a dark ash-color, with only a very faint trace of the double transverse wavy bands that are found in most species of *Agrotis;* the two ordinary spots are small and narrow, the anterior spot being oblong oval, and connected with the oblique kidney-shaped spot by a longitudinal black line. The hind wings are dirty brownish-white, somewhat darker behind. The head, the collar, and the abdomen are chestnut-colored. It expands one inch and three quarters. The wings, when shut, overlap on their inner edges, and cover the top of the back so flatly and closely that these moths can get into very narrow crevices. During the day they lie hidden under the bark of trees, in the chinks of fences, and even under the loose clapboards of buildings. When the blinds of our houses are opened in the morning, a little swarm of these insects, which had crept behind them for concealment, is sometimes exposed, and suddenly aroused from their daily slumber. This kind of moth has the form and general appearance of some species of *Pyrophila*, but not the essential characters of the genus. It differs also from *Agrotis* and *Graphiphora* in some respects, and therefore I have thought it best to leave it, for the present, in the old genus *Noctua*, under the specific name of *clandestina*, the clandestine owlet-moth.

Among the various remedies that have been proposed for preventing the ravages of cut-worms in wheat and corn fields, may be mentioned the soaking of the grain, before planting, in copperas-water and other solutions supposed to be disagreeable to the insects; rolling the seed in lime or ashes; and mixing salt with the manure. These may prevent wire-worms (*Iuli*) and some insects from destroying the seed; but cut-worms prey only on the sprouts and young stalks, and do not eat the seeds. Such stimulating applications may be of some benefit, by promoting a more rapid and vigorous growth of the grain, by which means the sprouts will the sooner become so strong and rank as to resist or escape the attacks of the young cut-worms. Fall-ploughing of sward-lands, which are intended to be sown with wheat or planted with corn the year following, will turn up and expose the insects to the inclemency of winter, whereby many of them will be killed, and will also bring them within reach of insect-eating birds. But this seems to be a doubtful remedy, against which many objections have been urged.*

* See Mr. Colman's Third Report of the Agriculture of Massachusetts, p. 62.

The only effectual remedy at present known has been humorously described by Mr. Asahel Foote in the "Albany Cultivator," and reprinted in the seventeenth volume of the "New England Farmer." After having lost more than a tenth part of the corn in his field, he "ordered his men to prepare for war, to sharpen their finger-ends, and set at once about exhuming the marauders. For several days it seemed as if a whole procession came to each one's funeral, but at length victory wreathed the brow of perseverance; and, the precaution having been taken to replace each foe dislodged with a suitable quantity of good seed-corn, he soon had the pleasure to see his field restored, in a good measure, to its original order and beauty, there being seldom a vacancy in a piece of four acres." Mr. Foote's statement, founded on an estimate of the time employed in digging up and killing the cut-worms, and the increased produce of the field, is conclusive in favor of this mode of checking the ravages of these insects.

Mr. Deane states that he "once prevented the depredations of cut-worms in his garden by manuring the soil with sea-mud. The plants generally escaped, though every one was cut off in a spot of ground contiguous." He acknowledges, however, that "the most effectual, and not a laborious remedy, even in field-culture, is to go round every morning, and open the earth at the foot of the plant, and you will never fail to find the worm at the root, within four inches. Kill him, and you will save not only the other plants of your field, but, probably, many thousands in future years." Mr. Preston, of Stockport, Pennsylvania, protected his cabbage-plants from cut-worms by wrapping a walnut or hickory leaf around the stem, between the roots and leaves, before planting it in the ground. The late Honorable Oliver Fiske, of Worcester, Massachusetts, says, that "to search out the spoiler, and kill him, is the very best course; but, as his existence is not known except by his ravages, I make a fortress for my cabbage-plants with paper, winding it conically and firmly above the root, and securing it by a low embankment of earth."

In the summer of 1851, one of our agricultural newspapers contained an account of certain naked caterpillars, that came out of the ground in the night, and, crawling up the trunks of fruit-trees, devoured the leaves, and returned to conceal themselves in the ground before morning.* Perhaps these depredators were the same as the following. Roses, currant-bushes, and other shrubs, and even young trees, often lose their tender shoots, by having them cut off and devoured during the night. This is the work of a naked caterpillar, which generally grows to a larger size than the common cut-worm, and, like the latter, may be found by digging at the root of the plant. One of these spoilers, which was turned out of his burrow early in June, measured an inch and a half in length. His body was livid or brownish and shining above, with a chestnut-colored head, and a horny spot of the same color on the top of the first and last rings. A few minute dots, producing very short inconspicuous hairs, were regularly disposed upon his body. This caterpillar changed to a chrysalis in the ground, and was transformed to a moth (Fig. 222) on the 1st of July. The moth very often enters houses in the evening, during the months of July and August, and, in its restrained flight, keeps bobbing against the ceiling and walls. When it alights, it sits with its wings sloping in the form of a steep roof. It is easily distinguished by its Spanish-brown upper wings, marked with a large pale kidney-spot, and a broad wavy blue-gray band near the end. Its eyes when living shine like coals of fire. It has been described by mistake

Fig. 222.

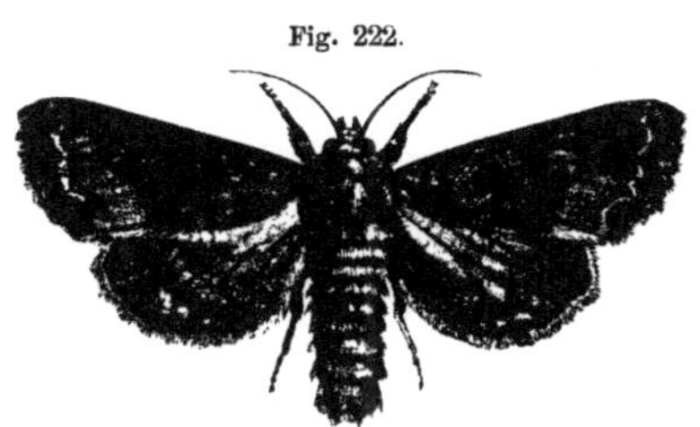

* See Massachusetts Ploughman for June 28, 1851.

as a British species, under the name of *Hadena amica*, or the barred arches-moth. The wings of this moth expand an inch and three quarters, or more, and are proportionally broader than those of the cut-worm moths. The general color of the fore wings, as already stated, is deep Spanish-brown, variegated with gray. The small ordinary oval spot is marked by a gray border. The kidney-spot is large, gray, and very conspicuous. There is a broad wavy band of a pearl-gray or blue-gray color near the outer hind margin, and a narrow wavy band between the oval spot and shoulder. The hind wings are pale ash-colored, shaded behind with brown, having a pale border, and a distinct central blackish spot beneath. The head and thorax are dark brown; the collar and tips of the shoulder-covers are edged with rust-red; and the hind body is ash-colored or pale brown, with a row of four rust-red tufts upon it. This common moth belongs to the same group or family as the following species, though differing therefrom in its caterpillar state.

There is another naked caterpillar (Fig. 223) which is often found to be injurious to cabbages, cauliflowers, spinach, beets, and other garden vegetables with succulent leaves. It does not conceal itself in the ground, but lives exposed on the leaves of the plants which it devours. When disturbed, it coils its body spirally. It is of a light yellow color, with three broad, longitudinal, black stripes, one on each side and the third on the top of the back; and the head, belly, and feet are tawny. The lateral black stripe is worthy of attentive examination. It consists of numerous transverse black marks somewhat like Runic letters, on a pure white ground; but the white ground, when seen without a glass, seems blue, by contrast with the black characters. Dr. Melsheimer calls this the zebra caterpillar, on account of its stripes. It comes to its full size here in

Fig. 223.

September, and then measures about two inches in length. Early in October it leaves off eating, goes into the ground, changes to a shining brown chrysalis (Fig. 224), and is transformed to a moth about the first of June. It is probable that there are two broods of this kind of caterpillar every summer, in some, if not all, parts of this country; for Dr. Melsheimer informs me that it appears in Pennsylvania in June, goes into the ground and is changed to a chrysalis towards the end of June or the beginning of July, and comes forth in the moth state near the end of August. The moth may be called *Mamestra picta*, the painted Mamestra, in allusion both to the beautiful tints of the caterpillar, and to the softly blended shades of dark and light brown with which the fore wings of the moth are colored. It is of a light brown color, shaded with purple-brown; the ordinary spots on the fore wings, with a third oval spot behind the round one, are edged with gray; and there is a transverse zigzag gray line, forming a distinct W in the middle, near the outer hind margin. The hind wings are white, and faintly edged with brown around the tip. It is evident that this insect cannot be included in either of the foregoing groups of the owlet-moths. It belongs to a distinct family, which may be called MAMESTRADÆ, or Mamestrians. The caterpillars in this group are generally distinguished by their bright colors; they live more or less exposed on the leaves of plants, and transform in the ground. The moths fly by night only; most of them have the thorax slightly crested; and they are easily known by the zigzag line, near the outer hind margin of the fore wings, forming a W or M in the middle.

Fig. 224.

As the caterpillar of the painted Mamestra does not seek concealment, it may easily be found, and destroyed by hand.

There is a small caterpillar which has been found injurious to the wheat-crop in England, by eating the grain before and after it is ripe. It is described and figured by

Mr. John Curtis, in the fifth volume of the "Journal of the Royal Agricultural Society of England" (pp. 477–481). Though unable to rear any of these caterpillars, which always shrivelled up and died, Mr. Curtis, for reasons stated by him, was impressed with the conviction that they were produced by a moth called *Noctua* (*Caradrina*) *cubicularis*. Our agricultural newspapers contain accounts of certain caterpillars, much like the foregoing in appearance and in habits, which devour the grains of wheat while growing and after being harvested. Their transformations have not been ascertained; and, on account of the diminutive size of these caterpillars, it remains uncertain whether they are the offspring of any species of *Noctua*. Nevertheless, this seems to be the most suitable place to record what has been said and seen of them. They have been called wheat-worms, gray worms, and brown weevils; and, although these different names may possibly refer to two or more distinct species, I am inclined to believe that all of them are intended for only one kind of insect. The name of grain-worms has likewise sometimes been applied to them; whereby it becomes somewhat difficult to separate the accounts of their history and depredations from those of the wheat-insect, called *Cecidomyia Tritici*. It may, however, very safely be asserted, that the wheat-worm of the western part of New York and of the northern part of Pennsylvania is entirely distinct from the maggots of our wheat-fly, and that it does not belong to the same order of insects.

Mr. Willis Gaylord described this depredator as a kind of caterpillar, or span-worm, from three to five eighths of an inch long, of a yellowish-brown or butternut color, provided with twelve legs, and having the power of spinning and suspending itself by a thread. He stated that it not only fed on the kernel in the milky state, but also devoured the germinating end of the ripened grain, without, however, burying itself within the hull; and that it was found, in great numbers, in the chaff, when the grain was threshed.

According to him, it had been known for years in the western part of New York; and it was not so much the new appearance of the insect, as its increase, which had caused alarm respecting it.* Mr. Nathaniel Sill, of Warren, Pennsylvania, has given a somewhat different description of it.† On threshing his winter-wheat, immediately after harvest, he found among the screenings a vast army of this new enemy. He says that it was a caterpillar, about three eighths of an inch in length when fully grown, and apparently of a straw color; but, when seen through a magnifier, it was found to be striped lengthwise with orange and cream-color. Its head was dark brown. It was provided with legs, could suspend itself by a thread, and resembled a caterpillar in all its motions.

This insect ought not to be confounded with the smaller worms found by Mr. Sill in the upper joints of the stems of the wheat, and within the kernels, until their identity has been proved by further observations. It appears highly probable that Mr. Gaylord's and Mr. Sill's wheat-caterpillars are the same, notwithstanding the difference in their color. Insects, of the same size as these caterpillars, and of a brownish color, have been found in various parts of Maine, where they have done much injury to the grain. Unlike the maggots of the wheat-fly, with which they have been confounded, they remain depredating upon the ears of the grain until after the time of harvest. Immense numbers of them have been seen upon barn-floors, where the grain has been threshed, but they soon crawl away and conceal themselves in crevices, where they probably undergo their transformations. Mr. Elijah Wood, of Winthrop, Maine, says that the chrysalis has been observed in the chaff late in the fall.‡ A gentleman from the southern part of Penobscot County informs me that he winnowed out nearly a bushel of these insects from his wheat, in the autumn of

* The Cultivator, Vol. VI. p. 43. † Ibid. p. 21.

‡ New England Farmer, Vol. XVII. p. 73.

1840; and he confirms the statements of others, that these worms devour the grain when in the milk, and also after it has become hard. In the autumn of 1838, the Rev. Henry Colman observed the same insect in the town of Egremont, in Berkshire County, Massachusetts. It was separated from the wheat, in great quantities, by threshing and winnowing the grain.*

On the 26th of September, 1846, my brother brought to me a sample of wheat-ears, from Dixmont, Maine, containing five of these insects, of different sizes. The largest measured five eighths of an inch in length, when fully extended. It was a very slender caterpillar, having sixteen legs, and was not a true span-worm either in structure or motions. It was of a pale reddish-brown color, with three longitudinal paler or colorless lines on the back, and a broader pale stripe on each side of the body. The head and the tops of the first and last segments were shining brown. A few minute black points (each furnishing a short inconspicuous hair) were regularly disposed on each segment. The body beneath and all the legs were pale brownish-red. Many of the kernels of wheat had been gnawed by these caterpillars; but they refused to eat any more, and died without change. In the summer of 1850, Dr. Ovid Plumb had the kindness to send to me some younger specimens of these caterpillars, from Salisbury, Connecticut, where they had long prevailed in the wheat-fields; and I saw them in the wheat at the same place, on the 25th of July, 1851. They had grown only to the length of three sixteenths or one fourth of an inch at most; but they resembled the larger specimens from Maine in all essential particulars. They were too young and delicate to survive the effects of a journey without fresh food, which could not be procured for them after my return. When disturbed, they readily suspended themselves by a slender thread, were very uneasy on being taken from the ears, and were quick in

* Second Report on the Agriculture of Massachusetts, p. 99.

all their motions. Previous accounts concerning their habits and depredations were fully confirmed by observations and information at Salisbury.

These wheat-worms, or wheat-caterpillars, as they ought to be called, if these accounts really refer to the same kind of insect, are supposed by some persons to be identical with the clover-worms, which have been found in clover, in various parts of the country, and have often been seen spinning down from lofts and mows where clover has been stowed away.* A striking similarity between them has been noticed by a writer in the "Genesee Farmer."† Stephen Sibley, Esq. informs me that he observed the clover-worms, in Hopkinton, New Hampshire, many years ago, suspended in such numbers by their threads from a newly gathered clover mow, and from the timbers of the building, as to be very troublesome and offensive to persons passing through the barn. He also states, that, if he recollects rightly, these insects were of a brown color, and about half an inch long.

I am sorry to leave the history of these wheat-worms unfinished; but hope that the foregoing statements, which have been carefully collected from various sources and compared with my own observations, will tend to remove some of the difficulties wherewith the subject has been heretofore involved. The contradictory statements and unsatisfactory discussions that have appeared in some of our papers, respecting the ravages of these worms and the maggots of the wheat-fly, might have been avoided, if the writers on these insects had always been careful to give a correct and full description of the insects in question. Had this been done, a crawling worm or caterpillar, of a brownish color, three eighths or one half of an inch in length, provided with legs, and capable of suspending itself by a silken thread of its own spinning, would never have been mistaken for a writhing maggot, of a deep yellow color, only one tenth

* New England Farmer, Vol. XVII. p. 73.

† Ibid., p. 164.

of an inch long, destitute of legs, and unable to spin a thread. As these destructive wheat-caterpillars may be separated from the wheat by threshing and winnowing, the chaff containing them may be put into large tubs, into which also a sufficient quantity of boiling-hot water may then be poured to kill all the insects. This will at least prevent their making their escape, completing their transformations, and laying the foundation of another brood.

At the end of the tribe of owlet-moths may be arranged certain insects, which, from the structure of their caterpillars and their manner of creeping, evidently seem to connect this tribe with the Geometers. Some of these caterpillars have the first, and sometimes also the second, pair of prop-legs, under the middle of the body, so short, that they cannot be used in creeping; others have only twelve or fourteen legs, the first pair of the prop-legs, or the second also, being entirely wanting in them. These caterpillars creep with a kind of halting gait, and arch up the middle of the body, more or less, with every step they take, thereby imitating the gait of the true geometers or span-worms. To this group belong the army-worms, or cotton-worms, which ravage the cotton-fields of the Southern States. They have sixteen legs; but the foremost prop-legs are shorter than the rest, and the caterpillars crook their backs in creeping, which has caused them to be mistaken for geometers by some writers. The cotton-worm is green, doubly striped with black on the back, and sprinkled with black dots. It grows to the length of an inch and a half, transforms in a kind of web or imperfect cocoon, and becomes an olive-brown moth, called *Noctua xylina* by Mr. Say. It is found only as far as the cotton-plant is cultivated, and never occurs in New England. The twelve-legged caterpillars are sometimes injurious to cultivated vegetables; but not enough so, in this country, to have attracted much notice. Their moths are distinguished by golden or silvery spots on their fore wings. The species, with the first and second pairs of prop-legs short and rudimentary, feed mostly on the leaves of shrubs and trees; their moths are of large size, with the hind wings often crimson, scarlet, or yellow, and traversed by black bands. But as these insects are not particularly interesting to the farmer, any further account of them, in this treatise, will be unnecessary.

3. Geometers. (*Geometræ.*)

The caterpillars of the Geometræ of Linnæus, earth-measurers, as the term implies, or geometers, span-worms, and loopers, have received these several names from their peculiar manner of moving, in which they seem to measure or span over the ground, step by step, as they proceed. Most of these caterpillars have only ten legs; namely, six, which are jointed and tapering, under the fore part of the body, and four fleshy prop-legs, at the hinder extremity; the three intermediate pairs of prop-legs being wanting. Consequently, in creeping, they arch up the back while they bring forward the hinder part of the body, and then, resting on their hind legs, stretch out to their full length, in a straight line, before taking another step with their hind legs. Some of the Geometers have twelve or fourteen legs; but the additional prop-legs are so short that the caterpillars cannot use them in creeping, and their motions are the same as those that have only ten legs. Some caterpillars with fourteen legs, and wanting only the terminal pair of prop-legs, are placed in this tribe, on account of the resemblance of their moths to those of the true Geometers.

The latter live on trees and bushes, and most of them undergo their transformations upon or in the ground, to reach which, by travelling along the branches and down the stem, would be a long and tedious journey to them, on account of the deficiency of their legs, and the slowness of their gait. But they are not reduced to this necessity; for they have the power of letting themselves down from any height, by means of a silken thread, which they spin from their mouths while falling. Whenever they are disturbed they make use of this faculty, drop suddenly, and hang suspended till the danger is past, after which they climb up again by the same thread. In order to do this, the span-worm bends back its head and catches hold of the thread above its head with one of the legs of the third segment, then, raising its head, it seizes the thread with its jaws and fore legs, and, by repeating the same operations with tolerable rapidity, it soon reaches its former station on the tree. These span-worms are naked, or only thinly covered with very short down; they are mostly smooth, but sometimes have warts or irregular projections on their backs. They change their color usually as they grow older, are sometimes striped, and sometimes of one uniform color, nearly resembling the bark of the plants on which they are found. When not eating, many of them rest on the two hindmost pairs of legs against the side of a branch, with the body extended from the branch, so that they might be mistaken for a twig of the tree; and in this position they will often remain for hours together.

When about to transform, most of these insects descend from the plants on which they live, and either bury themselves in the ground, or conceal themselves on the surface under a slight covering of leaves fastened together with silken threads. Some make more regular cocoons, which, however, are very thin, and generally more or less covered on the outside with leaves. The cocoons of the European, tailed Geometer (*Ourapteryx sambucaria*), which lives on the elder, and of our chain-dotted Geometer (*Geometra catenaria*), (Fig. 225, Fig. 226 cocoon, Fig. 227 larva,) which is found on the wood-wax, are made with regular meshes, like

Fig. 225

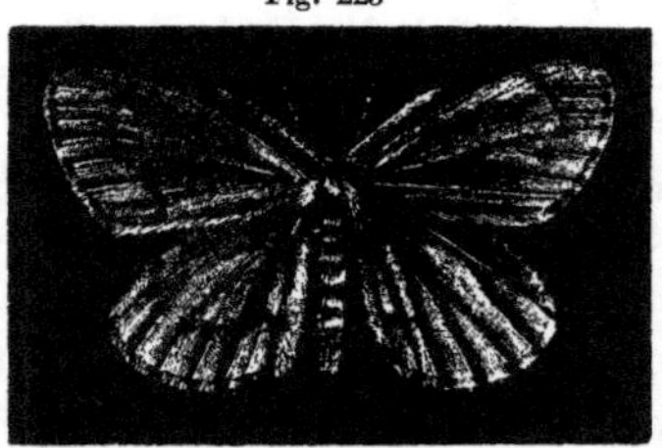

net-work, through which the insects may be seen. A very few of the span-worms fasten themselves to the stems of plants, and are changed to chrysalids, which hang suspended, without the protection of any outer covering.

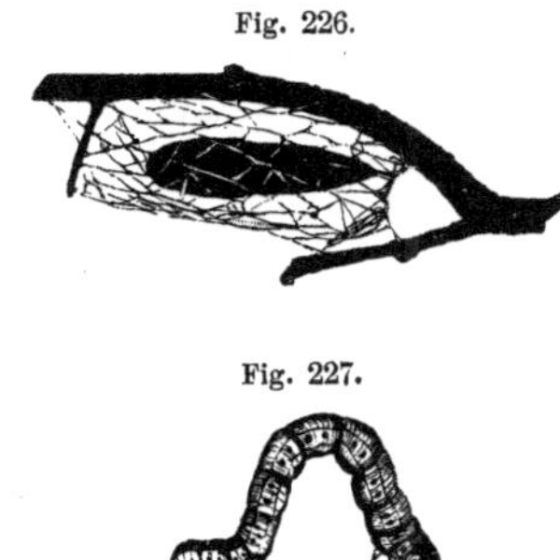

Fig. 226.

Fig. 227.

In their perfected state, these insects are mostly slender-bodied moths, with tapering antennæ, which are often feathered in the males. Their feelers are short and slender; the tongue is short and weak; the thorax is not crested; the wings are large, thin, and delicate, sometimes angular, and often marked with one or two dark-colored oblique bands. They generally rest with the wings slightly inclined, and almost horizontal; some with them extended, and others with the hind wings covered by the upper pair. A very few carry their wings like the Skippers. Some of the females are without wings, and are distinguished also by the oval and robust form of their bodies. These moths are most active in the night; but some of them may be seen flying in thickets during the day-time. They are very short-lived, and die soon after their eggs are laid.

Those kinds, whereof the females are wingless, or have only very short, scale-like wings, and naked antennæ, while the males have large, entire wings, and feathered or downy antennæ, seem to form a distinct group, which may be named Hybernians (HYBERNIADÆ), from the principal genus included therein. The caterpillars have only ten legs, six before and four behind; and they undergo their transformations in the ground. The insects called canker-worms in this country, are of this kind. The moths from which they are produced belong to the genus *Anisopteryx*,* so named because in some species the wings in the two sexes are very

* Literally *unequal wing*.

unequal in size, and in others the females are wingless. Among those whose females are wingless are the canker-worm moths. In the late Professor Peck's "Natural History of the Canker-Worm," which was published among the papers of "the Massachusetts Society for Promoting Agriculture," and obtained a prize from the Society, this insect is called *Phalæna vernata*, on account of its common appearance in the spring, and also to distinguish it from the winter moth (*Phalæna* or *Cheimatobia brumata*) of Europe. In the male canker-worm moth (Fig. 228) the antennæ have a very narrow, and almost downy edging, on each side, hardly to be seen with the naked eye. The feelers are minute, and do not extend beyond the mouth. The tongue is not visible. The wings are large, very thin, and silky; and, when the insect is at rest, the fore wings are turned back, entirely cover the hind wings, and overlap on their inner edges. The fore wings are ash-colored, with a distinct whitish spot on the front edge, near the tip; they are crossed by two jagged, whitish bands, along the sides of which there are several blackish dots; the outermost band has an angle near the front edge, within which there is a short, faint, blackish line; and there is a row of black dots along the outer margin, close to the fringe. The hind wings are pale ash-colored, with a faint blackish dot near the middle. The wings expand about one inch and a quarter.

Fig. 228.

This is the usual appearance of the male, in its most perfect condition; by which it will be seen that it closely resembles the *Anisopteryx Æscularia* of Europe. Compared with the latter, I find that our canker-worm moth is rather smaller, the wings are darker, proportionally shorter and more obtuse, the white bands are less distinct, and are often entirely wanting, in which case only the whitish spot near the tip remains, the hind wings are more dusky, and the feelers are gray instead of being white. Specimens of a rather smaller size are sometimes found, resembling the figure and description given by Professor Peck, in which the whitish bands and spot are wanting, and there are three interrupted dusky lines across the fore wings, with an oblique blackish dash near the tip. Perhaps they constitute a different species from that of the true canker-worm moth. Should this be the case, the latter may be called *Anisopteryx pometaria*, or the Anisopteryx of the orchard, while the former should retain the name originally given to it by Professor Peck. The female is wingless, and its antennæ are short, slender, and naked. Its body approaches to an oval form, but tapers and is turned up behind. It is dark ash-colored above, and gray beneath.

It was formerly supposed that the canker-worm moths came out of the ground only in the spring. It is now known that many of them rise in the autumn and in the early part of the winter. In mild and open winters I have seen them in every month from October to March. They begin to make their appearance after the first hard frosts in the autumn, usually towards the end of October, and they continue to come forth, in greater or smaller numbers, according to the mildness or severity of the weather after the frosts have begun. Their general time of rising is in the spring, beginning about the middle of March, but sometimes before, and sometimes after, this time; and they continue to come forth for the space of about three weeks. It has been observed that there are more females than males among those that appear in the autumn and winter, and that the males are most abundant in the spring. The sluggish females (Fig. 229) instinctively make their way towards the nearest trees, and creep slowly up their trunks. In a few days afterwards they are followed by the winged and active males, which flutter about and accompany them in their ascent, during which the insects pair. Soon after this, the

Fig. 229.

females lay their eggs (Fig. 230, natural size and magnified) upon the branches of the trees, placing them on their ends, close together in rows, forming clusters of from sixty to one hundred eggs or more, which is the number usually laid by each female. The eggs are glued to each other, and to the bark, by a grayish varnish, which is impervious to water; and the clusters are thus securely fastened in the forks of the small branches, or close to the young twigs and buds. Immediately after the insects have thus provided for a succession of their kind, they begin to languish, and soon die. The eggs are usually hatched between the first and the middle of May, or about the time that the red currant is in blossom, and the young leaves of the apple-tree begin to start from the bud and grow. The little canker-worms, upon making their escape from the eggs, gather upon the tender leaves, and, on the occurrence of cold and wet weather, creep for shelter into the bosom of the bud, or into the flowers, when the latter appear. As this treatise may fall into the hands of persons who are not acquainted with the habits and devastations of our canker-worms, it should be stated that, where these insects prevail, they are most abundant on apple and elm trees; but that cherry, plum, and lime trees, and some other cultivated and native trees, as well as many shrubs, often suffer severely from their voracity. The leaves first attacked will be found pierced with small holes; these become larger and more irregular when the canker-worms increase in size; and, at last, the latter eat nearly all the pulpy parts of the leaves, leaving little more than the midrib and veins.

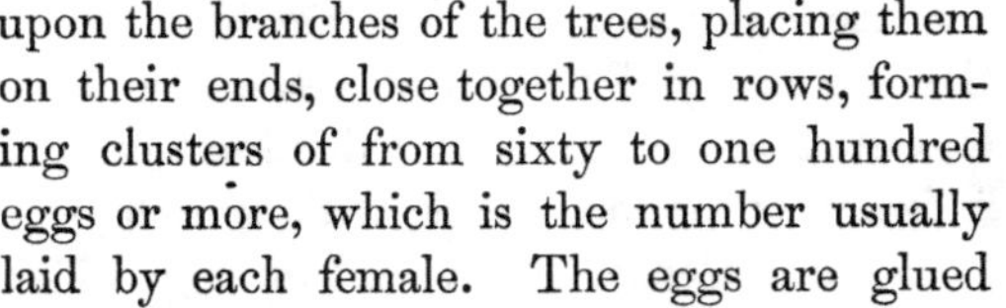

Fig. 230.

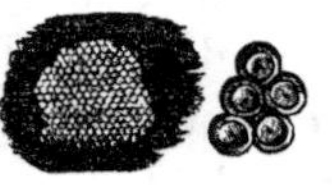

A very great difference of color is observable among canker-worms of different ages, and even among those of the same age and size. It is possible that some of these variations may arise from a difference of species; but it is also true that the same species varies much in color. When very young, they have two minute warts on the top of the last ring; and they are then generally of a blackish or dusky-brown color, with a yellowish stripe on each side of the body; there are two whitish bands across the head; and the belly is also whitish. When fully grown (Fig. 231), these individuals become ash-colored on the back, and black on the sides, below which the pale yellowish line remains. Some are found of a dull greenish-yellow and others of a clay color, with slender interrupted blackish lines on the sides, and small spots of the same color on the back. Some are green, with two white stripes on the back. The head and the feet partake of the general color of the body; the belly is paler. When not eating, they remain stretched out at full length, and resting on their fore and hind legs, beneath the leaves. When fully grown and well fed, they measure nearly or quite one inch in length. They leave off eating when about four weeks old,* and begin to quit the trees; some creep down by the trunk, but great numbers let themselves down by their threads from the branches, their instincts prompting them to get to the ground by the most direct and easiest course. When thus descending, and suspended in great numbers under the limbs of trees overhanging the road, they are often swept off by passing carriages, and are thus conveyed to other places. After reaching the ground, they immediately burrow in the earth, to the depth of from two to six inches, unless prevented by weakness or the nature of the soil. In the latter case, they die, or undergo their transformations on the surface. In the former, they make little cavities or cells (Fig. 232) in the ground, by turning round repeatedly and fastening the loose grains of earth about them with a few silken threads. Within twenty-four

Fig. 231.

Fig. 232.

* In the year 1841 the red currant flowered and the canker-worms appeared on the 15th of May. The insects were very abundant on the 15th of June, and on the 17th scarcely one was to be seen.

hours afterwards, they are changed to chrysalids (Fig. 233) in their cells.

Fig. 233.

The chrysalis is of a light-brown color, and varies in size according to the sex of the insect contained in it; that of the female being the largest, and being destitute of a covering for wings, which is found in the chrysalis of the males. The occurrence of mild weather after a severe frost stimulates some of these insects to burst their chrysalis skins and come forth in the perfected state; and this last transformation, as before stated, may take place in the autumn, or in the course of the winter, as well as in the spring; it is also retarded, in some individuals, for a year or more beyond the usual time. They come out of the ground mostly in the night, when they may be seen struggling through the grass as far as the limbs extend from the body of the trees under which they had been buried. As the females are destitute of wings, they are not able to wander far from the trees upon which they have lived in the caterpillar state. Canker-worms are therefore naturally confined to a very limited space, from which they spread year after year. Accident, however, will often carry them far from their native haunts, and in this way, probably, they have extended to places remote from each other. Where they have become established, and have been neglected, their ravages are often very great. In the early part of the season, the canker-worms do not attract much attention; but it is in June, when they become extremely voracious, that the mischief they have done is rendered apparent, when we have before us the melancholy sight of the foliage of our fruit-trees and of our noble elms[25] reduced to withered and lifeless shreds, and whole orchards looking as if they had been suddenly scorched with fire.

[[25] The insect which ravages the foliage of our "noble elm" in the South is the larva of a beetle, *Galeruca calmariensis*, and hence the precautions here recommended are inapplicable. The female flies upon the leaves to deposit her eggs,

In order to protect our trees from the ravages of canker-worms, where these looping spoilers abound, it should be our aim, if possible, to prevent the wingless females from ascending the trees to deposit their eggs. This can be done by the application of tar around the body of the tree, either directly on the bark, as has been the most common practice, or, what is better, over a broad belt of clay-mortar, or on strips of old canvas or of strong paper, from six to twelve inches wide, fastened around the trunk with strings. The tar must be applied as early as the first of November, and perhaps in October, and it should be renewed daily as long as the insects continue rising; after which the bands may be removed, and the tar should be entirely scraped from the bark. When all this has been properly and seasonably done, it has proved effectual. The time, labor, and expense attending the use of tar, and the injury that it does to the trees when allowed to run and remain on the bark, have caused many persons to neglect this method, and some to try various modifications of it, and other expedients.

Among the modifications may be mentioned a horizontal and close-fitting collar of boards, fastened around the trunk, and smeared beneath with tar; or four boards nailed together, like a box without top or bottom, around the base of the tree, to receive the tar on the outside. These can be used to protect a few choice trees in a garden, or around a house or a public square, but will be found too expensive to be applied to any great extent. Collars of tin-plate fastened around the trees, and sloping downwards like an inverted tunnel, have been proposed, upon the supposition that the moths would not be able to creep in an inverted position, beneath the smooth and sloping surface. This method will also prove too expensive for general adoption, even should it be

and does not crawl up the trunk like the apterous female of *Anisopteryx*. Some persons hearing of the New England method, and presuming that the insects were the same, adopted the plan here recommended, but of course it failed. They were taught better, and now squirt a decoction of tobacco-leaves on the trees, which is an effectual antidote, when the trees are not too high. — MORRIS.]

found to answer the purpose. A belt of cotton-wool, which it has been thought would entangle the feet of the insects, and thus keep them from ascending the trees, has not proved an effectual bar to them.

Little square or circular troughs of tin or of lead, filled with cheap fish-oil, and placed around the trees, three feet or more above the surface of the ground, with a stuffing of cloth, hay, or sea-weed between them and the trunk, have long been used by various persons in Massachusetts with good success; and the only objections to them are the cost of the troughs, the difficulty of fixing and keeping them in their places, and the injury suffered by the trees when the oil is washed or blown out and falls upon the bark. Mr. Jonathan Dennis, Jr., of Portsmouth, Rhode Island, has obtained a patent for a circular leaden trough to contain oil, offering some advantages over those that have heretofore been used, although it does not entirely prevent the escape of the oil, and the nails, with which it is secured, are found to be injurious to the trees. These troughs ought not to be nailed to the trees, but should be supported by a few wooden wedges driven between them and the trunks. A stuffing of cloth, cotton, or tow should never be used; sea-weed and fine hay, which will not absorb the oil, are much better. Before the troughs are fastened and filled, the body of the tree should be well coated with clay paint or whitewash, to absorb the oil that may fall upon it. Care should be taken to renew the oil as often as it escapes or becomes filled with the insects. These troughs will be found more economical and less troublesome than the application of tar, and may safely be recommended and employed, if proper attention is given to the precautions above named. Some persons fasten similar troughs, to contain oil, around the outer sides of an open box enclosing the base of the tree, and a projecting ledge is nailed on the edge of the box to shed the rain; by this contrivance, all danger of hurting the tree with the oil is entirely avoided.

In the "Manchester Guardian," an English newspaper, of the 4th of November, 1840, is the following article on the use of melted Indian rubber to prevent insects from climbing up trees. "At a late meeting of the Entomological Society, [of London?] Mr. J. H. Fennell communicated the following successful mode of preventing insects ascending the trunks of fruit-trees. Let a piece of Indian rubber be burnt over a gallipot, into which it will gradually drop in the condition of a viscid juice, which state, it appears, it will always retain; for Mr. Fennell has, at the present time, some which has been melted for upwards of a year, and has been exposed to all weathers without undergoing the slightest change. Having melted the Indian rubber, let a piece of cord or worsted be smeared with it, and then tied several times round the trunk. The melted substance is so very sticky, that the insects will be prevented, and generally captured, in their attempts to pass over it. About three pennyworth of Indian rubber is sufficient for the protection of twenty ordinary-sized fruit-trees." Applied in this way it would not be sufficient to keep the canker-worm moths from getting up the trees; for the first-comers would soon bridge over the cord with their bodies, and thus afford a passage to their followers. To insure success, it should be melted in larger quantities, and daubed with a brush upon strips of cloth or paper, fastened round the trunks of the trees. Worn out Indian rubber shoes, which are worth little or nothing for any other purpose, can be put to this use. This plan has been tried by a few persons in the vicinity of Boston, some of whom speak favorably of it. It has been suggested that the melted rubber might be applied immediately to the bark without injuring the trees. A little conical mound of sand surrounding the base of the tree is found to be impassable to the moths, so long as the sand remains dry; but they easily pass over it when the sand is wet, and they come out of the ground in wet as often as in dry weather.

Some attempts have been made to destroy the canker

worms after they were hatched from the eggs, and were dispersed over the leaves of the trees. It is said that some persons have saved their trees from these insects by freely dusting air-slacked lime over them while the leaves were wet with dew. Showering the trees with mixtures that are found useful to destroy other insects has been tried by a few, and, although attended with a good deal of trouble and expense, it may be worth our while to apply such remedies upon small and choice trees. Mr. David Haggerston, of Watertown, Mass., has used, for this purpose, a mixture of water and oil-soap (an article to be procured from the manufactories where whale-oil is purified), in the proportion of one pound of the soap to seven gallons of water; and he states that this liquor, when thrown on the trees with a garden engine, will destroy the canker-worm and many other insects, without injuring the foliage or the fruit. This application may be found useful in protecting grafts; for if canker-worms attack these, they will very much injure, if not entirely destroy them. Jarring or shaking the limbs of the trees will disturb the canker-worms, and cause many of them to spin down, when their threads may be broken off with a pole; and if the troughs around the trees are at the same time replenished with oil, or the tar is again applied, the insects will be caught in their attempts to creep up the trunks. In the same way, also, those that are coming down the trunks to go into the ground will be caught and killed. If greater pains were to be taken to destroy the insects in the caterpillar state, their numbers would soon greatly diminish.

Even after they have left the trees, have gone into the ground, and have changed their forms, they are not wholly beyond the reach of means for destroying them. One person told me that his swine, which he was in the habit of turning into his orchard in the autumn, rooted up and killed great numbers of the chrysalids of the canker-worms. Some persons have recommended digging or ploughing under the trees, in the autumn, with the hope of crushing some of the chrysalids by so doing, and of exposing others to perish with the cold of the following winter. If hogs are then allowed to go among the trees, and a few grains of corn are scattered on the loosened soil, these animals will eat many of the chrysalids as well as the corn, and will crush others with their feet. Mr. S. P. Fowler* thinks it better to dig around the trees in July, while the shells of the insects are soft and tender. He and Mr. John Kenrick, of Newton, Mass., advise us to remove the soil to the distance of four or five feet from the trunk of the trees, and to the depth of six inches, to cart it away and replace it with an equal quantity of compost or rich earth. In this way, many of the insects will be removed also; but unless the earth, thus carried away, is thrown into some pond-hole, and left covered with water, many of the insects contained in it will undergo their transformations and come out alive the next year.

Canker-worms are subject to the attacks of many enemies. Great numbers of them are devoured by several kinds of birds, which live almost entirely upon them during their season. They are also eaten by a very large and splendid ground-beetle (*Calosoma scrutator*), (Fig. 234,) that appears about the time when these insects begin to leave the trees. These beetles do not fly, but they run about in the grass after the canker-worms, and even mount upon the trunks of the trees to seize them as they come down. The potter-wasp (*Eumenes*

Fig. 234.

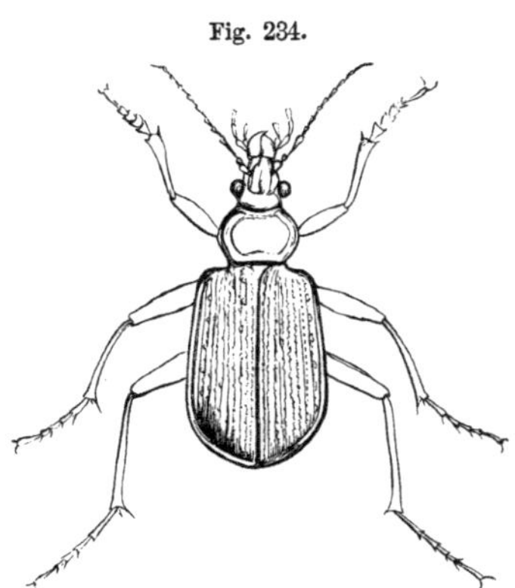

* See Yankee Farmer of July 18, 1840, and New England Farmer of June 2, 1841, for some valuable remarks by Mr. Fowler.

fraterna), an insect rather smaller than the common brown wasp, fills her clay cells with canker-worms, often gathering eighteen or twenty of them as food for her young.* A four-winged ichneumon-fly also stings them, and deposits an egg in every canker-worm thus wounded. From the egg is hatched a little maggot, that preys on the fatty substance of the canker-worm, and weakens it so much that it is unable to go through its future transformations. I have seen one of these flies sting several canker-worms in succession, and swarms of them may be observed around the trees as long as the canker-worms remain. Their services, therefore, are doubtless very considerable. Among a large number of canker-worms, taken promiscuously from various trees, I found that nearly one third of the whole were unable to finish their transformations, because they had been attacked by internal enemies of another kind. These were little maggots, that lived singly within the bodies of the canker-worms, till the latter died from weakness; after which the maggots underwent a change, and finally came out of the bodies of their victims in the form of small two-winged cuckoo-flies, belonging to the genus *Tachina*.

Mr. E. C. Herrick, of New Haven, Connecticut, has made the interesting discovery that the eggs of the canker-worm moth are pierced by a tiny four-winged fly (Fig. 235, greatly magnified), a species of *Platygaster*, which goes from egg to egg, and drops in each of them one of her own eggs. Sometimes every canker-worm egg in a cluster will be found to have been thus punctured and seeded for a future harvest of the *Platygaster*. The young of this *Platygaster* is an exceedingly minute maggot, hatched within the canker-worm egg, the shell of which, though only one thirtieth of an inch long, serves for its habitation, and the contents for its food, till it is fully grown; after which it

Fig. 235.

* See the history of this insect, and a figure of her cells, in the Boston Cultivator, for July 15, 1848.

becomes a chrysalis within the same shell, and in due time comes out a *Platygaster* fly, like its parent. This last transformation Mr. Herrick found to take place towards the end of June, from eggs laid in November of the year before; and he thinks that the flies continue alive through the summer, till the appearance of the canker-worm moths in the autumn affords them the opportunity of laying their eggs for another brood. As these little parasites prevent the hatching of the eggs wherein they are bred, and as they seem to be very abundant, they must be of great use in preventing the increase of the canker-worm. Without doubt such wisely appointed means as these were once enough to keep within due bounds these noxious insects; but, since our forests, their natural food, and our birds, their greatest enemies, have disappeared before the woodman's axe and the sportsman's gun, we are left to our own ingenuity, perseverance, and united efforts, to contrive and carry into effect other means for checking their ravages.

Between the years 1841 and 1847, canker-worms almost entirely disappeared in the vicinity of Boston. At the latter date, there was a visible increase of them here, and their numbers have rapidly augmented every subsequent year. In a few years more, unless checked by natural or artificial means, they will probably prove as destructive as at any former time. The writer of this work has given repeated warning of these facts in the public prints, and has pointed out the remedies to be applied.*

Apple, elm, and lime trees are sometimes injured a good deal by another kind of span-worm, larger than the canker-worm, and very different from it in appearance. It is of a bright yellow color, with ten crinkled black lines along the top of the back; the head is rust-colored; and the belly is paler than the rest of the body. When fully grown, it

* See Prairie Farmer, Vol. VIII. p. 172, for June, 1848. Massachusetts Ploughman, for June 24, 1848, Nov. 23, 1850, and May 17, 1851. Boston Cultivator, Nov. 24, 1849. New England Farmer, Vol. II. p. 252, for August, 1850.

measures about one inch and a quarter in length. It often rests with the middle of the body curved upwards a little, and sometimes even without the support of its fore legs. The leaves of the lime seem to be its natural and favorite food, for it may be found on this tree every year; but I have often seen it in considerable abundance, with common canker-worms, on other trees. It is hatched rather later, and does not leave the trees quite so soon as the latter. About or soon after the middle of June it spins down from the trees, goes into the ground, and changes to a chrysalis in a little cell five or six inches below the surface; and from this it comes out in the moth state towards the end of October or during the month of November. More rarely its last transformation is retarded till the spring.

The females are wingless and grub-like, with slender thread-shaped antennæ. As soon as they leave the ground they creep up the trees, and lay their eggs in little clusters, here and there, on the branches. The males have large and delicate wings, and their antennæ have a narrow feathery edging on each side. They follow the females, and pair with them on the trees. This kind of moth closely resembles the lime-looper or umber moth (*Hybernia defoliaria*) of Europe; but differs from it so much in the larva state, that I have not the slightest doubt of its being a distinct species, and accordingly name it *Hybernia Tiliaria* (Fig. 236), the lime-tree winter-moth, from *Tilia*, the scientific name of its favorite tree. The fore wings of the male are rusty buff or nankin-yellow, sprinkled with very fine brownish dots, and banded with two transverse wavy, brown lines, the band nearest the shoulders being often indistinct; in the space between the bands, and

Fig. 236.

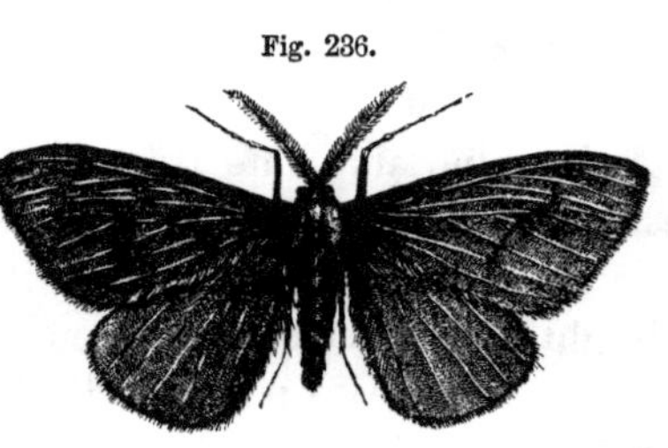

near to the thick edge of the wing, there is generally a brown dot. The hind wings are much paler than the others, and have a small brownish dot in the middle. The color of the body is the same as that of the fore wings; and the legs are ringed with buff and brown. The wings expand one inch and three quarters. The body of the female is grayish or yellowish white; it is sprinkled on the sides with black dots, and there are two square black spots on the top of each ring, except the last, which has only one spot. The front of the head is black; and the antennæ and the legs are ringed with black and white. The tail is tipped with a tapering, jointed egg-tube, that can be drawn in and out, like the joints of a telescope. Exclusive of this tube, the female measures about half an inch in length. The eggs are beautiful objects when seen under a microscope. They are of an oval shape, and pale yellow color, and are covered with little raised lines, like net-work, or like the cells of a honeycomb.

As these span-worms appear at the same time as canker-worms, resemble them in their habits, and often live on the same trees, they can be kept in check by such means as are found useful when employed against canker-worms.

Probably more than one hundred different kinds of Geometers may be found in Massachusetts alone. Seventy-eight are already known to me. Some of these are small, and are not otherwise remarkable; some are distinguished for their greater size and beauty in the moth state, or for the singularity of the forms and habits of their caterpillars. None of them, however, have become so notorious on account of their devastations as the species already described.

4. Delta-Moths. (*Pyralides.*)

The Pyralides of Linnæus are nearly akin to the Geometers. Latreille called them *Deltoides*, because the form of the moths, when their wings are closed, is triangular, like that of the Greek letter Δ. For the same reason I have called them Delta-moths. The body, in these moths, is long and slender. The fore wings are long and rather narrow, and cover the hind wings nearly horizontally when at rest. The feelers are generally very long, flattened sidewise, and more or less turned up at the end. The tongue in some is of moderate length, in others it is very small or invisible. The antennæ are long and generally simple or bristle-formed in both sexes; in some males, however, they are feathered, and in a few others they have a singular knot or crook in the middle. The legs are long and slender; and the first pair is often fringed with tufts of long hairs. Most of these moths fly at night; a few are on the wing in the daytime also. They generally prefer moist and shady places, where the long grass and thick foliage shelter them from the light and heat of the sun. Some of them frequent houses.

The meal-moth (*Pyralis farinalis*), (Plate VII. Fig. 8,) the caterpillar of which may be found in old flour-barrels, is often seen on the ceilings of rooms, sitting with its tail curved over its back. The fore wings of this pretty moth are light brown, crossed by two curved white lines, and with a dark chocolate-brown spot on the base and tip of each. The tabby, or grease-moth (*Aglossa pinguinalis*), the larva of which lives in greasy animal substances, is also to be found in houses, and is known by its narrow glossy wings, of a smoky gray color, crossed by wavy lighter-colored bands; its tongue is not visible. The motions of some of the day-flying kinds (*Simaethis*) are very curious. When they alight upon a leaf, they whirl round sidewise, in a circular direction, with the head in the centre of the circle, and then return in the contrary direction, and repeat these gyrations several times in succession.

The larvæ or caterpillars of the Delta-moths are long and slender, tapering at each end, and naked, or with only a few short hairs, which are rarely visible to the naked eye.

Some of them have sixteen legs, others have only fourteen. The latter creep very much like the span-worms, but are more active and quick in their motions. Most of them live exposed upon or under the leaves of plants, and, when they come to their full growth, they enclose themselves in cocoons formed of folded leaves thinly lined with silk, in which they undergo their transformations. Some kinds (*Hydrocampa* and *Petrophila*) live in the water upon aquatic plants, and secure themselves in cylindrical leafy cases, fitted to cover the whole of the body except the head and six fore legs, and made air-tight. These cases prevent the water from getting into the lateral breathing-holes of the caterpillars, and contain a sufficient quantity of air for them to breathe; and, with them, they can easily move about under the surface, upon the plants which serve them for food. Some of the aquatic kinds do not make these air-tight cases, for they do not need them, as they breathe through fringed gills, placed along the sides of their bodies. Thus we see that even aquatic plants are inhabited by peculiar tribes of insects, which keep in check their redundant vegetation, and which are fitted, by extraordinary and curious contrivances, for the element wherein they are appointed to live. These aquatic insects stand on the limits of the order, and connect the *Lepidoptera* with the *Neuroptera*, by means of the May-flies (*Phryganeadæ*) belonging to the latter order.

Those caterpillars of the Pyralides that have only fourteen legs may be called Herminians (HERMINIADÆ), after the principal genus in the group. The hop-vine is often infested by great numbers of these caterpillars. They eat large holes in the leaves, and thereby sometimes greatly injure the plant. Caterpillars of this kind have also been observed on the hop in Europe, from whence ours may have been introduced; but until specimens from Europe and this country are compared together, in all their states, it will be well to consider the latter as distinct. Our hop-vine caterpillars are false-loopers, bending up the back a little when they creep, because the first pair of prop-legs, found in other caterpillars, is wanting in them. The rings of their bodies are rather prominent, the cross-lines between them being deep. They are of a green color, with two longitudinal white lines along the back, a dark green line in the middle between them, and an indistinct whitish line on each side of the body. The head is green, and very regularly spotted with minute black dots, from each of which arises a very short hair. There are similar dots and hairs arranged in two transverse rows on each of the rings. When disturbed, they bend their bodies suddenly and with a jerk, first on one side and then on the other, each time leaping to a considerable distance, so that it is difficult to catch or hold them. They make no webs on the leaves, and do not suspend themselves by silken threads like the Geometers; but they are very active, creep fast, and soon get upon the leaves again after leaping off. When fully grown they are about eight tenths of an inch long. They then form a thin, imperfect, silky cocoon within a folded leaf, or in some crevice or sheltered spot, and are changed to brownish chrysalids, which present nothing remarkable in their appearance. Three weeks afterwards the moths come forth from these cocoons.

There are two broods of these insects in the course of the summer. The caterpillars of the first brood appear in May and June, and are transformed to moths towards the end of June, and during the early part of July. Those of the second brood appear in July and August, and are changed to moths in September. The insects of the second brood are much the most numerous usually, and do much more damage to the hop-vine than the others. The moth has been named *Hypena Humuli* (Fig. 237), the hop-vine Hypena, upon the supposition that it is distinct from the *Hypena rostralis*, or hop-vine snout-moth of Europe. These moths are readily known by

Fig. 237.

their long, wide, and flattened feelers, which are held close together, and project horizontally from the fore part of the head, in the manner of a snout. The antennæ in both sexes are naked, and bristle-formed. The wings vary in color, being sometimes dusky or blackish brown, and sometimes of a much lighter rusty-brown color. The fore wings are marbled with gray beyond the middle, and have a distinct oblique gray spot on the tip; they are crossed by two wavy blackish lines, one near the middle, and the other near the outer hind margin; these lines are formed by little elevated black tufts, and there are also two similar tufts on the middle of the wing. The hind wings are dusky brown or light brown, with a paler fringe, and are without bands or spots. The wings expand about one inch and a quarter.

The means for destroying the hop-vine caterpillars are showering or syringing the plants with strong soapsuds, or with a solution of oil-soap in water, in the proportion of two pounds of the soap to fourteen or fifteen gallons of water.

The foregoing is the only kind of Delta-moth that appears to be particularly injurious to any of our useful or cultivated plants.

5. Leaf-Rollers. (*Tortrices.*)

There are many caterpillars that curl up the edges of the leaves of plants into little cylindrical rolls, open at each end, and fastened together with bands or threads of silk. These rolls serve at once for the habitations and the food of the insects; and to the latter Linnæus gave the name of Tortrices, derived from a Latin word signifying to curl or twist. All the caterpillars now put in this tribe are not leaf-rollers. Some of them live in leaf and flower buds, and fasten the leaves together so that the bud cannot open, while they devour the tender substance within. Some live in a kind of tent formed of several leaves, drawn together and secured with silken threads. Others are found in the tender shoots or under the bark of plants. A few bore into young fruits, which they cause to ripen and fall prematurely. A still smaller number of kinds live on the leaves of plants, exposed to view, and without any kind of covering over them. Most of these insects, when disturbed, let themselves down by threads, like the Geometers. Very few of them make cocoons; the greater number transforming within the rolled leaves, or in the other situations wherein they usually dwell. They are furnished with sixteen legs, and their bodies are nearly or quite naked. Many of their chrysalids have two rows of minute prickles across each of the rings of the hind body, by the help of which they push themselves half-way out of their habitations, when the included moths are about to come forth.

The moths of this tribe are mostly of small size, very few of them expanding more than one inch. They carry their wings like a steep roof over their bodies when they are at rest. Their fore wings are very much curved, and are very broad at the shoulders, and hence these insects are called *Platyomides*, that is, broad shoulders, by the French naturalists. These wings are generally very prettily banded and spotted, and are sometimes ornamented with brilliant metallic spots. The hind wings are plain, and of a uniform dusky or grayish color, and the inner edge is folded like a fan against the side of the body. Their antennæ are naked or thread-like. Their feelers, two in number, are broad, of moderate length, or project like a short beak in front of the head, and are never curved upwards. The spiral tongue is mostly short, and sometimes invisible. The body is rather short and thick, and the legs are also much shorter in proportion than in the Delta-moths. These little moths fly only in the evening and night, and remain at rest during the day upon or near the plants inhabited by their caterpillars. They are most abundant in midsummer, but certain species appear in the spring or autumn. The habits of the Tortrices, in all their states, are not yet known well enough to enable us to group the insects together under family names.

The caterpillars of some of our largest species are found on the ends of the branches of various trees and bushes, in nests, made of the young leaves drawn together in bunches, and fastened with threads. In the middle of these nests the caterpillars live, either singly, or in companies of several individuals together. Nests of this kind, containing a large number of caterpillars, may often be seen on oak-trees in the summer. The chrysalids force their way partly out of the nests by the help of the transverse rows of prickles on their backs, when the moths are about to make their escape. The moths resemble in form and general appearance those of another species, the caterpillars of which live singly in much smaller nests, on apple-trees and rose-bushes. Early in May, or soon after the buds of the apple-tree begin to open, these little caterpillars begin their labors. They curl up and fasten together the small and tender leaves that supply them both with shelter and food; and in this way, they often do considerable damage to the trees. These caterpillars are sometimes of a pale green color, with the head and the top of the first ring brownish; and sometimes the whole body is brownish or dull flesh-red; they are rough to the touch, with minute warts, each of which produces a very short hair, invisible to the naked eye. They come to their full size towards the middle of June, and then measure nearly or quite half an inch in length. After this, they line the inner surface of the curled leaves composing their nests with a web of silk, and are then changed to chrysalids of a dark brown color. Towards the end of June, or early in July, the chrysalis pushes itself half-way out of its nest, and bursts open at the upper end, so that the moth may come out. The moth closely resembles the *Lozotœnia** *oporana* of Europe, but differs from it in having

* This word was probably an error of the press in the "Catalogue" of Mr. Stephens, by whom the genus was proposed. It has, however, been copied in

the fore wings broader at the base, more curved on the front edge, and more hooked at the tip, and its markings are also somewhat different. It may be called *Loxotœnia Rosaceana* (Fig. 238), the oblique-banded moth of the Rose tribe, for to the latter the apple-tree belongs as well as the rose. The fore wings of this moth are very much arched on their outer edge, and curve in the contrary direction at the tip, like a little hook or short tail. They are of a light cinnamon-brown color, crossed with little wavy darker-brown lines, and with three broad oblique dark brown bands, whereof one covers the base of the wing, and is oftentimes indistinct or wanting, the second crosses the middle of the wing, and the third, which is broad on the front edge and narrow behind, is near the outer hind margin of the wing. The hind wings are ochre-yellow, with the folded part next to the body blackish. It expands one inch or a little more.

Fig. 238.

Little caterpillars of another species are sometimes found in May and June in the opening buds and among the tender leaves of the apple-tree. They live singly in the buds, the leaves of which they fasten together and then devour. These caterpillars are of a pale and dull brownish color, warty and slightly downy like the foregoing kind, with the head and the top of the first ring dark shining brown; and a dark brown spot appears through the skin on the top of the eighth ring. They generally come to their growth by the middle of June, and are changed to shining brown chrysalids within the curled leaves, in a little web of silk, wherewith their retreats are lined. The chrysalis has only one row of prickles across the rings of the back. The moths come out early in July. They very closely

several other works by other authors, without correction or comment. *Loxotœnia*, meaning oblique band, seems to be the right name for the moths of this genus, which are distinguished by the oblique bands on their fore wings.

resemble the European *Penthina comitana*,* and perhaps may be merely a variety of it. The head and thorax are dark ash-colored. The fore wings are of the same color at each end, and grayish white in the middle, mottled with dark gray; there are two small eye-like spots on each of them; one near the tip, consisting of four little black marks, placed close together in a row, on a light brown ground, the inner marks being longer than the others; the second eye-spot is near the inner hind angle, and is formed by three minute black spots, arranged in a triangle, in the middle of which there is sometimes a black dot. The hind wings are dusky brown. This moth expands from one half to six tenths of an inch. It may be called *Penthina oculana*, the eye-spotted Penthina. My attention was called to the depredations of this bud-moth, and of the preceding species, by John Owen, Esq., of Cambridge, by whom the moths were raised from the caterpillars, and presented to me. It is difficult at first to conceive how such insignificant creatures can occasion so much mischief as they are found to do. This seems to arise from the number of the insects, and their mode of attack, whereby the opening foliage is checked in its growth or nipped in the bud. To pull off and crush the withered clusters of leaves containing the caterpillars or the chrysalids, is the only remedy that occurs to me. It were to be wished that some better way of putting a stop to the ravages of the leaf-rollers and bud-moths that infest many of our fruit-trees and flowering shrubs could be discovered.

Apricot, peach, and plum trees, when trained against walls in the open air, are said to suffer very much sometimes from the attacks of insects whose habits resemble those of the eye-spotted Penthina. But, as I have not yet seen them in the moth state, I cannot say whether they are of the same species as the bud-moth above named.

* *Spilonota comitana*, Stephens; *Pœcilochroma comitana*, Curtis; *Penthina luscana*, Duponchel.

Perhaps they are identical with the apricot-bud caterpillars (*Ditula angustiorana*) of Europe, the depredations of which have been described by Mr. Westwood in the fourteenth volume of the "Gardener's Magazine." Besides picking off the curled and confined clusters of leaves, when practicable, I would recommend thoroughly drenching the trees with Mr. Haggerston's remedy, a pound of oil-soap in from seven to ten gallons of water, in the hope that some of the mixture might penetrate the injured buds and leaves, and destroy the caterpillars concealed therein. A mixture of one gallon of the liquor expressed by tobacconists from tobacco, with five gallons of water, has been used to the same intent.

Roses are infested with several kinds of caterpillars belonging to this tribe. Mr. Westwood has described one of them, and mentions others that are found in Europe, in the thirteenth volume of the "Gardener's Magazine." Similar species are not uncommon in this country. Some of these spoilers fasten upon the leaves, and roll them up, or stick them together, to serve them for food and shelter; while others lurk unseen in the flower-buds, and canker them to the heart, before they can spread their lovely petals to the sun, and breathe out their fragrance to the air. A particular description of each of these insects would occupy too much space here; and I can only add, that the worm in the bud is to be destroyed only by hand.

Pine and fir trees are also injured by some of the *Tortrices*, that pierce the tender shoots and terminal buds. The seat of their depredations becomes known by the oozing of the resin and by the withering of the bud or shoot. The latter commonly dies in consequence of the injury, the upward growth is checked, and the stem only puts forth side shoots the following year. Some one of these side shoots, in time, takes the place of the leading shoot, and thus gives to the trunk an irregular and crooked appearance, and renders it unfit for timber. The history of several European *Tortrices* or turpentine-moths, that thus injure pines and firs, is given in Kollar's Treatise, wherein we are advised to search for the lumps of turpentine in the autumn, and destroy the caterpillars under them, or to cut off the injured shoots and burn them with their inhabitants. This advice it may be proper for us to follow, although it is not yet certain that our turpentine-moths are actually the same as those of Europe.

Among the insects that have been brought to America with other productions of Europe may be mentioned the apple-worm, as it is here called, which has become naturalized wherever the apple-tree has been introduced. This mischievous creature has sometimes been mistaken for the plum-weevil (*Rhynchænus* (*Conotrachelus*) *Nenuphar*), described in another part* of this treatise; but it may be easily distinguished therefrom by its shape, its habits, and its transformations. Although the plum-weevil prefers stone fruit, it is sometimes found in apples also; on the other hand, the apple-worm has never been found here in plums. It is not a grub, but a true caterpillar, belonging to the *Tortrix* tribe, and in due time is changed to a moth, called *Carpocapsa Pomonella* (Fig. 239),† the codling-moth, or fruit-moth of the apple. An anonymous writer, in the "Entomological Magazine" ‡ of London, has well remarked that this moth "is the most beautiful of the beautiful tribe to which it belongs; yet, from its habits not being known, it is seldom seen in the moth state; and the apple-grower knows no more than the man in the moon to what cause he is indebted for his basketfuls of worm-eaten windfalls in the stillest weather."

Fig. 239.

* Page 75.

† *Tinea Pomonella*, L.; *Pyralis Pomana*, F. If the modern name of the genus be correct, it was probably formed from two Greek words signifying to devour fruit. Perhaps the name should have been *Carpocampa*, that is, in English, fruit-caterpillar.

‡ Vol. I. p. 144.

The apple-worm has been long known in Europe, and its history has been written by Rösel, Réaumur, Kollar, Westwood,* and other European naturalists. A good account of it, and of its transformations, by Joseph Tufts, Esq., of Charlestown, Massachusetts, was published in the year 1819, in the fifth volume of "The Massachusetts Agricultural Repository and Journal"; and Mr. Joseph Burrelle, of Quincy, Massachusetts, has also made some remarks on the same insect, in the eighteenth volume of "The New England Farmer."† At various times, between the middle of June and the first of July, the apple-worm moths may be found. They are sometimes seen in houses in the evening, trying to get through the windows into the open air, having been brought in with fruit while they were in the caterpillar state. Their fore wings, when seen at a distance, have somewhat the appearance of brown watered silk; when closely examined, they will be found to be crossed by numerous gray and brown lines, scalloped like the plumage of a bird; and near the hind angle there is a large, oval, dark brown spot, the edges of which are of a bright copper-color. The head and thorax are brown mingled with gray; and the hind wings and abdomen are light yellowish brown, with the lustre of satin. Its wings expand three-quarters of an inch. This insect is readily distinguished from other moths by the large, oval brown spot, edged with copper-color, on the hinder margin of each of the fore wings. During the latter part of June and the month of July, these fruit-moths fly about apple-trees every evening, and lay their eggs on the young fruit. They do not puncture the apples, but they drop their eggs, one by one, in the eye or hollow at the blossom-end of the fruit, where the skin is most tender. They seem also to seek for early fruit rather than for the late kinds, which we

* Gardener's Magazine, Vol. XIV. p. 234.

† Page 398. See also some remarks on this insect in my "Discourse before the Massachusetts Horticultural Society, in 1832," page 42.

find are not so apt to be wormy as the thin-skinned summer apples. The eggs begin to hatch in a few days after they are laid, and the little apple-worms or caterpillars produced from them immediately burrow into the apples, making their way gradually from the eye towards the core. Commonly only one worm will be found in the same apple; and it is so small at first, that its presence can only be detected by the brownish powder it throws out in eating its way through the eye. The body of the young insect is of a whitish color; its head is heart-shaped and black; the top of the first ring or collar and of the last ring is also black; and there are eight little blackish dots or warts, arranged in pairs, on each of the other rings. As it grows older, its body becomes flesh-colored; its head, the collar, and the top of the last ring turn brown, and the dots are no longer to be seen. In the course of three weeks, or a little more, it comes to its full size, and meanwhile has burrowed to the core and through the apple in various directions. To get rid of the refuse fragments of its food, it gnaws a round hole through the side of the apple, and thrusts them out of the opening. Through this hole also the insect makes its escape after the apple falls to the ground; and the falling of the fruit is well known to be hastened by the injury it has received within, which generally causes it to ripen before its time.

Soon after the half-grown apples drop, and sometimes while they are still hanging, the worms leave them and creep into chinks in the bark of the trees, or into other sheltered places, which they hollow out with their teeth to suit their shape. Here each one spins for itself a cocoon or silken case, as thin, delicate, and white as tissue paper. Some of the apple-worms, probably the earliest, are said by Kollar to change to chrysalids immediately after their cocoons are made, and in a few days more turn to moths, come out, and lay their eggs for a second generation of the worms; and hence much fruit will be found to be worm-eaten in the autumn. Most of the insects, however, remain in their cocoons through the winter, and are not changed to moths till the following summer. The chrysalis is of a bright mahogany-brown color, and has, as usual, across each of the rings of its hind body, two rows of prickles, by the help of which it forces its way through the cocoon before the moth comes forth.

As the apple-worms instinctively leave the fruit soon after it falls from the trees, it will be proper to gather up all wind-fallen apples daily, and make such immediate use of them as will be sure to kill the insects, before they have time to escape. Mr. Burrelle says, that if any old cloth is wound around or hung in the crotches of the trees, the apple-worms will conceal themselves therein; and by this means thousands of them may be obtained and destroyed, from the time when they first begin to leave the apples, until the fruit is gathered. By carefully scraping off the loose and rugged bark of the trees, in the spring, many chrysalids will be destroyed; and it has been said that the moths, when they are about laying their eggs, may be smothered or driven away, by the smoke of weeds burned under the trees. The worms, often found in summer pears, appear to be the same as those that affect apples, and are to be kept in check by the same means. Cranberries are likewise affected by worms, altogether similar to apple-worms.

6. Tineæ.

The word moth was formerly used in a much more restricted sense than it now is. It was originally given to the caterpillars of certain insects, called Tineæ by Linnæus, and well known as the destroyers of clothing and of other household stuffs. In this sense we find it used in our version of the Scriptures, and in the works of old English writers. It occurs, with very little change, in other languages also, and seems to have been derived from a word signifying to gnaw or to eat.* Nearly all the moth-worms, or caterpillars belonging to the tribe of Tineæ, gnaw holes or winding paths in the substances wherein they live. Some of the fragments they devour, and the rest they fasten together, with a few silken threads, so as to shelter or clothe their tender bodies. With these materials some of them make cylindrical burrows, through which they can move freely, and carry on the destruction unseen; and others, with the same, shape for themselves various kinds of pods or cases, large enough to cover their bodies entirely when they are at rest, and so light that they can bear them about on their backs, as snails do their shells. Some moth-worms are dark-colored; but most of them are of a dirty white color, with a brownish head, and a brown spot on the top of the first ring. They are either wholly naked, or have only a few short hairs thinly scattered over the surface of their bodies. They generally have sixteen legs. Some, however, want the first pair of prop-legs, having only fourteen in all. They undergo their transformations in the burrows or cases that have served them for habitations, either with or without the additional covering of a cocoon spun within their places of abode. The chrysalids are of a brown color, and are rather more slender than those of other moths. In the winged state they vary greatly both in form and color. They all agree, however, in having the wings long and narrow, and folded or wrapped around the body, more or less closely, when they are at rest. Their antennæ are bristle-shaped, and very rarely feathered in either sex. Some of them have four feelers, others only two; and the spiral tongue is short. Most of these winged moths are very small; indeed, the least of the Lepidoptera belong to this tribe. They have been divided by some naturalists into two, and by others into three groups, namely, *Crambidæ*, *Yponomeutadæ*, and *Tineadæ*, the differences be-

* From the Gothic *maten*, to gnaw, and from *matjan*, to eat, we have the Anglo-Saxon word *moth*, as now used, and *matha*, a maggot.

tween which it is not necessary particularly to notice in this place.

Some moth-worms burrow into leaves, and make winding passages in the pulpy substance thereof, under the skin; some bore into the stems of plants; and a few are found only on the surface of leaves, or on roots. Living plants, however, form but a small part of the food of the Tineæ, most of which subsist on other substances; and, for this reason, they would have been passed by without further notice, were it not for the depredations of certain species on some of our most valuable possessions. Most of these pests are foreign insects, and have been introduced into this country from abroad; it will not, therefore, be in my power to offer anything absolutely new about them. Nevertheless, a few remarks on some of the most remarkable or destructive of these moths may not be wholly useless or unacceptable to those persons for whom this treatise was particularly designed.

The largest insects of this tribe belong to the group called Crambidæ, or Crambians, among which the bee-moth or wax-moth is to be placed. This pernicious insect was well known to the ancients, and we find it mentioned, under the name of *Tinea*, in the works of Virgil and Columella,* old Roman writers on husbandry. In the winged state, the male and female differ so much in size, color, and in the form of their fore wings, that they were supposed, by Linnæus and by some other naturalists, to be different species, and accordingly received two different names.† To avoid confusion, it will be best to adopt the scientific name given to the bee-moth by Fabricius, who called it *Galleria cereana* (Fig. 240), that is, the wax Galleria, because, in its caterpillar

Fig. 240.

* Virgil, Georgic IV. line 246. Columella, Husbandry, Book IX. chap. 14.

† *Tortrix cereana*, the male; *Tinea mellonella*, the female.

62

state, it eats beeswax. Doubtless it was first brought to this country, with the common hive-bee, from Europe, where it is very abundant, and does much mischief in hives. Very few of the *Tineæ* exceed or even equal it in size. In its perfect or adult state it is a winged moth or miller, measuring, from the head to the tip of the closed wings, from five eighths to three quarters of an inch in length, and its wings expand from one inch and one tenth to one inch and four tenths. The feelers are two in number; and the tongue is very short, and hardly visible. The fore wings shut together flatly on the top of the back, slope steeply downwards at the sides, and are turned up at the end, somewhat like the tail of a fowl. This resemblance probably suggested the name of the genus, *Galleria*, which seems to have been derived from the Latin word for a fowl. The male is of a dusty gray color; his fore wings are more or less glossed and streaked with purple-brown on the outer edge, they have a few dark brown spots near the inner margin, and they are scalloped or notched inwardly at the end; his hind wings are light yellowish-gray, with whitish fringes. The female is much larger than the male, and much darker-colored; her fore wings are proportionally longer, not so deeply notched on the outer hind margin, and not so much turned up at the end; they are more tinged with purple-brown, sprinkled with darker spots; and the hind wings are dirty or grayish white. There are two broods of these insects in the course of a year. Some winged moths of the first brood begin to appear towards the end of April, or early in May; those of the second brood are most abundant in August; but between these periods, and even later, others come to perfection, and consequently some of them may be found during the greater part of the summer. By day they remain quiet on the sides or in the crevices of the bee-house; but, if disturbed at this time, they open their wings a little, and spring or glide swiftly away, so that it is very difficult to seize or to hold them. In the evening they take wing, when the bees are at rest, and hover around the hive, till, having found the door, they go in and lay their eggs. Those that are prevented by the crowd, or by any other cause, from getting within the hive, lay their eggs on the outside, or on the stand, and the little worm-like caterpillars hatched therefrom easily creep into the hive through the cracks, or gnaw a passage for themselves under the edges of it.

These caterpillars, at first, are not thicker than a thread. They have sixteen legs. Their bodies are soft and tender, and of a yellowish-white color, sprinkled with a few little brownish dots, from each of which proceeds a short hair; their heads are brown and shelly, and there are two brown spots on the top of the first ring. Weak as they are, and unprovided with any natural means of defence, destined, too, to dwell in the midst of the populous hive, surrounded by watchful and well-armed enemies, at whose expense they live, they are taught how to shield themselves against the vengeance of the bees, and pass safely and unseen in every direction through the waxen cells, which they break down and destroy. Beeswax is their only food, and they prefer the old to the new comb, and are always found most numerous in the upper part of the hive, where the oldest honeycomb is lodged. It is not a little wonderful, that these insects should be able to get any nourishment from wax, a substance which other animals cannot digest at all; but they are created with an appetite for it, and with such extraordinary powers of digestion, that they thrive well upon this kind of food.

As soon as they are hatched they begin to spin; and each one makes for itself a tough silken tube, wherein it can easily turn around, and move backwards or forwards at pleasure. During the day they remain concealed in their silken tubes; but at night, when the bees cannot see them, they come partly out, and devour the wax within their reach. As they increase in size, they lengthen and enlarge their dwellings, and cover them on the outside with a coating of grains of wax mixed with their own castings, which resemble gunpowder. Protected by this coating from the stings of the bees, they work their way through the combs, gnaw them to pieces, and fill the hive with their filthy webs; till at last the discouraged bees, whose diligence and skill are of no more use to them in contending with their unseen foes, than their superior size and powerful weapons, are compelled to abandon their perishing brood and their wasted stores, and leave the desolated hive to the sole possession of the miserable spoilers. These caterpillars grow to the length of an inch or a little more, and come to their full size in about three weeks. They then spin their cocoons, which are strong silken pods, of an oblong oval shape, and about one inch in length, and are often clustered together in great numbers in the top of the hive. Some time afterwards, the insects in these cocoons change to chrysalids of a light brown color, rough on the back, and with an elevated dark brown line upon it from one end to the other. When this transformation happens in the autumn, the insects remain without further change till the spring, and then burst open their cocoons, and come forth with wings. Those which become chrysalids in the early part of summer are transformed to winged moths fourteen days afterwards, and immediately pair, lay their eggs, and die.

Bees suffer most from the depredations of these insects in hot and dry summers. Strong and healthy swarms, provided with a constant supply of food near home, more often escape than small and weak ones. When the moth-worms have established themselves in a hive, their presence is made known to us by the little fragments of wax, and the black grains scattered by them over the floor. Means should then be taken, without delay, to dislodge the depredators and invigorate the swarm. These are so fully described in Dr. Thacher's "Treatise on the Management of Bees," and in other works on the same subject, that I shall limit myself to a few remarks, and refer the reader for further particulars to these works. Kollar states that there is but one sure method of clearing bee-hives of the moth, and this is to look for and destroy the caterpillars or moth-worms and the chrysalids; and he advises that the hives should be examined, for this purpose, once a week, and that all the webs and cocoons, with the insects in them, should be taken out and destroyed. At all events, the examination ought to be made every year, early in September, when the cocoons will be found in greater numbers than at any other time, and should be carefully removed and burned. The winged moths are very fond of sweets; and if shallow vessels, containing a mixture of honey or sugar, with vinegar and water, are placed near the bee-house in the evening, the moths will get into them and be drowned. In this way great numbers may be caught every night. Several kinds of hives and bee-houses have been contrived and recommended, for the purpose of keeping out the bee-moth; but it does not appear that any of them entirely supersede the necessity for the measures above recommended.

The various kinds of destructive moths found in houses, stores, barns, granaries, and mills, are mostly very small insects; the largest of them, when arrived at maturity, expanding their wings only about eight tenths of an inch. The ravages of some of these little creatures are too well known to need a particular description. Among them may be mentioned the clothes-moth (*Tinea vestianella*), the tapestry or carpet-moth (*T. tapetzella*), the fur-moth (*T. pellionella*), the hair-moth (*T. Crinella*), and the grain-moth (*T. granella*), with some others, belonging to a group which may be called Tineans (TINEADÆ); also the pack-moth (*Anacampsis sarcitella*), which is very destructive to wool and fabrics made of this material, and the Angoumois grain-moth (*Butalis cerealella*), both of which are to be included among the Yponomeutians. In the cabinet of the Boston Society of Natural History, the cases containing the large and beautiful collection of shells were formerly lined with fine white flannel. In this some moths soon established themselves, multiplied very fast, and, in the course of a few years, did so much damage that it became necessary entirely to remove the moth-eaten linings. In their winged state these moths (Fig. 241) were of a light buff color, with the lustre of satin, and had a thick orange-colored tuft on the forehead; the wings were deeply fringed, and the first pair were lance-shaped, and expanded rather more than half an inch. This species agrees very well with the description given, by the old naturalists, of the *Tinea flavifrontella** (Fig. 242, larva, natural size and magnified), or the orange-fronted Tinea, and with Wood's figure of *Tinea destructor*, the destroyer. Should it prove to be different from these, it may be named the satin-buff moth. Objects of natural history are very apt to be injured by another moth, closely resembling the foregoing, and differing from it chiefly in being somewhat smaller, and in having the hind wings tinged with gray. Chocolate, as Réaumur has remarked, is devoured by another Tinea, whose little silken cases are often seen between the cakes, and I have also found them in chocolate put up in tin cases. Other articles of food are also devoured by some of these Tineæ, and even our books are not spared by them.

Fig. 241.

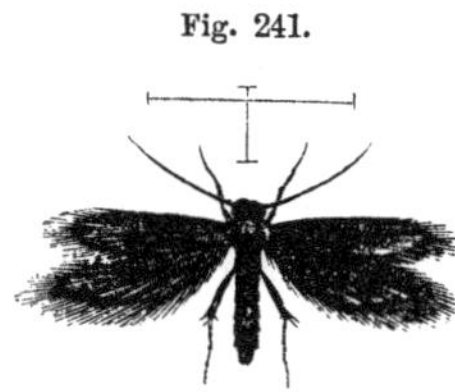

Fig. 242.

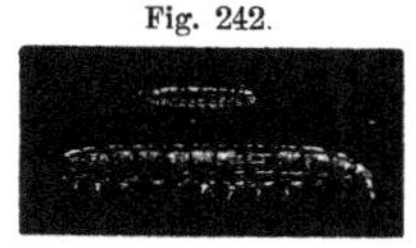

The Tineans, in the winged state, have four short and slender feelers, a thick tuft on the forehead, and very narrow wings, which are deeply fringed. They lay their eggs mostly in the spring, in May and June, and die immediately afterwards. The eggs (according to Latreille and Duponchel, from whose works the following remarks are

* Not the *Batia flavifrontella* of the English entomologists.

chiefly extracted) are hatched in fifteen days, and the little whitish caterpillars or moth-worms proceeding therefrom immediately begin to gnaw the substances within their reach, and cover themselves with the fragments, shaping them into little hollow rolls and lining them with silk. They pass the summer within these rolls, some carrying them about on their backs as they move along, and others fastening them to the substance they are eating; and they enlarge them from time to time by adding portions to the two open extremities, and by gores set into the sides, which they slit open for this purpose. Concealed within their movable cases, or in their lint-covered burrows, they carry on the work of destruction through the summer; but in the autumn they leave off eating, make fast their habitations, and remain at rest and seemingly torpid through the winter. Early in the spring they change to chrysalids within their cases, and in about twenty days afterwards are transformed to winged moths, and come forth, and fly about in the evening, till they have paired and are ready to lay their eggs. They then contrive to slip through cracks into dark closets, chests, and drawers, under the edges of carpets, in the folds of curtains and of garments hanging up, and into various other places, where they immediately lay the foundation for a new colony of destructive moth-worms.

Early in June the prudent housekeeper will take care to beat up their quarters and put them to flight, or to disturb them so as to defeat their designs and destroy their eggs and young. With this view wardrobes, closets, drawers, and chests will be laid open, and emptied of their contents, and all woollen garments, and bedding, furs, feathers, carpets, curtains, and the like, will be removed and exposed to the air, and to the heat of the sun, for several hours together, and will not be put back in their places without a thorough brushing, beating, or shaking. By these means, the moths and their eggs will be dislodged and destroyed. In old houses, that are much infested by moths, the cracks in the floors, in the wainscot, around the walls and shelves of closets, and even in the furniture used for holding clothes, should be brushed over with spirits of turpentine. Powdered black pepper, strewed under the edges of carpets, is said to repel moths. Sheets of paper sprinkled with spirits of turpentine, camphor in coarse powder, leaves of tobacco, or shavings of Russia leather, should be placed among the clothes, when they are laid aside for the summer.

Furs, plumes, and other small articles, not in constant use, are best preserved by being put, with a few tobacco-leaves, or bits of camphor, into bags made of thick brown paper, and closely sewed or pasted up at the end. Chests of camphor-wood, red cedar, or of Spanish cedar, are found to be the best for keeping all articles from moths and other vermin. The cloth linings of carriages can be secured for-ever from the attacks of moths by being washed or sponged on both sides with a solution of the corrosive sublimate of mercury in alcohol, made just strong enough not to leave a white stain on a black feather. Moths can be killed by fumigating the article containing them with tobacco-smoke or with sulphur, or by shutting it in a tight vessel and then plunging the latter into boiling water, or exposing it to steam, for the space of fifteen minutes, or by putting it into an oven heated to about one hundred and fifty degrees of Fahrenheit's thermometer.

Stored grain is exposed to much injury from the depredations of two little moths, in Europe, and is attacked in the same way, and apparently by the same insects, in this country. Not having had sufficient opportunity to examine these insects myself, I have been obliged to rely upon the accounts given by foreign writers for most of the following particulars respecting their history.

The European grain-moth (*Tinea granella*), in its perfected state, is a winged insect, between three and four tenths of an inch long, from the head to the tip of its wings, and expands six tenths of an inch. It has a whitish tuft on its forehead; its long and narrow wings cover its back like a sloping roof, are a little turned up behind, and are edged with a wide fringe. Its fore wings are glossy, like satin, and are marbled with white or gray, light brown, and dark brown or blackish spots, and there is always one dark square spot near the middle of the outer edge. Its hind wings are blackish. Some of these winged moths appear in May, others in July and August, at which times they lay their eggs; for there are two broods of them in the course of the year. The young from the first-laid eggs come to their growth and finish their transformations in six weeks or two months; the others live through the winter, and turn to winged moths in the following spring. The young moth-worms (Plate VII. Fig. 6) do not burrow into the grain, as has been asserted by some writers, who seem to have confounded them with the Angoumois grain-worms; but, as soon as they are hatched, they begin to gnaw the grain and cover themselves with the fragments, which they line with a silken web. As they increase in size, they fasten together several grains with their webs (Plate VII. Fig. 7), so as to make a larger cavity, wherein they live. After a while, becoming uneasy in their confined habitations, they come out, and wander over the grain, spinning their threads as they go, till they have found a suitable place wherein to make their cocoons. Thus wheat, rye, barley, and oats, all of which they attack, will be found full of lumps of grains cemented together by these corn-worms, as they are sometimes called; and when they are very numerous, the whole surface of the grain in the bin will be covered with a thick crust of webs and of adhering grains.

These destructive corn-worms are really soft and naked caterpillars, of a cylindrical shape, tapering a little at each end, and are provided with sixteen legs, the first three pairs of which are conical and jointed, and the others fleshy and wart-like. When fully grown, they measure four or five tenths of an inch in length, and are of a light ochre or buff color, with a reddish head. When about six weeks old, they leave the grain, and get into cracks, or around the sides of corn-bins, and each one then makes itself a little oval pod or cocoon, about as large as a grain of wheat. The insects of the first brood, as before said, come out of their cocoons, in the winged form, in July and August, and lay their eggs for another brood; the others remain unchanged in their cocoons, through the winter, and take the chrysalis form in March or April following. Three weeks afterwards, the shining brown chrysalis forces itself part way out of the cocoon, by the help of some little sharp points on its tail, and bursts open at the other end, so as to allow the moth therein confined to come forth.

From various statements, deficient however in exactness, that have appeared in some of our agricultural journals, I am led to think that this corn-moth, or an insect much like it in its habits, prevails in all parts of the country, and that it has generally been mistaken for the grain-weevil. Many years ago I remember to have seen oats and shelled corn (maize) affected in the way above described; and Dr. Asa Fitch has favored me with a grain-moth, obtained in a flour-mill at East Greenwich, New York, which agreed with the descriptions and figures of the European *Tinea granella.* In some remarks upon this insect in the Albany "Cultivator," for January, 1847, he states that the American insect was observed to make its cocoon within the webs among the grain, instead of retiring therefrom when about to undergo its transformations. The habits of the European grain-moth are probably sometimes varied; for, although most writers on its history agree in saying that the insect leaves the grain and conceals itself in crevices of the granary when preparing to make its cocoon, Olivier* expressly states that it undergoes its transformation in its web among the grain.

* Encyclopédie Méthodique, Insectes, Tom. IV. p. 114.

There is another grain-moth, which, at various times, has been found to be more destructive in granaries, in some provinces of France, than the preceding kind. It is the Angoumois moth, or *Anacampsis* (*Butalis*) *cerealella*, an insect evidently belonging to the family of Yponomeutadæ, or Yponomeutians. The winged moths of this group have only two visible feelers, and these are generally long, slender, and curved over their heads. Their narrow wings most often overlap each other, and cover their backs horizontally when shut. It is stated in the "Introduction to Entomology,"* by the Rev. Mr. Kirby and Mr. Spence, that the insect under consideration is not yet named. This, however, is a mistake; for it was named *Alucita cerealella*, by Olivier,† as long ago as the year 1789. Olivier's name for it appears also to have been overlooked by Latreille, who has given it that of *Œcophora granella*.‡ Moreover, the writers of the "Introduction" have extracted from the works of Réaumur § an account of the habits of this insect, which they call *Tinea Hordei* and *Ypsolophus granellus*,‖ without seeming to be aware that it is the same as the Angoumois moth. In the first edition of this treatise, I stated that "the Angoumois grain-moth probably belongs to the modern genus *Anacampsis*, a word derived from the Greek, and signifying recurved, in allusion to the direction of the feelers of the moths." To this genus, as understood by most English entomologists, it certainly does belong; but Mr. Curtis is disposed to place it in his genus *Laverna*, including certain species which he has separated from *Anacampsis*. The French naturalist Duponchel, who has described and figured it in the fourth volume of the Supplement to his "Histoire Naturelle des Lépidoptères de

* Fifth edition, Vol. I. p. 172.

† Encyclopédie Méthodique, Hist. Nat. Insectes, Tom. IV. p. 121. See also Guérin's edition of Tigny's Histoire Nat. des Insectes, Tom. IX. p. 301.

‡ Cuvier's Règne Animal, 2d edition.

§ Mémoires, Tom. II. p. 486.

‖ Introduction to Entomology, Vol. I. p. 174.

France," refers this insect to the genus *Butalis*, which name I have thought proper now to adopt.

For more than a century, this insect has prevailed in the western parts of France, and has gradually been extending in an easterly and northerly direction. In the year 1736, the French naturalist Réaumur published an interesting account of it, illustrated by rude figures, in the second volume of his instructive "Mémoires." He found it to be very injurious to stored barley, at Luçon, in the province of La Vendée, and ascertained that it destroyed wheat also. In the adjacent province of Angoumois, it continued to increase for many years, till at length the attention of government was directed to its fearful depredations. This was in 1760, when the insect was found to swarm in all the wheat-fields and granaries of Angoumois and of the neighboring provinces, and the afflicted inhabitants were thereby deprived not only of their principal staple, wherewith they were wont to pay their annual rents, their taxes, and their tithes, but were threatened with famine and pestilence from the want of wholesome bread. Two members of the Academy of Sciences of Paris, the celebrated Duhamel du Monceau and M. Tillet, were then commissioned to visit the province of Angoumois, and inquire into the nature of this destructive insect. The result of their inquiries was communicated to the Academy, in whose history and memoirs it may be found, and was also subsequently republished in a separate volume.* From this work, and from the "Mémoires" of Réaumur, the following particulars are derived.

The Angoumois grain-insect, in its perfected state, is a little moth, of a pale cinnamon-brown color above, having the lustre of satin, with narrow broadly fringed hind wings of an ashen or leaden color, two thread-like antennæ, con-

* "Histoire d'un Insecte qui dévore les Grains de l'Angoumois," 12mo, Paris, 1762. See also "Histoire de l'Académie Royale des Sciences," Année 1761, p. 66, and "Mémoires," p. 289, 4to, Paris, 1763.

sisting of numerous beaded joints, a spiral tongue of moderate length, and two tapering feelers, turned over its head. It lays from sixty to ninety eggs, placing them in clusters of twenty or more on a single grain. From these are hatched, in from four to six days, little worm-like caterpillars, not thicker than a hair. These immediately disperse, and each one selects for itself a single grain, and burrows therein at the most tender part, commonly the place whence the plumule comes forth. Remaining there concealed, it devours the mealy substance within the hull; and this destruction goes on so secretly, as only to be detected by the softness of the grain or the loss of its weight. When fully grown, this caterpillar is not more than one fifth of an inch long. It is of a white color, with a brownish head; and it has six small jointed legs, and ten extremely small wart-like prop-legs. Having eaten out the heart of the grain, which is just enough for all its wants, it spins a silken web or curtain to divide the hollow, lengthwise, into two unequal parts, the smaller containing the rejected fragments of its food, and the larger cavity serving instead of a cocoon, wherein the insect undergoes its transformations. Before turning to a chrysalis it gnaws a small hole nearly or quite through the hull, and sometimes also through the chaffy covering of the grain, through which it can make its escape easily when it becomes a winged moth.

The insects of the first, or summer brood, come to maturity in about three weeks, remain but a short time in the chrysalis state, and turn to winged moths in the autumn, and at this time may be found, in the evening, in great numbers, laying their eggs on the grain stored in barns and granaries. The moth-worms of the second brood remain in the grain through the winter, and do not change to winged insects till the following summer, when they come out, fly into the fields in the night, and lay their eggs on the young ears of the growing grain. Although there seem to be two principal broods in the course of a year, we are not to understand that these are the only ones; for French writers inform us, that others are produced during the whole summer, and that the production of the insects is accelerated or retarded by differences in the temperature of the air.* When damaged grain is sown, it comes up very thin; the infected kernels seldom sprout, but the insects lodged in them remain alive, finish their transformations in the field, and in due time come out of the ground in the winged form.

To the foregoing sketch must now be added an account of an American grain-insect, which, in the first edition of this treatise, I suggested would prove to be the same as the Angoumois grain-moth. Having since obtained some of these American insects from various quarters, and having had a colony of them living and increasing, for three years, under my own eye, I find them to agree, in all essential particulars, with the European species. Until, therefore, they are proved, by actual comparison with perfect specimens of the latter, to be absolutely distinct, I must consider it as next to certain that they are identical, and that they have been introduced into this country from Europe. Perhaps, hereafter, the mode of their introduction may be as satisfactorily ascertained as that of the Hessian fly. In the year 1768, Colonel Landon Carter, of Sabine Hall, Virginia, communicated to the American Philosophical Society at Philadelphia some interesting "Observations concerning the Fly-Weevil that destroys Wheat." These were printed in the first volume of the "Transactions" of the Society, and were followed by some remarks on the subject by "the Committee of Husbandry." This is the earliest authentic account of the insect that I have met with. The Committee stated, that "it was said the injury of wheat from these flies began in North Carolina about forty years before, — and that they had extended gradually from Carolina into Virginia, Maryland, and the lower counties of Delaware, but had not then

* Olivier, Encyclopédie Méthodique, Insectes, Tom. IV. p. 115.

penetrated into Pennsylvania or passed the Delaware." They remarked, moreover, that the insects "appeared to be of the same kind with those that do the like mischief in Europe, as described to Mr. Duhamel by a gentleman of Angoumois."

Mr. Louis A. G. Bosc, who was sent by the French government, in 1796, to this country, where he spent several years, found the *Alucita cerealella* "so abundant in Carolina as to extinguish a candle when he entered his granary in the night." * This fly-weevil, or little grain-moth, has spread from North Carolina and Virginia, where its depredations were first observed, into Kentucky, and the southern parts of Ohio and Indiana, and probably more or less throughout the wheat region of the adjacent States, between the thirty-sixth and fortieth degrees of north latitude. But these are not the extreme limits of its occasional depredations, as it has been found even in New England, where, however, its propagation seems to have been limited by the length and severity of the winter. Wheat, barley, oats, and Indian corn suffer alike from it, the last especially when kept unprotected more than six or eight months.

Several essays on this insect have appeared in agricultural journals, none of which, however, were known to me when my first account of the Angoumois moth was written. One of these is an elaborate article by Edward Ruffin, Esq., of Hanover County, Virginia, printed in "The Farmers' Register" for November, 1833. The object of the writer is to prove, by a series of experiments, that there is a continued reproduction of the insect, in stored grain, at short intervals, throughout the warm season, or from the latter part of June till further increase is checked by cold weather. Mr. Ruffin thinks that but very few eggs are deposited on corn in the field, that these do not ordinarily hatch till the following summer, and that then they are sufficient to stock the whole

* Encyclopédie Méthodique, Agriculture, Tom. V. p. 243. — Mr. Bosc, a contributor to this work, resided some time at Wilmington, North Carolina.

crop of stored grain with their progeny. Mr. Samuel Judah, of Vincennes, Indiana, in a short and very sensible article, published in "The Indiana Farmer and Gardener" for October 4, 1845, seems to have come to nearly the same conclusions. Mr. Richard Owen, of New Harmony, Indiana, has given a very good history of this insect, accompanied with wood-cuts, in "The Cultivator," for July and November, 1846. To this I may have occasion again to refer, as also to two other articles, on the same subject, by Edward Ruffin, Esq., in the sixth volume of "The American Agriculturist," pages 52 and 93, published in February and March, 1847.

In the summer of 1840, Mr. E. C. Herrick, of New Haven, Connecticut, sent to me a few grains of wheat, that had been eaten by moth-worms precisely in the same way as grain is attacked by the Angoumois insect; and a gentleman, to whom this moth-eaten wheat was shown, informed me that he had seen grain thus affected in Maine. Unfortunately, the insects contained in this wheat were dead when received, having perished in the chrysalis state. Had they lived to finish their transformations, they would have afforded me an opportunity of ascertaining their suspected identity with the fly-weevil of Virginia, and the Angoumois moth of France. All my attempts to obtain specimens of the fly-weevil from the South and West were unsuccessful, till the 10th of November, 1845, when I had the pleasure of receiving a parcel of damaged wheat and a bottle full of the moths from Richmond, Virginia, through the kindness of Mr. John Dunlop Osborne, then a student in the Law School of Harvard College. Living specimens, and the insects in the worm or larva state, were still wanting. These were most unexpectedly obtained nearer home.

The late Samuel M. Burnside, Esq., of Worcester, told me, in the summer of 1844, that he had a quantity of corn, grown the year before, which had become infested with insects, and that he found great numbers of the insects, on the wing, in the room where the corn was kept. He also brought to me two large ears of corn from the infected heap. At that time, I was not aware that the fly-weevil attacked Indian corn, at least in New England; and these ears, appearing sound externally, were rolled up in several sheets of strong brown paper, securely tied, and laid away for future examination. They were forgotten, however, till December, 1845, when, upon opening the parcel, I found a great quantity of dead moths, and several living ones, in the paper. Every kernel appeared to have been perforated, and many of the kernels had three or four holes in each of them. Some contained the insect in the worm state, and some the fully formed chrysalis. The moths differed from the Virginia fly-weevil only in being rather larger, with blackish fore legs, and in having a more conspicuous blackish spot near the tips of the feelers, showing them to be merely varieties of the same species. This remark seems to be confirmed by the now well-known fact, that the fly-weevil, at the South and West, attacks corn as well as wheat, and by the statement of Mr. Owen, that "the insect found in corn does not differ from that found in wheat; it is usually," says he, "somewhat larger than the specimens from wheat, but this may be owing to the greater amount of nourishment which the corn has afforded." Moreover, we learn from the works of Olivier and of Bonafous,* that maize also suffers from the Angoumois moth in France. It is related that Kalm, the Swedish traveller, on finding some bugs in pease that he had carried home from this country, was filled with alarm, "fearing lest he might thereby introduce so great an evil into his beloved Sweden." With something of the same feeling, on finding what the insects were that had been depredating in my friend's corn-bin, I put the two ears of corn into a large glass jar, and corked it tight, to prevent the escape of any moths that might be

* Encyclopédie Méthodique, Insectes, Tom. IV. p. 121. Histoire du Maïs, par M. Bonafous, p. 111.

developed from worms and chrysalids remaining in the kernels. The next June, a swarm of moths appeared in the jar, in which they continued to propagate three years, successively, producing moths in considerable quantities in June and in August, with a smaller number at various intermediate times, except during the depth of winter.

These corn-moths, as already stated, were rather larger than those from the wheat, the wings of some of them expanding nearly six tenths of an inch.* The head is smooth, and not tufted. The antennæ are thread-like, with distinctly marked joints. The feelers are long and curved upwards; the terminal joint naked, acute, and blackish near the tip; the second or middle joint rather shorter and thicker, hairy beneath, and blackish on the outer side; the basal joint very short and hairy. The tongue makes several spiral turns, and, when extended, is about half the length of the antennæ. The body and fore wings are of that tint of pale brownish-gray which the French call coffee and milk color, and they have the lustre of satin. The fore wings are long and narrow, and are pointed at the end; together with their wide fringes, they are more or less sprinkled with blackish dots, especially near the tips. The hind wings are blackish, with a leaden lustre; they are narrow, and are very suddenly and obliquely contracted to a point at the tips; they are entirely surrounded with a blackish fringe, which is wider on the inner margin than the wing itself. They are folded lengthwise, when at rest, beneath the upper wings. The fore legs are blackish, and the hindmost legs are fringed with long hairs on the inner side. The chrysalis is obtuse at each end; the tail surrounded with a few minute points, three of which are larger than the rest; the rings of the body are smooth, or not

* Mr. Curtis, probably through inadvertence, has stated that *Butalis cerealella* "expands rather more than one inch." Half an inch is the true measure. See Journal of the Royal Agricultural Society of England, Vol. VII. p. 86. Compare Duponchel, Hist. Nat. des Lépidoptères de France, Supplement, Tom. IV. pl. 85, fig. 3.

notched; and the wing-cases extend nearly to the hinder extremity. The chrysalis-skin generally remains within the grain when the moth comes out; in some few cases, however, it was found sticking out of the orifice in the kernel, and sometimes in the crevices between the kernels. The foregoing minute description, which is taken from perfectly fresh and uninjured specimens, will serve to remove any doubt as to the genus and species to which this corn-moth is to be referred.

It has been proved by experience, that the ravages of the two kinds of grain-moths whose history has been now given can be effectually checked by drying the damaged grain in an oven or kiln; and that a heat of one hundred and sixty-seven degrees, by Fahrenheit's thermometer, continued during twelve hours, will kill the insects in all their forms. Indeed, the heat may be reduced to one hundred and four degrees with the same effect, but the grain must then be exposed to it for the space of two days. Insect-mills, somewhat like coffee-roasters on a large scale, have been invented in France, for the purpose of heating and agitating the infested wheat, by which the eggs and larvæ of the little corn-moth, or *Butalis*, are destroyed. Fumigation in close vessels, with the gas of burning charcoal, is found to be an effectual remedy; and Dr. Herpin states that this process neither imparts any bad flavor to the grain, nor does it impair its power of vegetating. He recommends also the early threshing and winnowing of wheat, as tending to preserve it.* This, indeed, is advocated by the most experienced wheat cultivators in this country, particularly if done by machinery; and it should not be deferred later than the end of July. The concussion and agitation undergone by the wheat in being threshed and winnowed, as intimated by Dr. Herpin, Mr. Judah, and others, is supposed to dis-

* See Duponchel, Lépidopt. de France, Supplem., Tom. IV. pp. 450 – 453; and Mr. Curtis's paper in the Journ. Royal Society of Agricult. of England, Vol. VII. pp. 87 – 89.

lodge the eggs and kill the larvæ of the insect. With the same view, Mr. Owen recommends passing the new wheat through "a rubbing mill, such as is used in Virginia and other large wheat-growing districts, to insure first-rate flour"; after which the wheat may be kept in bulk, or may be immediately ground. If a large surface of grain be exposed in the barn, the granary, or the mill, during the season of the moth, it will assuredly become affected; for, in the night, when these insects are most active and on the wing, they will light upon the exposed surface and deposit their eggs, which, in a few months of hot weather, will produce numerous and successive broods of moth-worms. To secure it from attack, therefore, the grain should be deposited in tight bins or casks, after having been properly prepared by being dried in a kiln, or even by exposure to the heat of the sun.

Some persons have succeeded perfectly in preserving grain from the corn-weevil and from the corn-moth by putting it into casks heated and fumigated with burning charcoal. The charcoal may be burnt in a portable furnace, lowered into the cask by a chain; and the grain should be poured in while the cask is hot. It has been observed that a low temperature checks the propagation of the corn-moth, and that the larvæ, or moth-worms, in the grain cannot survive the winter in those places where the thermometer falls to zero. Hence, in the cool and well-ventilated corn-barns of New England, grain will ordinarily be exempt from attack. During the summer, however, grain that has been brought from infected districts, or that has otherwise become contaminated, will be likely to suffer to some extent, even here. From these facts we learn how important it is that wheat and corn, which are to be kept over winter, for use, for sale, or for seed, should be previously well prepared, and should be deposited in suitable vessels in cool apartments, no matter how cold, provided they are also dry. It has been observed that very little corn is attacked in the field, the husks or shucks protecting it from the moths, which find only a few ears, whose ends protrude beyond the husks, whereon to deposit their eggs. Hence some persons recommend keeping corn in the husks, to preserve it from the corn-moth and also from the corn-weevil. This method is objectionable on account of the trouble it occasions, and the increased bulk of the corn; and it is less sure than the means above described.

Mr. Owen has made the interesting discovery, that the larvæ of the wheat-moth are sometimes preyed upon by still smaller larvæ, which, having destroyed their victims, are transformed to minute black ichneumon-flies. These have not yet been obtained from any of the samples of infected wheat or corn that have come under my notice; but, from the figures given of them by Mr. Owen in "The Cultivator," for November, 1846, they appear evidently to be Chalcidian parasites, and belong perhaps to the genus *Pteromalus*. Of these parasitical flies he remarks, that "some farmers had noticed large numbers among the tailings of the winnowing machine." Where they prevail, they doubtless contribute, in no small measure, to check the increase of the moths.

The Angoumois moth is unknown in England. Hence specimens of the American insect, sent by me to my friend, the late Mr. Edward Doubleday, of the British Museum, in December, 1845, were not immediately recognized by him and by Mr. Curtis, the celebrated English entomologist. Afterwards, on consulting the work of Duponchel on the Lepidoptera of France, they identified my specimens as belonging to the *Butalis cerealella*, the true Angoumois grain-moth, described and figured in that work. This identification is the more interesting and satisfactory, from the circumstance that I had not communicated to these gentlemen my belief that the insects were the same, and had given to them no account of the habits of my specimens, being desirous of obtaining their opinion unbiased by my own. I am not aware that any attempt had been made by European naturalists, before the publication of the first edition of this treatise, to determine the modern genus to which the Angoumois moth belongs, or to clear up and make known the synonymy of this species. This labor seems to have been left to an American, remote from the scene of the early and long continued depredations of the insect, and deprived of the common facilities enjoyed by European naturalists.

7. Feather-winged Moths. (*Alucitæ.*)

The last tribe of Lepidopterous insects remaining to be noticed contains the Alucitæ of Linnæus, or feather-winged moths, called Pterophoridæ by the French naturalists. These moths are easily known by their wings being divided lengthwise into narrow, fringed branches, resembling feathers. The fore wings in the genus *Pterophorus* are split, nearly half-way, into two, and the hind wings are divided, to the shoulder-joint, into three feathers; and each of the wings, in *Alucita*, consists of six feathers, connected only at the joint. The antennæ of these moths are slender and tapering; the tongue is long; the feelers are two in number, and of moderate length; and the body and legs are very long and slender. When at rest, their wings do not cover the body, but stand out from it on each side, not spread however, but folded together like a fan, so that only the outer part of each of the fore wings is visible. They fly slowly and feebly, some of them by day, and others only at night, and, when on the wing, they somewhat resemble the long-legged gnats. Their caterpillars are rather short and thick, are clothed with a few hairs, and have sixteen short legs. Most of them live on the leaves of low or herbaceous plants, and, when about to change to chrysalids, they fasten themselves by the hind feet and by a loop over the back, like the Lycænians. Those which belong to the genus *Alucita* are said to live in buds, and undergo their transformations in thin, transparent cocoons. The number of species in this tribe is small; and those that are found in this country are so few, and of so little consequence, in an economical point of view, that a particular description of them will not be necessary in this treatise.

CHAPTER VI.

HYMENOPTERA.

STINGERS AND PIERCERS. — HABITS OF SOME OF THE HYMENOPTERA. — SAW-FLIES AND SLUGS. — ELM SAW-FLY. — FIR SAW-FLY. — VINE SAW-FLY. — ROSE-BUSH SLUG. — PEAR-TREE SLUG. — HORN-TAILED WOOD-WASPS. — GALL-FLIES. — CHALCIDIANS. — BARLEY INSECT AND JOINT-WORM.

BEES, wasps, ants, saw-flies, and ichneumon-flies, of many different kinds, together with other insects, unknown by any common names in the English language, belong to the order HYMENOPTERA. Their wings are four in number, are traversed by a few branching veins, and are more or less transparent, or of a thin and filmy texture, as expressed by the name of the order, which signifies membranaceous wings. They fly swiftly, and are able to keep on the wing much longer than any other insects, because their bodies are light and compact, and their wings very thin, narrow, and withal very strong. They have four nippers or jaws; the upper pair being horny, stout, and fitted for biting or cutting; the lower are longer and softer, and, with the lower lip, which they cover, form a kind of beak or sucker. Their antennæ vary in form and length; but are most often cylindrical, and of equal thickness to the end. The males have no weapons of offence or defence except their jaws. The females are armed with a venomous sting, concealed within the end of the hind body, or are provided with a piercer, of some sort, for boring or sawing the holes wherein their eggs are deposited. Hence the insects of this order may be divided into two groups, Stingers and Piercers. Though both of them undergo a complete transformation in coming to maturity, they differ from each other in the early states of their existence.

The young of all the stinging Hymenoptera are soft, white, and maggot-shaped, and are without legs; some of those of the Piercers have the same form, but the others more nearly resemble grubs and caterpillars, having a horny head, and six jointed legs, and some of them numerous fleshy prop-legs besides. The latter, when food fails them in one place, are able to creep to another, and can look out for themselves a proper place of shelter, wherein to go through with their transformations. The others are exceedingly helpless, and depend wholly upon the instinctive foresight of their parents, or the daily care of attentive nurses, for their food and habitations. When fully grown, nearly all of these young insects spin oblong oval cocoons, wherein they change to chrysalids, and finally to winged insects. A few, however, never obtain wings in the adult state; but these are mostly certain neuter and female ants, the males of which possess wings. With the exception of the white ants, belonging to another order, it is only among Hymenopterous insects that we find certain individuals constantly barren, and hence called neuters. These form the principal part of those communities of bees, of wasps, and of ants, that unite in making a habitation for the whole swarm, and in providing a stock of provisions for their own use, and for that of their helpless brood; and nearly or quite all the labor falls upon these industrious neuters, whose care and affection for the young, which they foster and shelter, could not be greater were they their own offspring.

Hymenopterous insects love the light of the sun; they take wing only during the daytime, and remain at rest in the night, and in dull and wet weather. They excel all other insects in the number and variety of their instincts, which are wonderfully displayed in the methods employed by them in providing for the comfort and the future wants of their offspring. In the introductory chapter some remarks have already been made on their habits and economy;

and the limits of this work will not allow me now to enlarge upon them. I shall not, therefore, attempt to show how admirably the Hymenoptera are fitted, in the formation of all their parts, for their appointed tasks. If any of my readers are curious to learn this, and to witness for themselves the various arts, resources, and contrivances resorted to by these insects, let them go abroad in the summer, and watch them during their labors. They will then see the saw-fly making holes in leaves with her double key-hole saws, and the horn-tail boring with her auger into the solid trunks of trees; — they will not fail to observe and admire the untiring scrutiny of the ichneumon-flies, those little busy-bodies, forever on the alert, and prying into every place to find the lurking caterpillar, grub, or maggot, wherein to thrust their eggs; — the curious swellings produced by the gall-flies, and inhabited by their young; — the clay cells of the mud-wasp, plastered against the walls of our houses, each one containing a single egg, together with a large number of living spiders, caught and imprisoned therein solely for the use of the little mason's young, which thus have constantly before them an ample supply of fresh provisions; — the holes of the stump-wasp, stored with hundreds of horse-flies for the same purpose; — the skill of the leaf-cutter bee in cutting out the semicircular pieces of leaves for her patchwork nest; — the thimble-shaped cells of the ground-bee, hidden, in clusters, under some loose stone in the fields, made of little fragments of tempered clay, and stored with bee-bread, the work of many weeks for the industrious laborer; — the waxen cells made by the honey-bee, without any teaching, upon purely mathematical principles, measured only with her antennæ, and wrought with her jaws and tongue; — the water-tight nests of the hornet and wasp, natural paper-makers from the beginning of time, who are not obliged to use rags or ropes in the formation of their durable paper combs, but have applied to this purpose fibres of wood, a material that the art of man has not yet been able to manufacture into paper;—the herculean labors of ants in throwing up their hillocks, or mining their galleries, compared wherewith, if the small size of the laborers be taken into account, the efforts of man in his proudest monuments, his pyramids and his catacombs, dwindle into insignificance. These are only a few of the objects deserving of notice among the insects of this order; many others might be mentioned, that would lead us to observe with what consummate skill these little creatures have been fashioned, and how richly they have been endowed with instincts that never fail them in providing for their own welfare, and that of their future progeny.

Comparatively speaking, there are not many of the Hymenoptera which are actually or seriously injurious to vegetation. Those which I propose now to describe are not provided with venomous stings, and, consequently, are to be included among the Piercers.

Such are the saw-flies (TENTHREDINIDÆ), insects that are found on the leaves of plants, and live almost entirely on vegetable food. They are the least active of the Hymenoptera, are sluggish in their habits, fly heavily and but little, and do not attempt to escape when touched. Most of them are rather short and somewhat flattened. They have a broad head, which, seen from above, appears transversely square. The hind body is not narrowed to a point where it joins the thorax, but is as broad as the latter, and is closely united to it. The antennæ are generally short; but they vary much in form; in many species they are thread-like and slightly tapering; in some, thickened or knobbed at the end; more rarely, they end suddenly with a few very small joints, much more slender than the rest; they are feathered in some males, and notched in the other sex; and sometimes they are forked, or divided into long branches. Their wings cross and overlap each other, and cover the back horizontally when closed. But the most striking peculiarity of these insects consists in the double saws wherewith the females are provided. These are lodged in a deep chink under the hinder part of the body, like the blade of a penknife in its handle, and are covered by two narrow, scabbard-like pieces. The saws are two in number, placed side by side, with their ends directed backwards, and are so hinged to the under side of the body that they can be withdrawn from the chink, and moved up and down when in use. They vary in their form, and in the shape of their teeth, in different kinds of saw-flies; but they generally curve upwards and taper towards the end, and are toothed along the lower or convex edges. Each of the saws, like a carpenter's fine saw, has a back to steady it; the blade, however, is not fastened to the back, but slides backwards and forwards upon it. Moreover, the saw-blade is not only toothed on the edge, but is covered, on one side, with transverse rows of very fine teeth, giving to it the power of a rasp, as well as that of a saw.

The female saw-flies use these ingeniously contrived tools to saw little slits in the stems and leaves of plants, wherein they afterwards drop their eggs. Some, it appears, lay their eggs in fruits; for Mr. Westwood discovered their young within apples that had fallen from the trees before they had grown to the size of walnuts. The wounds made in plants by some kinds of saw-flies swell, and produce galls or knobs, that serve for habitations and for food to their young. The eggs, themselves, of all these flies, are found to grow, and increase to twice their former size after they are laid, probably by absorbing the sap of the plant through their thin shells.

Most of the larvæ or young of the saw-flies strikingly resemble caterpillars, being usually of a cylindrical form, of a greenish color, and having several pairs of legs. Hence they are sometimes called false caterpillars. With the exception of such as belong to the genera *Lyda* and *Cephus*, in which the legs are only six, and the prop-legs are entirely wanting, these false caterpillars have a greater number of legs than true caterpillars, being provided with from eighteen to twenty-two; but their prop-legs have not the numerous little hooks that arm those of caterpillars. They have the means of spinning silk from their lower lips, but not often in any great quantity. They are mostly naked and without hairs; a few have forked prickles on their backs; some are covered with a white flaky substance, that easily rubs off; and others have a dark-colored slimy skin, which has caused them to be called slugs or slug-worms. They shed their skins about four times, and, after the last moulting, often materially change in appearance. Not only do these insects resemble caterpillars in their forms, but they have nearly the same habits. They are generally found on the leaves of plants, which they devour. Many kinds are altogether solitary; a few live together in swarms, under silken webs, which they spin for a common place of shelter; others are found also in swarms, but without any webs over them, and, when disturbed, they throw up their heads and tails, in a very odd way; some roll up leaves, and live in the hollow thus formed, like the Tortrices; others make portable cases of bits of leaves, which they carry about on their backs, like the Tineæ; certain kinds live within the stems of plants, and devour the pith; and wheat, in Europe, is said to suffer considerable injury from internal feeders (*Cephus pygmæus*) of this kind. When fully grown, most of them go into the ground, and enclose themselves in thin silken cocoons, of an oblong oval shape, coated with grains of earth. Some make much thicker cocoons, in texture resembling parchment, and fasten them to the plants on which they live, or conceal them in crevices, or under leaves and stones on the ground. They generally remain for a long time unchanged in their cocoons, most of them during the winter; are transformed to chrysalids, of a whitish color, in the spring, and come out in the winged form soon afterwards. Of some kinds there are two broods in the course of the summer, the false caterpillars of the first brood coming to their growth, and passing through all their transformations, within six or seven weeks from their first appearance.

The names of above sixty native species of saw-flies may be found in my "Catalogue of the Insects of Massachusetts." Some of these are very interesting in their appearance and habits in the caterpillar state. In what follows, an account will be given of one of the largest species, and of some smaller kinds, that have been found very injurious to cultivated plants.

Our largest saw-fly belongs to the genus *Cimbex* (Plate VIII. Fig. 12, *Cimbex Laportei*). This name was originally given by the Greeks to certain insects resembling bees and wasps, but not producing honey. It therefore applies very well to some kinds of saw-flies, such as the female of this species, which, at first sight, might be mistaken for a hornet. Her head and thorax are shining black. Her hind body is oval, and of a steel-blue or deep violet color, with three or four oval yellowish spots on each side. Her antennæ are buff-colored, except at the base, where they are dusky; they are short, and end with an egg-shaped knob. Her wings are smoky brown, and semitransparent. Her legs are blue-black, and her feet pale yellow. The length of her body varies from three quarters to seven eighths of an inch, and her wings expand an inch and three quarters or more. In the manuscript lectures of the late Professor Peck, she is called *Cimbex Ulmi*, because she inhabits the elm. The male is the *Cimbex Americana* of Dr. Leach, and differs so much from the female, that it might be taken for a different species. His body is longer and narrower than that of the female, and wants the white spots on the sides; and there is a transverse, oval hole, filled with a whitish film, behind the thorax, which is hardly perceptible in the other sex. His hind legs are very thick; the shins are bowed, and hairy within; and the first joint of his feet ends with a stout hook, curved inwards. He often measures an inch in length, and his wings expand about two inches.

These insects appear from the latter part of May to the middle of June, during which period the female lays her eggs upon the common American elm, the leaves whereof are the food of her young. (Fig. 243, larva.) The latter come to their growth in August, and then measure from one inch and a half to two inches in length. They are rather thick, and nearly cylindrical in form, and have twenty-two legs, or a pair to every ring except the fourth. They have a firm, rough skin, of a pale greenish yellow color, covered with numerous transverse wrinkles, with a black stripe, consisting of two narrow black lines, along the top of the back, from the head to the tail; and their spiracles, or breathing-holes, are also black. When at rest, they lie on their sides, curled up in a spiral form, and in this position look not much unlike some kinds of cockle or snail shells.

Fig. 243.

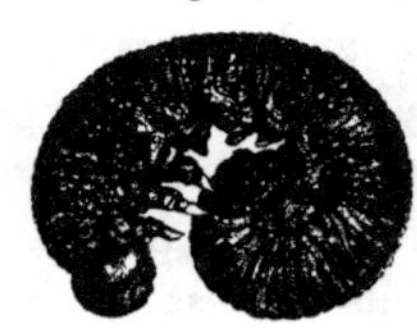

Like all the false caterpillars of the genus Cimbex, this insect, when handled or disturbed, betrays its fears or its displeasure by spirting out a watery fluid from certain little pores situated on the sides of its body just above its spiracles. After its feeding state is over, it crawls down from the tree to the ground, and conceals itself under fallen leaves or other rubbish, and there makes an oblong oval, brown cocoon (Plate VIII. Fig. 11), very closely woven, as tough as parchment, and about an inch in length. In this the false caterpillar remains unchanged throughout the winter, and is not transformed to a chrysalis till the following spring. At length the insect bursts its chrysalis skin, and, by pushing against the end of its cocoon, forces off a little circular piece like a lid, and through the opening thus made it comes forth in its winged form.

For some years past many of the fir-trees, cultivated for

ornament, in this vicinity, have been attacked by swarms of false caterpillars, and, in some instances that have fallen under my notice, have been nearly stripped of their leaves every summer, and in consequence thereof have been checked in their growth, and now seem to be in a sickly condition. These destructive insects agree in their habits and in their general appearance, in all their states, with the pine and fir saw-flies, described by Kollar,* by whose ravages whole forests of these trees have been destroyed in some parts of Germany. It is probable, however, that the American fir saw-flies are not identical with those of Europe, as they differ from them rather too much to have originated from the same stock; neither do they sufficiently agree with Dr. Leach's descriptions of *Lophyrus Americanus*, *Abbotii*, *compar*, &c.; and therefore I propose to name this apparently undescribed species *Lophyrus Abietis*, the Lophyrus of the fir-tree.

The following is a description of the insect in its winged state. The two sexes differ very much from each other in size and color, and still more remarkably in the form of their antennæ. The male (Plate VIII. Fig. 3) is the smallest, measures one quarter of an inch in length, and expands his wings about two fifths of an inch. His body is black above, and brown beneath; his wings are transparent, with changeable tints of rose-red, green, and yellow; and his legs are wholly of a dirty leather-yellow color. His antennæ (Plate VIII. Fig. 4, magnified) resemble very short, black feathers, wide at the end, and narrowed to a point, and are curled inwards on each edge, so as to appear hollow. The genus *Lophyrus* derives its name from the plume-like crest on the heads of the male insects. The body of the female (Plate VIII. Fig. 5) is about three tenths of an inch long, and her wings expand half an inch or more. She is of a yellowish-brown color above, with a short blackish stripe on each side of the middle of the

* Treatise, pp. 340 and 347.

thorax; her body beneath and her legs are paler, of a dirty leather-yellow color; and her wings resemble those of the male. Her antennæ are short, taper to a point, consist of nineteen joints, and are toothed on one side like a saw. My specimens of this kind of saw-fly, which were raised from the caterpillars in the summer of 1838, came out of their cocoons towards the end of July in the same year; but I have also found them on pines and firs early in May.

The European pine saw-flies lay their eggs in slits which they make with their saws in the edges of the leaves; and it is probable that our fir saw-flies proceed in the same way. In June and July the false caterpillars of the latter may be found on firs; and, according to notes made by me many years ago, the same insects, or some very much like them, were observed on the leaves of the pitch-pine also. They are social in their habits, living together in considerable swarms, and so thick that sometimes two may be seen feeding together on the same leaf, and sitting opposite to each other. In order to lay hold of the leaf more firmly, they curl the hinder part of the body around it; and, if they are disturbed, they throw up their heads and tails with a jerking motion. When fully grown, they are from five to six tenths of an inch in length; they are nearly cylindrical in form, thickest before the middle, and tapering behind, and have twenty-two legs. The head and the first three pairs of legs are black. The body is of a pale and dirty green color above, with a light stripe along the top of the back, separating two of a darker-green color; there are two dark green stripes on each side of the body; and the belly and prop-legs are yellowish. When young, the two stripes on the back are much darker, and those on the sides are nearly black. The skin, though covered with very fine transverse wrinkles, is not rough, and with a magnifying-glass a few short hairs may be seen scattered over it. After the last moulting their color fades, and they become almost yellow. The greater part of them then suddenly leave the trees, either by travelling down the trunks, or by falling from the branches to the ground. A few, either from weakness or from some other cause, remain on the trees, make their cocoons among the leaves, and rarely finish their transformations, most of them perishing from the internal attacks of ichneumon-grubs. Some creep into cracks in fences, and into other crevices; but most of those which reach the ground bury themselves under decayed leaves, or among the roots of the grass, and in such secure places make their cocoons. The latter are oblong oval cases, of tough grayish silk, and measure nearly three tenths of an inch in length. In due time the insects change to saw-flies, and come out of their cocoons, one end whereof separates, like a lid, to allow of their escape. Although some of them are found to finish their transformations in August, it is probable that the greater part of them remain unchanged in the ground till the following spring.

No means for the destruction of the caterpillars of the fir saw-fly have been tried here, except showering them with soapsuds, and with solutions of whale-oil soap, which has been found effectual. They may also be shaken off or beaten from the trees, early in the morning, when they are torpid and easily fall, and may be collected in sheets, and be burned or given to swine. For other means to check their depredations, the reader may consult the articles on the pine and fir saw-flies of Europe, contained in Kollar's "Treatise."

The following account of a kind of saw-fly which attacks the grape-vine is chiefly extracted from my Discourse before the Massachusetts Horticultural Society, in 1832, where the insect is named *Selandria Vitis*. The saw-fly of the vine is of a jet-black color, except the upper side of the thorax, which is red, and the fore legs and under side of the other legs, which are pale yellow or whitish. The wings are semitransparent, of a smoky color, with dark brown veins. The body of the female (Fig. 244) measures one quarter of an inch in length, that of the male is somewhat shorter. These flies rise from the ground in the spring, not all at one time, but at irregular intervals, and lay their eggs on the lower side of the terminal leaves of the vine.

Fig. 244.

In the month of July the false caterpillars, hatched from these eggs, may be seen on the leaves, in little swarms, of various ages, some very small, and others fully grown. They feed in company, side by side, beneath the leaves, each swarm or fraternity consisting of a dozen or more individuals, and they preserve their ranks with a surprising degree of regularity. Beginning at the edge they eat the whole of the leaf to the stalk, and then go to another, which in like manner they devour, and thus proceed, from leaf to leaf, down the branch, till they have grown to their full size. They then average five eighths of an inch in length, are somewhat slender and tapering behind, and thickest before the middle. They have twenty-two legs. The head and the tip of the tail are black; the body, above, is light green, paler before and behind, with two transverse rows of minute black points across each ring; and the lower side of the body is yellowish. After their last moulting they become almost entirely yellow, and then leave the vine, burrow in the ground, and form for themselves small oval cells of earth, which they line with a slight silken film. In about a fortnight after going into the ground, having in the mean time passed through the chrysalis state, they come out of their earthen cells, take wing, pair, and lay their eggs for a second brood. The young of the second brood are not transformed to flies until the following spring, but remain at rest in their cocoons in the ground through the winter.

For some years previous to the publication of my Discourse, I observed that these insects annually increased in number, and in the year 1832 they had become so numerous and destructive that many vines were entirely stripped of their leaves by them. Whether the remedies then proposed by me, or any other means, have tended to diminish their numbers, or to keep them in check, I have not been able to ascertain, and have had no further opportunity for making observations on the insects themselves. At that time, air-slacked lime, which was found to be fatal to these false caterpillars of the vine, was advised to be dusted upon them, and strewed also upon the ground under the vines, to insure the destruction of such of the insects as might fall. A solution of one pound of common hard soap in five or six gallons of soft water is used by English gardeners to destroy the young of the gooseberry saw-fly; and the same was recommended to be tried upon the insects under consideration.

All the young of the saw-flies do not so closely resemble caterpillars as the preceding; some of them, as has already been stated, have the form of slugs or naked snails. Of this description is the kind called the slug-worm in this country, and the slimy grub of the pear-tree in Europe. So different are these from the other false caterpillars, that they would not be suspected to belong to the same family. Their relationship becomes evident, however, when they have finished their transformations; and accordingly we find that the saw-flies of our slug-worms and those of the vine are so much alike in form and structure, that they are both included in the same genus. Moreover, there are certain false caterpillars intermediate in their forms and appearance between the slimy and slug-like kinds and those that more nearly resemble the true caterpillars; thus admirably illustrating the truth of the remark, that nature proceeds not with abrupt or unequal steps; * or, in other words, that, amidst the immense variety of living forms wherewith this earth has been peopled, there is a regular gradation

* "Natura saltus non facit." — Linnæus, Syst. Nat. I. 11.

and connection, which in particular cases if we fail to discover, it is rather to be attributed to our own ignorance and short-sightedness than to any want of harmony and regularity in the plan of the Creator. In considering the resemblances of species, we cannot fail to admire the care that has been taken, by almost insensible shades of difference among them, or by peculiar circumstances controlling their distribution, their habits of life, and their choice of food, to prevent them from commingling, whereby each species is made to preserve forever its individual identity.

The saw-fly of the rose, which, as it does not seem to have been described before, may be called *Selandria Rosæ* (Fig. 245), from its favorite plant, so nearly resembles the slug-worm saw-fly as not to be distinguished therefrom except by a practised observer. It is also very much like *Selandria barda*, *Vitis*, and *pygmæa*, but has not the red thorax of these three closely allied species. It is of a deep and shining black color. The first two pairs of legs are brownish-gray or dirty white, except the thighs, which are almost entirely black. The hind legs are black, with whitish knees. The wings are smoky, and transparent, with dark brown veins, and a brown spot near the middle of the edge of the first pair. The body of the male is a little more than three twentieths of an inch long, that of the female one fifth of an inch or more, and the wings expand nearly or quite two fifths of an inch. These saw-flies come out of the ground, at various times, between the 20th of May and the middle of June, during which period they pair and lay their eggs. The females do not fly much, and may be seen, during most of the day, resting on the leaves, and, when touched, they draw up their legs, and fall to the ground. The males are more active, fly from one rose-bush to another, and hover around their sluggish partners. The latter, when about to lay their

Fig. 245.

eggs, turn a little on one side, unsheathe their saws, and thrust them obliquely into the skin of the leaf, depositing, in each incision thus made, a single egg. The young (Fig. 246) begin to hatch in ten days or a fortnight after the eggs are laid. They may sometimes be found on the leaves as early as the first of June, but do not usually appear in considerable numbers till the 20th of the same month. How long they are in coming to maturity, I have not particularly observed; but the period of their existence in the caterpillar state probably does not exceed three weeks. They somewhat resemble young slug-worms in form, but are not quite so convex. They have a small, round, yellowish head, with a black dot on each side of it, and are provided with twenty-two short legs. The body is green above, paler at the sides, and yellowish beneath; and it is soft, and almost transparent, like jelly. The skin of the back is transversely wrinkled, and covered with minute elevated points; and there are two small, triple-pointed warts on the edge of the first ring, immediately behind the head.

Fig. 246.

These gelatinous and sluggish creatures eat the upper surface of the leaf in large irregular patches, leaving the veins and the skin, beneath, untouched; and they are sometimes so thick that not a leaf on the bushes is spared by them, and the whole foliage looks as if it had been scorched by fire, and drops off soon afterwards. They cast their skins several times, leaving them extended and fastened on the leaves; after the last moulting they lose their semitransparent and greenish color, and acquire an opaque yellowish hue. They then leave the rose-bushes, some of them slowly creeping down the stem, and others rolling up and dropping off, especially when the bushes are shaken by the wind. Having reached the ground, they burrow to the depth of an inch or more in the earth, where each one makes for itself a small oval cell (Fig. 247), of grains of earth, cemented with a little gummy silk.

Fig. 247.

Having finished their transformations, and turned to flies, within their cells, they come out of the ground early in August, and lay their eggs for a second brood of young. These, in turn, perform their appointed work of destruction in the autumn; they then go into the ground, make their earthen cells, remain therein throughout the winter, and appear, in the winged form, in the following spring and summer.

During several years past, these pernicious vermin have infested the rose-bushes in the vicinity of Boston, and have proved so injurious to them as to have excited the attention of the Massachusetts Horticultural Society, by whom a premium of one hundred dollars for the most successful mode of destroying these insects was offered, in the summer of 1840. In the year 1832, I first observed them in gardens in Cambridge, and then made myself acquainted with their transformations. At that time they had not reached Milton, my former place of residence, and they did not appear in that place till six or seven years later. They now seem to be gradually extending in all directions, and an effectual method for preserving our roses from their attacks has become very desirable to all persons who set any value on this beautiful ornament of our gardens and shrubberies. Showering or syringing the bushes with a liquor made by mixing with water the juice expressed from tobacco by tobacconists, has been recommended; but some caution is necessary in making this mixture of a proper strength, for if too strong it is injurious to plants; and the experiment does not seem, as yet, to have been conducted with sufficient care to insure safety and success. Dusting lime over the plants when wet with dew has been tried, and found of some use; but this and all other remedies will probably yield in efficacy to Mr. Haggerston's mixture of whale-oil soap and water, in the proportion of two pounds of the soap to fifteen gallons of water.

Particular directions, drawn up by Mr. Haggerston himself, for the preparation and use of this simple and cheap application, may be found in the "Boston Courier" for the 25th of June, 1841, and also in most of our agricultural and horticultural journals of the same time. The utility of this mixture has already been repeatedly mentioned in this treatise, and it may be applied in other cases with advantage. Mr. Haggerston finds that it effectually destroys many kinds of insects; and he particularly mentions plant-lice, red spiders, canker-worms, and a little jumping insect which has lately been found quite as hurtful to rose-bushes as the slugs or young of the saw-fly. The little insect alluded to has been mistaken for a Thrips or vine-fretter; it is, however a leaf-hopper, or species of *Tettigonia*, and is described in a former part of this treatise.

According to the plan to which I have found it necessary to limit this work, only one more species of saw-fly remains to be described. Of the habits and transformations of this insect the late Professor Peck has given us an admirable account, under the title of a "Natural History of the Slug-Worm," which was printed in Boston, in the year 1799, by order of the Massachusetts Agricultural Society, and obtained the Society's premium of fifty dollars and a gold medal. As my own observations on this insect agree perfectly with those of Professor Peck, in the following remarks I have merely abridged and condensed his "Natural History of the Slug-Worm," a work now out of print, and rarely to be met with. It will be proper to premise that Professor Peck was inclined to believe this slug-fly to be a variety of the *Tenthredo Cerasi* of Linnæus, an insect found more commonly on the pear-tree in Europe than on the cherry, although it has a specific name derived from the latter tree.

Most naturalists now reject the name given by Linnæus to the slimy grub of the pear-tree, because it is not strictly correct, and substitute a specific name imposed upon it by Fabricius. The European insect, therefore, is now called *Selandria* (*Blennocampa*) *Æthiops*; and a good account of it, by Mr. Westwood, may be found in the thirteenth volume of "The Gardener's Magazine." It is possible that our slug-fly may have been imported from Europe, and it may turn out to be really a mere variety of the European insect. Professor Peck was aware that it did not agree with the description given by Linnæus of the latter; and it appears to me that the difference between the two insects, in their winged state, is enough to entitle them to be considered as specifically distinct from each other. For this reason I shall retain for our insect the specific name adopted by Professor Peck, because this slug does really live upon the cherry, in this country, as well as on the pear-tree; and shall merely prefix to it the generical name which it should bear according to modern nomenclature. The fly of our slug-worm may therefore be called *Selandria* (*Blennocampa*) *Cerasi*. The meaning of the word *Selandria* is unknown to me. *Blennocampa* signifies slimy caterpillar, a name which, it will be seen, may be applied with great propriety to our slug-worm.

This slug-fly is of a glossy black color, except the first two pairs of legs, which are dirty yellow or clay-colored, with blackish thighs, and the hind legs, which are dull black, with clay-colored knees. The wings are somewhat convex and rumpled or uneven on the upper side, like the wings of the saw-flies generally. They are transparent, reflecting the changeable colors of the rainbow, and have a smoky tinge, forming a cloud or broad band across the middle of the first pair; the veins are brownish. The body of the female (Fig. 248) measures rather more than one fifth of an inch in length; that of the male is smaller. In the year 1828, I observed these saw-flies, on cherry and plum trees, in Milton, on the 10th of May; but they usually appear towards the end of May or early in June. Soon afterwards some of them begin to lay their eggs, and all of them finish this business and disappear within the space of three weeks. Their eggs

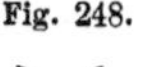
Fig. 248.

are placed, singly, within little semicircular incisions through the skin of the leaf, and generally on the lower side of it. The flies have not the timidity of many other insects, and are not easily disturbed while laying their eggs. On the fourteenth day afterwards, the eggs begin to hatch, and the young slug-worms (Fig. 249) continue to come forth from the 5th of June to the 20th of July, according as the flies have appeared early or late in the spring.

Fig. 249.

At first the slugs are white; but a slimy matter soon oozes out of their skin and covers their backs with an olive-colored sticky coat. They have twenty very short legs, or a pair under each segment of the body except the fourth and the last. The largest slugs are about nine twentieths of an inch in length, when fully grown. The head, of a dark chestnut color, is small, and is entirely concealed under the fore part of the body. They are largest before, and taper behind, and in form somewhat resemble minute tadpoles. They have the faculty of swelling out the fore part of the body, and generally rest with the tail a little turned up. These disgusting slugs live mostly on the upper side of the leaves of the pear and cherry trees, and eat away the substance thereof, leaving only the veins and the skin beneath untouched. Sometimes twenty or thirty of them may be seen on a single leaf; and in the year 1797 they were so abundant, in some parts of Massachusetts, that small trees were covered with them, and the foliage entirely destroyed; and even the air, by passing through the trees, became charged with a very disagreeable and sickening odor, given out by these slimy creatures. The trees attacked by them are forced to throw out new leaves, during the heat of the summer, at the ends of the twigs and branches that still remain alive; and this unseasonable foliage, which should not have appeared till the next spring, exhausts the vigor of the trees, and cuts off the prospect of fruit.

The slug-worms come to their growth in twenty-six days, during which period they cast their skins five times. Frequently, as soon as the skin is shed, they are seen feeding upon it; but they never touch the last coat, which remains stretched out upon the leaf. After this is cast off, they no longer retain their slimy appearance and olive color, but have a clean yellow skin, entirely free from viscidity. They change also in form, and become proportionally longer; and their head and the marks between the rings are plainly to be seen. In a few hours after this change, they leave the trees, and, having crept or fallen to the ground, they burrow to the depth of from one inch to three or four inches, according to the nature of the soil. By moving their body, the earth around them becomes pressed equally on all sides, and an oblong oval cavity is thus formed, and is afterwards lined with a sticky and glossy substance, to which the grains of earth closely adhere. Within these little earthen cells or cocoons the change to chrysalids takes place; and, in sixteen days after the descent of the slug-worms, they finish their transformations, break open their cells, and crawl to the surface of the ground, where they appear in the fly form. These flies usually come forth between the middle of July and the first of August, and lay their eggs for a second brood of slug-worms. The latter come to their growth, and go into the ground, in September and October, and remain there till the following spring, when they are changed to flies, and leave their winter quarters. It seems that all of them, however, do not finish their transformations at this time; some are found to remain unchanged in the ground till the following year; so that, if all the slugs of the last hatch in any one year should happen to be destroyed, enough from a former brood would still remain in the earth to continue the species.

The disgusting appearance and smell of these slug-worms do not protect them from the attacks of various enemies. Mice and other burrowing animals destroy many of them in their cocoons, and it is probable that birds also prey upon them when on the trees, both in the slug and the winged states. Professor Peck has described a minute ichneumon-fly, stated by Mr. Westwood to be a species of *Encyrtus*, that stings the eggs of the slug-fly, and deposits in each one a single egg of her own. From this, in due time, a little maggot is hatched, which lives in the shell of the slug-fly's egg, devours the contents, and afterwards is changed to a chrysalis, and then to a fly like its parent. Professor Peck found that great numbers of the eggs of the slug-fly, especially of the second hatch, were rendered abortive by this atom of existence.

Ashes or quicklime, sifted on the trees by means of a sieve fastened to the end of a pole, was recommended, by the late Hon. John Lowell, of Roxbury, for the destruction of the slugs; and it is found to answer the purpose. It is probable that Mr. Haggerston's almost universal remedy may prove to be still more effectual.

The saw-flies, though undoubtedly belonging to the order Hymenoptera, depart from the general characters thereof more than any other insects in it. They are more dull and heavy in all their motions; they have not the powerful jaws of the predaceous tribes, nor the long and slender lower jaws and tongue of those that subsist upon honey. They live but a short time, and their food appears to be pollen, the tender parts of leaves, and sometimes the plant-lice and other soft-bodied insects frequenting flowers. In the stiffness of their upper wings, and the heaviness of their flight, they somewhat resemble beetles, and, analogically, may be said to typify the Coleoptera, or, in other words, they may be called the beetles of the Hymenoptera. They will be found, on comparison, to have some features in common with the crickets, which, with the earwigs, are also the representatives of the Coleoptera. Although they differ essentially from butterflies and moths, the resemblance of most of their young to caterpillars, in form and in habits, is very striking and remarkable. Hence the saw-flies plainly show the relation existing between the orders Lepidoptera and Hymenoptera, and serve closely to connect them together.

The next piercing insects to be described belong to the family of Uroceridæ, or horn-tails, so called because they have a horny point at the end of the body. The Germans call them wood-wasps. Their antennæ are slender, and thread-like, or tapering. They have a large head, convex before, and flat behind where it joins the thorax. Their wings are long, narrow, and strong, and overlap on the top of the back, when closed. The body is very long, and nearly or quite cylindrical; the thorax and the after part of the body are of equal thickness, and are closely joined together. The horn, at the end, is short, and conical or triangular, in the males; longer, and sometimes spear-pointed, in the females. Moreover, the latter are provided with a long, cylindrical borer, hinged to the middle of the belly, which is furrowed to receive it. The borer usually extends some distance beyond the end of the body, and consists of five pieces. The two outermost are grooved within, and, when shut, form a hollow tube or scabbard to the others, one of which represents the two backs of the saws of the saw-flies, joined together, and encloses two needles for boring holes. The part serving for a back to these needles is notched on each side, and the needles themselves, which are as fine as a hair, and as strong and elastic as wire, have several small teeth along the lower side towards the end. These needles, and the back in which they play, are so connected as to appear to be only a single spear-pointed awl. With this complicated and powerful tool the females bore holes into the trunks of trees, wherein they drop their eggs. Their young are cylindrical and fleshy grubs, of a whitish color, with a small, rounded, horny head, and a pointed and horny tail. They have six very small legs under the fore part of the body, and are provided with strong and powerful jaws, wherewith they bore long holes in the trunks of the trees that they inhabit. Like other borers, these grubs are wood-eaters, and often do great damage to pines and firs, wherein they are most commonly found.

When fully grown, the grubs make thin cocoons of silk, interwoven with little chips, in their burrows, and in them go through their transformations. The chrysalis is somewhat like the winged insect in form, but is of a yellowish white color till near the time of its last change, and the wings and legs are folded under the breast; in all these respects it agrees with the chrysalids of other Hymenopterous insects. After the chrysalis skin is cast off, the winged insect breaks through its cocoon, creeps to the mouth of its burrow, and gnaws through the covering of bark over it, so as to come out of the tree into the open air. It is stated that the grubs of the large species come to their growth in seven weeks after the eggs are laid. If this be true, and it seems hardly possible, the chrysalis state must last a long time, for the perfected insects have been known to come out of timber that had been cut up and applied to mechanical uses by the carpenter. Some persons have supposed that they attacked only diseased and decayed trees, in which it must be admitted they are often found in great numbers. But many instances might be mentioned of their appetite for sound wood also, and it is probable that the presence of these insects, like that of many others, is the cause, and not the consequence, of the decay of the trees wherein they live.

It is stated in the London "Zoölogical Journal," that two hundred Scotch firs have been destroyed by the *Urocerus Juvencus*, in the woods of Henham Hall, the seat of the Earl of Stanhope, their trunks being bored through and through by the grubs of this insect. Mr. Westwood relates,* that a piece of wood, twenty feet in length, from

* Introduction to the Modern Classification of Insects, Vol. II. p. 118.

a fir-tree in Bewdley Forest, Worcestershire, England, was found to be so intersected by the burrows of these grubs, as to be fit for nothing but firewood; and that the winged insects continued to come out of it, at the rate of five, six, or more each day, for the space of several weeks. Mr. Marsham states, on the authority of Sir Joseph Banks, that several specimens of *Urocerus gigas* were seen to come out of the floor of a nursery in a gentleman's house, to the no small alarm and discomfiture of both nurse and children. The grubs must therefore have existed in the boards or timbers before they were employed in building, and these materials would not have been used if in a decayed state. The sexes of most of these insects differ considerably in size and color, and in the shape of their body and of their hind legs. There are not many different kinds, but they are very prolific, and abound in mountainous districts, and in temperate climates, where forests of pines and firs prevail. A new order was proposed for their reception by Mr. Macleay, and was named *Bomboptera*, on account of the humming sound that they make in flying. Their young partake of the nature of the wood-eating grubs of the capricorn beetles, which therefore they may be said to represent, as the saw-flies do some of the leaf-eating insects of the same order.

Eight of the UROCERIDÆ are enumerated in my "Catalogue of the Insects of Massachusetts," including two kinds of *Xiphydria*, which are now known to belong to the same family.

In the autumn of 1826, Major E. M. Bartlett, of Northampton, "found, on the body of one of his almost lifeless pear-trees, a dead insect, about one inch and a half long, attached to the tree by its awl or borer, of about the same length, near an inch of which was fast in the hard wood; and there were several deep punctures near it, evidently made by the same instrument, and in some of them eggs were deposited." Not long afterwards Major Bartlett found that the body of this tree, two or three feet from the ground, was pierced with many small holes, to the depth of an inch or more, and in these holes there were great numbers of larvæ, about one sixth of an inch in length, which he supposed were hatched from the eggs seen there before; and he came to the conclusion that the tree was "destroyed by the deadly needles of the winged insect" above mentioned.* The latter was subsequently sent to me for examination, and enabled me to furnish an account of it, which, with a description of the male insect, was published in January, 1827, in the fifth volume of the "New England Farmer."

The insect proved to be the *Sirex Columba* of Linnæus, or *Tremex Columba* of modern naturalists. *Sirex* is a corruption of the Greek name for a wild bee; *Tremex* signifies a perforator, or maker of holes; and *Columba* a pigeon. The body of the female (Fig. 250) is cylindrical, about as thick as a common lead-pencil, and an inch and a half or more in length, exclusive of the borer, which is an inch long, and projects three eighths of an inch beyond the end of the body. The latter rounds upwards, like the stem of a boat, and is armed with a point or short horn. The head and the thorax are rust-colored, varied with black. The abdomen, or hinder and longest part of the body, is black, with seven ochre-yellow bands across the back, all of them but the first two interrupted in the middle. The horned tail, and a round spot before it, impressed as if

Fig. 250.

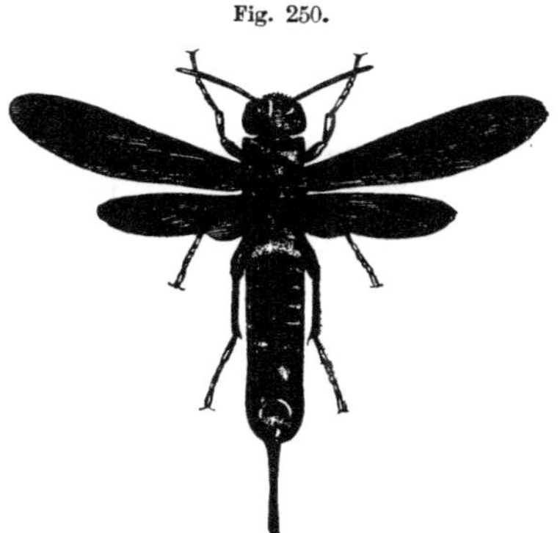

* See New England Farmer, Vol. V. pp. 167, 175, 186, and 211.

with a seal, are ochre-yellow. The antennæ are rather short and blunt, rust-colored, with a broad black ring in the middle. The wings expand two inches and a quarter, or more; they are smoky brown and semitransparent. The legs are ochre-yellow, with blackish thighs. The borer, awl, or needle is as thick as a bristle, spear-pointed at the end, and of a black color; it is concealed, when not in use, between two narrow rust-colored side-pieces, forming a kind of scabbard to it.

This insect is figured and described in the second volume of the late Mr. Say's "American Entomology." The male does not appear to have been described by any author; and, although agreeing, in some respects, with the two other species represented by Mr. Say, is evidently distinct from both of them. He is extremely unlike the female in color, form, and size, and is not furnished with the remarkable borer of the other sex. He is rust-colored, variegated with black. His antennæ are rust-yellow or blackish. His wings are smoky, but clearer than those of the female. His hind body is somewhat flattened, rather widest behind, and ends with a conical horn. His hind legs are flattened, much wider than those of the female, and of a blackish color; the other legs are rust-colored, and more or less shaded with black. The length of his body varies from three quarters of an inch to one inch and a quarter; and his wings expand from one inch and a quarter to two inches or more.

An old elm-tree in this vicinity used to be a favorite place of resort for the *Tremex Columba*, or pigeon Tremex; and around it great numbers of the insects were often collected, during the months of July and August, and the early part of September. Six or more females might frequently be seen at once upon it, employed in boring into the trunk and laying their eggs, while swarms of the males hovered around them. For fifteen years or more, some large buttonwood-trees in Cambridge have been visited by them in the same way. The female, when about to lay her eggs, draws her borer out of its sheath, till it stands perpendicularly under the middle of her body, when she plunges it, by repeated wriggling motions, through the bark into the wood. When the hole is made deep enough, she then drops an egg therein, conducting it to the place by means of the two furrowed pieces of the sheath. The borer often pierces the bark and wood to the depth of half an inch or more, and is sometimes driven in so tightly that the insect cannot draw it out again, but remains fastened to the tree till she dies. The eggs are oblong oval, pointed at each end, and rather less than one twentieth of an inch in length. The larva, or grub, is yellowish-white, of a cylindrical shape, rounded behind, with a conical, horny point on the upper part of the hinder extremity, and it grows to the length of about an inch and a half. It is often destroyed by the maggots of two kinds of ichneumon-flies (*Pimpla atrata* and *lunator* (Fig. 251) of Fabricius). These flies may frequently be seen thrusting their slender borers, measuring from three to four inches in length, into the trunks of trees inhabited by the grubs of the Tremex, and by other wood-eating insects; and, like the female Tremex, they sometimes become fastened to the trees, and die without being able to draw their borers out again.

Urocerus albicornis, of Fabricius, the white-horned Urocerus, has white antennæ, longer and more tapering than those of the pigeon Tremex, and black at each end. The female is of a deep blue-black color, with an oval white spot behind each eye, and another on each side of the hinder part of the abdomen. The horn on the tail is long, and shaped like the head of a lance. The wings are smoky brown, and semitransparent. The legs are black, with white joints. The body measures about an inch in length, and the wings expand nearly two inches. The male has a black head, with a white spot on each side, behind the eyes. His thorax and legs are black. His abdomen is flattened, and rust-colored, and ends with a flattened horny point. He

Fig. 251.

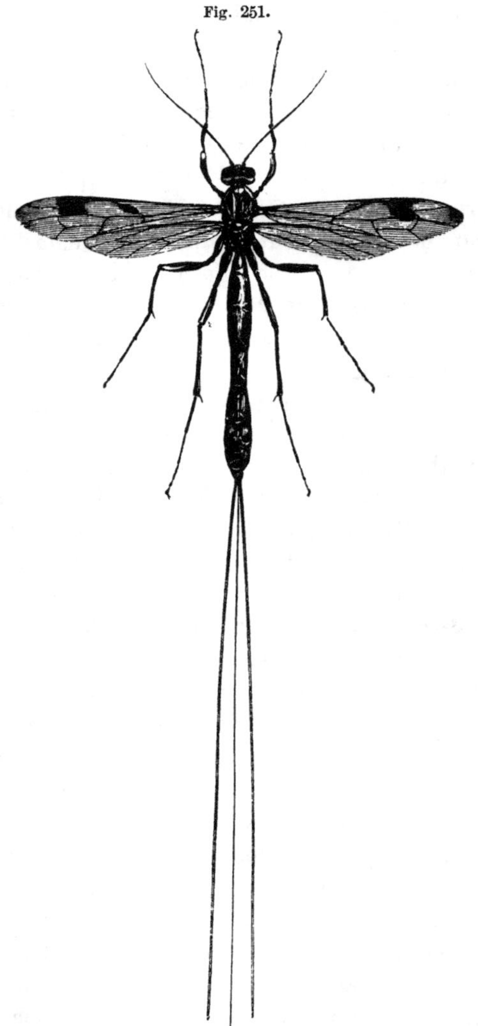

measures about an inch in length. This species, which is not common, has been found on pine-trees in July.

Urocerus nitidus,[1] the polished horn-tail, is an undescribed species, for which I am indebted to the Rev. L. W. Leonard. The male is not known to me. The female is of a deep blue color, downy on the head and thorax, smooth and highly polished on the abdomen, the end of which is armed with a flattened horny point. Her wings are clear and perfectly transparent, with brownish veins, and have only a faint smoky tinge towards the tip. Her legs are ochre-yellow. The body of this insect measures rather more than three quarters of an inch, exclusive of the horn on the tail. This insect differs from the European *Urocerus Juvencus* in the much greater brilliancy of its color, and in having shorter antennæ. The borer of this and of the preceding species resembles, in form and structure, that of the pigeon Tremex, and is used in the same way.

Urocerus abdominalis,* the black and orange horn-tail, of which only the male is known to me, has not been described before. It is black, with the four middle segments of the abdomen deep orange. There is a pale yellow spot behind each eye; the front corners of the thorax are pale brownish-yellow; and there are two minute yellowish scales on the back part of the thorax. The abdomen is flattened and widened behind, and ends with a flattened or triangular point. The antennæ are long and tapering, of a reddish brown color, with the two extremities black. The wings are transparent, with brown veins, and are a little smoky at the tips. The first four legs are ochre-yellow, with black thighs; and the hind legs are black, with yellow knees and feet. This insect varies in length from six tenths to more than three quarters of an inch. It is found in July, on the trunks of the white pine.

Mr. Westwood has ascertained that the grubs of the

[[1] *Urocerus nitidus.* This is the *cyaneus* of Fab. Syst. Piez. p. 50. — NORTON.]

* So named from the great contrast in the colors of the abdomen. In my "Catalogue" it stands under the genus *Sirex* of Linnæus, which is the same as *Urocerus* of Geoffroy.

insects belonging to the genus *Xiphydria* have the same form and habits as those of the horn-tailed wood-wasps. The name comes from a word signifying a small sword, in allusion to the borer of the female, which is shorter than in the preceding horn-tails. The winged insects have a rounded head, distant from the thorax, to the lower part of which it is joined by a slender conical neck. The body is nearly cylindrical, a little flattened, somewhat turned up behind, and ends with an obtuse point. The antennæ are short, curved, and tapering at the end.

Xiphydria albicornis of my "Catalogue," or the white-horned Xiphydria, has white antennæ with the two lowest joints black. The head is black, with a narrow white line around each of the eyes, forming a large oval, interrupted only in two places, on each side of the head. The body is black, with a spot on the front corners of the thorax, and six spots on each side of the abdomen, of a white color. The legs are reddish yellow or honey-yellow, with dusky feet. The wings are transparent, and have blackish veins. The body measures from six tenths to nearly three quarters of an inch in length. This insect is found on the trunks of trees of soft wood, in August.

Xiphydria mellipes of my "Catalogue" may be merely a variety of the preceding, from which it differs chiefly in having only four white spots on each side of the abdomen. It is four tenths of an inch long. I am indebted to the Rev. L. W. Leonard for specimens of these two species.

The name of the genus *Oryssus* comes from a Greek word signifying to dig holes. The insects belonging to it differ considerably from the other *Uroceridæ*, but, from what little is known respecting them, they appear to have the same habits. They have a cylindrical body, almost rounded behind, or bluntly pointed, and not distinctly horned. Their heads are large, and very rough on the front. Their antennæ appear to come out of the mouth, being inserted close to it, under the outer angles of the visor; are rather short, curved, and thread-like; and are unequal in the number and size of the joints, in the two sexes. They have a short and thick neck. Their borer is very slender, is entirely concealed in a deep and narrow chink under the hinder part of the body, and is coiled up at its base, so that it can be darted out to some distance when extended. The fore legs of the females are very thick, and have only three joints to the feet; while the rest, as well as all of the feet of the male, are five-jointed. Their wings have but few veins and meshes in them. These insects are active, fly quickly, and love to alight and run about on the sunny side of the trunks of trees, wherein they are supposed to lay their eggs.

For a long time, only two kinds of *Oryssus* were known to naturalists, and both of them were European insects. In the year 1833, three undescribed species were enumerated in my "Catalogue of the Insects of Massachusetts"; and these, in the second edition of the Catalogue, which was published early in 1835, received the following descriptive names, by means whereof an entomologist would find little or no difficulty in recognizing them; namely, *hæmorrhoidalis*, the red-tailed, *maurus*, the dark-colored, and *affinis*, the allied, so called from its near resemblance to the preceding species. These singular insects were taken upon a willow-tree, by my friend, the Rev. L. W. Leonard, and were presented to me many years ago.

The red-tailed Oryssus has been renamed and described, by Mr. Newman, in the October number of the fifth volume* of "The Entomological Magazine," published in London in 1838. It is his *Oryssus terminalis* (Fig. 252). The female only is known to me. Her body is black, rough before, and smooth behind, with the last three segments of a blood-red color.

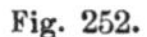
Fig. 252.

* Page 486.

The outer side of the fourth and fifth joints of her antennæ, her knees, and a line on the outer edge of her shins, are white. Her feet are dull red. Her wings are clear and transparent, with a broad, smoky-brown, transverse band, beyond the middle of the first pair. Her body measures nearly six tenths of an inch in length.

The dark-colored Oryssus is probably the same as one described by Mr. Westwood, in 1835, in the fifth volume* of "The Zoölogical Journal," under the name of *Oryssus Sayii*, in honor of the late Mr. Say, who sent him the insect. It is of a deep black color, rough before and smooth behind, and is marked with white on the antennæ and legs, like the red-tailed kind, with the addition of two short white lines on the forehead, between the lower corners of the eyes. The feet are black. The wings have a smoky band beyond the middle, which, however, fades away towards the inner margin. I have seen only females of this species, and they measure from four to five tenths of an inch in length.

It is possible that my *Oryssus affinis*, which is a male, may be the mate of the foregoing dark-colored species, from which it differs in having reddish feet, and in wanting the two white spots on the forehead. It measures four tenths of an inch in length.

From this somewhat extended account, it is evident that we have very little power over the insects of the foregoing family. The most that we can do towards checking their ravages will be to destroy the females, whenever they are found laying their eggs.

The four-winged gall-flies have very little outward resemblance to the saw-flies and horn-tailed wood-wasps. They agree with them, however, in boring into plants, and in laying their eggs therein. Vegetation does not often suffer much injury from their attacks, and it is only on account of the very singular productions, called galls, arising from

* Page 440.

the irritating punctures of these insects, that the attention of cultivators is at all likely to be drawn to them. There are some two-winged flies, and also some other insects, which produce various kinds of excrescences or galls on plants; but these now under consideration are very small, four-winged insects, belonging to the order *Hymenoptera*, and distinguished by the following peculiarities. The head is small; the antennæ are rather short, slender, and thread-like; and the thorax is thick and hunched. The abdomen or hind body, viewed sidewise, appears round or oval, but it is sharp-edged above and below, very thin or pinched up at the sides, and is hung to the thorax by a very short and slender stem. The fore wings are rather long, and have only a few veins in them; the hind wings are small, and seemingly veinless. The borer of the females is very long, and slender, concealed in the under side of the hind body, the curvature whereof it follows, and is capable of being straightened and thrust out of a narrow chink, which is covered by two little, grooved, sheath-like pieces, that serve to conduct the eggs into the holes made with the instrument.

The genus containing most of the gall-flies was called, by Geoffroy, *Diplolepis*, that is, double scales, on account of the two pieces that cover the opening for the borer in the hinder part of the abdomen. The same insects, however, had previously been placed by Linnæus in the genus *Cynips*, so called from a word used by ancient authors to designate some small piercing insect. The Linnæan name, though for some time rejected, has been restored to the gall-flies, which accordingly are now included in a family called CYNIPIDÆ. The punctures, made by these insects in the leaves, buds, stems, and roots of plants, are followed by swellings of the wounded parts, which increase rapidly in size, and become spongy or pulpy within. The thin-skinned eggs, dropped into the punctures, grow awhile, by absorbing the sap around them, and, when at length they are hatched, the little grubs, proceeding therefrom, find themselves comfortably bedded within the pulpy tumors, and plentifully supplied with food on every side. They feed on the vegetable substance immediately around them, come to their growth in due time, cast their skins, and appear first in the chrysalis and then in the winged form, and finally gnaw their way through the hard shell of the galls, and come out into the open air. There are a few of the grubs, however, that leave the galls when fully grown, and finish their transformations in the ground.

The grubs or young of the gall-flies are of a whitish color, and somewhat resemble maggots, but are shorter and thicker, and have a small, distinct head. They are without proper legs, and move only by means of the swollen edges of their rings, with the aid, it is said, of certain little contractile warts on their bodies, that serve them instead of feet. There are almost as many kinds of galls as there are species of gall-flies; and each species confines its attacks to some one sort of plant, and to some particular part thereof. It is wonderful that there should be such a diversity in the forms and texture of the galls of insects so nearly resembling each other in form and structure; and, on the other hand, that each species of gall-fly should invariably produce galls of the same kind. Many galls are very irregular and uneven, others are round and resemble fruits; some are smooth, others are beset with prickles, or covered with a woolly substance; some hang by little stems, others are perfectly flat, and adhere closely to the surface of leaves. At first they are soft or spongy within, but after some time they become hard, and almost or quite woody. The eggs of some gall-flies do not hatch till the galls begin to grow hard on the outside; this is the reason why we do not find any insects within certain kinds of galls, so long as they remain soft and unripe.

The round and hard Aleppo galls, or nutgalls of commerce, used in the making of ink, in coloring, and in medicine, are caused by the punctures of the *Cynips gallæ tinctoriæ* on a kind of oak growing in the western part of Asia; and the insect may often be found in those which are not pierced with holes. Some galls contain only a single insect, lodged in a little cavity in the centre; other kinds are inhabited by several grubs, each in a cell by itself, and the cells not unfrequently resemble numerous small seeds, clustered together in the middle of a fruit. Two or three different kinds of insects are often found to come from one gall, namely, a few gall-flies, which are the lawful proprietors thereof, and more numerous four-winged flies (CHALCIDIDÆ), with elbowed antennæ. The latter are bred from grubs, which devour the grubs of some of the gall-flies, or starve them by eating up their food, and thereby contribute to check the too great increase of the gall-flies.

The largest galls found in this country are commonly called oak-apples. They grow on the leaves of the red oak, are round and smooth, and measure from an inch and a half to two inches in diameter. This kind of gall (Plate VIII. Fig. 9) is green and somewhat pulpy at first, but when ripe it consists of a thin and brittle shell, of a dirty drab color, enclosing a quantity of brown spongy matter, in the middle of which is a woody kernel about as big as a pea. A single grub (Fig 253, magnified) lives in the kernel, becomes a chrysalis (Fig. 254) in the autumn, when the oak-apple falls from the tree, changes to a fly in the spring, and makes its escape out of a small round hole which it gnaws through the kernel and shell. This is probably the usual course, but I have known this gall-fly to come out in October. The name of this insect is *Cynips confluens*.* (Plate

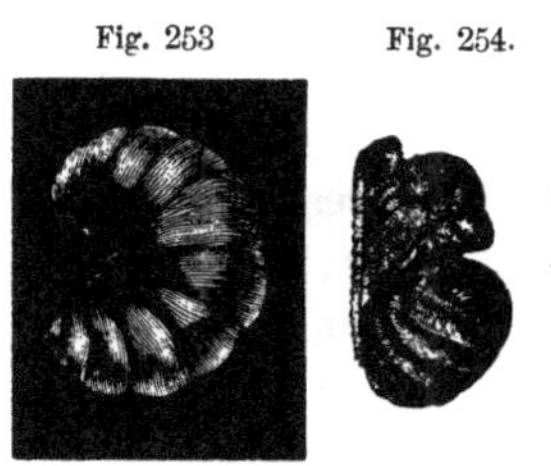

Fig. 253 Fig. 254.

* *Diplolepis confluentus* of my Catalogue, and so named by Mr. Say.

VIII. Fig. 10.) Its head and thorax are black, and are rough with numerous little pits and short hairs; the hind body is smooth, and of a shining pitch-color; the legs are dull brownish red; and the fore wings have a brown spot near the middle of the outer edge. Its body is nearly one quarter of an inch long, and its wings expand five eighths of an inch.

A dwarf oak (*Quercus infectoria*), growing on the borders of the Dead Sea, produces galls somewhat like the foregoing, which have been supposed to be the apples of Sodom, described by ancient writers as fruits fair to the view, but crumbling into dust when handled. A late writer,* however, has shown that these tempting and deceptive productions are the real fruits of a tree, the *Asclepias procera*, resembling our common silk-weed in its botanical characters.

Clusters of three or four round and smooth galls are often seen on the small twigs of the white oak. They are nearly as large as bullets, of a greenish color on one side, and red on the other. They approach in hardness to the Aleppo galls, and perhaps might be put to the same use. Each one is the nest of a single insect, which turns to a fly and eats its way out in June and July, having passed the winter as a chrysalis, within the gall, lodged in a clay-colored egg-shaped case, about three twentieths of an inch long, and with a brittle shell. These little cases appear to be cocoons, but are not made of silk or fibrous matter. Similar cocoons are found within many other galls, and I have some which were discovered under stones, and were not contained in galls, but produced gall-flies, the insects having left their galls, to finish their transformations in the ground. The gall-fly of the white oak varies in color. Sometimes it closely resembles the gall-fly of our oak-apple, differing from it only in size, and in wanting the brownish spot and dark-colored veins on the fore wings; and sometimes it is of a dull brownish-yellow color, with a brown

* Robinson's Biblical Researches in Palestine, Vol. II. p. 235.

spot on the back. It is three twentieths of an inch long, and its wings expand three tenths of an inch. It is the *Diplolepis*, or more properly *Cynips oneratus* of my Catalogue.

Galls of the size and color of grapes are found on the leaves of some oaks. Each one contains a grub, which finishes its transformations in June. The winged insect is my *Cynips nubilipennis*, or cloudy-winged Cynips, so named from the smoky cloud on the tips of its wings. Excepting in this respect, it closely resembles the dark-colored variety of *Cynips oneratus*, and very little exceeds it in size.

One of our smallest gall-flies may be called *Cynips seminator*, or the sower. She lays a great number of eggs in a ring-like cluster around the small twigs of the white oak, and her punctures are followed by the growth of a rough or shaggy reddish gall, as large sometimes as a walnut. When this is ripe, it is like brittle sponge in texture, and contains numerous little seed-like bodies, adhering by one end around the sides of the central twig. These seeming seeds have a thin and tough hull, of a yellowish-white color; they are egg-shaped, pointed at one end, and are nearly one eighth of an inch long. The gall-insects live singly, and undergo their transformations, within these seeds; after which, in order to come out, they gnaw a small hole in the hull, and then easily work their way through the spongy ball wherein they are lodged. They are less than one tenth of an inch long, are almost black, or of the color of pitch, highly polished, especially on the abdomen, and their mouth, antennæ, and legs are cinnamon-colored.

It has been observed that no tree in Europe yields so many different kinds of galls as the oak. Those which I have described are not all that are found on oaks in this country, and they seem to be sufficiently distinct from the galls of European oaks.

Round, prickly galls, of a reddish color, and rather larger than a pea, may often be seen on rose-bushes. Each of them contains a single grub, and this in due time turns to a gall-fly, which may be called *Cynips bicolor*, the two-colored Cynips. Its head and thorax are black, and rough with numerous little pits; its hind body is polished, and, with the legs, of a brownish-red color. It is a large insect compared with the size of its gall, measuring nearly one fifth of an inch in length, while the diameter of its gall, not including the prickles, rarely exceeds three tenths of an inch.

Cynips dichlocerus, or the gall-fly with two-colored antennæ, (Plate VIII. Fig. 6, Fig. 7 magnified,) is of a brownish-red or cinnamon color, with four little longitudinal grooves on the top of the thorax, the lower part of the antennæ red, and the remainder black. It varies in being darker sometimes, and measures from one eighth to three sixteenths of an inch in length. Great numbers of these gall-flies are bred in the irregular woody galls, or long excrescences, of the stems of rose-bushes (Plate VIII. Fig 8).

The small roots of rose-bushes, and of other plants of the same family, sometimes produce rounded, warty, and woody knobs, inhabited by numerous gall-insects, which, in coming out, pierce them with small holes on all sides. The winged insects closely resemble the dark varieties of the preceding species, in color, and in the little furrows on the thorax; but their legs are rather paler, and they do not measure more than one tenth of an inch in length. This species has been named *Cynips semipiceus*.

Monstrous swellings of buds, and various other kinds of excrescences, may often be seen on plants; but my specimens of the insects producing them are not in a condition to be described. The foregoing account, however, will serve to illustrate the habits of some of our most common gall-flies, and explain the origin, forms, and structure of their singular productions. Such excrescences, as soon as they are observed on plants of any value, should immediately be cut off, and put into the fire.

Gall-insects, as already stated, are often destroyed by little parasites belonging to the family Chalcididæ; and as these are liable to be mistaken for the former, especially when coming from the same gall, it may be well to point out the difference between them. The four-winged gall-flies have rather long, straight, thread-like, and ascending antennæ; the fore wings with a few veins, forming two triangular meshes, one of which is very small, and situated near the middle of the wing, the other mesh much larger, and near the base; the hind body roundish, but laterally compressed; and the piercer spiral or curved, and concealed. The Chalcidians have shorter, elbowed, and drooping antennæ, which are enlarged towards the end; a single vein, running from the shoulder near the outer margin of the fore wing, uniting with this margin near its middle, and emitting thence, towards the disk of the wing, a short oblique branch, which is enlarged or forked at the end; the hind body generally oval, pointed at the end in the females, and provided in this sex with a straight piercer, which is more or less visible beneath, and prominent at the extremity. By means of their piercers, the Chalcidians thrust their eggs into the galls made by various kinds of gall-insects, and the maggots hatched from these eggs devour the young of the gall-flies. (Fig. 255, larva of a Chalcidian, which attacks *Cynips dichlocerus;* Fig. 256, pupa of the same.) Nor do they destroy these alone; they prey upon many other larvæ, especially caterpillars, and also on pupæ or chrysalids. Some of them are egg-parasites, puncturing the eggs of other insects, and depositing therein their own tiny eggs. They are the minute ichneumons (*Ichneumones minuti*) of Linnæus, and, like the true ichneumon-flies, they are eminently useful in checking the increase of the noxious tribes.

Fig. 255. Fig. 256.

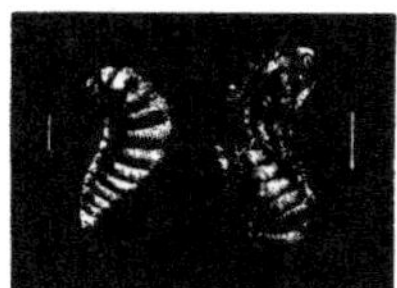

Such being the known habits and services of the greater

part of the Chalcidians, it may seem singular that any doubt should exist in regard to others of them. There are, however, some kinds which have been thought to produce galls themselves, instead of being the parasites of gall-insects; certain species wearing indeed the form of a Chalcidian, but appearing to have the habits of a Cynips. These species belong to the genus *Eurytoma*, which, though agreeing in structure with the Chalcidians, Professor Nees von Esenbeck was inclined to place among the CYNIPIDÆ, because he took them to be gall-makers. Mr. Westwood* controverts this opinion; and Dr. Ratzeburg† considers it as founded upon error. It may nevertheless be correct, if there be no mistake in the result of observations made upon the insects, called barley-straw insects and joint-worms, that produce gall-like swellings upon the stems of barley and of wheat in this country.

In the years 1829 and 1830 several communications were published in the eighth volume of Fessenden's "New England Farmer,"‡ respecting a disease of barley-straw, produced by the punctures of insects. The first account of this disease that has fallen under my notice is contained in an extract from a letter, dated August 16th, 1829, from the Honorable John Merrill, of Newburyport, to Mr. Fessenden; wherein it is stated, that the barley, in the neighborhood of Newburyport, yielded only a very small crop; on some farms, not much more than the seed sown. Most of the stalks were found to have a number of small worms within them, near to the second joint, and had become hardened in the part attacked, from the interruption of the circulation of the sap. During several years previous to this date, the barley crops, in various parts of Essex and Middlesex Counties, were more or less injured in the same way; and in some places the cultivation of this grain was

* Modern Classification of Insects, Vol. II. p. 161, note.

† Die Ichneumonen der Forstinsecten, I. 172.

‡ Pages 43, 138, 217, 299, 330, and 402. Also Vol. IX. p. 2, and Vol. X. p. 11.

given up in consequence thereof. It was supposed that the insects producing this disease were imported from Bremen, or some other port in the North of Europe, in some barley that was sown in the vicinity of Newbury, three or four years before 1829.* The worms or maggots were found, by John M. Gourgas, Esq., of Weston, Massachusetts, to be transformed to small flies, "about the make and size of a small black ant, with wings," which were thought, by some persons, to be the same as the Hessian flies.

In the summer of 1831, myriads of these flies were found alive in straw beds in Gloucester; the straw having been taken from the fields the year before. An opinion at that time prevailed, that the troublesome humors wherewith many persons were then afflicted were occasioned by the bites of these flies; and it is stated that the straw beds in Lexington, being found to be infested with the same insects, were generally burnt.† Mr. Gourgas observes,‡ that when the barley is about eight or ten inches high, the effects of the disease in it begin to be visible by a sudden check in the growth of the plants, and the yellow color of their lower leaves. If the buts of the straw are now examined, they will be found to be irregularly swollen, and discolored, between the second and third joints, and, instead of being hollow, are rendered solid, hard, and brittle, so that the stem above the diseased part is impoverished, and seldom produces any grain. Suckers, however, shoot out below, and afterwards yield a partial crop, seldom exceeding one half the usual quantity of grain. Dr. Andrew Nichols, of Danvers, states,§ that the worms are about one tenth of an inch in length, and of a yellow or straw color; and that, in the month of November, they appeared to have passed to the chrysalis state. They live through the winter unchanged in the straw, many of them in the stubble in the field, while others are carried away

* New England Farmer, Vol. VIII. p. 217.
† Ibid., Vol. X. p. 11.
‡ Ibid., Vol. VIII. p. 299.
§ Ibid., p. 138.

when the grain is harvested. When the barley is threshed, numerous small pieces of diseased straw, too hard to be broken by the flail, will be found among the grain. Some of these may be separated by the winnowing-machine, but many others are too large and heavy to be winnowed out, and remain with the grain, from which they can only be removed by the slow process of picking them out by hand.

In the winter of 1829, Cheever Newhall, Esq., furnished me with a few pieces of diseased barley-straw, each of which contained several small whitish maggots. Since that time this affection of the barley has only once fallen under my notice, though I have reason to think that it continues to prevail in many parts of Massachusetts. Each maggot was imbedded in the thickened and solid substance of the stem, in a little longitudinal hollow, of the shape of its own body; and its presence was known by an oblong swelling upon the surface. In some pieces of straw the swellings were so numerous as greatly to disfigure the stem, the circulation in which must have been very much checked, if not destroyed. Early in the following spring these maggots entered the pupa or chrysalis state, and on the 15th of June the perfected insects began to make their escape through minute perforations in the straw, which they gnawed for this purpose. Seven of these little holes were counted in a piece of straw only half an inch in length. The insects continued to release themselves from their confinement till the 5th of July, after which no more were seen. Much to my surprise, they proved to be minute, four-winged flies, belonging to the genus *Eurytoma*. Supposing these insects to be parasites, in accordance with the known habits of others of the same family, I described them as such, under the name of *Eurytoma Hordei* (so called from *Hordeum*, the Latin for barley), in the "New England Farmer," for July 23, 1830,* and in the first edition of this work. It was then my belief that the true culprits, or original cause

* Vol. IX. p. 2.

of the disease, would prove to be some species of *Cecidomyia*, allied to, but distinct from, the Hessian fly; and that they, while in the larva or pupa state, had been preyed upon and destroyed by the *Eurytoma*. The larvæ of the Hessian fly are often destroyed by a somewhat similar Chalcidian parasite, great numbers of which have been observed, in their winged form, in wheat-fields, and have then been mistaken for Hessian flies.

The body of the *Eurytoma Hordei* is jet-black, and slightly hairy. The head and thorax are opaque, and rough with dilated punctures. The hind body is smooth and polished. The thighs, shanks, and claw-joints are blackish; the knees, and the other joints of the feet, are pale honey-yellow. The females are twelve or thirteen hundredths of an inch long. The males are rather smaller, and are distinguished from the females by the following characters. They have no piercer. The joints of their antennæ are longer, and are surrounded with whorls of little hairs. The hind body is shorter, less pointed behind, and is connected with the thorax by a longer stem or peduncle. These insects are very active, and move by little leaps; but the hindmost thighs are not thickened. About eight years ago, some of these insects, that had come from a straw bed in Cambridge, were shown to me. They had proved very troublesome to children sleeping on the bed; their bites or stings being followed by considerable inflammation and irritation, which lasted several days. So numerous were the insects, that it was found necessary to empty the bed-tick and burn the straw. Since that time, I have heard nothing more either of the insects or of the disease of barley-straw in this part of the country.

My attention was again called to the history of the barley-straw insect by an article on the joint-worm, published at Albany in "The Cultivator," for October, 1851. The account given in this magazine, by Mr. Rives, of the ravages of the joint-worm in the wheat-fields of Virginia, and the remarks by Dr. Fitch on the peculiar affection of the wheat-straw produced by this worm, led me to suspect that the disease was identical with that which had been observed in barley-straw, and that it originated from the same cause. In the article above named, Dr. Fitch appears to have come to the conclusion that the disease was produced by some species of *Cecidomyia*. He found the disease of the wheat-straw to be situated immediately above the lower joint, in the sheathing base of the leaf, the substance of which, for a distance exceeding half an inch, was much swollen, and was changed to a more solid and wood-like texture, while the surface exhibited several long pale spots, slightly elevated like a blister. The hollow of the stem was entirely obliterated, at some parts, by the pressure of the enlarged portion of the sheath, and was hardly visible at others. Each of the blistered spots covered an elongated cavity, containing a footless worm or maggot, about ten hundredths of an inch long, of an oval form rather more tapering posteriorly than towards the head, and divided by slight constrictions into thirteen segments. The worm was soft, shining, of a uniform milk-white color, with a small V-shaped brown line marking the situation of the mouth. "So exactly," remarks Dr. Fitch, "does this worm in its form and appearance resemble the larvæ of the Hessian fly and other species of *Cecidomyia* which have fallen under my examination, that I entertain no doubt it pertains to the same genus of insects."

On the 16th of March, 1852, F. G. Ruffin, Esq., of Shadwell, Virginia, the editor of "The Southern Planter," sent to me that paper for July, 1851, containing some account of the joint-worm, and with it a few samples of diseased wheat-straw. A much larger quantity of the straw, soon afterwards received from him, was divided into two unequal portions, the larger of which was sent to Dr. Fitch, in the hope that between us something definite concerning the origin of the disease might be obtained. Upon examining my samples, I found that the disease was not invariably confined to the sheathing base of the leaf, but that, in many cases, it was seated in the joint itself, the whole substance of which became enlarged and distorted. In a smaller number of cases, it was found to occupy the culm or stem, above the joint, which was swollen so as to form an irregular gall-like tumor, while the leaf-sheath remained unaffected. These woody tumors had several little cells in them, varying in number from six to ten or more; and every cell contained an insect, in the pupa or chrysalis state. The samples of straw reserved for myself were put into a small glass jar, to secure the insects when they had completed their transformations. Early in May, winged insects began to perforate the tumors and come forth, and they continued to issue during ten days or more. Their appearance was probably hastened by the jar being kept in the house, instead of being exposed to the air abroad.

These insects so nearly resemble, in form, size, and color, the *Eurytoma* formerly obtained from the barley-straw, that I am persuaded they are at least mere varieties of the same species, if not absolutely identical. The only apparent difference between them consists in the color of the fore shanks; these, in the wheat-insects, being pale yellow, and faintly tinged with black only on the outer edges, in a few individuals. Among fifteen specimens only one male was found, and this did not appear till the month of June. Dr. Fitch obtained from his samples of straw above one hundred specimens of the same kind of *Eurytoma*, and all of them females. Among them he found another Chalcidian insect, a species of *Pteromalus*, probably a parasite of the *Eurytoma*, and has favored me with a description of it. The head and thorax are of a dark metallic green color; the abdomen is slightly depressed, polished, purplish black above, bright copper-colored beneath. The antennæ are black, except the basal joint, which is of a brilliant copper-color. The thighs are pale yellow; the shanks and feet blackish, the hind pair with a broad pale ring around the bottom of the shank and the contiguous part of the foot. The length of the body is ten hundredths of an inch, being somewhat less than that of the *Eurytoma.*

From my samples of the straw I have obtained another and a different parasite, belonging to the same family, but to the genus *Torymus.* The specimen is a female, and, like others of the same genus, it is provided with an exserted slender piercer, nearly as long as its own body. The latter is about as long as that of the *Pteromalus* above described, and is of a deep black color, slightly tinged with green on the face and thorax, both of which are rough and opaque, while the hind body is smooth and polished. The fore wings have an elongated cloudy spot near the middle, and the oblique branch is very short. The thighs, claws, and the antennæ except the basal joint, are blackish, the other parts of the legs and the base of the antennæ are pale yellow. The hindmost thighs are much thicker than the others, and are notched beneath the end. The eyes have a dull reddish tinge, perhaps not their true color in life. Professor Cabell has sent to me some specimens of this *Torymus*, including a male, which differs from the female in having all the joints of the antennæ black.

The ravages of the joint-worm in the wheat-fields of Virginia are said to have been first observed in Albemarle County, about four or five years ago. They have alarmingly increased from year to year, and have extended over many parts of the adjacent counties, becoming more aggravated each time that they are renewed in the same place. The loss occasioned thereby often amounts to one third of the average crop, and is sometimes much greater; and during the present season, "some farmers did not reap as much as they sowed." These statements are made chiefly on the authority of Professor J. L. Cabell, of the University of Virginia, who has given some attention to the natural history of the joint-worm, and has recently communicated to me the result of his interesting observations. He has come to the conclusion that the joint-worm is the larva of a Hymenopterous, and not of a Dipterous insect. He finds that the parts of its mouth are very different from those of the dormant larva of the Hessian fly (the latter extracted from its flax-seed case before it had undergone any change of form), and that the mouth of the former agrees essentially with that of the larvæ obtained from galls of the oak. In the mouth of the joint-worm he observed that "the mandibular hooks cross each other on the middle line," while in the Hessian fly larva the "two hooks are directed downwards." His samples of diseased wheat-straw of the previous year yielded him, in the spring, numerous specimens of the *Eurytoma*, and nothing else. A few specimens of the same insect were developed from the tumors on plants of the present season, thus showing that "a small proportion of the larvæ undergo their transformations during the summer." Among his specimens he obtained a very few Hymenopterous insects, differing from the *Eurytoma*, and probably parasites. In several instances Professor Cabell saw a small semitransparent whitish worm, scantily covered with hairs, in the same cell with a lifeless joint-worm, and adhering to its body. In other cases, the former kind of worm or larva "was found alone, but it was then of a larger size, and there were almost always some more or less unequivocal signs of the worm having fed on the joint-worm."

Having been favored by Professor Cabell with some samples of wheat-straw, containing living joint-worms, I have been able to verify his observations during the present summer, while this sheet is passing through the press. At my request, Professor Jeffries Wyman, of Harvard College, an accomplished anatomist and a skilful microscopical observer, has examined these larvæ, and also some of the parasitical worms, found in the straw, and has made for me several magnified sketches of them. Both kinds are found to differ essentially from the larvæ of the locust and of the willow gall-flies, with living specimens of which I have compared them. Their bodies are softer, and their skins more delicate and tender; and the form of the head and structure of the mouth are entirely unlike those of the Cecidomyian larvæ. The true joint-worm varies from one tenth to nearly three twentieths of an inch in length. It is of a pale yellowish white color, with an internal dusky streak, and is destitute of hairs. The head is round, and partially retractile. The jaws are lateral and hooked; they meet at the points, and are of a blackish color, and apparently of a horny texture; and they are distinctly to be seen even with a pocket microscope. It is evident, therefore, that these joint-worms are not the larvæ of any Dipterous insect; they are doubtless Hymenopterous larvæ, and probably, from their abundance, those of the foregoing *Eurytoma*. The other larvæ, few in number compared with the joint-worms, are distinguished therefrom by their inferior size, and whiter color, and by being sparingly covered with short hairs. Their heads are round, are provided with blackish hooked jaws, and have two little tubercles on the front. I judge them to be the young of one of the parasites, probably of the *Torymus*, described on a former page.

The foregoing account might be thought to afford conclusive evidence that the *Eurytoma* alone was the author of the mischief done to the wheat and barley, and that it is not a parasitical insect. In favor of this conclusion, we have the fact that hitherto no person has succeeded in obtaining from the diseased wheat-straw so much as a single specimen of *Cecidomyia;* while both the wheat and the barley straw have yielded to several observers, in repeated instances, numerous specimens of the same kind of *Eurytoma*, and nothing else, saving an extremely small number of lesser parasites. The determination of this difficult and interesting question is of much importance in a scientific and an economical point of view. The great amount of property that is at stake, and the serious losses already sustained by the ravages of the joint-worm, render it necessary to ascertain the true history of the insect before proceeding to take measures for the protection of our crops. We are to consider, in destroying the *Eurytoma*, whether we shall kill an enemy or a friend. If it be a parasite, as the almost universal opinion of entomologists would lead us to believe, it would be the height of folly to attempt to interfere with its operations. On the other hand, if we can show it to be a plant-eating insect, we may use such means as are in our power towards checking its career, not only with perfect safety, but with eminent advantage. In this case, in dealing with the joint-worm, we need not be restrained by the consideration that the diseased straw contains also some truly parasitical larvæ; for these, as already stated, are very few in number compared with the immense swarms of the *Eurytoma* that are annually produced. If we can succeed in exterminating these destroyers, we shall have no occasion for the services of the parasites.

Admitting the *Eurytoma* to be the sole cause of the mischief, the following suggestions will be found useful in arresting its ravages. As the disease is seated mostly near the base of the straw, in or near the second or the third joint, the greater part of the diseased portions will be left in the stubble when the grain is reaped. Most of the insects remain unchanged in the stubble till the following year. If, then, we can destroy the maggots in the stubble before they have acquired wings and made their escape, we shall, in great measure, restrain their further propagation and increase; for it is in the winged state alone that insects propagate their kind. It has been found in Massachusetts that ploughing in the stubble has little or no effect upon the insects, which continue alive and uninjured under the slight covering of earth, and easily make their way to the surface when they have completed their transformations.

The only practicable method of destroying the insects is to burn the stubble containing them. All the straw and refuse, which is unfit for fodder, should likewise be consumed, because it will be found occasionally to contain a small amount of diseased portions of the straw. Some of these may remain among the grain itself, being too heavy to be separated by the process of winnowing. These will have to be picked out by hand. Moreover, as some few of the insects are transformed to flies during the first summer, and these will suffice to continue the race, it becomes important that all the means above recommended should be continued during several successive years; and when these are universally, carefully, and thoroughly put in practice, they can hardly fail to exterminate the *Eurytoma*. A free use of manure and thorough tillage, by promoting a rapid and vigorous growth of the plant, may render it less liable to suffer from the attacks of the insect. Large fields, well seeded, will probably escape better than those that are smaller and thinner sown, in which the insects, when about to lay their eggs, can penetrate easily and to a greater distance.[2]

[[2] In the "American Agriculturist," New York, August, 1861, p. 235, Dr. Fitch reasserts the opinion that this is the "joint-worm," and enumerates four distinct species, viz.: —

"*Eurytoma Hordei*, Harris (the black-legged or Massachusetts barley-fly). It has the shanks of all the legs black."

"*Eurytoma fulvipes*, Fitch, Jour. N. Y. St. Ag. Soc., Vol. IX. p. 115 (the yellow-legged or New York barley-fly). It has all the shanks and thighs of a tawny-yellow or pale orange hue."

"*Eurytoma Tritici*, Fitch, Jour. N. Y. St. Ag. Soc., Vol. IX. p. 115 (the joint-worm fly). It has the shanks of the forward legs pale yellow, and the others black."

"*Eurytoma Secalis*, Fitch, new species (the rye-fly). It has the fore and hind shanks pale yellow, and the middle ones black." This is very abundant in Connecticut, and without doubt in all the Eastern States. — NORTON.]

CHAPTER VII.

DIPTERA.

GNATS AND FLIES. — MAGGOTS, AND THEIR TRANSFORMATIONS. — GALL-GNATS. HESSIAN FLY. — WHEAT-FLY. — REMARKS UPON AND DESCRIPTIONS OF SOME OTHER DIPTEROUS INSECTS. — RADISH-FLY. — TWO-WINGED GALL-FLIES, AND FRUIT-FLIES. — CONCLUSION.

UNDER the name of DIPTERA, signifying two-winged, are included all the insects that have only two wings, and are provided with two little knobbed threads in the place of hind wings, and a mouth formed for sucking or lapping.

Various kinds of gnats and of flies are therefore the insects belonging to this order. The proboscis or sucker, wherewith they take their food, is placed under the head, and sometimes can be drawn up and concealed, partly or wholly, within the cavity of the mouth. It consists of a long gutter, usually ending with two fleshy lips, and enclosing, in the channel on its upper side, several fine bristles, from two to six in number, which are sometimes as sharp as needles, and are then capable of inflicting severe punctures. These piercing bristles really take the place of the jaws of biting insects, and hence the wounds made therewith, by gnats and mosquitos, are very properly called bites. The saliva of these insects, flowing into the wounds, renders them more painful, and is the cause of the inflammation and itching that follow. The grooved sheath of the proboscis is usually very large and fleshy in the flies that only lap or sip their food. Two small, jointed feelers are commonly found attached to the base of the proboscis.

Gnats and flies have softer bodies than most other winged

insects. The head is large, and fastened to the thorax by a very slender neck. The eyes, especially in the males, are large, and occupy the whole of the sides of the head. The antennæ, in gnats and mosquitos, are rather long, slender, and many-jointed; in flies, they are short, consisting of only two or three thick joints, the last of which often bears a little bristle or delicate feather. The wings are filmy, like those of Hymenopterous insects, but usually have a greater number of veins in them. Just behind the wing-joints there are two little, convex scales, which open and shut with the motion of the wings; they are called the winglets. The two balancers or poisers are short threads, knobbed at the end, and placed on each side of the hindmost part of the thorax, immediately behind the winglets. The thorax is often the thickest and hardest part of the body; to it the hind body is more or less closely united, and the latter, in many females, ends with a tapering, retractile tube, wherewith the eggs are deposited. The legs are six in number, and each of the feet is provided with two claws, and two or three little cushions or skinny palms, by the help whereof the insects can walk on the smoothest surfaces, and on the ceilings of rooms, with the back downwards, as easily as when upright; for the palms act like suckers, and thus prevent them from falling.

Mosquitos and gnats are active both by day and night, but flies take wing only during the day. The life of these insects, even from the time when they are first hatched, is generally very short, seldom lasting more than a few weeks; but of some kinds several broods are produced in the course of a single summer, and often in the greatest profusion. In certain countries and seasons they multiply so fast, and appear in such immense swarms, as to become a serious annoyance both to man and beast.

The young insects, hatched from the eggs of gnats and of flies, are fleshy larvæ, usually of a whitish color, and without legs. They are commonly called maggots, and sometimes are mistaken for worms. They vary a good deal in their forms, structure, habits, and transformations, so that it is somewhat difficult to give any general description of them. Their breathing-holes are usually situated near the extremities of the body. Aquatic maggots often have a tubular tail, through which they breathe, and the orifice of this tube is sometimes surrounded with beautiful feather-formed appendages. The larvæ or maggots of the gnats, and of nearly all those flies which have four or six bristles in the proboscis, have a distinct head covered with a horny shell. Larvæ of this kind, when fully grown, cast off their skins to become pupæ or chrysalids. These pupæ are usually of a brown color, and somewhat resemble the chrysalids of certain moths, or more nearly those of Hymenopterous insects; for their short and imperfect legs and wings, though folded on the breast, are not immovably fastened to it. They commonly have several small thorns on each end of the body, and a row of smaller prickles across each of the rings of the back. By the help of these thorns and prickles they work their way out of the places wherein they had previously lived, just before they burst open their pupa-skins to come forth in the perfected or winged state. The pupæ of mosquitos are not prickly, but they possess the power of swimming or tumbling about in the water, by the help of two little fins on their tails.*

The larvæ of the Dipterous insects in general do not make cocoons; those of some gnats (*Mycetophilœ*), which live in tree mushrooms, or *boleti*, not only cover themselves with a silken web, under which they live, but also spin cocoons, wherein they undergo their transformations. Some of the Cecidomyians also make silken cocoons. The larvæ of the other flies are not so variable in their forms as the foregoing. They are commonly plump, whitish maggots, obtuse behind, and tapering before, with a small and soft head, that can be drawn within the fore part of the body. They take their

* See pages 4 and 5.

food almost entirely by suction, for their jaws are merely two little hooks, that enable them to fasten themselves upon the substances which serve for their nourishment. They increase rapidly in size, and when they are fully grown, they change their forms, without casting off their skins at all, merely by the gradual shortening of their bodies, which take an oblong oval shape, and turn hard and brown on the outside. The hardened skin of the larva thus becomes a shell or kind of cocoon, within which the insect is afterwards changed to a pupa, having its imperfect limbs folded on its breast, and from which, in due time, it comes forth in the form of a fly, by forcing off one end of the shell.*

The far-famed Hessian fly, and the wheat-fly of Europe and of this country, are small gnats or midges, and belong to the family called CECIDOMYIADÆ, or gall-gnats. The insects of this family are very numerous, and most of them, in the maggot state, live in galls or unnatural enlargements of the stems, leaves, and buds of plants, caused by the punctures of the winged insects in laying their eggs, or by the irritation of the maggots hatched therefrom. The Hessian fly, wheat-fly, and some others, differ from the majority in not producing such alterations in plants. The proboscis of these insects is very short, and does not contain the piercing bristles found in the long proboscis of the biting gnats and mosquitos. Their antennæ are long, composed of many little, bead-like joints, which are more distant in the males than in the other sex; and each joint is surrounded with short hairs. Their eyes are kidney-shaped. Their legs are rather long and very slender. Their wings have only two, three, or four veins in them, and are fringed with little hairs around the edges; when not in use, they are generally carried flat on the back. The hind body of the females often ends with a retractile, conical tube, wherewith they deposit their eggs. Their young are little, footless maggots, tapering at each end, and generally of a deep yellow or orange color.

* See page 5.

They live on the juices of plants, and undergo their transformations either in these plants, or in the ground.

The transformations of these insects offer some peculiarities that do not seem to have been described by European naturalists, and probably are not well understood by them. Three modifications in the process have been observed in this country, and examples of these are affordeed by *Cecidomyia Salicis*, *destructor*, and *Tritici*. In all of them the pupa has the limbs and wings free or unconfined, and becomes active shortly before its final change, being enabled to crawl out of the place where it had hitherto lodged, when about to take the winged form. It appears also that these Cecidomyians retain the larva-skin when the insect is changed to a pupa; this skin undergoing only certain alterations in the course of the process, without being thrown off. The abdominal part of the larva-skin remains with little or no change; the fore part of the body becomes swollen, shining, and apparently gelatinous, and allows the budding limbs and wings of the pupa to push outwards, each carrying with it an enveloping portion of the skin, which by extension or growth, or by both, is modified so as to suit the changed condition of the insect. This peculiarity was first made known to me by a letter from Dr. Asa Fitch, of Salem, New York, who has paid much attention to the natural history of the Cecidomyians, and has published several elaborate essays upon them in "The American Quarterly Journal of Agriculture and Science," and in "The Transactions of the New York State Agricultural Society." In these essays, however, the point under consideration is not so distinctly stated and described as in his letter. I am also indebted to him for galls, containing larvæ of the willow gall-fly. These, with specimens of the Hessian fly in the flax-seed state, received from him and from other correspondents, have enabled me to verify the result of his observations.

The willow gall-gnat, or gall-fly, is one of the largest of our species. It has been described and figured by Dr. Fitch, under the name of *Cecidomyia Salicis*.* On account of the size of the larva and the ease with which it may be raised, it is an excellent object for the observation of the transformation that is peculiar to it and to other species of the genus. It inhabits a small woody gall, growing at the ends of the slender twigs of the American basket-willow (*Salix rigida*), and other dwarf willows. This kind of gall is of an oval shape, about three quarters of an inch long, by three eighths of an inch thick, and is terminated by a brittle conical beak, which seems to me to consist of the unexpanded and dry terminal bud of the twig. Upon being cut open in the winter or spring, a longitudinal channel will be found in the middle, extending from the apex of the beak nearly to the base of the gall, and lined in the upper part with a delicate silken web. Within this hollow is lodged a single orange-colored maggot, about one fifth of an inch long. In the spring this maggot takes the pupa form, the approaching change being marked by an alteration of the color of the anterior segments, which from orange become red, shining, and swollen, as if distended with blood. Within a few hours after this change of color, rudimentary legs, wings, and antennæ, begin, as it were, to bud and put forth, and rapidly grow to their full pupal dimensions; and thus the transformation to the pupa is effected without any moulting of the skin of the larva. In a few days, the pupa works its way upwards, bursts through the silken film, and rests half-way out of the orifice of the beaked summit of the gall, where it casts off and leaves its pupa-skin, and appears in its winged form. This little gnat or fly is of a deep black color above, paler and downy beneath, with livid legs and smoky wings. The length of its body is a little over one fifth of an inch, and its wings expand rather more than three tenths of an inch.

The *Cecidomyia Robiniæ*, of Professor Haldeman,† is a

* American Quarterly Journal of Agriculture and Science, Vol. I. p. 263.

† American Journal of Agriculture and Science, Vol. VI. p. 193.

much smaller and more common species, inhabiting the locust-tree. During the month of August, some of the leaves of this tree will be found to have one edge thickened in substance and rolled over, so as to form an oblong cavity, cylindrical in the middle, and tapering at each end. This is the work of the larvæ or young Cecidomyians, two or three of which will sometimes be found in each cavity, where also they complete their transformations. The larva is a maggot of a whitish color, faintly tinged with orange, particularly towards the head. The pupa or chrysalis is not contained within a cocoon. The fly measures three twentieths of an inch in length. It is orange-colored, with dusky antennæ and wings, three dusky lines on the thorax, and two dusky spots on the sides of the body. An apparent interruption in one of the veins of the wings, noticed by Professor Haldeman, is not peculiar to this insect, but may be seen, more or less distinctly, in our other species of *Cecidomyia*.

The Hessian fly was scientifically described by Mr. Say, in 1817, under the name of *Cecidomyia destructor* * (Fig. 257). It obtained its common name from a supposition that it was brought to this country, in some straw, by the Hessian troops under the command of Sir William Howe in the war of the Revolution.† This supposition, however, has been thought to be erroneous, because the early inquiries made to discover the Hessian fly in Germany were unsuccessful; and, in consequence thereof, Sir Joseph Banks, in his report to the British government, in 1789, stated that "no such insect could be found to exist in Germany or any other part of Europe." ‡ It

Fig. 257.

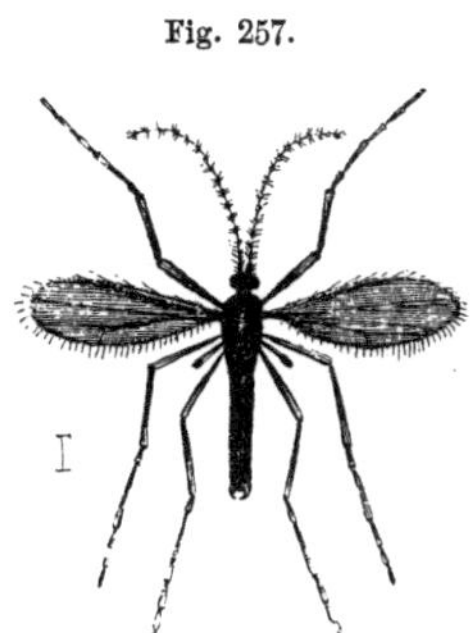

* Journal of the Academy of Natural Sciences of Philadelphia, Vol. I. p. 45.

† Dobson's Encyclopædia, Vol. VIII. p. 491.

‡ Encyclopædia Britannica, and Dobson's Encyclopædia, Vol. VIII., Article *Hessian Fly*.

appears, however, that the same insect, or one exactly like it in habits, had been long known in the vicinity of Geneva; an account of it may be found in Duhamel's "Practical Treatise of Husbandry," * and in a communication † made to the Duke of Dorset, in 1788, by the Royal Society of Agriculture of France.

In the year 1833 the wheat in Austria and in Hungary was considerably injured by an insect of the same kind, supposed to be the Hessian fly by the Baron Kollar.‡ Moreover, Mr. E. C. Herrick, of New Haven, Connecticut, has published an account § of the discovery of the true Hessian fly, by Mr. James D. Dana, in Minorca, near Toulon in France, and in the vicinity of Naples, in the year 1834. Nothing has yet been found relative to the existence of the Hessian fly in America before the Revolution. It was first observed in the year 1776, in the neighborhood of Sir William Howe's debarkation on Staten Island, and at Flat Bush, on the west end of Long Island. Having multiplied in these places, the insects gradually spread over the southern parts of New York and Connecticut, and continued to proceed inland at the rate of fifteen or twenty miles a year. They reached Saratoga, two hundred miles from their original station, in 1789. Dr. Chapman says, that they were found west of the Alleghany Mountains in 1797; from their progress through the country, having apparently advanced about thirty miles every summer. Wheat, rye, barley, and even timothy grass, were attacked by them; and so great were their ravages in the larva state, that the cultivation of wheat was abandoned in many places where they had established themselves.‖

* Page 90 (4to, Lond., 1759). See also his Elements of Agriculture, Vol. I. p. 269 (8vo, Lond., 1664).

† Encyclopædia Britannica, and Dobson's Encyclopædia, Vol. VIII., Article *Hessian Fly.*

‡ Treatise, pp. 118, 119.

§ Silliman's American Journal of Science, Vol. XLI. p. 153.

‖ Encyclopædia Britannica, and Dobson's Encyclopædia, Vol. VIII., Article *Hessian Fly.*

In a communication by Mr. J. W. Jeffreys, published in the sixth volume of Buel's "Cultivator," it is stated, that soon after the battle of Guilford, in North Carolina, the wheat crops were destroyed by the Hessian fly in Orange County, through which the British army, composed in part of Hessian soldiers, had previously passed. Although it is possible that, in this instance, the chinch-bug may have been mistaken for the Hessian fly, the remark shows how prevalent was the belief respecting the introduction of the latter. The foregoing statements, taken in connection with the habits of the Hessian fly, induce me to think that the common opinion relative to its origin is deserving of some credit.

The head, antennæ, and thorax of this fly are black. The hind body is tawny, more or less widely marked with black on each ring, and clothed with fine grayish hairs. The egg-tube of the female is rose-colored. The wings are blackish, except at the base, where they are tawny and very narrow; they are fringed with short hairs, and are rounded at the tip. The legs are pale red or brownish, and the feet are black. The body measures about one tenth of an inch in length, and the wings expand one quarter of an inch, or more. After death, the hind body contracts and becomes almost entirely black.

The Hessian fly is a true *Cecidomyia*, differing from *Lasioptera* in the shortness of the first joint of its feet, and in the greater length of its antennæ, the bead-like swellings whereof are also more distant from each other, especially in the males. According to Mr. Herrick, the number of the joints of the antennæ varies "from fourteen to seventeen, besides the basal joint, which appears double." As in other species of *Cecidomyia*, the form of the joints differs according to the sex; those of the male being globular, and those of the female, except at base, oblong oval. In both they are surrounded with whorls of short hairs. The difference in the antennæ of the sexes has been pretty well represented by Mr. Lesueur, in the plate designed to accompany Mr. Say's description of the insect.

The following brief history of the habits and transformations of the Hessian fly will be found to agree essentially with the excellent observations on this insect, written in the year 1797, by Dr. Isaac Chapman, and published in the fifth volume of the "Memoirs of the Philadelphia Society for Promoting Agriculture," and with the more full and equally valuable history of the insect, by Jonathan N. Havens, Esq., contained in the first volume of the "Transactions of the Society for the Promotion of Agriculture, &c., in New York." Mr. Herrick has kindly permitted me to make free use of his valuable account of this insect, contained in the forty-first volume of "The American Journal of Science," and of other information communicated by him to me in various letters. He has spent some time in carefully observing the habits of the fly, during many years in succession, after having fitted himself for the task by the study of the natural history of insects in general. His statements therefore may be relied upon, as in the main correct. Moreover, they are corroborated by the observations of many other persons, published in various works, which have been consulted in the course of my investigations.

Of this insect, two broods or generations are brought to maturity in the course of a year, and the flies appear in the spring and autumn, but rather earlier in the Southern and Middle States than in New England. The transformations of some in each brood appear to be retarded beyond the usual time, as is found to be the case with many other insects; so that the life of these individuals, from the egg to the winged state, extends to a year or more in length, whereby the continuation of the species in after years is made more sure. It has frequently been asserted, that the flies lay their eggs on the grain in the ear; but whether this be true or not, it is certain that they do lay their eggs on the young plants, and long before the grain is ripe; for many persons have witnessed and testified to this fact. In the New England States, winter wheat, as it is called, is usually sown about the first of September. Towards the end of this month, and in October, when the grain has sprouted, and begins to show a leaf or two, the flies appear in the fields, and, having paired, begin to lay their eggs, in which business they are occupied for several weeks.

The following interesting account of the manner in which this is done was written by Mr. Edward Tilghman, of Queen Ann County, Maryland, and was published in the eighth volume of "The Cultivator," in May, 1841. "By the second week of October, the first-sown wheat being well up, and having generally put forth its second and third blades, I resorted to my field in a fine warm forenoon, to endeavor to satisfy myself, by ocular demonstration, whether the fly did deposit the egg on the blades of the growing plant. Selecting a favorable spot to make my observation, I placed myself in a reclining position in a furrow, and had been on the watch but a minute or two before I discovered a number of small black flies alighting and sitting on the wheat plants around me, and presently one settled on the ridged surface of a blade of a plant completely within my reach and distinct observation. She immediately began depositing her eggs in the longitudinal cavity between the little ridges of the blade. I could distinctly see the eggs ejected from a kind of tube or sting. After she had deposited eight or ten eggs, I easily caught her upon the blade, and wrapped her up in a piece of paper. I then proceeded to take up the plant, with as much as I conveniently could of the circumjacent earth, and wrapped it all securely in a piece of paper. After that I continued my observations on the flies, caught several similarly occupied, and could see the eggs uniformly placed in the longitudinal cavities of the blades of the wheat; their appearance being that of minute reddish specks. My own mind being thus completely and fully satisfied as to the mode in which the egg was deposited, I proceeded directly to my dwelling, and put the plant with the eggs upon it in a large glass tumbler, adding a little water to the earth, and secured the vessel by covering it with paper, so that no insect could get access to the interior. The paper was sufficiently perforated with pin-holes for the admission of air. The tumbler with its contents was daily watched by myself to discover the hatching of the eggs. About the middle of the fifteenth day from the deposit of the eggs, I was so fortunate as to discover a very small maggot or worm, of a reddish cast, making its way with considerable activity down the blade, and saw it till it disappeared between the blade and stem of the plant. This, I have no doubt, was the produce of one of the eggs, and would, I presume, have hatched much sooner, had the plant remained in the field. It was my intention to have carried on the experiment, by endeavoring to hatch out the insect from the flax-seed state into the perfect fly again; but being called from home, the plant was suffered to perish. The fly that I caught on the blade of the wheat, as above stated, I enclosed in a letter to Mr. John S. Skinner, the editor of 'The American Farmer,' of Baltimore, who pronounced it to be a genuine Hessian fly, and identical in appearance with others recently received from Virginia."

Dr. Chapman agrees with this writer in saying, that the Hessian fly lays her eggs in the small creases of the young leaves of the wheat. Mr. Havens states, that the fly lays her eggs on the leaves. In the fortieth number of "The Connecticut Farmer's Gazette," Mr. Herrick says: "I have repeatedly, both in autumn and in spring, seen the Hessian fly in the act of depositing eggs on wheat, and have always found that she selects for this purpose the leaves of the young plant. The eggs are laid in various numbers on the upper surface of the strap-shaped portion (or blade) of the leaf." His remarks in Professor Silliman's Journal are to the same effect. Other authorities on this point might be mentioned; but the foregoing are sufficient, in my opinion, to establish the fact, that the Hessian fly lays her eggs on the leaves of wheat soon after the plants are up. "The number on a single leaf," says Mr. Herrick, "is often twenty or thirty, and sometimes much greater. In these cases many of the larvæ must perish. The egg is about a fiftieth of an inch long, and four thousandths of an inch in diameter, cylindrical, translucent, and of a pale red color. Mr. Tilghman was correct in supposing that the eggs would hatch in less than fifteen days, under favorable circumstances; for, if the weather be warm, they commonly hatch in four days after they are laid.

Fig. 258.

The maggots (Fig. 258, natural size), when they first come out of the shells, are of a pale red color. Forthwith they crawl down the leaf, and work their way between it and the main stalk, passing downwards till they come to a joint, just above which they remain, a little below the surface of the ground, with the head towards the root of the plant. Having thus fixed themselves upon the stalk, they become stationary, and never move from the place till their transformations are completed. They do not eat the stalk, neither do they penetrate within it, as some persons have supposed, but they lie lengthwise upon its surface, covered by the lower part of the leaves, and are nourished wholly by the sap, which they appear to take by suction. They soon lose their reddish color, turn pale, and will be found to be clouded with whitish spots; and through their transparent skins a greenish stripe may be seen in the middle of their bodies. As they increase in size, and grow plump and firm, they become imbedded in the side of the stem, by the pressure of their bodies upon the growing plant. One maggot thus placed seldom destroys the plant; but when two or three are fixed in this manner around the stem, they weaken and impoverish the plant, and cause it to fall down, or to wither and die. They usually come to their full size in five or six weeks, and then measure about three twentieths of an inch in length. Their skin now gradually hardens, becomes brownish, and soon changes to a bright chestnut-color. This change usually happens about the first of December.

The insect, in this form, has been commonly likened to a flax-seed (Fig. 259, natural size and magnified, larva on the left). Hence "many observers speak of this as the flax-seed state." Others regard it as the beginning of the pupa state, wherein the condition of the insect is analogous to the immature pupa (*boule allongée*) of common flies. Such indeed has been my own impression concerning it; and even so it seems to have been regarded by Mr. Herrick, although he was well aware of the actual form of the insect included within this "leathery" outer skin of the larva, and of all its subsequent changes. While this change of the color and texture of the skin is going on, the body of the insect, as remarked by Mr. Herrick, "gradually cleaves from the dried skin, and, in the course of two or three weeks, is wholly detached."

Fig. 259.

In a letter dated February 21, 1843, he alludes more explicitly to the condition of the insect, in these words: "In two or three weeks after this change of color, the animal within becomes entirely detached from the old larva-skin, and lies *a motionless grub*." Accordingly, when this dried skin or flax-seed case is opened, the insect will be found loose within it, and still retaining the maggot form, as stated by Mr. Herrick, Mr. Worth,* and Professor

* Mr. James Worth, writing on this insect in 1820, remarked that "as soon as it changes to the flax-seed color, by rolling it lightly with the finger, the tegument can be taken off; the *worm* will then appear with a greenish stripe through it, which is evidently the substance extracted from the plant." (American Farmer, Vol. II. p. 180.)

Cabell.* Kollar alludes to the unchanged condition of the insect within this case, in the European specimens which he had examined.† Mr. Westwood makes the following remarks upon some from Vienna that were in his possession: "The insects are enclosed in a leathery case, and on opening them I discovered the larvæ shrivelled up and dead." ‡ Referring to Mr. Say's account of the Hessian fly, and its flax-seed case, Mr. Westwood says, "It is not described in what manner this case is formed." That it really consists of the loosened outer skin of the maggot is evident from its shape and structure. It has nearly the same form and size, is convex on both sides, and retains traces of the former segments in the transverse lines wherewith it is marked. This flax-seed shell has been correctly called a *puparium*, or pupa-case, because the pupa is subsequently matured within it.

Dr. Chapman repeatedly alludes to the pupa, or chrysalis as he calls it, and to "the outward coat" of the larva "becoming a hard shell or covering for the chrysalis"; by which we perceive that he was acquainted with the origin and office of the one, and the condition of the other. But as the true figure of the included insect is concealed, and cannot be determined without opening the puparium, "it is customary," as stated by Messrs. Kirby and Spence,§ "in speaking of pupæ of this description, to refer solely to the exterior covering." Agreeably to this common usage, sanctioned by the best entomologists of our time, the flax-seed case, or puparium, has been commonly denominated the pupa, even by such writers as Mr. Say, to whom the real nature of its contents must have been well known.

In the letter before mentioned, Mr. Herrick thus continued his account of the transformations of the insect. "The

* See page 557.

† Kollar's Treatise, p. 121.

‡ Note in Kollar's Treatise, p. 121. See also Westwood's Modern Classification of Insects, Vol. II. p. 520.

§ Introduction to Entomology, Vol. III. p. 258.

process of growth goes on, and, by and by, on opening the leathery maggot-skin, now a puparium, you find the pupa so far advanced that some of the members of the future fly are discernible through the scarf which envelopes and fetters it on all sides." In his observations communicated to the Commissioner of Patents in 1844,* he referred to the same process in the following words: "Within this shell (the flax-seed case) the pupa gradually advances towards the winged state; it contracts in length, but not in breadth; and its skin appears covered with minute elevations. Just before evolution (of the fly), we find the pupa invested in a delicate membrane or scarf, which not long previous was its outer skin, through which many parts of the future fly may be distinctly seen."

From the foregoing passages, it appears that the transition of the insect, within the flax-seed case, from the form of a larva or maggot to that of a mature pupa, takes place only a short time before its final transformation to a fly, that is, towards the end of April or beginning of May; and that the scarf or proper skin of this pupa is the same as that wherein the body of the insect had been previously enveloped. In this respect, the Hessian fly agrees in its transformations with the willow gall-fly; and doubtless the transition in question is effected in the same way as in that insect. But the larva of the Hessian fly does not spin a silken web or cocoon like that of the willow gall-fly and some other Cecidomyians; and it differs from these insects also in being finally invested with two skins, the outer one, when detached, serving instead of a cocoon for the included insect; while the inner one, of a much thinner and more delicate texture, becomes the true skin of the matured pupa.

Towards the end of April and in the fore part of May, or as soon as the weather becomes warm enough in the spring, the insects are transformed to flies. They make

* Report, p. 163.

their escape from their winter quarters by breaking through one end of their shells and the remains of the leaves around them. In the "Observations on the Hessian Fly," written by Jonathan N. Havens, Esq., it is stated, that, "whenever the fly has been hatched in the house, it always comes forth from its brown case wrapt in a thin white skin, which it soon breaks, and is then at liberty"; and Mr. Havens supposes that the same thing occurs when the transformation takes place abroad. Mr. Herrick states, that this skin or "scarf," as he calls it, "splits on the thorax or back," and the fly is disengaged from it by working through the rent.

This process, and the appearance of the insect through the pupa-skin, is fully described in his letter of the 21st of February, 1843, from which the following extract is taken. It is from a memorandum made May 12, 1837. "On looking over culms of wheat, which ripened last July, I found a puparium of the Hessian fly; began to cut it open; found within a fly nearly matured. Opened only the anterior part of the puparium; but the animal soon squirmed itself out, enveloped in a thin scarf. The puparium was left entirely clean. — The animal worked its abdomen back and forth, and, in about twenty minutes, was detached from the scarf." In one instance, Mr. Herrick found the empty scarf-skin "attached to one end of the puparium." Ordinarily, however, the insect seems to crawl entirely out of the puparium, or flax-seed shell, before disengaging itself from the pupa-skin, as stated above by Mr. Havens. Upon examining a puparium after the escape of the insect, I could not discover any vestige of larva or pupa skin within it. It was left entirely empty.

Very soon after the flies come forth in the spring, they are prepared to lay their eggs on the leaves of the wheat sown in the autumn before, and also on the spring-sown wheat, that begins, at this time, to appear above the surface of the ground. They continue to come forth and lay their eggs for the space of three weeks, after which they entirely disappear from the fields. The maggots, hatched from these eggs, pass along the stems of the wheat, nearly to the roots, become stationary, and take the flax-seed form in June and July. In this state they are found at the time of harvest; and, when the grain is gathered, they remain in the stubble in the fields. To this, however, as Mr. Havens remarks, there are some exceptions; for a few of the insects do not pass so far down the side of the stems as to be out of the way of the sickle when the grain is reaped, and consequently will be gathered and carried away with the straw. Most of them are transformed to flies in the autumn, but others remain unchanged in the stubble or straw till the next spring. Hereby, says Mr. Havens, "it appears evident that they may be removed from their natural situation in the field, and be kept alive long enough to be carried across the Atlantic; from which circumstance it is possible that they might have been imported" in straw from a foreign country.

In the winged state, these flies, or more properly gnats, are very active, and, though very small and seemingly feeble, are able to fly to a considerable distance in search of fields of young grain. Their principal migrations take place in August and September in the Middle States, where they undergo their final transformations earlier than in New England. There, too, they sometimes take wing in immense swarms, and, being probably aided by the wind, are not stopped in their course either by mountains or rivers. On their first appearance in Pennsylvania, they were seen to pass the Delaware like a cloud. Being attracted by light, they have been known, during the wheat harvest, to enter houses in the evening in such numbers as seriously to annoy the inhabitants.*

Mr. Havens has alluded to "an opinion, entertained by

* British, and Dobson's Encyclopædia, and Colonel Morgan's letter in Carey's American Museum, Vol. II. p. 298.

many observers, that there are three generations of this insect in a year," "two" being completed "before harvest." This opinion was revived, in 1821, by Mr. James Worth, of Sharon, Pennsylvania.* According to him, the second brood of flies, which appears early in June, had been altogether overlooked, or confounded with the spring brood. Their "eggs were lain on the upper leaves of the weakest or stunted wheat, and the larvæ became lodged about the two upper joints, but most about the upper." Being very numerous, and crowded together, many of the larvæ perished for want of food, and many also were destroyed by parasites. Enough, however, remained alive to continue the race; and the flies were evolved from them at irregular intervals, and continued laying from the 15th of August till October, when the earliest of their progeny entered on the fly state; thus making, during the year, as remarked by Mr. Worth, "three complete broods, and partially a fourth."

Mr. Say, though he does not appear to have been fully acquainted with the history of the insect, has recorded the occurrence of the fly in June. His remarks are these: "The perfect fly appears early in June, lives but a short time, deposits its eggs, and dies; the insects from these eggs complete the history by preparing for the winter brood."

In the year 1833, Mr. Herrick saw a Hessian fly laying eggs on the 3d of June, another on the 5th, and a third on the 7th of the same month. The fact of the occasional appearance of the flies as late as the 12th of June, when Mr. Worth found the insects in all their stages, seems to be well established; while it is equally certain that ordinarily only two broods are brought to perfection in the course of one year. Various circumstances may contribute to accelerate or to retard a portion of each brood; and, hence, some of the flies may be found from the middle of April to the middle of June, and others from the beginning of August till December. These circumstances have been so fully consid-

* See American Farmer, Vol. III. p. 188.

ered by Mr. Havens, that it is unnecessary to repeat them here. The observations of Mr. Worth are interesting, as showing that the insect is not left without resources, although there are no young wheat-plants growing in June; the upper joints of those old plants that are late in ripening being found to yield sufficient nourishment for a portion, at least, of the progeny of the June flies. They show, also, how easily the insects might be imported from Europe in the straw containing them, in the flax-seed state, about the upper joints.

The old discussion, concerning the place where the Hessian fly lays her eggs, was revived in the year 1841, in consequence of a communication made by Miss Margaretta H. Morris, of Germantown, Pennsylvania, to "The American Philosophical Society," of Philadelphia. The following remarks upon it are extracted from a Report made to the same Society, and published in their "Proceedings" for November and December, 1840. "Miss Morris believes she has established that the ovum (egg) of this destructive insect is deposited in the seed of the wheat, and not in the stalk or culm. She has watched the progress of the animal since June, 1836, and has satisfied herself that she has frequently seen the larva within the seed. She has also detected the larva, at various stages of its progress, from the seed to between the body of the stalk and the sheath of the leaves. According to her observations, the recently hatched larva penetrates to the centre of the straw, where it may be found of a pale greenish-white semitransparent appearance, in form somewhat resembling a silkworm. From one to six of these have been found at various heights from the seed to the third joint."

Miss Morris's communication had not been published in full when the first edition of this work was prepared for the press; but, from the foregoing Report, we are led to infer, that the egg, being sowed with the grain, is hatched in the ground, and that the maggot afterwards mounts from the seed through the middle of the stem, and, having reached a proper height, escapes from the hollow of the straw to the outside, where it takes the pupa or flax-seed state. The fact that the Hessian fly does ordinarily lay her eggs on the young leaves of wheat, barley, and rye, both in the spring and in the autumn, is too well authenticated to admit of any doubt. If, therefore, the observations of Miss Morris are found to be equally correct, they will serve to show, still more than the foregoing history, how variable and extraordinary is the economy of this insect, and how great are the resources wherewith it is provided for the continuation of its kind.

The foregoing remarks were written in 1841. Since that time, the communication, to which they refer, has been printed,* and this has been followed by the publication of several other articles,* on the same subject, by Miss Morris. This ingenious and persevering lady has also favored me with letters concerning her investigations, and with some of the flies. The latter were sent, as she says, " to convince me, at least, that she had not mistaken a curculio, moth, or bee for a Cecidomyia." Miss Morris has come to the conclusion that this insect is a different species from the Hessian fly, for which it had previously been mistaken, and has given to it the name of *Cecidomyia culmicola.* According to her, the fly " deposits its eggs early in June on the grain, in or over the germ. The eggs remain unhatched until the grain germinates, but when the plant has grown about three or four inches, the worm may be seen, with the aid of a strong magnifying-glass, feeding above the top joint, in the centre of the culm, where it remains until it arrives at maturity. Should this occur before the culm has become hard, the worm eats its way through the joint, inside of the straw, and makes its escape at the root, as-

* Transactions of the American Philosophical Society, Philadelphia, New Series, Vol. VIII. p. 48. Proceedings of the Academy of Natural Sciences, Philadelphia, Vol. I. p. 66; Vol. III. p. 238; and Vol. IV. p. 194.

cends the straw on the outside, where it attaches itself firmly, and awaits its change; the outer skin becomes the puparium. In the pupa or flax-seed state, it closely resembles the *C. destructor*. Should the culm of the wheat become prematurely hard before the worm has finished feeding, as is often the case, the insect will remain imprisoned for life, passing through its changes inside the straw, and there perish without the power to escape, unless some accidental passage be made for it. I have liberated," she adds, "hundreds with my penknife, and thousands make their escape after the grain has been reaped and carried into the barn. When the insect is thus unnaturally retarded, the time of its perfect development is uncertain"; and she has "found them on the straw, and in spiders' webs, from June until September."

Four of the specimens sent to me by Miss Morris were males. Another subsequently received was a female. The former were not more than half the size of the latter, and indeed were smaller even than the wheat-fly, which they seem somewhat to resemble. The female was evidently much darker-colored originally than the males. These insects were genuine specimens of *Cecidomyia*, and apparently of a different species from the Hessian fly. The condition of the specimens, which had suffered by compression and by being badly preserved, was such, that an accurate comparison and description of them could not be made. I understand that the species has disappeared from Germantown and the vicinity, and hence no opportunity for obtaining living or recent specimens has occurred since the year 1843.

Various means have been recommended for preventing or lessening the ravages of the Hessian fly; but they have hitherto failed, either because they have not been adapted to the end in view, or because they have not been universally adopted; and it appears doubtful whether any of them will ever entirely exterminate the insect. It is stated in the before mentioned Report of the Philosophical Society, that Miss Morris advises obtaining "fresh seed from localities in which the fly has not made its appearance," and that "by this means the crop of the following year will be uninjured; but, in order to avoid the introduction of straggling insects of the kind from adjacent fields, it is requisite that a whole neighborhood should persevere in this precaution for two or more years in succession." "This result," Miss Morris says, "was obtained, in part, in the course of trials made by Mr. Kirk, of Bucks County, Pennsylvania, with some seed-wheat from the Mediterranean, in and since the year 1837. His first crop was free from the fly; but it was gradually introduced from adjacent fields, and, in the present year (1840), the mischief has been considerable." In other hands this course has proved of no use whatever.

Not to mention other instances, the following appears to be conclusive on this point. About fifty years ago, Mr. Garret Bergen, of Brooklyn, New York, procured two bushels of wheat from the Genesee country, then an uninfected district, which he sowed in a field adjoining a piece seeded with grain of his own gathering. Both pieces were severely damaged by the Hessian fly, which could not have happened, in the same season, if the eggs of the insect are laid only on the grain. A few years ago he soaked his seed-wheat in strong pickle, and the crop was comparatively free from the fly. In 1839 he tried this experiment again, but not with similar success. In 1840 he sowed without previously soaking the grain, and his crop was uninjured. He says, moreover, that he has uniformly found the grain most affected in spots, usually near the edges of the field, where long grass and weeds grew, which afforded shelter and protection to the fly. This fact, he thinks, affords another proof that the egg is not deposited in the grain. I regret that my limits will not permit me to extract the whole of Mr. Bergen's interesting remarks, which may be found in number eight of the eighth volume of "The Cultivator," published in Albany in August, 1841.

The best modes of preventing the ravages of the Hessian fly are thus stated by Mr. Herrick.* "The stouter varieties of wheat ought always to be chosen, and the land should be kept in good condition. If fall wheat is sown late, some of the eggs will be avoided, but the risk of winter-killing the plants will be incurred. If cattle are permitted to graze the wheat-fields during the fall, they will devour many of the eggs. A large number of the pupæ may be destroyed by burning the wheat-stubble immediately after harvest, and then ploughing and harrowing the land. This method will undoubtedly do much good. As the Hessian fly also lays its eggs, to some extent, on rye and barley, these crops should be treated in a similar manner." On mature reflection, I am confident that burning the stubble, as originally recommended by Mr. Havens, and advised by Mr. Herrick, is the very best method of getting rid of the Hessian fly. It is true that, by so doing, many of the numerous parasites of the insect will also be destroyed. But this need not give us any concern; for if we can succeed in putting a stop to the ravages of the Hessian fly, by these or any other means, we shall not have occasion to mourn the loss of the parasites. It is found that luxuriant crops more often escape injury than those that are thin and light. Steeping the grain and rolling it in plaster or lime tend to promote a rapid and vigorous growth, and will therefore prove beneficial. Sowing the fields with wood ashes, in the proportion of two bushels to an acre, in the autumn, and again in the first and last weeks in April, and as late in the month of May as the sower can pass over the wheat without injury to it, has been found useful.† Favorable reports have been made upon the practice of allowing sheep

* American Journal of Science, Vol. XLI. p. 158.

† Cultivator, Vol. V. p. 59.

to feed off the crop late in the autumn, and it has also been recommended to turn them into the fields again in the spring, in order to retard the growth of the plant till after the fly has disappeared.* Too much cannot be said in favor of a judicious management of the soil, feeding off the crop by cattle in the autumn, and burning the stubble after harvest; a proper and general attention to which will materially lessen the evils arising from the depredations of this noxious insect.

Fortunately our efforts will be aided by a host of parasitical insects, which are found to prey upon the eggs, the larvæ, and the pupæ of the Hessian fly. Mr. Herrick states,† that, in this part of the country, a very large proportion, probably more than nine tenths, of every generation of this fly is thus destroyed. One of these parasites was made known by Mr. Say, in the first volume of the "Journal of the Academy of Natural Sciences of Philadelphia"; and the interesting discovery of three more kinds is due to the exertions of Mr. Herrick. They are all minute Hymenopterous insects, similar in their habits to the true Ichneumon-flies.

The chief parasite of the pupa is the *Ceraphron destructor* ‡ of Say, a shining black four-winged fly, about one tenth of an inch in length. This has often been mistaken for the Hessian fly, from being seen in wheat-fields, in vast numbers, and from its being found to come out of the dried larva-skin of that fly. In the month of June, when the maggot of the Hessian fly has taken the form of a flax-seed, the *Ceraphron* pierces it, through the sheath of the leaf, and lays an egg in the minute hole thus made. From this

* Cultivator, Vol. IV. p. 110, and Vol. V. p. 49.

† American Journal of Science, Vol. XLI. p. 156.

‡ It is evident, from Mr. Say's description, and from Mr. Lesueur's figures, that this insect is not a *Ceraphron*. Neither does it belong to the genus *Eurytoma*, to which I formerly referred it. It certainly comes very near to *Pteromalus*, as suggested by Mr. Westwood; but I apprehend that it should be placed in the genus *Rhaphitelus* of Walker, or *Storthygocerus* of Ratzeburg.

egg is hatched a little maggot, which devours the pupa of the Hessian fly, and then changes to a chrysalis within the shell of the latter, through which it finally eats its way, after being transformed to a fly. This last change takes place both in the autumn and in the following spring. Some of the females of this or of a closely allied species come forth from the shells of the Hessian fly, without wings, or with only very short and imperfect wings, in which form they somewhat resemble minute ants.

Two more parasites, which Mr. Herrick has not yet described, also destroy the Hessian fly, while the latter is in the flax-seed or pupa state. Mr. Herrick says, that the egg-parasite of the Hessian fly is a species of *Platygaster*, that it is very abundant in the autumn, when it lays its own eggs, four or five together, in a single egg of the Hessian fly. This, it appears, does not prevent the latter from hatching, but the maggot of the Hessian fly is unable to go through its transformations, and dies after taking the flax-seed form. Meanwhile its intestine foes are hatched, come to their growth, spin themselves little brownish cocoons within the skin of their victim, and, in due time, are changed to winged insects, and eat their way out. Such are some of the natural means, provided by a benevolent Providence, to check the ravages of the destructive Hessian fly. If we are humiliated by the reflection, that the Author of the universe should have made even small and feeble insects the instruments of His power, and that He should occasionally permit them to become the scourges of our race, ought we not to admire His wisdom in the formation of the still more humble agents that are appointed to arrest the work of destruction?

The wheat crops in England and Scotland often suffer severely from the depredations of the maggots of a very small gnat, called the wheat-fly, or the *Cecidomyia Tritici* of Mr. Kirby. This insect seems to have been long known in England, as appears from the following extract from a letter by Mr. Christopher Gullet, written in 1771, and published in the "Philosophical Transactions" for 1772. "What the farmers call the yellows in wheat, and which they consider as a kind of mildew, is, in fact, occasioned by a small yellow fly, with blue wings, about the size of a gnat. This blows in the ear of the corn, and produces a worm, almost invisible to the naked eye; but, being seen through a pocket microscope, it appears a large yellow maggot, of the color and gloss of amber, and is so prolific that I distinctly counted forty-one living yellow maggots in the husk of one single grain of wheat, a number sufficient to eat up and destroy the corn in a whole ear. One of those yellow flies laid at least eight or ten eggs, of an oblong shape, on my thumb, only while carrying by the wing across three or four ridges."

In 1795, the history of this insect was investigated by Mr. Marsham,* and since that time Mr. Kirby,† Mr. Gorrie, and Mr. Shirreff ‡ have also turned their attention to it. The investigations of these gentlemen have become very interesting to us, on account of the recent appearance in our own country, and the extensive ravages, of an insect apparently identical with the European wheat-fly. The following account of the latter will serve to show how far the European and American wheat-flies agree in their essential characters and in their habits. §

The European wheat-fly somewhat resembles a mosquito in form, but is very small, being only about one tenth of an inch long. Its body is orange-colored. Its two wings are transparent, and changeable in color; they are narrow at the base, rounded at the tip, and are fringed with little hairs on the edges. Its long antennæ, or horns, consist, in the female, of twelve little bead-like joints, each encircled with

* Transactions of the Linnæan Society, Vol. III. p. 142, and Vol. IV. p. 224.

† Ibid., Vol. IV. p. 230, and Vol. V. p. 96.

‡ Loudon's Magazine of Natural History, Vol. II. pp. 323 and 448.

§ See also my article on wheat insects in the New England Farmer, for March 31, 1841, Vol. XIX. p. 306.

minute hairs; those of the male will probably be found to have a greater number of joints.

Towards the end of June, or when the wheat is in blossom, these flies appear in swarms in the wheat-fields during the evening, at which time they are very active. The females generally lay their eggs before nine o'clock at night, thrusting them, by means of a long, retractile tube in the end of their bodies, within the chaffy scales of the flowers, in clusters of from two to fifteen or more. By day they remain at rest on the stems and leaves of the plants, where they are shaded from the heat of the sun. They continue to appear and lay their eggs throughout a period of thirty-nine days. The eggs are oblong, transparent, and of a pale buff color, and hatch in eight or ten days after they are laid. The young insects produced from them are little footless maggots, tapering towards the head, and blunt at the hinder extremity, with the rings of the body somewhat wrinkled and bulging at the sides. They are at first perfectly transparent and colorless, but soon take a deep yellow or orange color. They do not travel from one floret to another, but move in a wriggling manner, and by sudden jerks of the body, when disturbed. As many as forty-seven have been counted in a single floret. It is supposed that they live at first upon the pollen, and thereby prevent the fertilization of the grain. They are soon seen, however, to crowd around the lower part of the germ, and there appear to subsist on the matter destined to have formed the grain. The latter, in consequence of their depredations, becomes shrivelled and abortive; and, in some seasons, a considerable part of the crop is thereby rendered worthless. The maggots, when fully grown, are nearly one eighth of an inch long. Mr. Marsham and Mr. Kirby found some of them changed to pupæ within the ears of the wheat, and from these they obtained the fly early in September. The pupa represented by them is rather smaller than the full-grown maggot, of a brownish yellow color, and of an oblong oval form, tapering at each end. The pupæ found in the ears were very few in number, scarcely one to fifty of the maggots. Hence Mr. Kirby supposes that the latter are not ordinarily transformed to flies before the spring. Towards the end of September he carefully took off the skin of one of them, and found that the insect within still retained the maggot form, and conjectures that the pupa is not usually complete until the following spring.

It is evident, from these observations, that the English naturalists above named regarded the insect as having entered upon the pupa state when it ceased feeding and became quiescent, at which time Mr. Kirby found it generally to adhere somewhat to the grain. In applying to it, in this condition, the name of chrysalis or pupa, and describing it as such before it exhibited any trace of "the lineaments of the future fly," and while "still in the form of the larva," they followed the common usage of naturalists, as stated in my account of the Hessian fly. They cannot, therefore, be said to have mistaken the larva for the matured pupa; the remarks of Mr. Kirby prove that he was well aware of the difference between them. Mr. Kirby, however, was mistaken in his conjecture that "the insect enclosed itself in a thin membrane to protect itself from the cold of the winter"; the membrane referred to being merely the outer skin of the larva, loosened previously to being cast off entirely, — a process which he did not observe.

According to Mr. Gorrie, the maggots quit the ears of the wheat by the first of August, descend to the ground, and go into it to the depth of half an inch. That they remain here unchanged through the winter, and finish their transformations, and come out of the ground in the winged form, in the spring, when the wheat is about to blossom, is rendered probable from the great number of the flies found by Mr. Shirreff, in the month of June, in all the fields where wheat had been raised the year before. The increase of these flies is somewhat checked by the attacks of three different parasites, which have been described by Mr. Kirby. An excellent summary of the history of this insect, illustrated with figures, was published by Mr. Curtis, in the year 1845, in the sixth volume of the "Journal of the Royal Agricultural Society of England."

An insect, resembling the foregoing in its destructive habits, and known, in its maggot form, by the name of "the grain-worm," and "the weevil," has been observed, for several years, in the northern and eastern parts of the United States, and in Canada. It seems by some to have been mistaken for the grain-weevil, the Angoumois grain-moth, and the Hessian fly; and its history has been so confounded with that of another insect, also called the grain-worm, in some parts of the country, that it is difficult to ascertain the amount of injury done by either of them alone. The wheat-fly is said to have been first seen in America about the year 1828,* in the northern part of Vermont, and on the borders of Lower Canada. From these places its ravages have gradually extended, in various directions, from year to year. A considerable part of Upper Canada, of New York, New Hampshire, and of Massachusetts, have been visited by it; and, in 1834, it appeared in Maine, which it has traversed, in an easterly course, at the rate of twenty or thirty miles a year. The country over which it has spread has continued to suffer more or less from its alarming depredations, the loss by which has been found to vary from about one tenth part to nearly the whole of the annual crop of wheat; nor has the insect entirely disappeared in any place, till it has been starved out by a change of agriculture, or by the substitution of late-sown spring wheat for the other varieties of grain.

Many communications on this destructive insect have

* Judge Buel's Report in the Cultivator, Vol. VI. p. 26; and New England Farmer, Vol. IX. p. 42. Mr. Jewett says, that its first appearance in Western Vermont occurred in 1820. See New England Farmer, Vol. XIX. p. 301.

appeared in "The Genesee Farmer," and in "The Cultivator," some of them written by the late Judge Buel, by whom, as well as by the editors of "The Yankee Farmer," rewards were offered for the discovery of the means to prevent its ravages. Premiums have also been proposed, for the same end, by the Kennebec County Agricultural Society, in Maine, which were followed by the publication, in "The Maine Farmer," of three "Essays on the Grain-Worm," presented to that Society. These essays were reprinted in the seventeenth volume of the "New England Farmer," wherein, as well as in some other volumes of the same work, several other articles on this insect may be found. From these sources, and, more especially, from some interesting letters wherewith I was favored in the years 1838 and 1841, by Mrs. N. G. Gage, formerly of Hopkinton, New Hampshire, the history of the wheat-fly in America, published in the first edition of this work, was chiefly derived. It will be found to contain a circumstantial relation of the moulting of the maggot, a process which hitherto does not appear to have been understood in Europe, and which later writers on the history of the wheat-fly in this country have failed to describe. Personal observations on this insect in Maine and New Hampshire, and in the western parts of Massachusetts and of Connecticut, together with information gathered there from intelligent farmers, confirm the general correctness of my former statements, and enable me to add thereto some further particulars.

The American wheat-insect, which I have seen alive, in its winged form, in Maine and in New Hampshire, and which I have also reared from the larva, agrees exactly with the descriptions and figures of the European wheat-fly, or *Cecidomyia Tritici* of Mr. Kirby. It is a very small orange-colored gnat, with long, slender, pale-yellow legs, and two transparent wings, reflecting the tints of the rainbow, and fringed with delicate hairs. Its eyes are black and prominent. Its face and feelers are yellow. Its antennæ are long and blackish. Those of the male are twice as long as the body, and consist of twenty-four joints, which, excepting the two basal ones, are globular, surrounded by hairs, and connected by slender portions, like beads on a string.* The antennæ of the females are about as long as the body, and consist of only twelve joints, which, except two at the base, are oblong oval, narrowed somewhat in the middle, and surrounded by two whorls of hairs. These insects vary much in size. The largest females do not exceed one tenth of an inch in length; and many are found, towards the end of the season, less than half this length. The males are usually rather smaller than the females, and somewhat paler in color. Among hundreds that I have examined in the living state, I have never found one specimen with spotted wings.

The time of their appearance in the winged form varies according to the season and the situation, from the beginning of June to the end of August. In Salisbury, Connecticut, they had entirely disappeared before the 25th of July, 1851; but during the same year I found them still in some numbers at North Conway, in New Hampshire, on the 17th of August; and, three days later, near the base of the White Mountains. In most parts of New England where wheat is cultivated, immense swarms of these orange-colored gnats infest fields of grain towards the last of June. While the sun shines they conceal themselves among the leaves and weeds near the ground. They take wing during the morning and evening twilight, and also in cloudy weather, when they lay their eggs in the opening flowers of the grain. New swarms continue to come forth in succession, till the end of July; but Mr. Buel says that the principal deposit of eggs is made in the first half of July, when late-sown winter-wheat and early-sown spring-wheat are in the blossom

* These joints seem to me to be somewhat approximated in pairs.

or milk; and this statement agrees with the observations of Mrs. Gage.

The flies are not confined to wheat alone, but deposit in barley, rye, and oats, when these plants are in flower at the time of their appearance. I have found the maggots within the seed-scales of grass, growing near to wheat-fields. The eggs hatch in about eight days after they are laid, when the little yellow maggots or grain-worms may be found within the chaffy scales of the grain. Being hatched at various times during a period of four or five weeks, they do not all arrive at maturity together. Mrs. Gage informs me that they appear to come to their growth in twelve or fourteen days. They do not exceed one eighth of an inch in length, and many, even when fully grown, are much smaller. From two to fifteen or twenty have been found within the husk of a single grain, and sometimes in every husk in the ear. In warm and sheltered situations, and in parts of fields protected from the wind by fences, buildings, trees, or bushes, the insects are said to be much more numerous than in fields upon high ground or other exposed places, where the grain is kept in constant motion by the wind. Grain is commonly more infested by them during the second than the first year, when grown on the same ground two years in succession; and it suffers more in the immediate vicinity of old fields, than in places more remote. These insects prey on the wheat in the milky state, and their ravages cease when the grain becomes hard. They do not burrow within the kernels, but live on the pollen and on the soft matter of the grain, which they probably extract from the base of the germs.

It appears, from various statements, that very early and very late wheat escape with comparatively little injury; the amount of which, in other cases, depends upon the condition of the grain at the time when the maggots are hatched. When the maggots begin their depredations soon after the blossoming of the grain, they do the greatest injury; for the kernels never fill out at all. Pinched or partly filled kernels are the consequence of their attacks when the grain is more advanced. The hulls of the impoverished kernels will always be found split open on the convex side, so as to expose the embryo. This is caused by the drying and shrinking of the hull, after a portion of the contents thereof has been sucked out by the maggots.

Towards the end of July and in the beginning of August, the full-grown maggots leave off eating, and become sluggish and torpid, preparatory to moulting their skins. This process, which has been alluded to by Judge Buel and some other writers, has been carefully observed by Mrs. Gage, who sent to me the maggots before and after moulting, together with some of their cast skins. It takes place in the following manner. The body of the maggot gradually shrinks in length within its skin, and becomes more flattened and less pointed, as may easily be seen through the delicate transparent skin, which retains nearly its original form and dimensions, and extends a little beyond the included insect at each end. The torpid state lasts only a few days, after which the insect casts off its skin, leaving the latter entire, except a little rent in one end of it. Mrs. Gage observed many of the maggots in the very act of emerging from their skins. The cast skins are exceedingly thin, and colorless, and, through a microscope, are seen to be marked with eleven transverse lines. Great numbers of the skins are to be found in the wheat-ears immediately after the moulting process is completed. Sometimes the maggots descend from the plants, and moult on the surface of the ground, where they leave their cast skins, as described by Mr. J. W. Dawson, of Pictou, Nova Scotia.* Late broods are sometimes harvested with the grain, and carried into the barn without having moulted. This seems to have often happened in England, where the insect has been repeatedly noticed in the transition state, still enclosed within its loosened filmy skin.

* Proceedings of the Academy of Natural Sciences of Philad., Vol. IV. p. 210.

It is somewhat remarkable that the true nature of this covering of the maggot should not have been ascertained by English naturalists. Mr. Kirby, as before stated, supposed it to be a thin membrane, formed by the insect for the protection of its body from the cold of winter. According to Professor Henslow's account, the larvæ "spin themselves up in a very thin and transparent web, which is often attached to a sound grain, or to the inside of one of the chaff-scales."* Mr. Curtis observed on the backs of some of the shrivelled grains "a long narrow filmy sac, on opening which a bright orange granulated maggot came out alive; and when shut up in a tin box, many voluntarily left their cases and wandered about."† Having carefully watched the insect during the moulting period, I am convinced that what these gentlemen called a "membrane," "web," or "sac," is really the loosened outer skin of the maggot, which is subsequently thrown off in the ears of the wheat, or is cast upon the surface of the ground.

After shedding its skin, the maggot recovers its activity, and writhes about as at first, but takes no food. It is shorter, somewhat flattened, and more obtuse than before, and is of a deeper yellow color, with an oblong greenish spot in the middle of the body. Within two or three days after moulting, the maggots either descend of their own accord, or are shaken out of the ears by the wind, and fall to the ground. They do not let themselves down by threads, for they are not able to spin. Nearly all of them disappear before the middle of August; and they are very rarely found in the grain at the time of harvest. Mrs. Gage stated, in one of her letters, that she had not observed "how and when the insects issue from the grain," but that it was "apparent they go in company," and "perhaps they crawl out upon the heads during a rain, and are washed down to the ground, where they remain through

* Journal of the Royal Agricultural Society of England, Vol. II. p. 22.

† Ibid., Vol. VI. p. 145.

the winter." On the 14th of August, 1841, she visited again the field of wheat where, on the 25th of July, she had found great numbers of the maggots, and observed that "a very few of all that multitude were left. On rubbing the ears, their silvery coverings glistened in the sunshine, and floated away on the breeze. A warm rain had fallen between these visits."

In an account of the damage done by these insects in Vermont, in the summer of 1833, it is stated that, "after a shower of rain, they have been seen in such countless numbers on the beards of the wheat, as to give the whole field the color of the insect."* Mr. Elijah Wood, of Winthrop, Maine, in a short communication, written in the summer of 1837, made the following remarks: "This day, 9th of August, a warm rain is falling, and a neighbor of mine has brought me a head of wheat which has become loaded with the worms. They are crawling out from the husk or chaff of the grain, and were on the beards, and he says he saw great numbers of them on the ground."† From these observations, and from remarks to the same effect made to me by intelligent farmers, it appears that the descent of the insects is facilitated by falling rain and heavy dews.

Having reached the ground, the maggots soon burrow under the surface, sometimes to the depth of about an inch, those of them that have not already moulted casting their skins before entering the earth. Here they remain, without further change, through the following winter. During the month of May, I have seen specimens still in the larva form, in the earth wherein they had been kept during the winter. It is not usually till June that they are transformed to pupæ. This change is effected without another moulting of the skin; not the slightest vestige of the larva-skin being found in the earth in which some of these insects had undergone their

* New England Farmer, Vol. XII. p. 60.

† Ibid., Vol. XVI. p. 61.

transformations. Moreover, the pupa is entirely naked, not being enclosed either in a cocoon or in the puparium formed of this outer skin of the larva, and it has its limbs and wings free or unconfined. The pupa state lasts but a short time, a week or two at most, and probably, in many cases, only a few days. Under the most favorable circumstances, the pupa works its way to the surface, before liberating the included fly; and when the insect has taken wing, its empty pupa-skin will be seen sticking out of the ground. In other cases, the fly issues from its pupa-skin in the earth, and comes to the surface with flabby wings, which soon expand and dry on exposure to the air. This last change occurs mostly during the months of June and July, when great numbers of the flies have been seen, apparently coming from the ground, in fields where grain was raised the year before. Some persons have stated that the insects are transformed to flies in the ears of the grain, having probably mistaken the cast-skins of the maggots found therein for the shells of the chrysalis or pupa.

Several cases of the efficacy of fumigation in preventing the depredations of these insects are recorded in our agricultural papers.* For this purpose brimstone has been used, in the proportion of one pound to every bushel of seed sown. Strips of woollen cloth, dipped in melted brimstone, and fastened to sticks in different parts of the field, and particularly on the windward side, are set on fire, for several evenings in succession, at the time when the grain is in blossom; the smoke and fumes thus penetrate the standing grain, and prove very offensive or destructive to the flies, which are laying their eggs. A thick smoke from heaps of burning weeds, sprinkled with brimstone, around the sides of the field, has also been recommended. Lime or ashes, strown over the grain when in blossom, has, in some cases, appeared to protect the crop; and the Rev. Henry Colman, the Commissioner for the Agricultural Survey of

* Among others, see The Cultivator, Vol. V. p. 136.

Massachusetts, says that this preventive, if not infallible, may be relied on with strong confidence.* For every acre of grain, from one peck to a bushel of newly slacked lime or of good wood-ashes will be required; and this should be scattered over the plants when they are wet with dew or rain. Two or three applications of it have sometimes been found necessary.

Whether it be possible to destroy the maggots after they have left the grain, and have betaken themselves to their winter quarters, just below the surface of the ground, remains to be proved. Some persons have advised ploughing up the ground, soon after the grain is harvested, in order to kill the maggots, or to bury them so deeply that they could not make their escape when transformed to flies. I am inclined to think that deep ploughing will prove to be the best and most practicable remedy. Perhaps thoroughly liming the soil before it is ploughed may contribute to the destruction of the insects. The chaff, dust, and refuse straw should be carefully examined, and, if found to contain any of the maggots, should be immediately burnt. It is stated that our crops may be saved from injury by sowing early in the autumn or late in the spring. By the first, it is supposed that the grain will become hard before many of the flies make their appearance; and by the latter, the plants will not come into blossom until the flies have disappeared. In those parts of New England where these insects have done the greatest injury, the cultivation of fall-sown or winter grain has been given up; and this, for some years to come, will be found the safest course. The proper time for sowing in the spring will vary with the latitude and elevation of the place, and the forwardness of the season. From numerous observations made in this part of the country, it appears that grain sown after the 15th or 20th of May generally escapes the ravages of these destructive insects. Late sowing has almost entirely banished the wheat-flies

* Third Report on the Agriculture of Massachusetts, p. 67.

from those parts of Vermont where they first appeared; and there is good reason to expect that these depredators will be completely starved out and exterminated, when the means above recommended have been generally adopted and persevered in for several years in succession.

In the introductory chapter * a short account has already been given of the habits of the various kinds of gnats and flies, belonging to the principal families of this order. Besides the species that are injurious to vegetation, which have been now described, there still remain some of our native flies that deserve a passing notice, on account of their size, or of peculiarities in their forms, structure, and habits, although few of them are to be included among the insects which are hurtful to plants.

Among our long-legged gnats there is no one more singular in its appearance and graceful in its motions than the *Ptychoptera clavipes* of Fabricius, or club-footed Ptychoptera. A new genus, called *Bittacomorpha*, on account of the fancied resemblance of this insect to the Neuropterous genus *Bittacus*, has lately been made for its reception, by Mr. Westwood.† This pretty gnat is of a black color, with a broad, white stripe on the face, a short, white line on the fore part of the thorax, and three broad, white rings on the legs. The sides of the thorax are silvery white, and the hind body is dusky brown, with a narrow white line on the edges of each of the rings. The head is small, and almost hidden under the thick and hunched thorax; the antennæ are many-jointed, slender, and tapering; the hind body is long, narrow, and somewhat flattened; the legs are very slender next to the body, and increase in thickness towards the end, and the first joint of the feet is swollen, oblong oval, and very downy. The length of the body is about half an inch, and the wings expand nearly three quarters of an inch. It appears in July, and takes wing by day. As it flies slowly

* Pages 16 and 17.

† Philosophical Magazine, Vol. VI. p. 281. Lond. 1835.

along, it seems almost to tread the air, balancing itself horizontally with its long legs, which are stretched out, like rays, from the sides of its body.

There are exceptions to almost all general rules. Thus we find, among Dipterous insects, some kinds that never have wings. One of these is the thick-legged snow-gnat, or *Chionea valga*[1] (Fig. 260). This singular insect looks more like a spider than a gnat. Its body is rather less than one fourth of an inch long, and is of a brownish yellow or nankin color. The legs are rather paler, and are covered with short hairs. The head is small and hairy. The first two joints of the antennæ are thick, the others slender and tapering, and beset with hairs. Although the wings are wanting, there is a pale yellow poiser on each side of the hinder part of the thorax. The hindmost thighs are very thick, and somewhat bowed, in the males, which suggested the name of *valga*, or bow-legged, given to the insect in my Catalogue. The body of the female ends with a sword-shaped borer, resembling that of a grasshopper. These wingless gnats live on the ground, and the females bore into it to lay their eggs. They are not common here. Mr. Gosse found considerable numbers of them in Canada, crawling on the snow, in pine woods, during the month of March.*

Fig. 260.

Travellers and new settlers, in some parts of New England and Canada, are very much molested by a small gnat, called the black fly (*Simulium molestum*), swarms of which fill the air during the month of June. Every bite that they make draws blood, and is followed by an inflammation and

[1 Mr. Walker has described two American species of this singular genus; one of them, *Chionea aspera*, seems to be identical with Dr. Harris's *C. valga*. I do not undertake to decide which name should be preferred. Dr. Harris's has the priority, but the few words he mentions about this insect can hardly be called a description. (Compare Walker's List of Diptera of British Museum, Vol. I. p. 82.) — OSTEN SACKEN.]

* Canadian Naturalist, p. 51.

swelling which last several days. These little tormentors are of a black color; their wings are transparent; and their legs are short, and have a broad whitish ring around them. The length of their body rarely exceeds one tenth of an inch. They begin to appear in May, and continue about six weeks, after which they are no more seen. They are followed, however, by swarms of midges, or sand-flies (*Simulium nocivum*), called no-see-'em, by the Indians of Maine, on account of their minuteness. So small are they, that they would hardly be perceived, were it not for their wings, which are of a whitish color, mottled with black. Towards evening these winged atoms come forth, and creep under the clothes of the inhabitants, and by their bites produce an intolerable irritation, and a momentary smarting, compared * to that caused by sparks of fire. They do not draw blood, and no swelling follows their attacks. They are most troublesome during the months of July and August.

The most common of our large gad-flies, or horse-flies, appears to be the *Tabanus atratus* of Fabricius. It is of a black color, and the back is covered with a whitish bloom, like a plum. The eyes are very large, and almost meet on the top of the head; they are of a shining purple-black or bronzed black color, with a narrow jet-black band across the middle, and a broad band of the same hue on the lower part. The body of this fly is seven eighths of an inch or more in length, and the wings expand nearly two inches. The *Tabanus cinctus* (Fig. 261) of Fabricius, or orange-belted horse-fly, is not so common and is rather smaller. It is also black, except the first three rings of the hind body, which are orange-colored. The most common of our smaller horse-flies is the *Tabanus lineola* (Fig. 262), so named by Fabricius, because it has

Fig. 261.

* See Gosse's Canadian Naturalist, p. 100.

a whitish line along the top of the hind body. Besides these flies, we have several more kinds of *Tabanus*, some of which do not appear to have been described. These blood-thirsty insects begin to appear towards the end of June, and continue through the summer, sorely tormenting both horses and cattle with their sharp bites. Their proboscis, though not usually very long, is armed with six stiff and exceedingly sharp needles, wherewith they easily pierce through the toughest hide. It is stated that they will not touch a horse whose back has been well washed with a strong decoction of walnut-leaves. The eyes of these flies are very beautiful, and vary in their colors and markings in the different species.

Fig. 262.

The golden-eyed forest-flies are also distinguished for the brilliancy of their spotted eyes, and for their clouded or banded wings. They are much smaller than the horse-flies, but resemble them in their habits. Some of them are entirely black (*Chrysops ferrugatus*, Fabricius[2]), others are striped with black and yellow (*Chrysops vittatus*, Wiedemann). They frequent woods and thickets, in July and August.

The bee-flies, or Bombylians (BOMBYLIADÆ), have a very slender proboscis, sometimes exceeding the length of their body. They are met with in sunny paths in the woods, in April and May. They fly with great swiftness, stop suddenly every little while, and, balancing themselves with their long, horizontal spread wings, seem to hang suspended in the air. They often hover, in this way, over the early flowers, sucking out the honey thereof, like humming-birds, with their long bills. Our largest bee-fly is the *Bombylius*

[[2] *Chrysops ferrugatus*, Fab., is a *Tabanus*, and not a *Chrysops*. Besides, it has much more *ferruginous* and *cinereous* than *black* in its coloring (compare its description in Wied. Auss. Zw., Vol. I. p. 186). Dr. Harris means probably the *Chrysops niger*, Meig, which, next to *C. vittatus*, Wied., is the common *Chrysops* of this country. — OSTEN SACKEN.]

æqualis[3] (Fig. 263), so named by Fabricius, because the wings are divided lengthwise, in their color, into two equal parts, the outer part being brownish black, and the inner half colorless and transparent. The body of this insect is short, rounded, and covered with yellowish hairs, like a humble-bee. It measures three eighths of an inch in length, and the wings expand rather more than seven eighths of an inch.

Fig. 263.

There are some flies that prey on other insects, catching them on the wing or on plants, and sucking out their juices. Some of them have thick and hairy bodies and legs, and bear a striking resemblance to our biggest humble-bees. Such are the *Laphria thoracica* (Fig. 264) of Fabricius, which is black, with yellow hairs on the top of the thorax, and measures eight or nine tenths of an inch in length; another species, which may be called *Laphria flavibarbis*,[4] differing from the former in having the face and sides of the head covered with a yellow beard, and in being an inch or more long; and the *Laphria tergissa* of Say, which is somewhat like the last, but has yellow hairs on the three middle segments of the hind body, and on the shanks of the anterior and middle pairs of legs, and measures about an inch in length.

Fig. 264.

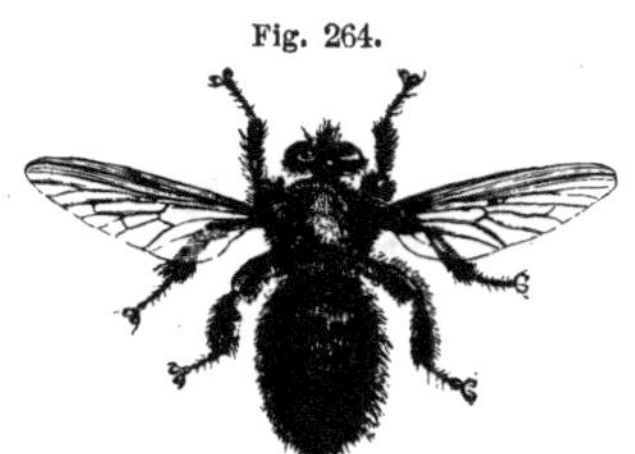

[[3] What the true *Bombylius æqualis*, Fab. is, remains as yet uncertain, and can be determined only by comparisons with his typical specimen. The insect I remember having seen in Dr. Harris's collection under this name belonged to the group which includes the European *B. major*, which group contains several species in this country. It might be either *B. albipectus*, Macq., or *B. fratellus*, Wied. (syn. *B. vicinus*, Macq.). — OSTEN SACKEN.]

[[4] *Laphria flavibarbis* is probably nothing but a variety of *L. tergissa*. I possess specimens of the latter, having *one*, *two*, or *three* segments of the abdomen covered *with yellow hair*. Similar varieties, as to the width of the abdominal yellow band, occur also in *L. thoracica*, Fab. — OSTEN SACKEN.]

These insects belong to a family called ASILIDÆ, from *Asilus*, the principal genus. In the larva state, those of the Asilians whose habits are known live in the ground upon the roots of plants, and sometimes do considerable mischief, as proved to be the case with some that were sent to me last May, by the Rev. Thomas Hill, of Waltham, who found them devouring the roots of the tart rhubarb. They were yellowish-white maggots, about three quarters of an inch long, not perfectly cylindrical, but a little depressed, and tapering at each end. The head was small, brown, and partially drawn within the first ring, and was provided with two little horny brown hooks. There was a pair of breathing-pores on the first ring, and another pair on the last but one. These maggots were transformed in the earth to naked pupæ, having the limbs free. The pupa was brown, and had a pair of short horns on the forehead, three spines on each side of the head, a forked tail, and a transverse row of little teeth across the middle of each ring of the hind body. When about to undergo their last transformation, the pupæ work their way to the surface of the ground by the help of the little teeth on their rings. I have repeatedly seen the empty pupa-shells sticking half-way out of the ground around rhubarb plants. In the fore part of July, there issued from these pupæ some long-bodied flies, which proved to be of the species called *Asilus sericeus* (Fig. 265) by Mr. Say. The body of this insect is slender and tapering, and measures from eight tenths of an inch to one inch and one tenth in length. It is of a brownish-yellow color, covered with a short silky down, varying in different lights from golden yellow to brown, and with a broad brown stripe on the top of the thorax. The wings are smoky brown, with broad

Fig. 265.

brownish-yellow veins, and expand one inch and a quarter, or more. We have several other kinds of *Asilus*, some larger and others smaller than the foregoing, of whose history nothing is known, except their predaceous habits in the winged state, which have been often observed. There are also several slender kinds of *Laphria;* but these are easily distinguished from every species of *Asilus* by their antennæ, which are not, as in the latter, tipped with a slender point, but are blunt at the end.

Besides the foregoing, there are many other rapacious flies, some of which are of great size. The largest one found here is the orange-banded Midas (*Midas filatus**[5]), (Fig. 266,) specimens of which are sometimes found measuring an inch and a quarter in length, with wings expanding two inches and a quarter. It is black, with an orange-colored band on the second ring of the hind body; and the wings are smoky brown, with a metallic lustre. It receives its scientific name, *filatus*, signifying thread-like, from its antennæ, which are long and slender, but they end with an oblong oval knob. Its generical name was also given to it on account of its long antennæ; *Midas*, in mythology, being the name of a person fabled to have had the long ears of an ass. The orange-banded Midas may often be seen flying in the woods in July and August, or resting and basking in the sun upon fallen trees. Its transformations

Fig. 266.

* Incorrectly named *Mydas filata* by Fabricius.

[[5] *Midas filatus* is now generally called *M. clavatus*, Drury, which is the older name. — OSTEN SACKEN.]

have never been described. Its larva and pupa almost exactly resemble those of the rapacious Asilians. The larva is a cylindrical, whitish maggot, tapering before, and almost rounded behind; it has only two breathing-holes, which are placed in the last ring but one; and it grows to the length of two inches. It lives and undergoes its transformations in decayed logs and stumps. The pupa measures about an inch and a quarter in length; it is of a brown color, and nearly cylindrical shape; its tail is forked; there are eight thorns on the fore part of its body; and each ring of the abdomen is edged with numerous sharp teeth, like a saw, all these teeth pointing backwards except those on the back of the first ring, which are directed forwards. The pupa pushes itself half-way out of the stump when the fly is about to come forth, and the latter makes its escape by splitting open the back of the pupa-skin.

In the month of June, there may sometimes be seen, resting on the grass or on rotten stumps, in open woods, a large light-brown or drab-colored fly, somewhat like a horse-fly in form, but easily distinguished therefrom by two little thorns on the hinder part of the thorax; and by the wings, which do not spread so much when the insect is at rest. It is heavy and sluggish in its motions, and does not attempt to fly away when approached. This insect was called *Cœnomyia pallida*, the pale Cœnomyia, by Mr. Say, in the Appendix to Keating's "Narrative," and in the second volume of the "American Entomology," where it is figured. The generical name, signifying a common fly, is rather unfortunate, for this is a rare insect. The only specimens known to Mr. Say were found by him in a small forest of scattered trees, on the Pecktannos River, in Wisconsin. A few have been taken in Massachusetts, one of them on Blue Hill, in Milton; and Mr. Gosse found three specimens, in as many years, in Canada. In its transformations this insect is more nearly related to the gad-flies and the Asilians than to the soldier-flies, near which it has generally been placed;

though it approaches the latter in its structure, and in its sluggish habits. The larvæ or maggots, though not yet discovered, undoubtedly live in the ground, or in decayed vegetable substances, like those of the horse-flies and other predatory insects; for Mr. Gosse found one of his specimens, on the grass, in the act of emerging from the pupa-skin. He has also figured * the pupa, which is of a chestnut-brown color, and has transverse rows of spines on the abdominal rings.

Most of the soldier-flies (STRATIOMYADÆ) are armed with two thorns or sharp spines on the hinder part of the thorax. They form the first family of the flies that undergo their transformations within the hardened skin of the larva, which is not thrown off till they break through it to come out in the winged state. Their proboscis contains, at most, only four bristles, is not fitted for piercing, but ends with large fleshy lips, by means whereof these flies suck the sweet juices of flowers. Most of them are found in wet places, where their larvæ live; some of the latter being provided with a tube, in the hinder extremity, which they thrust out of the water in order to breathe. The skin of these larvæ is merely shortened a little, without wholly losing its former shape, when the enclosed insects change to pupæ; thereby showing that this family is truly intermediate between the preceding flies, which cast off their larva-skins, and those which retain them, and take an oblong oval shape, when they become pupæ. Some of the soldier-flies (*Stratyomys*) have a broad oval body, ornamented with yellow triangles or crescents on each side of the back, and their antennæ are somewhat like those of Midas and of the gad-flies; others (*Sargus*) are slender, often of a brilliant brassy-green color, with a bristle on the tip of their antennæ. The maggots of the latter live in rich mould.

The Syrphians (SYRPHIDÆ) have a fleshy, large-lipped proboscis, elbowed near the base, and enclosing only four

* Canadian Naturalist, p. 199.

slender bristles. They live on the honey of flowers. The last joint of their short antennæ bears a bristle, which is sometimes feathered. Their heads are large and hemispherical. Many of these flies are often mistaken for bees or wasps, and some of them lay their eggs in the nests of the insects they so closely resemble. Others drop their eggs among plant-lice, which their young afterwards destroy in great numbers. The larvæ of a few are aquatic, and are provided with very long, tubular tails, through which they breathe, and have been called rat-tailed maggots. Some of the largest and most beautiful of these flies live, in the maggot state, in rotten wood. One of these rat-tailed flies is often seen on windows, in the autumn. It flies with a buzzing noise. Its eyes are very large, and of a bright copper-color; its body is brassy green; and there are five gray stripes on the thorax. It measures about four tenths of an inch in length. It is the *Eristalis sincerus*[6] of my Catalogue. The *Milesia excentrica*[7] (Fig. 267), named in the same work, strikingly resembles a hornet; its hind body being banded with black and yellow in the same way. Its head and thorax are black, the former margined around

Fig. 267.

[[6] *Eristalis sincerus*, Harris, is identical with *E. æneus*, Linn., so common in Europe. — OSTEN SACKEN.]

[[7] The description of this species is too short to have a scientific value. In order to prevent its being superseded by a subsequent description, under some other name, I subjoin a full description.

Milesia excentrica, Harris. Thorace nigro, flavomaculato, *scutello æneo;* abdomine flavo, nigro fasciato; pedibus fulvis; tibiis, tarsisque anticis nigris; Length, $\frac{6}{10}$ to $\frac{8}{10}$ inch.

Male : Hypostoma golden-yellow, sericeous, with a shining black stripe in the middle. *Antennæ* pale ferruginous; the space between the eyes has the form of a triangle, which is golden-yellow at the tip, and black on the *vertex.* The hind part of the head is sericeous, yellow along the borders. *Thorax* black; a yellow spot on the shoulder, another behind it, from which runs a narrow groove, yellow at the bottom; a third yellow, elongated spot, pointed upwards, near the corner of the scutellum. A gray stripe runs along the middle of the thorax; it becomes very faint beyond the centre and ends in a triangular yellowish spot near the *scu-*

the eyes, and the latter spotted, with yellowish white. The legs are ochre-yellow, except the shanks and feet of the first pair, which are black. Its body measures nearly three quarters of an inch in length. My *Sphecomyia undata* (Fig. 268) has the slender form of a *Sphex* or mud-wasp. It is of a light-brown color, darker on the back, and on the middle of the thighs and shanks; its head is conical, and bears the antennæ on the tip of the cone; its wings are brown on the outer part, with a small transparent spot near the edge, and the inner part is transparent in two large wavy spaces. It is about five eighths of an inch long, and its wings expand one inch and a quarter, or more. It is possible that this singular fly may be the *Pyrgota undata* of Wiedemann.[8] An insect closely

Fig. 268.

tellum. *Pleuræ* with an elongated yellow spot. *Scutellum* brassy green, metallescent. *Poisers* pale ferruginous. *Legs* yellowish ferruginous, fore *tibiæ* (excepting the knees) and *tarsi* black. *Hind thighs* unarmed. *Wings* tinged with gray and brown; a pale stripe along the latter part of the cubital vein; another runs along the pobrachial area and reaches the posterior margin. *Abdomen* yellow; first segment black; the three following segments have more or less black at the incisures along the fore border, and a black band, attenuated at both ends, and not reaching the lateral borders, in the middle; the band on the second segment is the broadest, and is sometimes connected with the black of the first segment; those of the third and fourth segments are interrupted in the middle. *Venter* yellow, with large black spots on the middle of the segments; *sexual organs* pale ferruginous.

Female: shows the following differences: *front* yellow, with a black stripe extending towards the black *vertex*. Beside the three pairs of spots already mentioned on the thorax, there is a fourth pair in the middle of the disk, each of the spots communicating by the groove with the second lateral spot. The black bands on the segments of the abdomen are broader, and connect at the fore and hind borders with the black incisures; the fifth segment has a similar band. The *wings* are more tinged with ferruginous than with gray; the pale stripes less apparent. Base of *thighs* more or less blackish.

I have before me two males from Illinois, collected by Mr. Kennicott, and one female from Maine, collected by Mr. Packard.

This species is very much like the European *M. vespiformis*, Linn. — OSTEN SACKEN.]

[[8] *Sphecomyia undata* is the *Pyrgota undata*, Wied. *S. valida* is likewise a *Pyrgota*, but a new species. *Myopa nigripennis*, Gray, might belong to the same group, as the figure in Griffith's Animal Kingdom seems to prove. — OSTEN SACKEN.]

resembling it is figured in Griffith's translation of Cuvier's "Animal Kingdom," under the name of *Myopa nigripennis*. It is found on fences around gardens in May and June. It sits with its wings half spread, moves slowly, and flies heavily. My *Sphecomyia valida*, though rather shorter than the preceding, has a thicker body. Its color is brownish yellow, and it is striped with brown. The wings are transparent, and are mottled with small, dusky spots.

Some of the Conopians (*Conopidæ*) still more closely resemble slender-bodied wasps than the preceding Sphex-flies. *Conops sagittaria* (Fig. 269) of Say (*nigricornis*, Wiedemann) might almost be mistaken for a species of *Eumenes*. Its hind body is very slender and cylindrical next to the thorax, and swells out behind. Its antennæ are long, and thickened towards the end. Its proboscis is very long and slender, elbowed at the base, and extends far beyond the head. This fly is of a black color; the rings of the hind body are edged with white; the face is yellow; the legs are brownish yellow, shaded with black on the thighs; and the wings are black, with two uncolored and wavy spaces on the inner margin. Its body is five eighths of an inch long, and its wings expand rather more than three quarters of an inch. This fly may be found sucking the honey of flowers in June and July. The Greeks gave the name of *Conops* to some stinging fly or gnat. The Conopians undergo their transformations in the bodies of humble-bees, their young subsisting on the fat contained within the abdomen of their luckless victims.

Fig. 269.

A host of flies, forming nearly one third of the whole number of species in the order Diptera, will be found to have a short and soft proboscis, ending with large fleshy lips, enclosing only two bristles, and capable of being drawn up within the cavity of the mouth. Their antennæ are generally short, hang down over the face, and end with a large oval joint, bearing a little bristle. Their larvæ, or young, are fleshy, whitish maggots, which never cast their skins, but, when the pupa state comes on, shorten, take the oblong-oval form of an egg, and become brown, dry, and hard on the outside. This immense tribe includes the various kinds of flesh-flies, blow-flies, house-flies, dung-flies, flower-flies, fruit-flies, two-winged gall-flies, cheese-flies, and many others, for which we have no common names, but all composing the tribe of Muscans, or MUSCADÆ. Some of these flies do not strictly conform to the foregoing characters of the tribe, in all respects; but the exceptions are few in number, and the most remarkable of them will be noticed in the following pages.

Many flies of this tribe are parasitic in their larva state, their young living and undergoing their transformations within the bodies of other insects, particularly in caterpillars, which they thereby destroy. These flies belong chiefly to the family of TACHINADÆ, a name applied to them on account of the swiftness of their flight. In form they somewhat resemble house-flies; like them, they have very large winglets, and their wings spread apart when they are at rest. They are easily distinguished, however, by the stiff hairs wherewith they are more or less covered, and by the bristles on their antennæ, which are not usually feathered. A large fly of this kind, the *Tachina vivida* (Plate VIII. Fig. 1) of my Catalogue, is often seen on fences, and on plants, and sometimes in houses, towards the end of June and during the month of July. Its large, oval hind body is of a clear and light red color, with two or three black spots, in a row, on the top of it, and a thick row of black bristles across each ring. The face is grayish white, like satin, and the eyes are copper-colored. The thorax is gray, with brownish lines upon it. The antennæ, proboscis, and legs are light red. Its body is short and thick, and is about half an inch long, and its wings expand rather more than nine tenths of an inch.

Most insects are hatched from eggs which are laid by the mother on the substances that are to serve for the food of her young. Some flesh-flies produce their young alive, or already hatched, and drop them on the dead and putrefying animal matter, which they are to consume and remove in the shortest possible time. An exception from the usual course among insects appears therefore to have been made in favor of these viviparous flesh-flies, to enable their young promptly to perform their appointed tasks. These insects produce an immense number of young, as many as twenty thousand having been observed by Réaumur in a single fly.* Our largest viviparous flesh-fly is the *Sarcophaga Georgina* of Wiedemann. It appears towards the end of June, and continues till the middle of August, or perhaps later. Its face is silvery white, and there is an oblong square black spot between the eyes, which are copper-colored. The thorax is light gray, with seven black stripes upon it. The hind body is nearly conical, has the lustre of satin, and is checkered with square spots of black and white, shifting or interchanging their colors according to the light wherein they are seen. The legs are black, and the hindmost pair are very hairy in the males. The female is about half an inch long; the male is rather smaller. In the Sarcophagans, or flesh-eaters, as the name implies, the bristles on the antennæ are feathered.

The flies that abound in stables in August and September, and sometimes enter houses on the approach of rain, might be mistaken for house-flies, were it not for the severity of their bites, which are often felt through our clothing, and are generally followed by blood. Upon examination they will be found to differ essentially from house-flies in their proboscis, which is very long and slender, and projects horizontally beyond the head. The bristles on their antennæ are feathered above. Cattle suffer sorely from the piercing bites of these flies, and horses are sometimes so much tor-

* Mémoires, Vol. IV. p. 417.

mented and enraged by them as to become entirely ungovernable in harness. The name of this kind of fly is *Stomoxys calcitrans* (Fig. 270); the first word signifying sharp-mouthed, and the second kicking, given to the fly from the effect it produces on horses. It lays its eggs in dung, where its young are hatched, and pass through their transformations. The larvæ and pupæ do not differ much in appearance from those of common house-flies.

Fig. 270.

The next three flies have feathered bristles on their antennæ. The first of them, a large, buzzing, and stinking meat-fly, named *Musca* (*Calliphora*) *vomitoria* (Fig. 271), is of a blue-black color, with a broad, dark blue, and hairy hind body. It is found all summer about slaughter-houses, butchers' stalls, and pantries, which it frequents for the purpose of laying its eggs on meat. The eggs are commonly called fly-blows; they hatch in two or three hours after they are laid, and the maggots produced from them come to their growth in three or four days, after which they creep away into some dark crevice, or burrow in the ground, if they can get at it, turn to egg-shaped pupæ, and come out as flies, in a few days more; or they remain unchanged through the winter, if they have been hatched late in summer. A smaller fly, of a brilliant blue-green color, with black legs, also lays its eggs on meat, but more often on dead animals in the fields. It seems hardly to differ from the *Musca* (*Lucilia*) *Cæsar* of Europe.

Fig. 271.

The house-fly of this country has been supposed to be the same as the European *Musca domestica;* but I cannot satisfy myself on this point for the want of specimens from Europe. It is possible that our sharp-biting stable-flies, the meat-flies, and the house-fly, may really be distinct species from those which are found in Europe. Our house-fly is the *Musca Harpyia*, or Harpy-fly, of my Catalogue. It begins to appear in houses in July, becomes exceedingly abundant in September, and does not disappear till killed by cold weather. It is probable that, like the domestic fly of Europe, it lays its eggs in dung, in which its larvæ live, and pass through their changes of form. The Americans are accused of carelessness in regard to flies, and apparently with some reason. But if these filthy, dung-bred creatures swarm in some houses, covering every article of food by day, and absolutely blackening the walls by night, in others comparatively few are found; for the tidy housekeeper takes care not to leave food of any kind standing about, uncovered, to entice them in, and makes a business of driving out the intruders at least once a day. If a plateful of strong green tea, well sweetened, be placed in an outer apartment accessible to flies, they will taste of it, and be killed thereby, as surely as by the most approved fly-poison. In the first volume of "The Transactions of the Entomological Society of London," Mr. Spence gives an account of a mode of excluding flies from apartments, which has been tried with complete success in England. It consists of netting, made of fine worsted or thread, in large meshes, or of threads alone, half an inch or more apart, stretched across the windows. It appears that the flies will not attempt to pass through the meshes, or between the threads, into a room which is lighted only on one side; but if there are windows on another side of the room they will then fly through; such windows should therefore be darkened with shutters or thick curtains.

The Anthomyians, or flower-flies (ANTHOMYIADÆ), are easily distinguished from the preceding flies, which they otherwise resemble, by the smaller size of their winglets, and by the mesh in the middle of their wings, which is long, narrow, and open at the end. They are smaller insects than the foregoing, their flight is more feeble, their wings, when at rest, do not spread so much, and the bristle on the last joint of their antennæ is not often feathered. Most of them frequent flowers, and are sometimes seen sporting together, in large swarms, in the air, like certain kinds of gnats. In the larva state some of them live in manure, and in rotten vegetable substances; others are found in the roots of living plants, such as onions, radishes, turnips, and even in the pulpy parts of leaves and of stems, which they devour. The latter have nearly the same form as the maggots of common flies; some of the former are shorter, flattened, and fringed on the sides with feathery hairs.

Many instances are recorded of these fringed maggots having been discharged from the human body. They are supposed to be the young of a fly named *Anthomyia* (*Homalomyia*) *scalaris*.* Flies closely resembling this are sometimes seen in privies, and a friend has presented me with one of them, together with the dried larva-skin out of which it came. The larva was found in excrement. The fly is grayish black, and hairy, with large copper-colored eyes, which are surrounded by a narrow silvery white line. It measures one quarter of an inch in length. The larva-skin has two rows of hairs on the back, and two more on each side. Another fly, sometimes seen on windows in the autumn, is produced, if I mistake not, from a hairy maggot that lives in rotten turnips. This fly strikingly resembles the *Anthomyia canicularis* of Europe, and is possibly identical with it. It is of a dark gray color, with copper-colored eyes, encircled by a silvery white line, and with a large, semitransparent, yellowish spot on each side of the first three rings of the hind body. It measures rather less than one quarter of an inch in length. The fringed maggots of the *canicularis* are stated by some naturalists to have been obtained from the human body. It is not impossible that they may have been swallowed with turnips, or other vegetables, eaten when going to decay.

Radishes, while growing, are very apt to be attacked by

* For an account of the transformations of the fly of privies, with figures, see Swammerdam's "Book of Nature," translated by Hill, Part II. p. 38, plate 38.

maggots, and rendered unfit to be eaten. These maggots are finally transformed to small, ash-colored flies, with a silvery-gray face, copper-colored eyes, and a brown spot on the forehead of the females; they have some faint brownish lines on the thorax, and a longitudinal black line on the hind body, crossed by narrower black lines on the edges of the rings. They vary in size, but usually measure rather more than one fifth of an inch in length. They finish their transformations, and appear above ground, towards the end of June. The radish-fly is called *Anthomyia Raphani*, in my Catalogue, from the botanical name of the radish, on the root of which its larvæ feed. It closely resembles the root-fly (*Anthomyia radicum*) of Europe.

Onions, soon after they come up in the spring, and until they are grown to a considerable size, are often observed to turn yellow and die. Many years ago I remember to have seen them extensively affected in this way, so that there was a failure of three fourths of the plants in a large bed. The cause of their death was not suspected at the time, and no examination was made for the discovery of insects in them. Since then, I have been favored by Mr. Westwood with copies of two articles* by him, on the onion-fly (*Anthomyia Ceparum*), (Fig. 272, pupa and imago,) which, in the maggot state, lives in the roots of onion-plants in Europe, and causes them to wither and perish exactly in the same way as young onions do here. Hence there is good reason to believe that the failure of our onion crop is caused by the ravages of maggots similar to those of the European onion-fly. The latter lays its eggs on the leaves of the onion, close to the earth, so that the maggots, when hatched, readily make their way to the heart of the onion. The maggots come to their growth in about two weeks, turn to

Fig. 272.

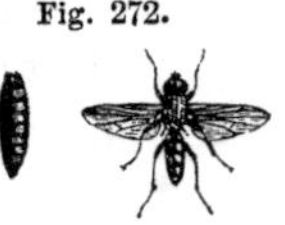

* See the Magazine of Natural History, Vol. VII. p. 425, and the Gardener's Magazine, Vol. XIII. p. 241. The same insect is also described and figured in Kollar's Treatise, p. 157.

pupæ within the onions, and come out as flies a fortnight afterwards.

We have a kind of fly corresponding almost exactly with the description of the onion-fly. This strengthens my belief that our onions suffer from the depredations of the maggots of this or of a similar insect. The fly to which I allude is often found on windows in the spring. It is ash-colored, with black hairs sparingly scattered on its body. It has a rust-colored forked spot on the top of its head, and three rust-red lines on the thorax; and the wings are tinged with yellow near the shoulders. It measures one fourth of an inch in length. It is stated that there are two or three generations of the European onion-flies during the summer, and that the late broods pass the winter in the pupa state, and are ready to burst forth at the first warmth of the following spring. It is stated that the onion crop may be preserved from the attacks of this fly, by sowing the seed on ground upon which a quantity of straw has been previously burnt.

The peculiar disease that has affected potatoes within the last ten years has been attributed, by many persons, to the depredations of insects. In the course of this work, several of these insects have been described. Another is now to be added to them, as will be seen by the following extract from a letter, received from a correspondent in July, 1851. "A new potato-rot theory has recently appeared in Brattleborough, Vermont. The mischief is referred to a fly, of which an authentic specimen is enclosed. It is said that the species first appeared simultaneously with the potato-rot; and the flies are accused of hovering about the manure, and depositing their eggs, so that the larvæ infect the potatoes." The specimen proved to be a common dung-fly, which may be found in abundance upon manure when carted into the field in the spring. The male is easily distinguished from other flies by its yellow and very hairy hind body and legs, and by its long and narrow wings. It is about half as large as a honey-bee; and it measures, from the face to the tips of the closed wings, from two fifths to one half of an inch, or more. The females are smaller, olive-colored, and sparingly clothed with short whitish hairs, with legs and wings like those of the male. The maggots or young, with the parent insects, live wholly upon dung, and are innocent of any injury to plants. The accusation brought against this insect entitles it to notice in this work, and to the distinction of a name and character by which it may hereafter be known. It may, therefore, be called *Scatophaga furcata*,* the forked dung-eater.

The dung-flies, or Scatomyians (SCATOMYZADÆ), in some of their characters, resemble the flower-flies, having similar wings, and very small ringlets; but their eyes are wide apart, and are of the same size in both sexes. The fly in question keeps its body remarkably clean, notwithstanding its dirty habits, and is neither offensive to the eye nor to the smell. The general color of the male is a bright ochre-yellow. The antennæ are pale red, and there is a wide forked red spot on the top of the head. The thorax is obscurely striped with brown above, and is lead-colored below the scutel. The hind body is oblong oval, and covered with long ochre-yellow hairs. The wings are ochre-yellow at the base and on the outer margin; and the two little transverse veins upon each of them are very conspicuous from their dark color and dusky borders. The legs are reddish yellow, and covered with long ochre-yellow hairs, intermixed with which there are a few black bristles; and there is a faint blackish line on the top of the first pair of thighs. A few black bristles are scattered upon the head and the top of the thorax. The bristle of the antennæ, when viewed with a powerful magnifier, is found to be covered with very minute hairs.

* *Pyropa furcata*, Say. Journ. Acad. Nat. Sciences, Vol. III. p. 98. To an imperfect specimen of this insect, Mr. Say gave the name of *Scatophaga postilena*, which it bears in the Catalogue of the Insects of Massachusetts.

Some two-winged flies deposit their eggs in the stems, buds, and leaves of plants, thereby producing large tumors or galls, wherein their young reside. Others lay their eggs in fruits, on the pulp of which their maggots live. These gall and fruit flies belong to a family called Ortalidians (ORTALIDIDÆ), from a word signifying to flap or shake the wings; for they keep their wings in motion nearly all the time, jerking them up and down, and twisting them round so that the thick outer edges often come together. Some of them are in the habit of suddenly raising their wings perpendicularly above their backs, and running along a few steps with them spread like the tail of a peacock. These insects, together with several other groups of flies, differ from all the foregoing in many respects, although they agree with them in their transformations. The forehead is broad in both sexes; their winglets are very small or entirely wanting; their powers of flight are feeble; and they are rarely found sporting on flowers in the sunshine, but seem generally to prefer shady and damp places.

The wings of the Ortalidians are often beautifully variegated, striped, or spotted with shades of brown or black. The hind body in the female generally ends with a pointed tube, wherewith the eggs are deposited. The little white maggots often found in over-ripe whortleberries, raspberries, cherries, and other fruits, are the young of some of these insects. Swellings, or galls, as large as a walnut, are often seen on the stems of some of our native asters or starworts. They are caused by the punctures of a fly, which lays its eggs, singly, in the stem, when the latter is tender. The puncture is followed by a spongy swelling, wherein the maggot, hatched from the egg, lives, and passes through its transformations. The insect finally comes out in the fly state, through a small hole previously made in the gall by the maggot. This fly may be called the gall-fly of the starwort (*Tephritis Asteris*). Its body is about one fifth of an inch long; it is of a light yellowish-brown color, with paler legs; the wings are broad, rounded at the tip, and clouded with brown in large spots, forming three wide, irregular bands across them.

Many of the smallest flies, belonging to several other groups, are placed near the end of the order. One of them has a head like a hammer-headed shark, short and very wide, with large globular eyes on each side of it. This little insect has been found in considerable numbers, flying near the ground, on the edges of banks. It is the *Sphyracephala brevicornis* of Mr. Say, and is figured and described in the third volume of his "American Entomology." The well-known cheese-maggots are the young of a fly (*Piophila casei*), not more than three twentieths of an inch long, of a shining black color, with the middle and hinder legs mostly yellowish, and the wings transparent like glass.

Some minute flies, belonging to a family called OSCINIDÆ, are found to be very injurious to wheat, rye, and barley, in Europe. One of them (*Oscinis frit*), a shining black fly, with yellowish feet, and measuring about one tenth of an inch in length, lays its eggs in the blossoms of barley, the grains of which afterwards perish in consequence of the depredations of the maggots of this fly; and Linnæus states that a tenth part of the produce of the barley in Sweden is thereby annually destroyed. The larvæ or maggots of *Oscinis lineata*, *Chlorops pumilionis*, *Chlorops glabra*, and other flies allied to them, live within the lower part of the stems of wheat, rye, and barley, thereby impoverishing the plants, and causing them to become stinted in their growth. They are rather larger insects than the frit-fly, and have black and yellow stripes on the thorax.

It is highly probable that some of these species, or other Oscinians, with similar habits, may be found in the stems of wheat and other grains in this country, and perhaps also in the ears. Several kinds of small flies, evidently different from the Hessian and wheat flies, have often been observed here, in fields of grain, when the plants are in flower; but their history has not yet been investigated, and the insects have not been scientifically examined and described. From the somewhat vague accounts that have been given of them, it is evident that they are much too large for any of the parasitical insects which attack the larvæ and eggs of the Hessian and wheat flies; and they appear sometimes to have been mistaken for the latter. In an extract from a paper by Mr. Worth, on the Hessian fly, mention is made of a pale yellow worm (maggot), about three sixteenths of an inch long, having been found by him within the stalks of wheat near the root, where its presence was detected by a swelling of the part attacked. This was perhaps the larva of one of the Oscinians. A careful examination of all the insects that inhabit our fields of grain is very much wanted.

The various insects, improperly called bot-bees, are two-winged flies, and belong to the order Diptera, and the family ŒSTRIDÆ, so named from the principal genus in it. Bot-flies do not seem to have any mouth or proboscis; for although these parts do really exist in them, the opening of the mouth is extremely small, and the proboscis is very short, and is entirely concealed in it; so that these insects, while in the winged state, do not appear to be able to take any nourishment. They somewhat resemble the Syrphians in form and color, and in the large size of their heads; but the eyes are proportionally small, and there is a large space between them. The face is swollen or puffed out before. The antennæ are very short, and almost buried in two little holes, close together, on the forehead. The winglets are large and entirely cover the poisers. The hind body of the females ends with a conical tube, bent under the body, and used for depositing the eggs, which the insect lays whilst flying. The larvæ or young of bot-flies live in various parts of the bodies of animals. They are thick, fleshy, whitish maggots, without feet, tapering towards the head, which is generally armed with two hooks; and the rings of the body are surrounded with rows of smaller hooks or prickles. When they are fully grown, they drop to the ground and burrow in it a short distance. After this, the skin of the maggot becomes a hard and brownish shell, within which the insect turns to a pupa, and finally to a fly, and comes out by pushing off a little piece like a lid from the small end of the shell.

More than twenty different kinds of bot-flies are already known, and several of them are found in this country. Some of them have been brought here with our domesticated animals from abroad, and have here multiplied and increased. Three of them attack the horse. The large bot-fly of the horse (*Gasterophilus equi*), (Plate VIII. Fig. 2,) has spotted wings. She lays her eggs about his knees; the small red-tailed species (*G. hæmorrhoidalis*) on his lips; and the brown farrier bot-fly (*G. veterinus*) under his throat, according to Dr. Roland Green. By rubbing and biting the parts where the eggs are laid, the horse gets the maggots into his mouth, and swallows them with his food. The insects then fasten themselves, in clusters, to the inside of his stomach, and live there till they are fully grown. The following are stated to be the symptoms shown by the horse when he is much infested by these insects. He loses flesh, coughs, eats sparingly, and bites his sides; at length he has a discharge from his nose, and these symptoms are followed by a stiffness of his legs and neck, staggering, difficulty in breathing, convulsions, and death. No sure and safe remedy has yet been found sufficient to remove bots from the stomach of the horse. The only treatment to be recommended is copious bleeding, and a free use of mild oils, in the early stages of the attack. The preventive means are very simple, consisting only in scraping off the eggs or nits of the fly every day.*

* See Dr. Green's "Natural History of the Horse-Bee," in Adams's Medical and Agricultural Register, Vol. I. p. 53; and the same in the New England Farmer, Vol. IV. p. 345.

Bracy Clark, Esq., who has published some very interesting remarks* on the bots of horses and of other animals, maintains that bots are rather beneficial than injurious to the animals they infest. His principal work on this subject I have not yet seen. The maggots (Fig. 273) of the *Œstrus bovis*, or ox bot-fly, live in large open boils, sometimes called wornils or wurmals, that is, worm-holes, on the backs of cattle. The fly is rather smaller than the horse bot-fly, although it comes from a much larger maggot. The sheep bot-fly (*Cephalemyia ovis*) lays its eggs in the nostrils of sheep, and the maggots crawl from thence into the hollows in the bones of the forehead. Deer are also afflicted by bots peculiar to them. Our native hare, or rabbit, as it is commonly called, sometimes has very large bots, which live under the skin of his back. The fly (*Œstrus buccatus*) is as big as our largest humble-bee, but is not hairy. It is of a reddish-black color; the face and the sides of the hind body are covered with a bluish white bloom; there are many small black dots on the latter, and six or eight on the face. This fly measures seven eighths of an inch or more in length, and its wings expand about three quarters of an inch. It is rarely seen; and my only specimen was taken in the month of July, many years ago.

Fig. 273.

At the very end of this order is to be placed a remarkable group of insects, which seems to connect the flies with the true ticks and spiders. Some of these insects have wings; but others have neither wings nor poisers. Of the winged kinds there is one (*Hippobosca equina*) that nestles in the hair of the horse; others are bird-flies (*Ornithomyia*), and live in the plumage of almost all kinds of birds. The wing-

* "Observations on the Genus Œstrus," in the Transactions of the Linnæan Society, Vol. III. p. 289, with figures; "On the Insect called Oistros by the Ancients," in Vol. XV. of the same work; and "An Essay on the Bots of Horses and other Animals," 1 vol. 4to (Lond., 1815).

less kinds have sometimes been called spider-flies, from their shape; such are sheep-ticks (*Mellophagus ovis*), and bat-ticks (*Nycteribia*). These singular creatures are not produced from eggs, in the usual way among insects, but are brought forth in the pupa state, enclosed in the egg-shaped skin of the larva, which is nearly as large as the body of the parent insect. This egg-like body is soft and white at first, but soon becomes hard and brown. It is notched at one end, and out of this notched part the enclosed insect makes its way, when it arrives at maturity.

The flea (*Pulex*) may almost be considered as a wingless kind of fly. Its proboscis seems to be intermediate in its formation between that of flies and of bugs; its antennæ are concealed in holes in the sides of its head, like those of certain water-bugs (*Nepa* and *Belostoma*), and somewhat resemble them in shape; while the transformations of the flea are not very much unlike those of the flies, whose maggots cast off their skins on becoming pupæ.

Having now arrived at the end of my work, I have only to add a few remarks by way of conclusion. It has been my design to present to the reader a sketch of the scientific arrangement of the principal insects which are injurious to vegetation, not only in New England, but in most of the United States. The descriptions of the insects, being drawn up in familiar language, will enable him to recognize them, when seen abroad, in all their forms and disguises. The hints and practical details, scattered throughout the work, it is hoped, will serve as a guide to the selection and the application of the proper remedies for the depredations of the insects described. I regret that it has not been in my power to do full justice to this important subject, which is far from having been exhausted. My object, however, will have been fully attained, if this treatise, notwithstanding its many faults and imperfections, should be found to afford any facilities for the study of our native insects, and should lead to the discovery and general adoption of efficient means for checking their ravages.

APPENDIX.

IT having been thought desirable, in consequence of the increased ravages of the "army-worm" during the past year (1861), to give a description and illustrations of it, although not specifically referred to in the original manuscript of the author of this treatise, the following account has been compiled from various authentic sources.

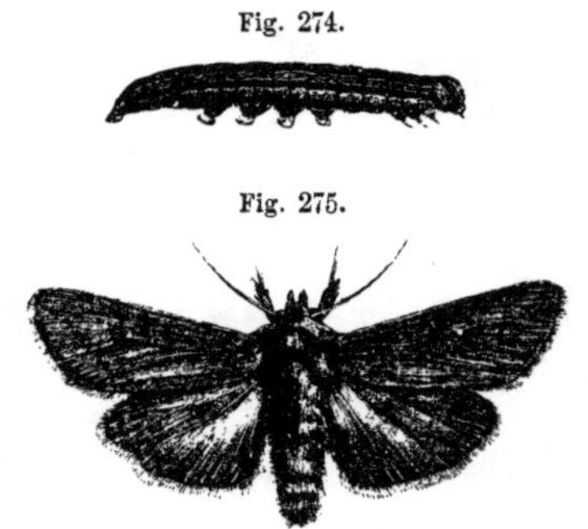

Fig. 274.

Fig. 275.

The army-worm (Fig. 274) is the larva of a night-flying moth, *Leucania unipuncta*, Haworth (Fig. 275). (Synonymes, *L. extranea*, Guénee; *L. impuncta*, Stephens.) The Imago "is very plain and unadorned in its appearance. The eye, on first glancing at it, only recognizes it as an ordinary-looking moth of a tarnished yellowish drab-color, inclining to russet, with a small white dot near the centre of its fore wings, and a dusky oblique stripe at their tips. On coming to look at it more particularly, we find it to be rather less than an inch long, to the end of its closed wings, or, if these are extended, it is about an inch and three quarters in width, different specimens varying somewhat in size. Its fore wings are sprinkled with blackish atoms, and a short distance forward of their hind edge they are crossed by a row of black dots, one on each of the veins. Outside of the middle of the wing this row of dots suddenly curves forward, and from this curve a dusky streak runs to the tip of the wing, the ground-color being more pale and clearer yellow along the outer side of this streak. . . . . .

Nearly in the centre of the wing is a milk-white dot placed upon the mid-vein. This dot is surrounded more or less by a dusky cloud, and this duskiness is frequently extended forward upon the mid-vein to its base, forming a faint darker streak along the middle of the wing. Contiguous to this dot on its outer side may be discerned a roundish spot of a slightly paler yellow color than the ground, and a very short distance forward of this is a similar spot, but smaller, both these spots often showing a more tarnished centre. On the hind part of the wing the veins are marked by slender, whitish lines, and between their tips on the hind edge of the wing is a row of minute black dots. The hind wings are smoky brown, with a purplish gloss, and are nearly transparent, with the veins blackish. The fringe of both pairs of wings is pale yellowish, with a dusky band on the middle. On the under side, the wings are much more glossy and paler, opalescent whitish inwardly, and smoky gray toward their outer and hind sides, where they are also freckled with blackish atoms. The smoky color on the hind wings has on its anterior edge a row of short blackish lines, one on each of the veins, and in a line with them on the fore wings is a faint dusky band, becoming more distinct toward its outer end, or sometimes only represented by a dusky dot on the outer margin forward of the tips. The veins are whitish, and also the hind edge, on which is a row of black dots placed between the tips of the veins. The hind wings have also a blackish crescent-shaped spot a little forward of their centre. The abdomen or hind body is smoky gray above, and on its under side ash-gray, freckled with black scales, and usually showing a row of black dots along each side."*

The LARVA, or "army-worm," varies considerably in color and size, owing to age and locality, but its characteristic markings are so constant, as to make it readily distinguished. As it appeared in New Hampshire and Massachusetts, it varied in length from less than one inch to one and three quarters, was of a dark gray, with three narrow yellowish stripes above, and a broader one of the same color or slightly darker on each side, thinly clothed with short hairs, which were longer and somewhat thicker on and about the head, the latter of a polished honey-yellow, with a net-work of fine dark brown lines, and a black line on the front like the letter V reversed. The following

* Dr. Asa Fitch, in "The Country Gentleman," Vol. XVIII. p. 66.

descriptions, with methods of destruction, most of which were tried with good success in the Eastern States, are taken from an "Essay" in the Ohio Agricultural Report for 1860, by Mr. J. Kirkpatrick, of Cleveland, Ohio.

"The caterpillar of the army-worm moth, when fully grown, measures from one and three fourths to two inches in length, — when well fed, the latter, — and is about as thick as a goose-quill; color a lighter or darker gray, two lateral stripes, the upper with a yellowish central line, the lower a reddish one; legs, sixteen, six true or pectoral, eight pro-legs or abdominal, and two anal, at the extremity of the body; the head brownish, sometimes marbled, and shining; a few short hairs on the body and longer straggling ones on the head. The pupa is of a mahogany-brown color, nine lines in length, and tipped at the end with a short spine."

Fig. 276.

The Pupa (Fig. 276) of the army-worm in the Eastern States is of the same color, but varies in size, being usually smaller, and the spine is double or cleft at the apex.

"When the army-worms are numerous, it is desirable to arrest their ravages; the most common, and probably the easiest, method of doing this is that commonly practised, — ploughing a double furrow around the field, or across any part of a field that they are marching toward. It is necessary to have the steep side of the furrow next the unharmed crop, so that when the worm attempts to climb over, it may fall back into the furrow. Running the plough once in the furrow is not sufficient; twice and even three times is better, and it requires to be renewed if washed down by rains. If the soil is stiff or stony, the worms will climb over the steepest ridge; it is on light, friable soil only that the ridge will suffice to protect the field. The foothold of the worm must give way, thus rolling it back into the furrow. And even under the best conditions of soil it is best to have two furrows, one about the width of a row of corn from the other. The worms thus trapped should be destroyed either by fire or hogs; laying dry straw in the furrows and then setting fire to it is a good way, for by this means the soil of the furrow is made more friable, and of course efficient. Where, however, there is dry grass or grain near by, this is a dangerous method, as the fire may extend to the field, and do more harm than the army-worm could if let alone.

Thousands of army-worms may be easily destroyed in a meadow, by running a heavy iron roller over it. A very small pressure is sufficient to burst their skins, and the slightest injury of this kind will kill them. If the meadow or field is level, few worms can escape the roller, and thus a stop will be put to them at once. Of course this cannot succeed as well where the ground is rough and uneven, for the worms always take shelter in the hollows. The best time to roll is while the worms are feeding and up among the leaves, for then they are sure to be crushed; at other times they are sheltered, and a great number will escape. . . . . . As soon as the crop is removed from an infested field, turn in all the hogs you can, and also chickens and turkeys; ducks will do a great deal of good by searching for and eating the caterpillars. All these animals will eat the fallen grain, and thus save it in making flesh and fowl. Sheep turned into the field will kill many of the worms by trampling upon them; especially will this be the case if the flock is large. If crows and blackbirds are visiting the fields, do not let them be disturbed; there never was a crow or blackbird, that would not prefer a fat caterpillar to a grain of corn, oats, or wheat."

Fig. 277 Fig. 278.

Figures 277 and 278 represent two species of ichneumon which destroy great numbers of the army-worm. Several species of two-winged flies belonging to the *Tachinadæ* are also destructive to them. (See page 612; also, Plate VIII. Fig. 1.)

INDEX.

Abdomen, 7, 9.
Achemon hawk-moth, 325, 326.
Acheta abbreviata, A. nigra, A. Pennsylvanica, 152. A. vittata, 153.
Achetadæ, 148.
Acridium hemipterum, 182. A. sulphureum, 177. A. tuberculatum, 176.
Acronycta, Acronyctians, 435, 436.
Acrydium, 172, 173, 175. A. alutaceum, A. Americanum, 173. A. femorale, 175. A. femur-rubrum, 169, 174. A. flavo-vittatum, 173. A. laterale, 187. A. marginatum, 182. A. Milberti, 174. A. ornatum, 186. A. peregrinum, 168. A. viridi-fasciatum, 182.
Ægerians, 319, 329.
Ægeria Cucurbitæ, 331. Æ. exitiosa, 331, 333. Æ. pyri, 335. Æ. tipuliformis, 334.
Aglossa pinguinalis, 475.
Agrilus, 116.
Agrion basalis, 12.
Agriotes, A. obesus, 56.
Agrotidians (Agrotididæ), 441.
Agrotis æqua, A. Agricola, 444. A. aquilina, 442, 445. A. devastator, A. inermis, 444. A. latens, 445. A. messoria, 444. A. ocellina, 445. A. segetum, 442, 444. A. suffusa, 443. A. telifera, 444. A. tessellata, 445. A. tritici, 442.
Ahaton skipper, 317.
Alaus, 54.
Alder sphinx, 328.
Alope butterfly, 305.
Alucita, 510. A. cerealella, 499, 503.
Alucitæ, 340, 510.
American blight, 242.
American copper-butterfly, 273, 274.
Amisomorpha buprestoides, 146.
Anacampsis, A. cerealella, 499. A. sarcitella, 493.
Ancylonycha, 30.
Angoumois grain-moth, 493, 499.
Anisopteryx, 460. A. Æscularia, 461. A. pometaria, 462. A. vernata, 461.
Anomala atrata, 35. A. cœlebs, A. lucicola, A. varians, 34. A. vitis, 29.
Antennæ, 7, 8.
Anthomya canicularis, 616. A. Ceparum, A. radicum, A. Raphani, 617. A. scalaris, 616.
Anthomyians (Anthomyiadæ), 615.
Antiopa butterfly, 296.
Ants, 237, 240, 513, 515.
Apate basillaris, 92.
Apatela Aceris, A. Americana, 436.
Aphaniptera, 19.
Aphidians (Aphididæ), 203, 230.
Aphis, 235. A. Brassicæ, A. Caryæ, 238. A. lanigera, 242. A. mali, 235. A. Persicæ, 241. A. radicum, 239. A. radicis, 240. A. Rosæ, 238. A. Salicis, A. Salicicola, A. Salicti, 239.
Aphrophora, 225.
Apion rostrum, A. Sayi, 67.
Apple-tree borer, 108; bud-moths, 480, 481; sphinx, 327, 328.
Apple-trees, injured by American blight, 242; bark-lice, 252, 254; canker-worms, 463, 472; caterpillars, 268, 357, 366, 370, 375, 377, 379, 388, 425, 429. Other insects attacking, 54, 120.
Apple-worm, 484.
Apples, attacked by plum-weevils, 76, 77.
Apples of Sodom, 547.
Apricot bud-moth, 482.
Aptera, 19.
Archippus butterfly, 280.
Arctia Acrea, 351, 354. A. Americana, 347, 348. A. Arge, 346. A. Caja, 348. A. Capratina, 354. A. fuliginosa, 357. A. Isabella, 356. A. luctifera, 360. A. phalerata, A. Phyllira, 347. A. punctatissima, 358. A. rubricosa, 356. A. textor, 358. A. Urticæ, 350. A. Virginica, 350. A. virgo, 345.
Arctians (Arctiadæ), 343.
Areoda lanigera, 24.
Argynnis, 284. A. Aphrodite, 285. A. Bellona, 287. A. Idalia, 285. A. Myrina, 286.
Army-worms, 457, 627.
Arrhenodes, 68.
Arthemis butterfly, 283.
Ash-tree borers, 330.
Asilians (Asilidæ), 605.

Asilus, 605, 606. A. æstuans, 17. A. sericeus, 605.
Astasia torrefacta, 437.
Aster gall-flies, 620.
Asterias butterfly, 265.
Attaci, 340.
Attacus Atlas, 338, 390. A. Cecropia, 385, 387, 389. A. Luna, 382 - 384. A. Polyphemus, 384, 386. A. Prometheus, 390, 391.
Attelabians (Attelabidæ), 65.
Attelabus analis, 65. A. bipustulatus, 66.
Azalea sphinx, 328.

Babia quadriguttata, 135.
Bacteria arumatia, 146. B. phyllina, B. rubispinosa, 146.
Bacunculus femoratus, B. Sayi, 146.
Balaninus, 74.
Baridius trinotatus, 81.
Bark-beetles, 85.
Bark-lice, 194, 204, 248, 250, 252, 256.
Barley injured by insects, 551, 621.
Basket worms, 415.
Bathyllus skipper, 312.
Batia flavifrontella, 494.
Bat-ticks, 625.
Beach-grass, use of, 57.
Beans attacked by insects, 138, 229.
Bear-caterpillars, 343, 349.
Bee-flies, 603.
Bee-moths, 489.
Bees, 512, 514.
Beeswax devoured by worms, 491.
Beetles, 10, 23, 28, 29.
Bellona butterfly, 287.
Bird-flies, 624.
Bittacomorpha, 600.
Bittacus, 600.
Blackberry-bush borer, 114.
Black fly, 601.
Black weevil, 83.
Blatta orientalis, 145.
Blennocampa, 528.
Blight beetle, 90.
Blight of pear-trees, 88.
Blistering beetles, 135.
Boisduval's butterfly, 305.
Bomboptera, 535.
Bombyces, 340.
Bombylians (Bombyliadæ), 603.
Bombylius æqualis, B. major, B. albipectus, B. fratellus, B. vicinus, 604.
Bombyx, 340. B. cossus, 412. B. grata, 427. B. mori, 380.
Bordered skipper, 308.
Borers, 46, 92, 331, 410.
Bostrichians (Bostrichidæ), 91.
Bot-bees, Bot-flies, 622.
Bots, 624.
Brenthians (Brenthidæ), 67.
Brenthus maxillosus, B. septemtrionis, B. septemtrionalis, 68.
Brizo skipper, 309.
Bruchians (Bruchidæ), 62.
Bruchus pisi, 61, 62.
Bud-moths, 480, 482.
Bug, chinch, 198; plant, 200; squash, 194.
Bugs, 193, 194.
Bulla, 185.
Buprestians (Buprestidæ), 45, 116.
Buprestis characteristica, B. dentipes, 49. B. divaricata, 48. B. Drummondi, B. femorata, B. fulvoguttata, 50. B. lurida, B. obscura, 49. B. Virginica, 48.
Butalis, 499, 507. B. cerealella, 506, 509.
Butterflies, 262, 263 ; four-footed, 279, 282.
Button-wood trees attacked by caterpillars, 363, 404; by wood-wasps, 537.

Cabbage butterfly, 270. C. caterpillar, 270, 451. C. cutworm, 443, 445, 449. C. plant-louse, 238.
Calandra, 69, 82. C. granaria, 83. C. Oryzæ, *ib.*
Callidium, 99. C. antennatum, C. bajulus, C. violaceum, 100.
Callimorpha Carolina, 344. C. Lecontei, 345. C. militaris, 344.
Calliphora vomitoria, 614.
Caloptenus, 174.
Calosoma scrutator, 470.
Camel-crickets, 146.
Canephoræ, 414.
Canker-worm, 461, 470.
Cantharides (Cantharididæ), 135.
Cantharis atrata, 139. C. cinerea, 138. C. marginata, 137. C. vittata, 136, 139.
Capricorn beetles, 92, 94, 96, 116.
Capsus oblineatus, 200.
Caradrina cubicularis, 453.
Carolina sphinx, 322.
Carpenter moths, 412.
Carpet-moth, 493.
Carpocapsa Pomonella, 484.
Carrot-caterpillars, 263.
Cassida aurichalcea, 122.
Cassidadæ, 121.
Catalogue, reference to, 32.
Caterpillars are the young of butterflies and moths, 257. C. described, 258. C. false, 516. C. food of, 258. C. habits of, 259. C. injurious to gardens, 349. C. numbers of, 257. C. spiny, 284. C. transformations of, 5, 260. C. wheat, 452, 455.
Cecidomyia, 554, 568, 570, 583. C. culmicola, 582. C. destructor, 566, 568, 583. C. Robiniæ, 567. C. Salicis, 566, 567. C. Tritici, 453, 566, 587, 592.
Cecidomyiadæ, 565.
Cedar (red), insects attacking, 87.
Celtis sphinx, 328.
Cephalemyia ovis, 624.
Cephus, 516. C. Pygmæus, 517.
Cerambycidæ, 92, 94, 96.
Cerambyx, 94. C. cinctus, 97. C. palliatus, 115. C. violaceus, 100.

Ceraphron destructor, 586.
Cerasphorus, 97.
Ceratocampa regalis, 399.
Ceratocampians (Ceratocampadæ), 398.
Ceratomia quadricornis, 323.
Cercopididæ, 224.
Cercopis ignipecta, C. obtusa, C. parallela, C. quadrangularis, 225.
Ceresa, 221.
Cernes skipper, 316.
Cerura, C. borealis, C. furcula, 423.
Cetonia barbata, 40. C. eremicola, 43. C. Inda, 40.
Cetonians (Cetoniadæ), 39.
Chalcidians (Chalcididæ), 550, 556.
Chafers, 26.
Chalcophora, 48.
Charæas graminis, 442.
Cheese-maggots, 621.
Cheimatobia brumata, 461.
Chelonia, 345.
Cherry-tree slug, 529. C. sphinx, 328. C. (wild), its borer, 48. C. caterpillars, 370, 421, 425.
Chinch-bug, 198.
Chionea aspera, C. valga, 601.
Chlamys gibbosa, 135.
Chloëaltis, 184.
Chlorops glabra, C. pumilionis, 621.
Chœrocampa, 326. C. Chœrilus, 328. C. Pampinatrix, 327. C. versicolor, 328.
Chrysalids, 6.
Chrysobothris, 49, 50, 51.
Chrysomela, 123. C. cæruleipennis, C. Polygoni, 133. C. scalaris, C. trimaculata, 132. C. vitivora, 130.
Chrysomelians (Chrysomeladæ), 123, 131.
Chrysopa euryptera, C. perla, 247.
Chrysops ferrugatus, C. niger, C. vittatus, 603.
Cicada, 141, 204. C. auletes, 219. C. canicularis, 217. C. hieroglyphica, 219. C. pruinosa, 218. C. Rosæ, 229. C. septendecim, 206, 211. C. seventeen-year, 204, 206. C. tibicen, 219. Cicadians (Cicadadæ), 203, 204.
Cimbex Americana, C. Laportei, C. Ulmi, 518, 519.
Cinara, 238.
Clastoptera proteus, 225.
Clear-winged sphinx, 328, 329.
Clematis attacked by insects, 137.
Climbing crickets, 154.
Clisiocampa Americana, 373. C. castrensis, 372, 373, 376. C. Neustria, 371, 373, 376. C. silvatica, 375.
Clostera Americana, 431, 433. C. anastomosis, C. inclusa, 433.
Clothes-moths, 494, 495.
Clover-worms, 456.
Clymene moth, 344.
Clypeus, 24.
Clythra dominicana, C. quadriguttata, 135.
Clytus flexuosus, 103. C. Hayii, 102. C. pictus, 103, 411. C. speciosus, 101.
Coccidæ, 204, 248.
Coccinella, C. novemnotata, 246.
Coccus, 248. C. Adonidum, 250. C. arborum linearis, 252. C. Cacti, 250. C. conchiformis, 252. C. cryptogamus, 254. C. Hesperidum, 250.
Cochenille, 249, 250.
Cock-chafer, 27.
Cockroaches, 143, 145.
Codling-moth, 484.
Cœnia butterfly, 293.
Cœnomyia pallida, 607.
Coleoptera, 10, 23.
Colias Philodice, 272.
Colona moth, 344.
Columbine roots destroyed by caterpillars, 440.
Comma butterfly, 300.
Conocephalus dissimilis, 164. C. ensiger, 163. C. uncinatus, 165.
Conopians (Conopidæ), 611.
Conops nigricornis, C. sagittaria, 611.
Conotrachelus Nenuphar, 75, 484. C. variegatus, 75.
Coreus lineolaris, 200, 201. C. linearis, 200. C. mœstus, 194. C. ordinatus, 195. C. rugator, C. tristis, 194.
Corn attacked by spindle-worms, 438, 439. C. destroyed by caterpillars, 347, 349. C. destroyed by cut-worms, 441, 443. C. moths, 506. C. weevil, 83, 84. C. worms, 497, 505.
Coscinoptera, 135.
Cossus ligniperda, 412. C. Robiniæ, 412, 413.
Cotalpa, 24.
Cotton-worms, 457.
Crambidæ, 488, 489.
Cranberry-worms, 487.
Crantor hawk-moth, 325.
Creeper attacked by insects, 223, 325, 337.
Crickets, 148, 154.
Criocerians (Crioceridídæ), 118.
Crioceris asparagi, 118. C. bipustulata, C. striolata, 129. C. trilineata, 118. C. vittata, 124.
Cryptocephalians (Cryptocephalidæ), 134. Cryptocephalus luridus, 135.
Cuckoo-spit, 220, 225.
Cucuio, 55.
Cucumber-beetle, 124, 127. Cucumbers, insects attacking, 124, 125, 127.
Curculio, 69, 75. C. granarius, 83. C. hilaris, 70. C. Nenuphar, 75. C. Oryzæ, 83. C. pales, 70.
Curculionidæ, 69.
Currant-bush borers, 334, 335.
Cut-worms, 441.
Cynipidæ, 544.
Cynips, 544. C. bicolor, 549. C. confluens, 546. C. dichlocerus, 549. C. gallæ tinctoriæ, 546. C. nubilipennis,

C. oneratus, C. seminator, 548. C. semipiceus, 549.
Cynthia Atalanta, 294. C. Cardui, 291. C. Huntera, 292, 293. C. Lavinia, 293.
Cyrtophillus, 158.

Dahlia attacked by spindle-worms, 439.
Danais Archippus, 280.
Deilephila Chamænerii, D. lineata, 328.
Deiopeia bella, 342. D. pulchella, 343.
Delta-moths, 474, 475.
Dermaptera, 20.
Desmocerus palliatus, 115.
Diapheromera, 146.
Dicerca, 48, 49.
Dicranura, 423.
Dictyotoptera, 20.
Diplolepis, 544, 546, 548.
Diptera, 16, 562.
Disippe butterfly, 281.
Ditula angustiorana, 483.
Dog's-bane beetle, 134.
Dominula moth, 344.
Donna moth, 344.
Dor-bug, 30, 31.
Dors, 26, 28.
Drop-worms, 415.
Dryocampa bicolor, 408. D. imperialis, 402, 403. D. pellucida, 407. D. rubicunda, 408. D. senatoria, 406. D. stigma, 407.
Dung-flies, 619.

Earwigs, 143, 144.
Elaphidion, 98.
Elater appressifrons, E. brevicornis, 56. E. cinereus, E. communis, 55. E. mancus, 56. E. noctilucus, 55. E. obesus, 56. E. oculatus, 54.
Elateridæ, 51.
Elder, its borer, 115.
Elm caterpillars, 297, 299, 302, 357. E. other insects attacking, 112, 124, 133, 465. E. false caterpillars on it, 519. E. sphinx-caterpillars, 323, 324. E. bored by wood-wasps, 537. E. injured by canker-worms, 465, 472.
Empoa, 229.
Encyrtus, 532.
Entilia, 220.
Erebus Strix, 338.
Eriosoma, 242, 245.
Eristalis sincerus, E. æneus, 609.
Ermine moths, 345, 350, 354.
Erythroneura, 229.
Euchætes Egle, 359, 360.
Euchenopa, 221.
Eudamus, 310. E. Bathyllus, 312. E. Tityrus, 310, 311.
Eudryus grata, 427. E. unio, 428.
Eumenes, 611. E. fraterna, 470, 471.
Eumolpus auratus, 134.
Euplexoptera, 20.
Euprepia, 345.
Euryomia, 40.
Eurytoma, 551, 559, 586. E. fulvipes, 561. E. Hordei, 553, 554, 561. E. Secalis, E. Tritici, 561.
Eurytris butterfly, 306.
Eyes of insects, 7, 258.

False caterpillars, 516.
Feather-winged moths, 510.
Fir saw-fly, 520.
Fir-trees attacked by moths, 483. F. destroyed by wood-wasps, 534.
Fire-beetle, 55.
Flea-beetles, 126.
Flea tribe, 19, 625.
Flesh-fly, 613.
Flies, 562. F. flower, 615. F. golden-eyed, 603. F. how excluded from houses, 615. F. parasitic, 612.
Flower-beetles, 24, 39.
Flower-flies, 615.
Fly, dung, 618. F. flesh, 613. F. golden-eyed lace-winged, 247. F. hammer-headed, 621. F. Hessian, 565, 568. F. house, F. meat, 614. F. radish, F. onion, 617. F. stable, 613, 614. F. viviparous, 613. F. wheat, 565, 587.
Fly-weevil that destroys wheat, 502.
Forest-flies, 603.
Forficula, 144.
Forficuladæ, 19.
Frit-fly, 621.
Frog-hoppers, 220, 224.
Fruit, weevils in, 75, 77, 484. F. flies, F. maggots, 620. F. moth, 484. F. trees injured by beetles, 29, 33, 36, 41; by canker-worms, 463; by cicadas, 212.
Fur-moths, 493.

Gad-flies, 602.
Galeruca Calmariensis, 124, 133, 465. G. vittata, 124.
Galerucians (Galerucadæ), 123.
Galleria cereana, 489.
Gall-flies, four-winged, 543, 550. G. two-winged, 620.
Gall-gnats, 566.
Galls, 543, 546, 567, 620.
Gasterophilus equi, G. hæmorrhoidalis, G. veterinus, 623.
Gastropacha Americana, G. Ilicifolia, G. occidentalis, 377. G. Velleda, 378, 379.
Geometers (Geometræ), 340, 458.
Geometra catenaria, 459, 460.
Glaucopidians (Glaucopididæ), 319, 336, 341.
Glaucopis Pholus, 341.
Gnat, snow, 601. G. wheat, 587, 592. G. wingless, 601.
Gnats, 562, 563, 564, 565. G. long-legged, 600. G. gall, 566.
Gnophria rubricollis, G. vittata, 342.

Goat-moths, 413.
Goldsmith-beetle, 24, 25.
Golden-rod, insects on, 103, 139.
Goliah-beetle, 40.
Goniloba, 310.
Gonocerus, 196.
Gortyna flavago, G. leucostigma, 440. G. Zeæ, 439.
Grain-moth, 493, 496.
Grain-weevil, 83.
Grain-worms, 453.
Grape-vine caterpillars, 325, 326, 426. G. leaf-hopper, 227. G. Procris, 336. G. sphinx, 325, 326. G. injured by bark-lice, 256. G. by false caterpillars, 523. G. by other insects, 25, 29, 34, 36, 129.
Grapta, 298.
Graspers, 143, 146.
Grasshopper, its growth and changes, 5. G. See Locust.
Grasshoppers, 148, 155. G. are locusts, 141.
Gray-worm, 453.
Grease-moth, 475.
Ground-beetles, 24, 470.
Grouse-locust, 185.
Grubs, 9, 23.
Gryllidæ, 148, 155.
Gryllotalpa brevipennis, G. borealis, 149. G. didactyla, 149, 150.
Gryllus, 148. G. chrysomelas, 182. G. equalis, 178. G. erythropus, 175. G. maculatus, 155. G. migratorius, 168. 175. G. sulphureus, 177. G. Virginianus, 182.
Gymnodus scaber, 41.

Hackberry sphinx, 328.
Hag-moth, 421, 422.
Hair-moth, 493.
Halesidota, 362.
Haltica bipustulata, H. chalybea, 129. H. Cucumeris, H. fuscula, H. pubescens, 127. H. striolata, 129.
Halticadæ, 126.
Hare bot-fly, 624.
Harlequin caterpillars, 359.
Harnessed moth, 347.
Harpy-fly, 614.
Harvest-flies, 141, 194, 203, 204. H. dog-day, 217. H. frosted, 218. H. leaping, 219.
Hawk-moths, 262, 318.
Hazel-nut weevil, 74.
Hedge-hog caterpillar, 355.
Hegemon Goliatus, 40.
Hemiptera, 11, 192. H. heteroptera, 193, 194. H. homoptera, 194, 203.
Hemiptycha, 221.
Hepialidæ, 408.
Hepiolus argenteomaculatus, 410. H. Humuli, 409.
Hera moth, 344.
Herminians (Herminiadæ), 476.
Hesperians (Hesperiadæ), 307. H. Ahaton, 317. H. Cernes, 316. H. Hobomok, 313, 314. H. Leonardus, 314. H. Metacomet, 317. H. Peckius, H. Sassacus, 315. H. Wamsutta, 318.
Heteropterus marginatus, 308.
Heterocera, 262.
Hickory borers, 46, 49, 97. H. caterpillars, 361, 362, 399. H. plant-louse, 238.
Hipparchians, 303. H. Alope, H. Boisduvallii, 305. H. Eurytris, H. Nephele, 306. H. semidea, 304.
Hippobosca equina, 624.
Hippoboscidæ, 20.
Hispadæ, Hispa marginata, H. quadrata, H. rosea, 120. H. suturalis, 121.
Hobomok skipper, 313, 314.
Hog-caterpillar, 326, 327, 328.
Homalomyia scalaris, 616.
Homaloptera, 20.
Homoptera, 20, 194, 203.
Honey-dew, 237.
Hop-vine caterpillars, 276, 299, 301, 476. H. Hepiolus, 409.
Horn-bugs, 43.
Horn-tailed wood-wasps, 14, 533.
Horse-bot, 623.
Horse-chestnut caterpillars, 367.
Horse-flies, 602.
House-fly, 614. H. crickets, 150.
Humming-bird moths, 319, 329.
Hybernians (Hyberniadæ), 460. H. defoliaria, H. Tiliaria, 473.
Hydrocampa, 476.
Hylecœtus Americanus, 59.
Hylobius, H. picivorus, 71.
Hylurgus dentatus, 87. H. terebrans, 86.
Hymenoptera, 14, 512.
Hypena Humuli, H. rostralis, 477.
Hyphantria, 358.
Hypogymna dispar, 366.

Iassus Rosæ, 229.
Ichneumones minuti, 550.
Ichneumon-flies, 73, 372, 532, 538, 550, 586, 630.
Idalia butterfly, 285.
Insects, structure of, 3. I. produced from eggs, 4. I. transformations, 5.

Jacobææ moth, 344.
Joint-worm, 551, 554.
Jumpers, 144, 148.
Juvenal's skipper, 309.

Kalmia sphinx, 328.
Katy-did, 158.
Kermes, 249.
Knot-grass beetle, 133.

Lachnosterna, 30.
Lachnus, 238.
Lackey caterpillars, 371.
Lady-birds, 246, 247.

Lamia titillator, 105.
Laphria, 606. L. flavibarbis, L. tergissa, L. thoracica, 604.
Lappet caterpillars, 377.
Larva, 6.
Lasiocampa Dumeti, 369. L. processionea, 394. L. Quercus, L. Roboris, L. Rubi, L. Trifolii, 369.
Lasiocampians (Lasiocampadæ), 369.
Lasioptera, 570.
Laurel sphinx, 328.
Laverna, 499.
Lavinia butterfly, 293.
Leaf-beetles, 117. L. hoppers, 220, 225. L. rollers, 478.
Leconte, Dr. John L., notes by, 19, 24, 30, 33, 40, 50, 56, 88, 96, 98, 100, 118, 135, 136, 138.
Lema, 118.
Leonard's skipper, 314.
Lepidoptera, 12, 257, 261.
Leptura, 94. L. picta, L. Robiniæ, 103.
Lepturians (Lepturadæ), 94, 114.
Leucania extranea, L. impuncta, L. unipuncta, 627.
Lilac sphinx, 328.
Limacodes, 419. L. cippus, L. Delphinii, 420. L. pithecium, 421. L. scapha, 420.
Lime or linden tree insects, 110, 133, 299, 436, 473.
Limenitis, 282.
Linnæus, anecdote respecting, 57.
Liparians (Liparidæ), 365.
Liparis, 365.
Lithosia miniata, 342. L. quadra, 343.
Lithosians (Lithosiadæ), 341.
Locust, 167. L. grouse, 185. L. Cicada, 204.
Locusts, 141, 144, 165, 167, 175.
Locust-tree butterfly, 311. L. boring caterpillars, 410. L. other insects upon it, 62, 103, 138, 221. L. (honey) attacked by insects, 138.
Locusta, 165, 168, 172, 175. L. abortiva, 184. L. æqualis, 178. L. agilis, 162. L. Carolina, 176. L. chrysomelas, 182. L. conspersa, 184. L corallina, 176. L. curtipennis, 184. L. curvicauda, 158, 161. L. eucerata, 180. L. fasciata, 163. L. fenestralis, 180. L. infuscata, 181. L. latipennis, 179. L. laurifolia, 159. L. leucostoma, L. maritima, 178. L. marmorata, 179. L. migratoria, 168, 175. L. nebulosa, 181. L. oblongifolia, 159. L. perspicillata, 158. L. radiata, 183. L. sulphurea, 177. L. viridi-fasciata, 180.
Locustadæ, 148, 165.
Loopers, 458.
Lophocampa Caryæ, 362. L. maculata, 363. L. tesselaris, 343, 364.
Lophyrus Abbotii, L. Abietis, L. Americanus, L. compar, 520.
Loxotænia Rosaceana, 481.
Lozotænia oporana, 480.
Lucanians (Lucanidæ), 43. L. Capreolus, L. Dama, 44.
Lucilia Cæsar, 614.
Ludius, 56.
Lycæna Americana, 273. L. Epixanthe, 274. L. Phlæas, 273.
Lycenians (Lycænadæ), 273.
Lyda, 516.
Lygæus leucopterus, 198.
Lymexylidæ, 58.
Lymexylon navale, 57. L. sericeum, 58.
Lytta, 136. L. atrata, 139. L. cinerea, 138. L. Fabricii, 138. L. marginata, 138. L. vittata, 136.

Macrodactylus subspinosus, 35.
Maggot, its transformations, 5.
Maggots, 612. M. in cattle, 624. M. in cheese, 621. M. in fruit, 620. M. in the human body, 616. M. in meat, 613, 614. M. in radishes, turnips, and onions, 616, 617, 618. M. in roots, 605, 617. M. rat-tailed, 609. M. wheat, 574, 575, 581, 592, 594 – 599.
Mamestra picta, 451, 452.
Mantes, 143, 146.
Maple caterpillars, 436.
Maple (sugar), its borer, 101.
Marshes, salt, insects injuring, 168, 175, 351.
May-beetles, 26, 30, 441.
May-flies, 12, 476.
Meadows injured by insects, 28, 31, 56.
Meal-moth, 475.
Mealy-bug, 250.
Melanophila, 50.
Melanotus, 55.
Melitæa, 287. M. Phaëton, M. Ismeria, 288. M. Pharos, 289.
Mellophagus ovis, 625.
Meloe angusticollis, 139, 140.
Melolontha, 26, 33. M. subspinosa, 35. M. variolosa, 33.
Melolonthians (Melolonthadæ), 26.
Membracidæ, 220.
Membracis acuminata, 221. M. Ampelopsidis, 220, 223. M. bimaculata, 221. M. binotata, *ib.*, 224. M. bubalus, 221. M. camelus, 220. M. Cissi, 223. M. concava, 220. M. diceros, 221. M. emarginata, 220. M. latipes, 221. M. sinuata, 220. M. taurina, 221. M. univittata, *ib.*, 223. M. vau, 220.
Metacomet skipper, 317.
Metamorphoses, 4.
Midas filatus, M. clavatus, 606.
Midges, 565, 602.
Milesia excentrica, 609. M. vespiformis, 610.
Milk-weed beetle, 132. M. caterpillars, 359.
Millers, 338, 350.
Misippus butterfly, 281.

Mole-cricket, 149, 150.
Monohammus, 105.
Morris, Rev. J. G., notes by, 257, 258, 261, 262, 265, 270, 276, 296, 298, 300, 310, 320, 325, 327, 328, 331, 342, 350, 358, 359, 362, 377, 433, 436, 465.
Mosquito, its transformations, 5.
Mosquitos, 563.
Moth, origin of the word, 488.
Moths, 262, 338. M. in houses, how destroyed, 496. M. worms, 488, 489, 495.
Mountain-butterfly, 304.
Muck-worm, 31.
Musca Cæsar, M. domestica, M. Harpyia, M. vomitoria, 614.
Muscans (Muscadæ), 612.
Mustard-butterfly, M. caterpillar, 270.
Mutilla coccinea, 15.
Mycetophilæ, 564.
Mydas filata, 606.
Myrina butterfly, 286.
Myopa nigripennis, 610, 611. Myrtle bark-louse, 250.

Nemeophila plantaginis, 345.
Nemobius, 153.
Nepa, 625. N. apiculata, 12.
Nettle-butterfly, 294, 303.
Neuroptera, 12, 20.
Neuter insects, 513.
Noctua clandestina, 448. N. devastator, 445. N. xylina, 457.
Noctuæ, 340, 434.
Nonagrians (Nonagriadæ), 437.
Norton, Mr. Edward, notes by, 540, 561.
Notodonta concinna, 425, 426. N. unicornis, 424.
Notodontians (Notodontadæ), 418.
Nut-galls, 545. Nut-weevils, 74.
Nycteribia, 625.
Nymphalis Arthemis, 283. N. disippus, 281. N. Ephestion, 283.

Oak-apples, 546.
Oak-caterpillars, 283, 375, 384, 397, 405, 412.
Oak gall-flies, 546, 548.
Oak-pruner, 98.
Oaks, other insects attacking, 49, 68, 211, 223.
Oberea, 114.
Ocelli, 7.
Œcanthus niveus, 154.
Œceticus, 414.
Œcophora granella, 499.
Œdipoda, 175-181. Œ. discoidea, 176. Œ. fenestralis, 177, 180.
Œstrians (Œstridæ), 622.
Œstrus bovis, Œ. buccatus, 624.
Oiketicus, 414, 415.
Oil-beetles, 140.
Omaloplia sericea, 34. O. vespertina, 33.
Onion-fly, 617.
Onions destroyed by maggots, 617.
Opsomala, 172.
Orchelimum gracile, 163. O. vulgare, 162.
Orgyia antiqua, 368. O. leucostigma, 367.
Ornithomyia, 624.
Ortalidians (Ortalididæ), 620.
Orthoptera, 11, 141. O. ambulatoria, 144, 146. O. cursoria, 143, 144. O. raptoria, 143, 146. O. saltatoria, 144, 148.
Orthosoma, 96.
Oryssus, 541. O. affinis, 543. O. hæmorrhoidalis, O. maurus, 542. O. Sayii, 543. O. terminalis, 542.
Oscinians (Oscinidæ), 621, 622.
Oscinis frit, O. lineata, 621.
Osmoderma eremicola, 43. O. scaber, 41.
Osten Sacken, Baron R., notes by, 601, 603, 604, 606, 609, 610.
Ourapteryx Sambucaria, 459.
Owl-moth, great, 338.
Owlet-moths, 434.
Ox bot-fly, 624.
Oxya, 174.

Pack-moth, 493.
Palpi, 8.
Pamphila, 312.
Pandeleteius, 70.
Papiliones, 262.
Papilio Ajax, P. Asterias, P. polyxenes, 265. P. Troilus, 266. P. Turnus, 268.
Parsley-caterpillars, 263.
Parthenice moth, 345.
Pea-weevil, 61, 62.
Peach-tree borers, 48, 331. P. plant-lice, 234, 240. P. Thrips, 234.
Pear-tree borer, 335. P. slug, 528. P. bored by wood-wasps, 535. P. injured by bark-lice, 254. P., other insects attacking, 25, 88, 214, 231, 233.
Pears, worms in, 487.
Peas, insects attacking, 61.
Peck's skipper, 315.
Pectinated antennæ, 336, 338.
Pelidnota punctata, 25.
Penthina comitana, P. luscana, P. oculana, 482.
Perophora Melsheimerii, 415, 416.
Petrophila, 476.
Phalæna Aceris, 436. P. anastomosis, 434. P. brumata, P. vernata, 461.
Phalænæ, 262, 338.
Phalangopsis lapidicola, P. maculata, 155.
Phaneroptera angustifolia, P. curvicauda, P. myrtifolia, P. septentrionalis, 161.
Phasma, 146.
Philampelus Achemon, P. satellitia, 325.

Philyra moth, 347.
Phloiotribus, 88.
Phryganeadæ, 20, 476.
Phyllium pulchrifolium, P. siccifolium, 146.
Phyllophaga, 30. P. fraterna, P. Georgicana, P. hirticula, 32. P. pilosicollis, 33. P. quercina, 30.
Phylloptera, 161. P. oblongifolia, 158, 159.
Phytocoris linearis, P. lineolaris, 200, 201.
Piercer, 8, 512.
Piercers, 513.
Pieris, P. casta, 270.
Pigeon Tremex, 537.
Pimpla atrata, P. lunator, 538, 539.
Pine saw-flies, 520, 521.
Pine-tree sphinx, 328. P. attacked by moths, 483; by wood-wasps, 534; by other insects, 48, 51, 70-72, 85-88, 99, 100, 115.
Piophila casei, 621.
Pissodes, 72.
Plant-bug, 200, 201. P. lice, 194, 203, 230, 235. P. cabbage, 238. P. downy, 242. P. hickory, 238. P. leaping, 230. P. peach-tree, 234, 240. P. rose, 238. P. willow, P. on roots, 239. P. how to destroy, 244, 245. P. their enemies, 246, 247.
Platygaster, 471, 587.
Platyomides, 479.
Platyphyllum concavum, P. perspicillatum, 158.
Plum-tree caterpillars, 424, 425. P. slug, 529. P. warts, 478, 479. P. weevil, 75, 484.
Pœcilochroma comitana, 482.
Polistes fuscata, 18.
Polyommatus Comyntas, P. Lucia, 275. P. Pseudargiolus, 274.
Polyphylla, 33.
Pontia oleracea, 270.
Poplar-tree caterpillars, 282, 297, 422, 431. P. other insects attacking, 96, 106.
Porthesia auriflua, P. chrysorrhea, 366.
Potato-fly, 136, 139. P. rot, 81, 118, 128, 137, 200, 246, 440, 618. P. vines, insects attacking, 81, 118, 128, 136, 138, 199, 320, 440. P. worm, 320. P. sweet, insects on, 122.
Potter-wasp 470.
Prionians (Prionidæ), 95.
Prionus brevicornis, 95. P. cylindricus, 96. P. laticollis, 95. P. unicolor, 96.
Procris Americana, P. ampelophaga, P. vitis, 336.
Progne butterfly, 301.
Psilura monacha, 366.
Psyche, 414.
Psychians (Psychadæ), 414.
Psylla, 231. P. Pyri, 233.
Pterology, 261.
Pteromalus, 509, 556, 586. P. Vanessæ, 300.
Pterophoridæ, 510.
Pterophorus, 510.
Ptychoptera clavipes, 600.
Pudica moth, 345.
Pulex, 625.
Pulicidæ, 19.
Pupa, 6.
Puparium, 576.
Purslane sphinx, 328.
Pygæra ministra, 430.
Pyralides, 340, 474.
Pyralis farinalis, 475. P. Pomana, 484.
Pyrgota undata, 610.
Pyropa furcata, 619.
Pyrophorus, 55.

Radish-fly, 617. R. injured by maggots, 616.
Rhaphitelus, 586.
Rhagium lineatum, 116.
Rhipiphoridæ, 19.
Rhipiptera, 18.
Rhopalocera, 262.
Rhubarb-root maggots, 605.
Rhynchænus, 69. R. Argula, 75. R. Cerasi, 78. R. nasicus, 74. R. Nenuphar, 75, 484. R. Strobi, 72.
Rhynchites bicolor, 66.
Rhynchophoridæ, 60.
Rhyparochromus devastator, 198, 199.
Rice-weevil, 83.
Romalea, 172.
Rose-bud moths, 480, 483.
Rose-bug, 35.
Rose-bush galls, 549. R. leaf-hopper, 229. R. plant-louse, 238. R. slug, 525. R. attacked by beetles, 33, 35, 36, 66.
Runners, 143, 144.
Rustic moths, 441.
Rutilians (Rutiladæ), 25.

Sack-bearer, 414.
Salt-marsh caterpillar, 351.
Sand-flies, 602.
Saperda, 106. S. bivittata, 107. S. calcarata, 106. S. candida, S. carcharias, 107. S. tridentata, 111. S. tripunctata, 114. S. vestita, 109.
Sarcophagans, 613.
Sarcophaga Georgina, 613.
Sargus, 608.
Sassacus skipper, 315.
Sassafras-tree caterpillars, 266, 389.
Satellitia hawk-moth, 325.
Saturnia Hera, 398. S. Io, 395. S. Maia, S. Proserpina, 396.
Saturnians (Saturniadæ), 381.
Satyrus Canthus, 306.
Saw-flies, 14, 515.
Saw-horned beetles, 45.
Scarabæians (Scarabæidæ), 10, 23, 88.
Scarabæus Indus, 40. S. relictus, 31.

Scarlet in grain, 248.
Scatomyians (Scatomyzadæ), 619.
Scatophaga furcata, S. postilena, 619.
Scientific names useful, 20, 21.
Scolytidæ, 86.
Scolytus destructor, 86. S. Pyri, 90. S. terrebrans, 86.
Scutel, 23.
Scymnus, 247.
Selandria barda, S. pygmæa, S. Rosæ, 525. S. vitis, 522, 525. S. (Blennocampa) Æthiops, 528. S. B Cerasi, 529.
Semicolon butterfly, 298, 299.
Serica, 33.
Sericaria, 364.
Sesia diffinis, 328. S. Pelasgus, *ib.*, 329.
Sesiæ, 319.
Sheep bot-fly, S. ticks, 624.
Silk, native, 380, 392.
Silk-worm, 259, 380.
Simaëthis, 475.
Simulium molestum, 601. S. nocivum, 602.
Siphonaptera, 19.
Sirex columba, 536.
Sitophilus, 83.
Skippers, 263, 307.
Slug-caterpillars, 419, 421. Slug-worm, 528.
Slugs, 517, 524.
Smerinthi, 319.
Smerinthus, 327. S. excæcata, 327, 328. S. Juglandis, S. myops, 328.
Snout-beetles, 60, 67.
Snow-gnat, 601.
Sodom, apples of, 547.
Soldier-flies, 608.
Soothsayers, 143, 146.
Spanish-flies, 135.
Span-worms, 458.
Spectrum bivittatum, 146. S. femoratum, *ib.*, 147.
Sphecomyia undata, 610. S. valida, *ib.*, 611.
Sphex, 610.
Sphex-flies, 611.
Sphinges, 262, 319, 328.
Sphinx Carolina, 322. S. cinerea, 328. S. cnotus, 327. S. coniferarum, S. drupiferarum, S. Gordius, S. Hylæus, S. Kalmiæ, 328. S. Myron, 327. S. quinquemaculatus, 321, 322.
Sphyracephala brevicornis, 621.
Spider-flies, 624.
Spilonota comitana, 482.
Spilosoma, 356, 358, 359.
Spindle-worm, 438, 439.
Spinners, 340.
Spring-beetles, 51.
Squash-bug, 194. S. vine Ægeria, 331.
Stag-beetles, 43.
Star-wort gall-fly, 620.
Stenocorus, S. cinctus, 97. S. cyaneus, 115. S. garganicus, 97. S. lineatus, 116. S. putator, S. villosus, 98.
Sting, 9, 512.
Stingers, 512.
Stinging caterpillars, 393.
Stomoxys calcitrans, 614.
Storthygocerus, 586.
Stratyomyadæ, 608.
Stratyomys, 608.
Strepsiptera, 18, 19.
Stylops, 19.
Stylopidæ, 19.
Syrphians (Syrphidæ), 608.
Syrphus, 248.

Tabanus, 603. T. atratus, T. cinctus, T. lineola, 602.
Tachinadæ, 612.
Tachina, 471. T. vivida, 612.
Tapestry-moth, 493.
Tarsi, 9.
Tenebrio molitor, 10, 11.
Tent-making caterpillars, 370.
Tenthredinidæ, 515.
Tenthredo Cerasi, 528.
Tephritis Asteris, 620.
Tetrix, 172, 185. T. bilineata, T. dorsalis, 186. T. lateralis, 187. T. ornata, 186. T. parvipennis, 187. T. quadrimaculata, 186. T. sordida, 187.
Tettigonia, 226. T. Fabæ, 229, 230. T. Rosæ, 229. T. Vitis, 227.
Tettigonians (Tettigoniadæ), 225.
Tettix, 226.
Thanaos Juvenalis, 309. T. Brizo, 310.
Thecla Auburniana, 277. T. Augusta, 279. T. Damon, 277. T. Favonius, T. Falacer, T. Humuli, T. Liparops, T. Melinus, 276. T. Mopsus, T. Niphon, 278. T. Simaëthis, T. Smilacis, 277. T. strigosa, 276.
Thistle butterfly, 291.
Thola, 249.
Thorn hedges injured by caterpillars, 369.
Thrips, 20, 234. T. cerealium, 235.
Thysanoptera, 20.
Tibia, 9.
Tiger-moths, 343, 345.
Timber-beetles, 58.
Tineæ, 340, 483.
Tineans (Tineadæ), 488, 493.
Tinea crinella, 493. T. destructor, T. flavifrontella, 494. T. granella, 493, 496. T. Hordei, 499. T. mellonella, 489. T. pellionella, 493. T. Pomonella, 484. T. tapetzella, T. vestianella, 493.
Tityrus skipper, 310, 311.
Tomicus exesus, 87. T. liminaris, T. Pini, 88. T. Pyri, 91.
Tortoise-beetles, 121.
Tortrices, 340, 478.
Tortrix cereana, 489.
Torymus, 557, 559.
Trachypteris, 50.

Tragocephala, 181.
Trama, 240.
Transformations, 4. T. imperfect, 142.
Tree-beetles, 25, 26. T. hoppers, 220.
Tremex Columba, 537.
Trichius scaber, 40.
Trichoptera, 20.
Trochilium, 331. T. denudatum, 330.
Troilus butterfly, 266.
Truxalis, 172.
Turnip fly, 126. T. other insects attacking, 130, 270.
Turnus butterfly, 268.
Turpentine-moths, 484.
Tussock-moths, 362, 364.

Uhler, Mr. P. R., notes by, 146, 149, 150, 152, 155, 158, 161, 173, 174, 176, 177, 178, 179, 180, 181, 187, 200, 219, 220, 221, 239, 240, 247.
Unicorn moth, 424.
Uroceridæ, 533.
Urocerus abdominalis, 540. U. albicornis, 538. U. gigas, 535. U. Juvencus, 534, 540. U. nitidus, 540.

Vanessa, 295. V. Antiopa, 296, 297. V. C. argenteum, V. C. album, 301. V. Comma, 300. V. furcillata, 302. V. Interrogationis, 298, 299. V. J. album, 298. V. Milberti, 302. V. Progne, 301.
Vanessians (Vanessiadæ), 284.
Vaporer-moths, 367.
Vine saw-fly, 522.
Virgin's bower, insects on, 137.
Virguncula moth, 345.
Visor, 24.

Walkers, 144, 146.
Walking leaves, 146.
Walnut-tree beetles, 92. W. caterpillars, 362, 382, 401. W. sphinx, 328.
Wamsutta skipper, 318.
Wasps, 512, 513, 514.
Wax-moth, 489.
Wax-work plant attacked by insects, 224.
Web-worms, 357.
Weevils, 21, 59, 60. W. black, 83. W. brown, 453.
Wheat injured by insects, 83, 197, 235, 453, 496, 499, 551, 555, 557, 570, 584. W. caterpillar, 452, 454. W. flies, 583, 588, 591. W. moths, 496, 502, 504. W. weevil, 83.
Whortleberry sphinx, 328.
Willow caterpillars, 282, 422. W. gall-gnat, 566. W. plant-louse, 239.
Willow-herb sphinx, 328.
Windsor bean attacked by insects, 138, 229.
Wire-worms, 52, 441, 448.
Wood-wasps, 533.
Woolly bears, 343.
Wurmals or worm-holes, 624.

Xenos Peckii, 18.
Xiphicera, 172.
Xiphydria, 535, 541. X. albicornis, X. mellipes, 541.
Xyleutes Cossus, X. Robiniæ, 412.

Yellow butterfly, 272.
Yponomeutadæ, 488, 499.
Ypsolophus granellus, 499.

Zebra caterpillar, 451.
Zeuzera, 409.
Zeuzerians (Zeuzeradæ), 408.

THE END.

Cambridge: Printed by Welch, Bigelow, & Co.

www.ingramcontent.com/pod-product-compliance
Lightning Source LLC
LaVergne TN
LVHW021054110826
845150LV00001B/66

* 9 7 8 1 4 2 5 5 6 7 3 0 9 *